# THE ROUTLEDGE HANDBOOK OF OCEAN SPACE

Invisible as the seas and oceans may be for so many of us, life as we know it is almost always connected to, and constituted by, activities and occurrences that take place in, on and under our oceans. *The Routledge Handbook of Ocean Space* provides a first port of call for scholars engaging in the 'oceanic turn' in the social sciences, offering a comprehensive summary of existing trends in making sense of our water worlds, alongside new, agenda-setting insights into the relationships between society and the 'seas around us'. Accordingly, this ambitious text not only attends to a growing interest in our oceans, past and present; it is also situated in a broader spatial turn across the social sciences that seeks to account for how space and place are imbricated in socio-cultural and political life.

Through six clearly structured and wide-ranging sections, *The Routledge Handbook of Ocean Space* examines and interrogates how the oceans are environmental, historical, social, cultural, political, legal and economic *spaces*, and also zones where national and international security comes into question. With a foreword and introduction authored by some of the leading scholars researching and writing about ocean spaces, alongside 31 further, carefully crafted chapters from established as well as early career academics, this book provides both an accessible guide to the subject and a cutting-edge collection of critical ideas and questions shaping the social sciences today.

This handbook brings together the key debates defining the 'field' in one volume, appealing to a wide, cross-disciplinary social science and humanities audience. Moreover, drawing on a range of international examples, from a global collective of authors, this book promises to be the benchmark publication for those interested in ocean spaces, past and present. Indeed, as the seas and oceans continue to capture world-wide attention, and the social sciences continue their seaward 'turn', *The Routledge Handbook of Ocean Space* will provide an invaluable resource that reveals how our world is a water world.

**Kimberley Peters** leads the Marine Governance Research Group at the Helmholtz Institute for Functional Marine Biodiversity (HIFMB), a collaboration between the University of Oldenburg and Alfred Wegener Institute (AWI), Germany.

**Jon Anderson** is Professor of Human Geography in the School of Geography and Planning, Cardiff University, UK.

**Andrew Davies** is Senior Lecturer in Human Geography at the University of Liverpool, and is Co-Director of the Centre for Port and Maritime History, a collaborative Centre run by the University of Liverpool, Liverpool John Moores University, UK and Merseyside Maritime Museum.

**Philip Steinberg** is Professor of Political Geography at Durham University, UK where he is Director of IBRU: Durham University's Centre for Borders Research and the Durham Arctic Research Centre for Training and Interdisciplinary Collaboration (DurhamARCTIC).

# THE ROUTLEDGE HANDBOOK OF OCEAN SPACE

*Edited by Kimberley Peters, Jon Anderson, Andrew Davies and Philip Steinberg*

Routledge
Taylor & Francis Group
LONDON AND NEW YORK

Cover image: © Pixabay

First published 2023
by Routledge
4 Park Square, Milton Park, Abingdon, Oxon OX14 4RN

and by Routledge
605 Third Avenue, New York, NY 10158

*Routledge is an imprint of the Taylor & Francis Group, an informa business*



*British Library Cataloguing-in-Publication Data*
A catalogue record for this book is available from the British Library

*Library of Congress Cataloging-in-Publication Data*
A catalog record has been requested for this book

ISBN: 978-1-138-08480-3 (hbk)
ISBN: 978-1-032-25920-8 (pbk)
ISBN: 978-1-315-11164-3 (ebk)

DOI: 10.4324/9781315111643

Typeset in Bembo
by Newgen Publishing UK

# CONTENTS

# FIGURES

# TABLES

# ABOUT THE CONTRIBUTORS

## About the Editors

**Kimberley Peters** leads the Marine Governance Research Group at the Helmholtz Institute for Functional Marine Biodiversity (HIFMB), a collaboration between the University of Oldenburg and Alfred Wegener Institute (AWI), Germany. Within this interdisciplinary centre for ocean work Kim uses spatial frames for understanding how watery spaces are organised and managed, and takes a critical approach to interrogating operations of power at sea. She is a socio-cultural and political geographer by training and has explored these interests in contexts ranging from offshore radio piracy, to prison transportation, deep-sea mining politics, to ship routeing. Kim's work on these topics appears in the edited books: *Water Worlds: Human Geographies of the Ocean* (Ashgate, 2014), *The Mobilities of Ships* (Routledge, 2015), *Carceral Mobilities* (Routledge, 2017) and *Territory Beyond Terra* (Rowman and Littlefield, 2018), as well as the monograph *Sound, Space, Society: Rebel Radio* (Palgrave, 2018). She is the author of the discipline-wide text-book *Your Human Geography Dissertation: Designing, Doing, Delivering* (Sage, 2017).

**Jon Anderson** is Professor of Human Geography in the School of Geography and Planning, Cardiff University, UK. His research interests focus on the relations between identity, culture, and place, in particular the actions, practices and politics that such relations produce (see www.spatialmanifesto.com). His work has focused extensively on cultural geographies, including the places of oceanic worlds, the practices of surfing, and literary geographies. He has published widely, including the books *Understanding Cultural Geography: Places and Traces* (Routledge, Third Edition, 2021), *Water Worlds: Human Geographies of the Ocean* (Ashgate, 2014), and *Page and Place: Ongoing Compositions of Plot* (Rodophi, 2014). He will soon publish a monograph on *Surfing Spaces* (Routledge, forthcoming) and recently completed an innovative mapping project of English language novels set in Wales (funded by the Arts and Humanities Research Council). Full details on this Literary Atlas project can be found here: www.literaryatlas.wales

**Andrew Davies** is Senior Lecturer in Human Geography at the University of Liverpool, and is Co-Director of the Centre for Port and Maritime History, a collaborative Centre run by the University of Liverpool, Liverpool John Moores University and Merseyside Maritime Museum. His research sits at the intersection of political, postcolonial and cultural geographies, and particularly maritime intersections with transnational anticolonial and imperial politics. He has authored *Geographies of Anticolonialism: Political Networks across and beyond South India, c. 1900–1930* (RGS-IBG Book Series, Wiley-Blackwell, 2019) and is a Co-Editor of the Liverpool radical sailor George Garrett's autobiography *10 Years on the Parish* (Liverpool University Press, 2017). He is a contributor of the Writing on the Wall's Great War to Race Riots project (www.greatwar-to-raceriots.co.uk/), including giving regular walking tours related to the events of the Port City Riots of 1919 in Liverpool.

**Philip Steinberg** is Professor of Political Geography at Durham University where he is Director of IBRU: Durham University's Centre for Borders Research and the Durham Arctic Research Centre for Training and Interdisciplinary Collaboration (DurhamARCTIC). His research focuses on the projection of social power onto spaces whose geophysical and geographic characteristics make them resistant to state territorialisation – spaces that include the world-ocean, the universe of electronic communication, and the Arctic. His publications include *The Social Construction of the Ocean* (Cambridge University, 2001), *Managing the Infosphere: Governance, Technology, and Cultural Practice in Motion* (Temple University Press, 2008), *What Is a City? Rethinking the Urban after Hurricane Katrina* (University of Georgia Press, 2008), *Contesting the Arctic: Politics and Imaginaries in the Circumpolar North* (I.B. Tauris/Bloomsbury, 2015), and *Territory beyond Terra* (Rowman and Littlefield, 2018).

## About the contributors

**Antony Adler** is Lecturer in the Department of History at Carleton College, USA, where he teaches courses on the history of science, the history of exploration, the Pacific World, and Public History. His research centres on the history of marine science and ocean exploration. He serves as an outreach coordinator for the International Commission for the History of Oceanography and he is the author of *Neptune's Laboratory: Fantasy, Fear, and Science at Sea* (Harvard University Press, 2019).

**Edward H. Allison** is Director of Science and Research at WorldFish, Penang, Malaysia, and Research Director of the Nippon Foundation's Oceans and Equity Nexus Program. He has held faculty positions in Development Studies at the University of East Anglia, UK, and in Marine Affairs at the University of Washington, Seattle, USA. He has visiting and adjunct professorships at the Lancaster Environment Centre, University of Lancaster, UK, and the School of Marine and Environmental Affairs, University of Washington, Seattle WA, USA. His research is informed by spells working with development agencies, including the UK's Department for International Development in Malawi and the UN Food and Agriculture Organization in central and West Africa. His current research addresses the equity implications of 'blue economy' initiatives.

**Sophie Bond** lectures in Human Geography at the University of Otago, Aotearoa New Zealand. She teaches and researches in areas of social and environmental justice. In particular, she is interested in how dissent, open public debate, and social action is enabled as a crucial part of democratic engagement. Her current work explores how action for climate justice is enabled or constrained by neoliberalism and contemporary forms of governance.

**Maria Borovnik** is Senior Lecturer in Development Studies at Massey University, Aotearoa New Zealand. She is working in the intersection of mobilities, development, and geographies.

Her long-term research, since 1999, explored a number of different perspectives on seafarers. She has looked at the living realities of seafarers and their families in Kiribati and Tuvalu, and has increasingly become interested in non-representational approaches to her work. Her most recent publication explored seafarers in the context of weather worlds, in her book on *Weather: Spaces, Mobilities and Affects*, co-edited with Tim Edensor and Kaya Barry (Routledge, 2021)

**Mike Brown** PhD is Associate Professor of Outdoor Learning at Auckland University of Technology, Aotearoa, New Zealand. He has an interest in place-responsive learning with a particular focus on marine based experiences. He has co-edited *Seascapes: Shaped by the Sea* (Routledge, 2015) and *Living with the Sea: Knowledge, Awareness and Action* (Routledge, 2018). He holds both recreational and commercial maritime qualifications and is involved in water-based activities including sailing.

**Christian Bueger** is Professor of International Relations at the University of Copenhagen, Denmark, honorary professor at the University of Seychelles and a research fellow at the University of Stellenbosch, South Africa. His areas of research include international practice theory, the sociology of expertise, ocean governance and maritime security. Further information is available at http://bueger.info

**Charmaine Chua** is Assistant Professor of Global Studies at the University of California, Santa Barbara. Her research focuses on critical political economy, in particular on the social and political economic relations between state and corporate actors, labour, and urban communities along global supply chains and across the transpacific maritime world. She is currently writing a book about the rise of the logistics industry as a 'counterrevolutionary' terrain of contestation under US empire. Her work has been published or is forthcoming in *Antipode, Historical Materialism, Political Geography*, and *Environment and Planning D: Society and Space*, among other venues.

**Greer Crawley** is Honorary Research Fellow in the Department of Drama, Theatre and Dance at Royal Holloway University, UK. She received a MAS (Masters of Advanced Studies in Scenography) from the University of the Arts, Zurich and a DrPhil from the University of Vienna for her dissertation *Strategic Scenography – Staging the Landscape of War*. Her research and professional practice is in landscape design, scenography and exhibition. She curated *Traces of the Future: Archaeology of Modern Science in Africa*, The Nunnery Bows Arts in 2017 and is a member of the Making | Art | Science | Environment (MASER) group at Bath Spa University: www.bathspa.ac.uk/research-and-enterprise/research-centres/art-research-centre/material-art-science-environment-research/. Working with artists, designers and curators, her aim is to facilitate the integration of scenographic perspectives in collaborative projects between performance, arts and science. Her publications include: 'The Generous Deceit' in: Neudecker M *Every Landscape is A State of Mind* (Dulwich Picture Gallery, 2019); 'The Scenographer as Camoufleur' in: Emeljanow V (ed) *War and Theatrical Innovation* (Palgrave Macmillan, 2017) and she is editor for *Sediment*, the work of Mariele Neudecker (Anomie Publishing, 2021).

**Emma Critchley** is an artist who uses a combination of photography, film, sound and installation to continually explore the human relationship with the underwater environment as a political, philosophical and environmental space. She is Royal College of Art alumni and has developed works funded by organisations including the National Media Museum, Arts

Council England, British Council, Singapore International Foundation, British Academy and the European Regional Development Fund. Her work has been shown extensively nationally and internationally in galleries and institutions including the Australian Centre of Photography, the ICA Singapore, the National Portrait Gallery, the Royal Academy, BALTIC Centre for Contemporary Art and Tate St Ives. In 2019 she completed a short film funded by the Jerwood Charitable Foundation called 'Common Heritage', about the imminent threat of deep-sea mining for rare earth minerals. For this she worked with experts in deep sea ecology and law at the National Oceanography Centre and Universities of Plymouth and Southampton. In collaboration with artist Lee Berwick, Emma has developed a large-scale public soundscape about underwater acoustic pollution, working with a number of organisations including the British Antarctic Survey, the Californian Ocean Alliance and the National Maritime Museum. This had its initial installation in the Greenwich Foot Tunnel at the beginning of 2020. In 2019, Emma was the winner of the Earth Water Sky residency programme with Science Gallery Venice, where she worked with the Ice Memory Project. The resulting film 'Witness' premiered in the official Italian Pavilion of the seventeenth Venice Architecture Biennale, 2021.

**Duncan Depledge** is Lecturer in Geopolitics and Security at Loughborough University, UK, and an Associate Fellow of the Royal United Services Institute. He is the author of *Britain and the Arctic* (Palgrave, 2018) and more than a dozen academic articles and book chapters on the changing geopolitics of the Arctic Ocean. He has also served as a consultant to the UK Ministry of Defence and as a special adviser to the House of Commons Defence Committee.

**Gradon Diprose** is a human geographer who works as a Researcher in Environmental Social Science at Manaaki Whenua Landcare Research, Aotearoa New Zealand. He is interested in how communities coalesce around shared concerns, respond to climate change and socio-economic inequalities, manage resources, and build more diverse and sustainable livelihoods.

**Klaus Dodds** is Professor of Geopolitics at Royal Holloway University of London, UK, and Director of Research for the School of Life Sciences and Environment. He is interested in the cultural and geopolitical significance of ice and cold. He published *Ice: Nature and Culture* (Reaktion, 2018) and a forthcoming co-edited (with Professor Sverker Sorlin) collection on *Ice Humanities* will appear with Manchester University Press in 2022. He has acted as specialist adviser to the UK Parliament on two parliamentary inquiries into the Arctic and is an Honorary Fellow of British Antarctic Survey.

**Leesa Fawcett** is Associate Professor in the Faculty of Environmental and Urban Change and the Director of the Graduate Program in Environmental Studies at York University, Canada. Originally trained as a marine biologist she specialises in animal studies, decolonial environmental pedagogies and ethics, natural history and conservation issues. She was Principal Investigator (with A. Zalik and E. Havice), on 'Ocean frontiers: An interdisciplinary workshop on changing contours of marine space and resource access', funded by a Social Sciences and Humanities Research Council Connections Grant.

**David Featherstone** is Reader in Human Geography at the University of Glasgow. He is the author of *Resistance, Space and Political Identities: The Making of Counter-Global Networks* (Wiley-Blackwell, 2008) and *Solidarity: Hidden Histories and Geographies of Internationalism* (Zed Books, 2012). He is co-editor with Christian Høgsbjerg of *The Red and the Black: the Russian Revolution and the Black Atlantic* (Manchester University Press, 2021). He is currently working on a monograph with the provisional title of *Politicising Race and Labour: Seafarers' Struggles for Equality and the Anti-Colonial Left, 1919–1953.* He is a member of the editorial collectives of *Antipode: A Radical Journal of Geography* and *Soundings: A Journal of Politics and Culture.*

**Ronan Foley** is Associate Professor in Health Geography and GIS at Maynooth University, Ireland, with specialist expertise in therapeutic landscapes and geospatial planning within health and social care environments. His current research focuses on relationships between water, health and place, including two authored/co-edited books and journal articles on auxiliary hospitals, holy wells, spas, social and cultural histories of swimming and 'blue space'. He co-edited a special issue on healthy blue space for *Health & Place* (2015) and is the Editor of *Irish Geography*. He collaborates on a number of water/health projects in Ireland, the UK, Spain, Germany, New Zealand and Australia.

**Basil Germond** is Senior Lecturer and Director of Research Training for the Faculty of Arts and Social Sciences at Lancaster University, UK. His research expertise is in naval affairs, the concept of seapower, maritime security, ocean narratives, Global Maritime Britain, and the maritime dimension of European security. His research is cross-disciplinary (International Relations, Human Geography, History) and aims at understanding human, social and political interactions at, from, within, and with the sea.

**Paul Griffin** is Senior Lecturer in the Department of Geography and Environmental Sciences at Northumbria University, UK. His research works across labour geography and labour history with papers published in *Progress in Human Geography, Geoforum, Area* and *Political Geography*. His research interests have most recently moved towards a project around the organising of 'unemployed workers' through a historical study of Unemployed Workers' Centres in the UK.

**Madeleine Gustavsson** is Researcher at Ruralis – Institute for Rural and Regional Research in Trondheim, Norway. Before joining Ruralis, she was a Research Fellow at the University of Exeter, holding an Economic and Social Research Council New Investigator fellowship, researching the changing lives of women in small-scale fishing families in the UK and Newfoundland, Canada. More broadly, her research focuses on marine, coastal and rural issues drawing on social science methods to understand the lifeworlds of people living and working with the sea.

**Maria Hadjimichael** is based in Cyprus and conducts research on the fields of political ecology, environmental politics and governance of the Commons with a focus on the sea and the coastline; how is the understanding of the sea and the coastal space as a 'Common' or 'Common Heritage' affected by international agreements or national law, and how are such institutional arrangements being instrumentalised to expand the authority of the State. She has published widely in this area, including in the journals *Sustainability Science, Marine Policy, Ocean and Coastal Management*, and *Political Geography*. She is co-editor (Marine Governance & Political Ecology) of the *Island Studies Journal.*

**Elizabeth Havice** is Professor of Geography at the University of North Carolina at Chapel Hill, USA. She uses the lens of governance to explore distributional outcomes in marine spaces, food systems, and global value chains. She is a co-founder of the Digital Oceans Governance

Lab (with L. Campbell) that explores intersections of data technologies and oceans governance and co-editor (with M. Himley and G. Valdivia) of *The Routledge Handbook of Critical Resource Geography* (Routledge, 2021).

**Matthew Heins** is an independent scholar specialising in infrastructure, urban history and urban design. He is the author of *The Globalization of American Infrastructure: The Shipping Container and Freight Transportation* (Routledge, 2016). Matthew obtained his PhD degree in architecture from the University of Michigan, and has been a faculty member at Wayne State University, Northeastern University and Rhode Island School of Design. He currently works as an urban planner near Boston, USA.

**Charles Heller** is a researcher and filmmaker whose work has a long-standing focus on the politics of migration. In 2015, he completed a PhD in research architecture at Goldsmiths, University of London, where he continues to be affiliated as a research fellow, focusing on migration and its control across the Mediterranean Sea. He is currently Research Associate at the Centre on Conflict, Development and Peacebuilding (CCDP), Graduate Institute, Geneva. Together with Lorenzo Pezzani, in 2011 he cofounded Forensic Oceanography, a collaborative project that critically investigates the militarised border regime in the Mediterranean Sea, as well as the WatchTheMed platform. Together with a wide network of NGOs, scientists, journalists, and activist groups, he has produced maps, video animations, installations, and human rights reports that attempt to document and challenge the ongoing death of migrants at sea. His work has been used as evidence in courts of law, published across different media and academic outlets, and exhibited widely.

**Stefan Helmreich** is Professor of Anthropology at the Massachusetts Institute of Technology, Boston, USA. He is author of *Alien Ocean: Anthropological Voyages in Microbial Seas* (University of California Press, 2009) and *Sounding the Limits of Life: Essays in the Anthropology of Biology and Beyond* (Princeton University Press, 2016). His essays have appeared in *Critical Inquiry, Representations, American Anthropologist, The Wire, Cabinet,* and *Public Culture.*

**Stephen Jay** is Reader at the University of Liverpool in the UK, where he specialises in the study of marine spatial planning (MSP). He has researched and published widely on its development, and is particularly interested in the interface between contemporary spatial theory and the practical implementation of MSP. Stephen has led or contributed to a number of European projects on transboundary cooperation in MSP and has engaged internationally with the uptake of MSP. He is co-founder of the MSP Research Network, which brings together researchers, practitioners and policy-makers. He is also Director of the Liverpool Institute for Sustainable Coasts and Oceans.

**Elizabeth R. Johnson** is Assistant Professor of Geography at Durham University, UK. Her work explores how emerging ties between the biosciences and technological innovation change Western understandings of environmental precarity. She is currently researching the interface between marine science and policy with a focus on tensions between marine conservation and the emerging Blue Economy. The project, titled 'Circulatory Entanglements', examines how marine organisms and the materials extracted from them pass through, across, and into the different ecological and epistemological worlds to figure in the production of healthy publics, ecological futures, and promissory economies.

**Stephanie Jones** is Associate Professor in English at the University of Southampton, and a Co-Director of the Southampton Institute for Arts and Humanities. She has published work on East African literature and law; Indian Ocean fictions; the poetics and metaphors of international maritime law and lore; fictional and historical piracy and privateering; and on literary and legal 'belonging' and 'commons'. Her research focuses on narratives about water and waterlands, and particularly concentrates on stories and poetry about oceanic worlds. Her research engages with ideas, methods and debates from across the interdisciplinary fields of law and literature; postcolonial and decolonial studies; and the environmental humanities.

**Isaac Land** is Professor of History at Indiana State University, USA. He is the author of *War, Nationalism, and the British Sailor, 1750–1850* (Palgrave Macmillan, 2009). Recent and forthcoming publications include 'Port towns and the "Paramaritime"' in *The Routledge Companion to Marine and Maritime Worlds, 1400–1800*, and 'Sea visibility and the anxious coastal gaze' in the journal *Global Environment: A Journal of Transdisciplinary History*. Currently, he is a co-editor of the new interdisciplinary journal *Coastal Studies & Society*, published by Sage.

**Jessica Lehman** is Assistant Professor of Geography (Human-Environment) at Durham University, UK. Her research focuses primarily on international environmental politics. She is especially interested in marine geographies, environmental knowledge production, and resource politics. She has published in a variety of journals including *Political Geography*, *Annals of the Association of American Geographers* and the *International Social Science Journal*.

**Emma McKinley** is a Research Fellow at Cardiff University, with expertise in marine social sciences, specifically topics relating to ocean literacy, marine citizenship and public perceptions research. Her research focuses on understanding the complex relationships between society and the sea, taking account of diverse perceptions, attitudes and values held by different communities and audiences, and considers how this insight can be used to support effective ocean governance. Emma is the founder of the Marine Social Science Network, a global, interdisciplinary community of marine social science researchers and practitioners and is Chair of the Royal Geographic Society's Coastal and Marine Research Group. She is the Co-Chair of the Marine Social Science Task Group of the UK's Marine Science Coordinating Committee, sits on the International Science Advisory Group for MEOPAR and the IOC-UNESCO's Ocean Literacy Research Community.

**Mariele Neudecker** (b. 1965, Düsseldorf, Germany) undertook a BA at Goldsmiths College, London (1987–90), and an MA in sculpture at Chelsea College of Art and Design, London (1990–1). Neudecker often uses technology's virtual capabilities in order to reproduce a heightened understanding of landscape, thus addressing the subjective and mediated condition of any first-hand encounter. She has worked with scientists and engineers and their methods and research. For her, technology both enables and limits the perception and experience of the worlds we inhabit. She has shown widely in international solo and group exhibitions. Neudecker is Professor of Fine Art at Bath School of Art, where she runs the research cluster Making | Art | Science | Environment. She is on the Arts at CERN's guest programme, the European Commission's JRC SciArt advisory panel and the steering committee of Centre of Gravity, UK. Neudecker works with Galeria Pedro Cera, Lisbon; In Camera Gallery, Paris; and Thomas Rehbein Galerie, Cologne.

**Lorenzo Pezzani** is an architect and researcher. He is currently lecturer at Goldsmiths, University of London, where he leads the MA studio in forensic architecture. His research deals with the spatial politics and visual cultures of migration, with a particular focus on the geography of the ocean. Together with Charles Heller, in 2011 he cofounded Forensic Oceanography, a collaborative project that critically investigates the militarised border regime in the Mediterranean Sea, as well as the WatchTheMed platform. Together with a wide network of NGOs, scientists, journalists, and activist groups, he has produced maps, video animations, installations, and human right reports that attempt to document and challenge the ongoing death of migrants at sea. His work has been used as evidence in courts of law, published across different media and academic outlets, and exhibited widely.

**Satya Savitzky** is Research Scientist at the Helmholtz Institute for Functional Marine Biodiversity in Germany. Prior to this he was a postdoctoral fellow at St Andrews University, UK. He completed his thesis, 'Icy Futures: Carving the Northern Sea Route', at Lancaster University in 2016. He has published several peer-reviewed articles and book chapters which examine the creation, maintenance and contestation of routes – on land and at sea. He continues to research the Indian Ocean as site, and ships and ballast tanks as spaces, of turbulent economic and ecologic circuits.

**Jeremy J. Schmidt** is Associate Professor of Geography at Durham University, UK. He is the author of *Water: Abundance, Scarcity, and Security in the Age of Humanity* (NYU Press, 2017) and, with Peter Brown, co-editor of *Water Ethics: Foundational Readings for Students and Professionals* (Island Press, 2010).

**Rachael Squire** is Lecturer in Human Geography at Royal Holloway University of London UK. Her research explores the historical and contemporary geographies and geopolitics of the oceans with a particular focus on ideas surrounding volume, territory, and terrain. She has published in journals including *Area, Political Geography* and *Environment and Planning D: Society and Space*. She is the author of the monograph, *Undersea Geopolitics: Sealab, Science, and the Cold War* (Rowman and Littlefield, 2021).

**Maurice Stierl** leads the research group "The Production of Knowledge on Migration" at the Institute for Migration Research and Intercultural Studies, Osnabrück University. Before, he was a lecturer in International Relations at the University of Sheffield. He has also taught at the University of Warwick and the University of California, Davis. His research focuses on migration struggles in contemporary Europe and (northern) Africa and is broadly situated in the fields of International Political Sociology, Political Geography, and Migration, Citizenship and Border Studies. His book *Migrant Resistance in Contemporary Europe* was published by Routledge in 2019. He is a member of the activist network WatchTheMed Alarm Phone.

**Elizabeth R. Straughan** is a cultural geographer interested in the connections between embodiment and wellbeing. Using qualitative methods her research focus is on practices that augment and manipulate the body as well as embodied practices of everyday mobile lives. Drawing on social theory and mobility studies Elizabeth considers the social, cultural and political implications of these practices through a focus on their emotional and affective geographies. She has published in international journals such as *Transactions of the Institute of British Geographers, Cultural Geographies* and *Emotion, Space and Society.*

**Amanda Thomas** is a feminist political geographer interested in environmental democracy. She takes a grounded approach to research, and seeks to do decolonising work and cultivate social justice. She is a senior lecturer in Environmental Studies at Te Herenga Waka Victoria University of Wellington, Aotearoa New Zealand.

**Jordan Te Aramoana Waiti (Ngāti Pikiao, Ngaati Maahanga, Te Rarawa)** is a Senior Lecturer within the School of Health at the University of Waikato. He has been researching aspects of Māori health and wellbeing for the past 20 years, with a particular focus on the benefits of reaffirming cultural identity. In 2015, he completed his PhD at Massey University in Māori Public Health, investigating the protective factors and coping strategies of resilient Māori whānau. Jordan is a member of the NZ Dietitians Governance Board, Water Safety NZ Māori Advisory Board, and is a co-founder of Aotearoa Water Patrol.

**Belinda Wheaton** is Professor in Te Huataki Waiora – School of Health University of Waikato, Aotearoa/New Zealand. She is a cultural sociologist, with research interests across leisure and sport, and a focus on identity, inclusion and inequality. Belinda is best known for her research on informal and lifestyle sport cultures which includes a monograph (*The Cultural Politics of Lifestyle Sports*, Routledge, 2013) and three edited collections. She is co-editor of *The Palgrave Handbook of Feminism and Sport, Leisure and Physical Education* (2017), and Managing Editor of *Annals of Leisure Research*. She has been fortunate to spent much of her life living by the sea, and spends much of her leisure time in, on and by the water.

**Anna Zalik** is Associate Professor in Environmental and Urban Change at York University, Canada, with areas of specialisation in global political economy and ecology, particularly with relation to industrial extraction. This has involved longstanding research on the offshore oil industry and, since 2013, on mining in the deep seabed beyond state jurisdiction, the zone referred to as the Area under the United Nations Convention on the Law of the Sea.

# FOREWORD

## Ocean space and the marine social sciences

*Emma McKinley*

Our global ocean has acted as connector across time and place for generations – ocean spaces surround and are inextricably connected to almost every aspect of contemporary society, with the ocean, coasts and seas, our water spaces, increasingly recognised as 'peopled' spaces (Bennett, 2019). Nevertheless, many of us live our lives in ways that are quite disconnected from the ocean. This sentiment, and indeed its challenges, are beautifully and quite hauntingly articulated by Rose George in her 2013 book, *Deep Sea and Foreign Going*:

> There are no ordinary citizens to witness the workings of an industry that is one of the most fundamental to their daily existence… they have fuelled if not created globalization… but who looks beyond a television now and sees the ship that carried it? Who cares about the men (sic) who brought your breakfast cereal through the winter storms? How ironic that the more ships have grown in size and consequence, the more their place in our imagination has shrunk.
>
> *George, 2013: 2*

Despite playing such a fundamental role in the development and growth of civilisations throughout history, understandings of the relationships between much of society and the ocean are often fragmented and disconnected (Potts et al., 2016). Indeed, a 2020 report from the High Level Panel for a Sustainable Ocean Economy begins by stating that, "[o]ver a third of the world's population lives within 100 kilometres of the ocean. Despite this, the role the ocean plays in sustaining human life and the global economy is often underappreciated and overlooked" (Northrop et al., 2020: 1). This societal 'sea blindness' is gradually garnering more recognition as one of the most significant challenges facing the ocean (Pascual et al., 2017). While, historically, there has been limited focus on these water dimensions of our blue planet, with the majority of studies planting themselves firmly on land, recent years have witnessed a growing call for us to turn back to the sea and to reconnect individuals and communities with the ocean (McKinley et al., 2020; see also Anderson and Peters, 2014; Steinberg, 1999). This call to arms has been echoed in the goals of the United Nations Decade of Ocean Science for Sustainable Development (2021–2030), which sets out aspirations which look towards a 'transformational relationship between society and the ocean', and more focus on the role of the ocean in individual and community sense of place and identity, and place attachment, through aspirations of enhanced ocean literacy (seen for example in recent work from the UK – McKinley and Burdon [2020] alongside evidence from the Canadian Ocean Literacy Coalition) and stewardship, sustainable, equitable blue economies (Bennett et al., 2019), and opportunities to support global ocean recovery for the benefit of people, ocean and place. Achieving these goals will require change across various scales and communities and will demand a concerted effort to build on this recent upswell of momentum to really develop our in-depth understanding of societal interactions with ocean spaces. The disciplines which fall under the umbrella of marine social sciences, crucially including disciplines from across geography, provide us with a diverse range of tools, approaches and methodologies through which we can come to better understand the multiple dimensions of human relationships with the watery part of our world. In particular, the lenses of geographical inquiry and their particular focus on the interconnectivity between space and spatiality can provide us with critiques and insight which will be invaluable to how we live, work and play in our ocean spaces, both now and in the future.

This *Handbook of Ocean Space* presents a collection of chapters full to the brim of essential insights into the diversity of relationships and connections between people, ocean and place, building on a growing discourse around the role of social sciences in understanding these relationships. As we look to building sustainable, equitable, and inclusive ways of living and working with our ocean spaces, the chapters in this book explore the complexities of these relationships through a number of geographical lenses. These include topics relating to knowledge of the ocean, cultural connections, mapping and planning for the ocean, blue economies, seafaring communities and workers, security, politics and migration, as well as more emergent areas of geographical inquiry drawing on arts and literature. Crucially, the authors frame the discourse through the various dimensions of space and place, and explore, interrogate and interpret the myriad of ways in which to be in, on, under and around the ocean. This collection of chapters is a signal of the ongoing shift in how ocean spaces are viewed and studied. There is no aspect of our lives that is untouched by the ocean, and this interconnectivity between people, ocean, space and place is encapsulated by the various geographical lenses applied by the authors. In this way, the *Handbook of Ocean Space* presents a valuable, and indeed much needed, addition to the ways in which we understand our global ocean, coasts and seas.

## References

Anderson J and Peters K (eds) (2014) *Water Worlds: Human Geographies of the Ocean*. Farnham: Ashgate Publishing.

Bennett NJ (2019) Marine social science for the peopled seas. *Coastal Management* 47(2): 244–252.

Bennett NJ, Cisneros-Montemayor AM, Blythe J, Silver JJ, Singh G, Andrews N, Calò A, Christie P, Di Franco A, Finkbeiner EM, Gelcich S, Guidetti P, Harper S, Hotte N, Kittinger JN, Le Billon P, Lister J, Lopez de la Lama R, McKinley E, Scholtens J, Solås A-M, Sowman M, Talloni-Álvarez N, Teh LCL, Voyer M and Sumaila UR (2019) Towards a sustainable and equitable blue economy. *Nature Sustainability* 2: 991–993.

George R (2013) *Deep Sea and Foreign Going*. London: Portobello Books.

McKinley E and Burdon D (2020) Understanding ocean literacy and ocean climate-related behaviour change in the UK: An evidence synthesis. Final report produced for the Ocean Conservation Trust and Defra. October 2020.

McKinley E, Acott TG and Yates K (2020) Marine Social Sciences: Looking towards a sustainable future. *Environmental Science and Policy* 108: 85–92.

Northrop E, Konar M, Frost N and Hollaway E (2020) *A Sustainable and Equitable Blue Recovery to the COVID-19 Crisis*. Washington, DC: World Resources Institute. Available at: www.oceanpanel.org/bluerecovery

Pascual U, Balvanera P, Díaz S, Pataki G, Roth E, Stenseke M, Watson RT, Başak Dessane E, Islar M, Kelemen E, Maris V, Quaas M, Subramanian SM, Wittmer H, Adlan A, Ahn SE, Al-Hafedh YS, Amankwah E, Asah ST, Berry P, Bilgin A, Breslow SJ, Bullock C, Cáceres D, Daly-Hassen H, Figueroa E, Golden CD, Gómez-Baggethun E, González-Jiménez D, Houdet J, Keune H, Kumar R, Ma K, May PH, Mead A, O'Farrell P, Pandit R, Pengue W, Pichis-Madruga R, Popa F, Preston S, Pacheco-Balanza D, Saarikoski H, Strassburg BB, van den Belt M, Verma M, Wickson F and Yagi N (2017) Valuing nature's contributions to people: The IPBES approach. *Current Opinion in Environmental Sustainability* 26–27: 7–16.

Potts T, Pita C, O'Higgins T and Mee L (2016) Who cares? European attitudes towards marine and coastal environments. *Marine Policy* 72: 59–66.

Steinberg PE (1999) Navigating to multiple horizons: Toward a geography of ocean-space. *The Professional Geographer* 51(3): 366–375.

# ACKNOWLEDGEMENTS

Our sincere thanks go to Egle Zigaite and Andrew Mould at Routledge for their persistence and encouragement during this project and to Claire Maloney for assistance as we neared the end. Our appreciation furthermore goes to those who assisted with careful pre- and post-submission checks (with special thanks to Jennifer Turner), and to those who worked with us through to the final production of this book (with thanks to Sara Marchington, Priyanka Mundada, Megan Smith and Helen Strain). Wholeheartedly, we must thank our contributors for their commitment, patience and the inspiration they've provided in writing for the volume.

Like any collection written over a period of time, much has happened during the compilation of this volume. In the academic world, theories have been advanced, methodologies have been developed, debates have emerged. Early career scholars are researching and writing about the oceans in novel but also necessary ways. Old hands have retired. Alongside the ever-increasing realities wrought through sea-level rise and the climate crisis, new events, within and beyond the ocean's depths, have attracted the attention of researchers. A globally altering pandemic has changed oceanic trade relations and the lifeworlds of seafarers. A devastating war has deployed the sea as a crucial access point for landed invasions. This book is thus not complete. It is a *work* in progress that reflects on a *world* in progress, an ocean that is ever re-forming. It is a handbook of ideas and provocations. It is also a handbook of omissions and shortcomings. As editors, these are our own. However, it is also a collection that, we hope, suggests the breadth of perspectives that might be reached through thinking the oceans spatially.

# SECTION I

# Ocean approaches, ocean perspectives

# 1
# INTRODUCTION
## Placing and situating ocean space(s)

*Jon Anderson, Andrew Davies, Kimberley Peters and Philip Steinberg*

### Oceanic (re)turns: Placing and situating

The oceans are all around us, so says the famous book title about the exploration of ocean environs by Rachel Carson (1989 [1951]). It is certainly also true that ocean scholarship – writ through a spatial lens – is now all around us. This spatial scholarship has (re)shaped and (re) defined geography as a discipline, but also the wider socio-cultural and political sciences and humanities that have taken interest in themes of space, place, territory and time. Geography – the disciplinary 'home' of spatial studies – is no longer wholly terra-centric, where a firm earthy bias exists (see also Peters and Squire, 2019: 101). Although historically lagging behind physical geographies, which have long taken to coastal and near shore spaces, human geography has now established an extensive catalogue of watery work (see Steinberg, 2001, through to Anderson and Peters, 2014; and for reviews see Peters, 2010, 2017; Steinberg, 2009, 2013, 2017). There is now a wealth of work from subfields including historical geographies (for example, Anim-Addo, 2011; Davies, 2019; Lambert et al., 2006; Legg, 2020; Lehman, 2020; Stafford, 2017), cultural geographies (such as Anderson, 2022; Choi, 2020; Satizábal and Dressler, 2019; Spence, 2014; Walsh and Döring, 2018), more-than-human geographies (see Bear, 2017; Gibbs and Warren, 2014; Johnson, 2015; Squire, 2020; Wang and Chien, 2020 to name but a few), as well as specific areas such as carceral geographies (Dickson, 2021; Stierl, 2021; Peters and Turner, 2015). The works listed, which 'take to the seas', are not exhaustive. Much work has also focused on the (geo)political dimensions of oceans (Childs, 2020; Dittmer, 2018; Dunnavant, 2021; Squire, 2021, among others) and the geographical aspects of ocean management and conservation (for example, Fairbanks et al., 2019; Gray et al., 2020; Jay, 2018).

The sea offers an empirical space that departs from the landed spaces of the city, the street, or even the underground or the air. In turn it offers a conceptual space for rethinking socio-cultural, political, economic and environmental spatial relations; for reflecting on notions of space, time and motion (Anderson and Peters, 2014; Steinberg and Peters, 2015; Peters and Steinberg, 2019); and for problematising the often-stark differentiations made between 'land' and 'sea' (Hau'ofa, 1995, 2008; George and Wiebe, 2020; Glissant, 1997; Pugh, 2016; Underhill-Sem, 2020). Indeed, the past two to three decades have witnessed an 'oceanic turn' in disciplines from geography to anthropology, from art to literature studies, and beyond (see, for example, Blum, 2010; DeLoughrey, 2019). Whilst history has long-held associations with

DOI: 10.4324/9781315111643-2

maritime pasts, histories of ocean worlds are increasingly shifting away from dominant narratives of state conquest and technological mastery to deploy perspectives from the Global South, writing sea stories that suggest different affinities between the ocean, its forces, and its various more-than-human inhabitants (Ingersoll, 2016). In these ways, the ocean, it seems, has shifted from a marginal to a central concern. In some areas, though – and it is important to note – the ocean has *always* been central to thinking and being (Hau'ofa, 1995, 2008). It is western scholarship itself which can be accused of emptying the ocean for its own filling.

This book is placed and situated within this 'oceanic turn' (or 'turns', acknowledging multiple knowledges of oceanic import), and is situated at the intersections of various disciplines engaging ocean space. In turning to ocean *space*, we signal an understanding of space that is increasingly prevalent within academic geography: space understood not simply as bounded area or metric calculation but as a force that shapes and is shaped by histories, scientific endeavour, cartographies, art practice, political action and so on. The 'oceanic turn' is thus part of a broader spatial turn, whose impacts extend far beyond the discipline of geography (Warf and Arias, 2009). Although space is a central concern of geography, space – as this book goes on to show – is also fundamental to ways of thinking through literary practice, international relations, anthropology, leisure studies, and beyond. This is because space is both a commonplace word as well as a specific tool. It pertains, most generally, to a dimension in which we live (the other dimension being time). Yet theories of space – how space is understood, grappled with and deployed – matters. Space has been understood as a geometric and abstract plane – separate from social life, a mere backdrop or surface on which lived experience occurs (see Cresswell, 2014 for an overview). However, space is now better understood as co-constituted through practice and performance, forged and formed through relation (Massey, 2005). These spatial ontologies are important. Earlier theorisations of space were fundamentally mathematical – an inanimate and unchanging surface of grid-like dimensions. The sea has oftentimes been read and constructed through such a framing – a backdrop to movement, a space to be traversed, a zone devoid of social interest – a mere plain of blue (see Mack, 2013; Steinberg, 2001 for discussion). Such understandings have, it has been argued, relegated the sea to the 'background' or the 'outside' in the social and political imagination (Steinberg, 2009). Yet, with post-structural, postcolonial and wider relational theories of connection, the ocean is understood as a space of dynamism, flux and flow – a lively agent with its own material force, and one also made meaningful through past, present and future (in)actions.

This book understands space in the latter framing: space as co-constituted through human and more-than-human engagement; space as an active agent; space as relational. The book hence understands the ocean as a space that is not an empty void to be filled with stories of its significance, but a space already rich, full, varied – ever emergent, ever becoming, ever in flux. Ocean space here is not singular, given or static. Indeed, it might be better to speak of ocean *spaces*, rather than an ocean space. The book sees ocean spaces critically examined as constructions, deconstructed through practices and performances, interrogated as sites of politics, and engaged as zones of possibility. The chapters offer provocations on a number of ideas related to ocean spaces/spaces as oceanic (empire, culture, discipline, solidarities, ice, mappings, science, depths). They explore ocean spaces through these frames reflecting on their spatial shapes and significances. Whilst the book plots concepts and topics, activities and events, phenomena and things in relation to ocean spaces, the book is not an encyclopaedia or dictionary of key terms. It will not provide a definitive companion to ocean space(s). Rather, it is an intervention by authors to consider ocean spaces in their various guises, and an invitation to readers to reflect on what such relations, between space, the ocean and the chapters at hand mean.

## Route laying: Outlining the book structure

There is often – almost always – more than one option when plotting a route across the ocean. Similarly, there is – and was – more than one way for us, as editors, to plot a route through this book by placing and situating chapters in relation to each other. Indeed, given the fluid, wet, churning nature of the ocean and also the ocean's non-linear histories, there is no perfect or smooth way to arrange the chapters that follow. Chapters have been loosely arranged into themes: 'frameworks and knowledges'; 'economies and labour'; 'histories and politics'; 'experiences and engagements'; and 'environments and worlds'. This is just one mode of organising that was possible. Another option might have been to structure the book around different world oceans (from the Indian, Atlantic to Arctic Oceans and beyond), or to have vertically split the book between surface practices of engaging oceans and those pertaining to the deep. Other arrangements would have likewise been feasible.

The structure here, though, has been designed to allow readers to readily identify chapters of interest, but it should be noted it is in no way intended to fix the chapters in place. There is overlap between the sections and the chapters, whereby themes leak, spill and even flood into others. For example, Waiti and Wheaton's chapter on culture and leisure practices in the opening section on 'frameworks and knowledges' could just as easily have been placed into the section on ocean 'engagements and encounters', or 'power and politics', whereby the Māori oceanic experiences described are lived, embodied and affectual but also never outside of colonial power dynamics past and present. Similarly, chapters by Griffin and Featherstone on processes of protest and acts of solidarity (respectively) are as much centred on ocean labour as they are focused on understanding histories and power. Accordingly, there is more than one way for readers to traverse the structure and it is hoped that the book's format will allow for easy navigation of key ideas from guideposting single-word titles and the possibility of finding more specific theoretical concepts and empirical studies from the subtitles.

**Section II** follows this introduction by setting out framings by which the ocean has been encountered and understood and, in turn, interrogating the knowledges of the ocean(s) that are constructed and perpetuated, but also challenged and transformed. The section begins with an exploration of mapping: one of the central frameworks by which oceans have been represented and 'known'. In this chapter, Jessica Lehman charts the operation of mapping and its politics in inscribing oceans with meanings for their use, as well as shifting this to other forms of ocean knowledge, which inform and also complement mapping practices – modelling and measuring (see also Helmreich on wave science, this volume). Following this, Antony Adler's chapter, from the perspective of history of science, relates the ways in which, again, particular representations or imaginations of oceans – particularly deep ocean spaces – in the ocean sciences, such as oceanography, create particular interpretations, shaping knowledges of the watery environment. Shifting not only from the horizontal shoreline to the vertical water column and the seafloor, and from the late eighteenth century to the present, Adler also moves from the cartographic practices discussed by Lehman, building from her scientific overviews of models and measurements of the ocean to show how the science of our oceans has been developed. Common to Lehman and Adler is an interest in how technologies play a role in the constructions of ocean knowledge. Indeed, technology is a theme that repeats throughout the book, specifically in chapters by Fawcett et al. (in a discussion of deep-sea mining), Squire (in relaying geographies of ocean depth) and Crawley et al. (in their artistic engagements with ocean worlds).

Continuing along a critical thread of how particular knowledges of the ocean are constructed (developing from Steinberg's foundational text, 2001), Basil Germond's chapter next examines the role of representation of the ocean – instead through a framework of nation-state seapower and maritime security, deriving from international relations and security studies perspectives. Here the section shifts from the role of science to more explicitly political, state and military roles in oceanic construction (with an appreciation that the nation state is also crucially part of mapping and scientific practices, as described in preceding chapters). Germond explores the long-held assumption foregrounding the approach of 'modern' states to security practice: the idea of an empty ocean (see also Hadjimichael, this volume). He then tracks the concept of *Mare Liberum*, or 'free seas', and how this has facilitated ocean (ab)uses that have fed into contemporary modes of ocean governance. In demonstrating these constructions, Germond points to the stability of narratives about 'good' and 'bad', and 'acceptable' and 'unacceptable' ocean engagements (e.g. the dominant discourse of the state as the upholder of governance, and the pirate, illegal fisher and so on, as the 'rogue' in need of governance). Although such perspectives endure in some IR scholarship, Germond problematises these 'dominant' representations made through state political and military practice.

Further rethinking the relationship between the state and the ocean, the next chapter expands from the western readings of seapower described by Germond to examine the workings of empire and imperialism more explicitly. Through a framing of the workings of anti-colonialisms and 'heterodox' readings of ocean space shaped by decolonial practice, Andrew Davies demonstrates, vitally, the ways in which frameworks for working with ocean space(s) have been limited, often by the roots of academic thinking and scholarship in imperial or Eurocentric thought. He works through a variety of spaces – the port, the port city, the ship, and military and carceral spaces – to show the complexity of power relations wrought through imperial practice, challenging dominant 'top down' readings and showing how imperialism was not just perpetuated but is also perpetually contested.

Following from this necessary engagement with oceans beyond dominant western and global north framings, the last two chapters of the section continue to engage with the construction of ocean knowledge through decolonial approaches and lenses. These are not alternative or counter constructions in a negative binary to the dominant western discourse but, rather, unsettle and unseat those narratives to reveal greater democracy in ways of knowing and working with the oceans. Leesa Fawcett, Elizabeth Havice and Anna Zalik's chapter advances Davies' attention towards empire to the more specific spatial frame of the frontier. The frontier looms large in ocean knowledge regimes. Like Germond's explanation of an 'empty ocean', the frontier, relatedly, posits the notion of an ocean expanse ripe for exploration and exploitation. Linking back to Adler and Lehman, Fawcett, Havice and Zalik show how frontiers are epistemological – known through particular regimes, tools and technologies that legitimise practices such as imperialism and capital appropriation and extraction. Focusing on three examples – the data-shaping deep seabed mining prospects; data technologies that increasingly make visible and available ocean space for democratising engagements and use; and decolonised, anti-anthropocentric, indigenous knowledges concerning oceans – the authors consider the role of frontier thinking in the exploration, extraction, conservation and commodification of the ocean, as well as necessarily challenging it. Their chapter links forward to the next section (see Thomas, Bond and Diprose) in a discussion of deep-sea mining, but also in the politics related to spaces of economic resource.

Completing the section, Jordan Te Aramoana Waiti and Belinda Wheaton zoom in further than the examples of the previous two chapters by presenting a post- and de-colonial account of ocean culture departing again from western constructions of the empty ocean or the blank frontier. Contrasting another dominant narrative – that the land and sea are separate and distinctive spaces – the authors examine how ocean cultures are known, understood and lived by indigenous Māori, where there is no neat line between liquid and solid worlds. They show the importance of focusing on specific cultural practices in knowing the ocean, which, in turn, upend powerful constructions that write-out localised, traditional and Indigenous knowledges of, and practices in, ocean space. Indeed, Waiti and Wheaton demonstrate how oceans are a space of cultural and political contestation for Māori, wrought through colonial constructions. Yet they also show how everyday leisure practices – waka hourua (double-hulled voyaging canoe), waka ama (outrigger canoe), and heke ngaru (surfing) – represent and embody ways of living with the ocean that are vital for promoting Māori cultural traditions and self-determination.

Having worked through various ways the ocean is constructed (and deconstructed) and knowledge is forged, formed, shared and challenged, **Section III** turns to the broad theme of ocean economies (economics being one of the predominant drivers of frontier politics and state interventions in and across the oceans, as shown by Fawcett, Havice and Zalik, and also chapters by Germond, and Davies – see also Campling and Colás, 2018, 2021). The section also explores, relatedly, ocean labour – the very visceral and felt processes of work at sea. It begins with an intervention considering the spatialities of fishing and, notably, fisheries governance. Madeleine Gustavsson and Edward Allison explore the spatial dimensions and territorialising logics that tend to drive fisheries management but that sit uncomfortably with the livelihoods and lived experiences of fisherfolk. Important to Gustavsson and Allison's chapter is a reading of maritime work that goes beyond the economic to pay attention to how fishing is a way of life. Echoing the preceding chapter, this contribution voices ocean knowledge not from the 'top down' but rather from localised and indigenous perspectives. The chapter ends by exploring how policy needs to be done differently to reflect the worldviews of fisher communities and to push back against the discriminations, injustices and harms that existing policies and approaches to management can enact (see also Satizábal and Batterbury, 2018). In this chapter, Gustavsson and Allison link some of their geographic discussion of fishing to processes of Marine Spatial Planning (MSP).

In the chapter that follows, planning scholar Stephen Jay introduces MSP as a uniquely spatial innovation for organising and coordinating uses in marine space, where there are often competing economic goals for use – fishing being one, but offshore wind energy, oil and gas extraction, shipping and aquacultural developments, as others. MSP remains a relatively novel intervention in planning, but is increasingly being advocated for in the governance and management of national waters in order to optimise and order marine resources. Jay's chapter explores the ways in which planning practices (much like the governance regimes explored earlier by Germond) are built from landed foundations (see Peters, 2020; Peters et al., 2018) whilst also signalling attention to the material qualities of the ocean – its fluidity and flux – in the shortcomings of MSP processes. Indeed, Jay stresses the need for 'fuzzy boundaries' in the planning of marine space that better reflect the oceans' geographies (see Jay, 2018). This materiality of the ocean – its depth, texture, shifting states, qualities and character – is further explored in later chapters (see, in particular, Section VI, Ocean environments, ocean worlds).

Shipping, as a key industry enfolded in MSP activities, is the next focus of the section. As Borovnik, Chua and Heins note in each of their chapters, shipping is central to the functioning of the global economy as we know it (see also Cowen, 2014; George, 2013; Khalili, 2021). And, as the geographer David Harvey has also lamented, the shipping container – the core mode of moving goods by vessel, at sea – is the singular most important (yet deceptively simple) technology "without which we would not have had globalisation" (Harvey, 2010: 6, cited in Martin, 2013: 1022). In this part of the section, Charmaine Chua examines the place of docks and the process of docking, illustrating, as in earlier chapters, the fluid relations between land and sea – in this case those enabled through the technology of the port and its associated infrastructure and labour force. Chua demonstrates not only the development of the dock through time but how port governance has likewise developed by shaping global spatial economic structures through public and private investments and an increasing corporatisation of port services which see "ports insert themselves into global intermodal networks defined by the imperatives of global capital accumulation, rather than by local or public interests" (Chua, this volume). Chua's assessment of docking and ports demonstrates the complex economic and infrastructural geographies that drive economies whereby basic port location is less important than entire chains and networks of connections within which particular ports are embedded. This critical reading also allows us to understand the inequalities wrought through ports and docking practices, not least for workers imbricated into the neoliberal workings of global logistics and commodity chains (see also Featherstone's discussions of seafarer solidarities in Section IV, Ocean histories, ocean politics).

From the dock and its place in the global circuit of goods, the next chapter turns to the specific place of the 'box' or container in understanding connections between economy, space and the sea. Matthew Heins does not tell a straightforward history of the shipping container but, rather, shows how this essential technology of global economic trade has been built on spatial standards – linked to the very microarchitecture of the box itself (see also Heins, 2016; Martin, 2016). He explains how the success of the box lay in the standardisation of its spatial dimensions to enable its intermodal capabilities. He further shows how this spatial standard is further embedded – or *imposed* – into wide-reaching infrastructural systems, with wide-reaching effects, reshaping space and blurring traditional spatial boundaries (such as between land and sea, and ship and shore).

Maria Borovnik next completes a trio of chapters attentive to economics, labour and shipping, with an explicit focus on seafarers. Borovnik's longstanding work on seafarer worlds has offered necessary insights into the often-hidden world of the global oceanic workforce (see Borovnik, 2007, 2011, 2017, for example). In this chapter, she draws from both existing literature and her own ethnographic and interview work to explore the intersections between seafarers' mobilities in space and the inequalities that can result in offshore workplaces. Here Borovnik examines seafarers through two lenses – their place as facilitators of the global economy, and their more localised place on board ships. She demonstrates, in both cases, the difficult working environments seafarers face and negotiations they must navigate. Central to Borovnik's chapter is an account that alerts us to, and gives voice to, those often forgotten in the servicing of the capitalist systems. She shows how, for example, international and national regulations and stipulations on contracts create seafarer uncertainties and precarities in terms of time onboard. She likewise explains how hierarchies and everyday discriminations onboard place some seafarers in the most challenging of circumstances – such as when it is impossible for seafarers to 'jump ship' on the job. Borovnik's attention to seafarers' lives also alerts us to seafarer rights, not least where she examines the influence of the Covid-19 pandemic on seafarers' health and wellbeing.

Scoping back out from shipping as a focal point, the final two chapters of this section consider how space and the economy interrelate in broader examples. Maria Hadjimichael, working in the nascent field of Marine Social Science through a social and political science framework, offers a reflection on the so-called 'Blue Economy' – the use of the ocean, in sustainable ways, for economic benefit. Hadjimichael complicates and problematises the 'blue' growth strategies that have been lauded for their potential to bring sustainable development to the ocean (see also Schlüter et al., 2020). Using the European Union's Blue Growth Strategy as an explicit thinking tool on the topic, Hadjmichael shows how growth raises questions of rights – rights that are inherently spatialised. Drawing from Henri Lefebvre and David Harvey, she argues that the blue economy constructs "injustices" over the "ownership, use and exploitation" of the seas (Hadjimichael, this volume; see also chapters by Fawcett at al. and Thomas et al., both this volume). Using the radical frameworks of Marxism, Hadjmichael raises vital questions related to blue economic growth, which concern "who is affected and how: who becomes dispossessed over their rights to this space (or resource) and who 'wins'" (Hadjimichael, this volume). Closing the chapter, Hadjimichael raises the possibilities of *de*growth in the oceanic commons to challenge regimes of enclosure and privatisation.

Completing this third section, Amanda Thomas, Sophie Bond and Gradon Diprose explore intersections between resources, the economy and politics. Focusing on processes of resource extraction in Aotearoa New Zealand, this chapter links back to the discussion of frontiers in Section II, and forward to debates on 'Security' by Bueger in Section IV. The authors explain how resource geographies are more-than-economic and pertain to questions of sovereignty, rights, and Indigenous and climate justice (also linking back to Hadjimichael, previous chapter). Central to Thomas, Bond and Diprose's chapter is a focus on practices of enclosure and appropriation – themes raised in earlier chapters by Fawcett et al. and Waiti and Wheaton on the role of colonial practices in such processes. This chapter shows the capacity to push back on dominant ocean discourses and the expressions of power articulated through assertions of autonomy by Indigenous, as well as environmental, groups. This chapter continues to demonstrate the necessity for studies to engage post-, de- and anticolonial thinking, not just for understanding how ocean spaces are shaped unequally but also for envisioning how they could be democratised in the future. Also central to Thomas, Bond and Diprose's chapter is the question of security. Drawing from a critical feminist perspective, they consider resource geographies as part of complex questions over future in/securities, which are embodied, felt and lived. They highlight the unevenness of security for different groups of people, dependent upon "how security is defined, by who, and for who" (Thomas et al., this volume). This take on security is a vital one for shifting matters of security from the state to the complex affectual politics of security on 'different' bodies.

The focus on security is taken up further in **Section IV** as attention turns to 'Ocean histories, ocean politics', starting with a more classic take on security and ocean space offered by Christian Bueger. He offers a critical start point to the section by demonstrating how the ocean is made a security space through new spatial configurations, wrought by constructed maritime threats (see Germond, Section II). Here, Bueger explores the political work of marking new spaces such as the High-Risk Area (HRA), established in response to Somali piracy; the so-called Southern Route for Afghan Heroin; and, finally, Areas of Interest and Common Operating Pictures as they are established in recent maritime domain awareness structures. He shows how particular problems are defined in the ocean and the spatial responses that arise, often based on regimes of spatial surveillance that transcend traditional state boundaries. He coins the notion of 'pragmatic' spaces to show how space becomes constructed around particular, constructed security 'problems'.

Continuing this attention to the securisation of the ocean, the next chapter turns to dominant agents of security in sea-space: navies. Duncan Depledge tracks a history of navies, pointing to the global inequality in their distribution and thus countries' abilities to assert naval power across the globe. From this starting point, Depledge explores how (largely Anglophone) naval thinking and practice have evolved since the 1500s, with particular attention to how naval strategists have conceptualised and, in turn, spatialised the sea. Although, Depledge argues, the agendas or objectives of navies in the assertion of power has not changed much over the centuries, the ocean environment has. He demonstrates how geographic interests in oceans' volumes and materialities, and how engaging with spatial ideas of geo-power, hybridity and assemblage could assist in deepening understandings of naval operations beyond descriptive historical or political accounts.

In the following chapter, Isaac Land homes in on the operations of power within the space of the ship via an historical account of the spatial politics of discipline at sea, expressed through the 'total institution' of the ship (following Aubert, 1982 and Rediker, 1987). In a series of examples, Land tracks the complexities of power expressed on ships – merchant and naval, and between different seafarers: captains, first officers, sailors, enslaved people – in respect of discipline and punishment. He exemplifies that, whilst discipline was often hierarchically exercised, there were often expressions of concealment, indiscipline and resistance on board ships. Land shows how the layout and internal spaces of ships mattered to discipline and punishment – to its operation and to the ways those at sea pushed back against it (see also Peters and Turner on discussions of the convict ship, 2015). Although this chapter contributes to our understanding of dominant shipboard modes of politics/power (maritime security, navies, onboard ship discipline regimes), it also marks a transition to the latter chapters in this section that are devoted to resistance and solidarity.

Next, Paul Griffin continues to complicate maritime hierarchies and dominant expressions of power, with a focus on the place of protest in ocean spaces and the spaces connected to ocean worlds. Drawing from radical geographic approaches, Griffin considers acts of protest *at* sea; protests constructed through movement *across* the sea; and *landed* protests articulating grievances of the sea. Continuing themes introduced in Land's chapter, Griffin demonstrates the spatial operation of power 'from below' in forms of protest onboard ships and in the lived and dynamic space of vessels. He shows how protest is constitutive of the making of subaltern identities (see also Featherstone, 2005) and how the seas can become distinct spatialities of disobedience and activism, dissent and resistance. Drawing also from his own work on maritime protest in Glasgow, UK, Griffin charts everyday struggles and exceptional moments. He points to the sea as a productive space in understanding resistances and explains that grievances and protest are ever tied to broader processes of colonialism, slavery and capitalism. For Griffin, exploring protest is a necessary task for acknowledging "alternative and resistant visions that similarly illuminate wider controlling and structural influences" (Griffin, this volume).

David Featherstone follows Griffin's discussion of protest, taking on the topic of 'Solidarities' as they relate to seafaring and maritime labour (see also Chua and Borovnik, in the previous section). Featherstone interrogates the ways in which solidarities take place on connected spaces of land (such as the port), as well as at sea, and the networks that connect seafaring solidarities across space. Indeed, solidarities are not always expressed in specific locations but traverse time and space, linking disparate communities around shared politics. Featherstone shows how solidarities are shaped by space and constitutive of its shaping. He also demonstrates the diversity of solidarities, from maritime solidarities stemming from 'white labourism' and their discrimination towards racialised minorities, to anti-colonial internationalisms, as well as contemporary articulations of maritime solidarity constituted in opposition to the rise of far-right politics. Featherstone's chapter concludes with reference to the solidarities that formed around the German captain, Carola Rackete of the Sea-Watch rescue group, in relation to her activities rescuing migrants from the Mediterranean – a theme taken up in the section's final chapter, by Charles Heller, Lorenzo Pezzani and Maurice Stierl.

In their chapter, which combines the critical work of political science, architecture and film studies, Heller, Pezzani and Stierl examine the role of overlapping spatial jurisdictions in creating particular political geographies of the Mediterranean, which result in creations of humanitarian and *de*-humanitarian politics in relation to saving lives at sea. Indeed, Heller et al. complicate the assumptions of state and inter-state (EU) roles in humanitarian action and stress the role of non-governmental organisations in the vital work of preventing migrant deaths at sea. Like other chapters, Heller and colleagues highlight the fluid relationships that span land and sea in the form of ocean governance (see Bueger, and Germond, this volume), ocean political contestation (Featherstone, and Griffin, this volume) and also the lived experiences of those crossing the oceans (see Chua, and Borovnik, also this volume). This chapter also continues a thread running through the book regarding the ways that turning attention to ocean space allows an exposure of the limitations, and even the violences, of 'traditional' 'master narratives' (Lambert et al., 2006) that, when left uninterrogated, crowd out radical politics and positions that upend dominant modes of understanding and knowing and that are alert to modes of justice and care (see Hadjimichael, and Thomas et al., this volume).

Next, in Stephanie Jones' wide-ranging chapter, she relays the work of writing as a space-making medium for engaging and encountering the seas and oceans, as well as methods of creating and constructing particular ocean spaces for readers. Leading from the previous section and Heller et al.'s examination of migration at sea, Jones reflects upon literature including Nam Le's *The Boat* (2008) and Behrouz Boochani's *No Friend but the Mountains* (2018) to show how such texts "narrate oceans as spaces of unfreedom and freedom" and, as "an enquiry into the 'necropolitics' of the nation state, and what liberty might, can and can't mean" (Jones, this volume, citing Mbembe, 2003). Jones' chapter marks the start of **Section V**, Ocean engagements, ocean encounters, which explores both literary and artistic relations with ocean space, as well as affectual, embodied and sensory explorations of the ocean through the acts of swimming, sailing, surfing and diving. Jones sets out by considering what constitutes literature at sea (from the ship's log to natural history observations), which can be deployed to critically explore the ocean as a way of reading literature. The chapter also charts through literary approaches to Indian, Atlantic, Pacific and comparative regional sea-studies to interests in the submarine, the deep ocean and icy seas (see also chapters by Adler, Squire, and Dodds who further attend to these themes respectively). Notably, Jones demonstrates the post- and de-colonial agencies spun through ocean space (echoing other chapters in this volume: Davies; Fawcett et al.; Featherstone; Thomas et al.; Waiti and Wheaton) this time in writing practice, where she powerfully concludes that, "world literature is being energetically reconceptualised as a decolonising idea" (Jones, this volume).

In the next chapter, Crawley, Critchley and Neudecker likewise demonstrate the capacities of creative practice for encountering the ocean, through their focus on visual arts. The chapter starts by outlining how ocean space has been subject to imaginings in art – not least as a space often hard to access and distanced from the land (see also the chapters by Adler and Squire). The authors then creatively and reflexively turn to examining the creative practices of Critchley and Neudecker – sharing their artistic works and the processes behind their work in capturing elements of ocean space for reflection. They show how art can be a mode of "responding to the effects of technological, ecological and economic exploitation of the oceans", demonstrating the claim of Neimanis et al. (2015: 11) that art can be "a catalyst for new kinds of engagements" which might "*in a very real and political sense*, produce the world we seek to live in" (Neimanis et al., 2015: 10, emphasis in the original). Here the authors also grapple with the ways in which ocean spaces are abstracted through representational artistic practice, but are simultaneously also material, wet, geophysical spaces: a point that is elaborated on throughout Section VI, Ocean environments, ocean worlds. They take note of the ways the imagined ocean is not separate from an ocean space that is stubbornly felt. This is picked up further in the four chapters to follow, which are alerted to the ocean's material form and engagements with the voluminous, liquid, salty, treacherous, sublime space of the sea through practices and performances of swimming, surfing, sailing and diving.

Ronan Foley explores the contemporary zeitgeist for wild swimming and the ways swimming offers an immersive and embodied encounter with the ocean, shaped by the ocean's various characters and qualities. Indeed, he notes the importance of accounting for relations between bodies and (blue) spaces as swimmers traverse dry to liquid worlds and negotiate more-than-human encounters with "sharks, jellyfish, dolphins, jetskis, boats, surfboards", the list goes on (Foley, this volume). Foley is also attentive to the inequalities of swimming – to the social geographies and "in and out of placeness" (Cresswell, 1996) that swimming spaces reveal through politics of access linked to gender, race, age, ability and intersections of those and other identities. Finally, building from his extensive research, Foley expands on the relations between ocean space and health geographies and the therapeutic affordances of swimming, shifting beyond linear biomedical accounts linking healthy bodies to healthy seas to a rather more critical understanding that situates the potentials of ocean space and health within questions of access, belonging, dignity and equality.

Foley's chapter on swimming is followed by Jon Anderson's chapter that explores the leisure pursuit of surfing. Surfing has been a niche, yet longstanding, interest shaping studies under the banner of 'geographies of the sea' – because of the very fact that surfing unlocks complex socio-cultural, political and environmental geographies, all the while complicating them in the context of the mobile, dynamic space of the sea (see Anderson, 2022 and Evers, 2009; Olive, 2019; Waitt, 2008). In this chapter, Anderson continues to understand the ocean as an immersive space in charting the 'what', 'where', 'how', 'why' and 'who' of surfing. Anderson takes a critical approach that resonates with Waiti and Wheaton's earlier chapter on culture and leisure practice that, while alert to the colonial politics of surfing (surfing as socio-cultural spatial appropriation) and its inequalities – particularly around the gendered access to space – also recognises surfing's potentials for resisting masculinist forms of sea-engagement and 'mastery' of the waves. Indeed, Anderson shows how surfed practices territorialise pockets of water, with perpetuated spatial practices determining the politics of line ups and drop ins, conversely, creating moments of spatial rupture. Importantly, Anderson shows how the body and ocean coalesce or converge in surfed practice, further complicating ideas on the relations between land and sea, body and water, ship and shore, which span this book.

In the following chapter, outdoor education scholar and professional sailor Mike Brown examines the platform of the boat in active engagements and encounters with the ocean. In an account that draws heavily on ethnographic methodologies, Brown explores embodied connections with the sea and how practices of sailing lead to particular oceanic knowledges. Through drawing out examples of sailing-as-practice, Brown reflects upon the material qualities of ocean space that come to define it as a space of alternate experience from the land, but he also uses these qualities to challenge the often-held western perception of the sea as empty (and in turn, bound-able for governance, or claimable for resources; see earlier chapters by Bueger; Fawcett et al.; Gustavsson and Allison; Hadjimichael; Thomas et al.). As Brown relays, "[f]or the sailor the sea is neither empty nor featureless" and this could be considered as "a crude and lazy shorthand, a way of saying 'I'm too busy to look, to see this as it is in itself'" (Brown citing Dorgan, 2004: 94, this volume). Looking to, and *feeling* the sea, Brown reflects on the materiality of the ocean (including its surprising solidness [see also Dodds, this volume]), the place of skill in navigation, and the emotional dis/connections with water (through feelings of grace, or through the affectual qualities of water, see also Anderson's discussion, this section, on the sensation of stoke whilst surfing). Building from his previous work on the sea (Brown and Humberstone, 2015; Brown and Peters, 2018), Brown concludes with a reminder of the difference between writing *about* the sea and writing *with* the sea through encounters, and what this spatial difference could evoke for environmental citizenship.

Also taking an immersive approach, Elizabeth Straughan's chapter on diving concludes the section. She takes the section full circle through exploring artistic practice and representation but also hints towards the final section of the book and its coverage of oceanic spaces as three-dimensional, volumetric zones of deep ecological crisis. Drawing from ethnographic accounts and conversations, Straughan's chapter considers engagements through diving that are both touristic (recognising the necessary addition of tourism to accounts of ocean space) and part of working practice (the underwater being a workplace, adding further nuance to understandings of labour and the ocean explored by Borovnik and others in Section III). Straughan's chapter further points towards the more-than-human underwater world, providing a connection to chapters in the section to follow on the topic of the deep (Squire) and ocean life (Johnson).

The final section of the book – **Section VI** – turns to Ocean environments, ocean worlds, and pays attention to the geophysical properties, material shapes and state-shifting capacities of oceans, and their role in earth systems. It also brings to the fore the forces and impacts of climate change that are submerged, but present, in other chapters of the volume. The section begins where the previous left off by exploring ocean depths. Here Squire's chapter complements Adler's contribution on the role of understanding deep space in the history of ocean science, as well as Crawley et al.'s and Straughan's chapters on underwater encounters and engagements. However, Squire takes a more firmly geopolitical approach in tracking ocean depths, including early oceanic representations and scientific work, submarine cable-laying and technological communications development, attempts at living underwater trialled during the Cold War by 'aquanauts' (see also Squire, 2021) and the 'gold rush' touted to emerge with deep-sea mining (see also Fawcett et al., this volume). Squire's chapter concludes by thinking of depth in relation to rising seas and climatic emergencies, arguing that this demands a reimagining of ocean depths as well as the development of ocean platforms (surface technologies) for countering the increasing depths around us.

Staying with the deep sea, Elizabeth Johnson's chapter explores more-than-human ocean life. Whilst this theme has arisen in previous chapters (see Crawley et al. and also Straughan), Johnson's chapter critically examines the relations of human and more-than-human worlds at sea through three important frames: biopolitics and ethics; consumption and extraction; and geopolitics and militarisms. Johnson begins by reflecting on ethical questions related to how people and life at sea relate: reframing simple questions of use and overuse of sea life as resource to instead ask critical questions of how marine life is positioned and what this means in relation to acts of care, protection, stewardship and even grief in relation to biodiversity loss. How life is calculated determines how it is valued and the ethical practices associated with it. Here, Johnson reminds us that colonial politics has also shaped how marine life is treated – in public perception and policy-scapes. In the second section she complicates 'blue economy' understandings and economic readings of ocean space and the place of marine life in entanglements of consumption (see also Hadjimichael, this volume), before closing the chapter by looking at the ways in which marine life is enrolled in various military activities linked to geopolitical strategy. The latter attention (see also Squire, 2020) notably provides a necessary reading of intersections between ocean space and military practices that extend the naval and seapower discussions of earlier chapters (see Depledge, and Germond).

Continuing a thread of connections between science and ocean space, Anthropologist Stefan Helmreich next takes on consideration of a specific element of ocean space – the wave. Departing from engagements of waves as described through surfing practice (see Anderson, this volume), Helmreich's chapter tracks back to Lehman's opening chapter on methods of modelling and measuring oceans, and forward to the close of the book in considering the more-than-human potentials of waves and how wave science helps us to grapple with anthropocenic change and sea-level rise (see also Squire, and Savitzky, this section). Helmreich's chapter demonstrates that, even as waves are stubbornly material, they are also "thinkable as media" (Helmreich, this volume), where technologies help scientists 'read' the tangibility of waves in specific ways, constituting particular knowledges of the oceans' form, mobilities and reach. Like the previous chapters, Helmreich connects histories of wave science to understandings of colonial practice, war and military action, in and through spaces at sea. Waves also, as Helmreich notes, are "hybrid forms that mix the phenomenological, mathematical, technological, legal, and more" (this volume). This idea of the connections, mixings and fluid relations that occur in (and beyond) the ocean leads to the next chapter where Jeremy Schmidt examines the Hydrosphere, "the combined mass and movement of all water on Earth in all its forms" (Schmidt, this volume).

In this expansive chapter, Schmidt does the necessary work of exploring an 'ocean in excess' (Peters and Steinberg, 2019), an ocean that is not simply oceanic, in occupying the distinct bounded 'blue' spaces between land on the map, but rather the ocean as part of the water cycle, and wider Earth System. Charting understandings of the hydrosphere from the late-nineteenth century through to the twenty-first century, Schmidt's chapter, like earlier ones, interrogates how ocean spaces are 'known' through scientific endeavour and geopolitical strategy. Most vitally, in merging 'ocean geographies' with broader geographies of water, water cycles, environment and geology, Schmidt provides a critical consideration of how the histories of bodies of water (the histories of oceans and oceans long disappeared) are linked to ways of thinking geologically, to global readings of the hydrosphere and ocean space particularly that understand it as part of integrated Earth systems.

The turn towards geology and to *earthly*, or *grounded*, ways of thinking about ocean space is followed by the penultimate chapter on ocean spaces that exist beyond their often-assumed liquidity. Here, Klaus Dodds explores sea ice and the shifting geographies and properties of oceans (solid to liquid, liquid to air) and, in turn, the spatial engagements that arise when the ocean is icy. Dodds draws from longstanding work (notably see Dodds, 2018) to explore sea ice as imagined, elemental, geopolitical – in other words, as something representational, material and practiced. He explores how ice is enrolled in both Indigenous and popular depictions of the ocean in ways that create conflicting knowledges of maritime space – as a site of conquest, as a political opportunity, as the environment of one's everyday lifeworld. Dodds further highlights the legal geographies that emerge as sea ice complicates where international conventions largely fail to attend to the particularities/peculiarities of ocean as ice, sea as solid. Dodds' chapter also considers the ever-present spectre of ice melt and the potentials of this for reshaping global geographies and mobilities.

The final chapter of the collection deals with this knowledge of an increasingly oceanic world of sea-level rise through exploring the place of islands, and particularly the strategies of island nations such as Singapore, to cope with an encroaching ocean. Drawing on, and problematising, themes of land–sea relations, this concluding chapter complicates (and inverts) those relations by showing how the sea is made land through processes of dredging and reclamation, and how islands adapt through vertical construction to the seas around them. In this chapter, Satya Savitzky builds from his formative work on how climate is forging new spatialities through the emergence of Arctic sea routes with increased ice melt (see Savitzky, 2016) to examine the production of new geographies through island building and its associated politics. Returning full circle to the very start of this book, Savitzky's chapter demonstrates how maps of the ocean – maps of the world – are ever in flux.

## From ocean geographies to ocean *spaces*: Themes, limitations and potentials

What is clear from the structure of the book, previously outlined, is that this is a book about ocean *spaces*, not necessarily ocean *geographies*. The distinction is important. Although geography is present (in a classical sense of absolute and relative location) – as are *geographies* (in the sense that the book presents multiple frames for understanding locale and senses of place) – the key contribution of the volume, as noted earlier, is to present a provocation on *spatialities* in relation to the ocean. In focusing on the spaces that are made in, and made from, the ocean, at times we turn away from the discipline of geography *per se* to consider how scholars from a range of fields engage space in making sense of relations with the ocean. The book thus reaches far beyond the formal discipline of geography to engage a wide range of individuals working with the ocean – including disciplines such as history, sociology, security studies, border studies, international relations, literature, politics, anthropology, architecture, health and leisure studies, education and film studies, as well as those working in the nascent field of marine social sciences and those working outside of the academy as professional artists, or those conducting research as independent scholars. It collates voices across career stages and across various oceans. Within the areas outlined, authors also work within and across sub-disciplinary settings – such as mobilities studies or the history of science, or within geography, for example, as part of political, socio-cultural or economic approaches to the discipline.

What is striking is that, as our chapter synopses have highlighted, in spite of various disciplinary or interdisciplinary starting points, there are key threads running through the chapters. Regardless of their disciplinary orientation, authors in this volume display a keen interest in the various constructions of ocean spaces; how particular ways and regimes of 'making' the oceans shape how they are understood and used; and how those understandings and uses can also be undermined. Linked to this, the book grapples, throughout, with identifying the work of 'dominant' oceanic discourses but likewise highlights the power and possibilities of knowing ocean spaces beyond western imaginaries. Indeed, running through a breadth of chapters are post-, de-, and anti-colonial approaches that are vital to understanding the ways in which oceans are spatial, and that *space writ large* is shaped by various ocean ontologies. That said, the book can be accused of lacking a wider diversity of authors, whereby a greater representation of Global South scholars and decolonial scholarship is needed. Patricia Noxolo's reminder should be heeded: "decolonisation begins from the scholarship of black and indigenous peoples, and should be led by that scholarship" (2017a: 318). The radical nature of such work, the necessary discomforts and ruptures it brings to academic spaces – and, in this case, ocean spaces – is further demanded, or else such ideas lose agency as they become "harnessed and domesticated in Western academic spaces" (Noxolo, 2017b: 342). It is not enough to say future work must do more. We as editors must do more to step forwards and *back* in more fully engaging decolonial perspectives on ocean spaces. Our focus on the open-ended, relational politics of ocean spatialities, rather than the bounded world of ocean geographies, opens space for these voices but it does not, on its own, fill it, and thus certain critical voices, even in this volume, remain unheard.

As Fawcett et al. note in their chapter, decolonial lenses also alert us to ocean spaces that are constituted by non- and more-than-human life, where indigenous peoples have relations to environments, animals, elements and sealife that differ from the stark lines drawn in global northern and western epistemologies. This is a reminder of another core thread of the book – one that is alert to ocean spaces complicated by the agencies of marine life, and the very geophysicality of oceans themselves and their relationality to other earthly spaces. Indeed, the handbook reflects on the qualities, characters and properties of ocean spaces as well as the animate and agential capacities of life within. However, the book does not exhaust all of these threads and, again, further contributions may have considered broader planetary–oceanic connections (e.g. the linkages between air, atmosphere and oceans, or skies, surveillance and drones and the oceans). Likewise, the book attends to the vertical depths and volumes of water within 3D articulations of geopolitics but could have expanded to think about the vertical aerially (following work on satellite observations, mentioned in Lehman's chapter, and further work 2016, 2018), as well as expanding more on the concept of ocean surfaces – mobilities across them, and the flattening of routes, plans and projections for policy (see Peters, 2020).

Indeed, perhaps a final shortcoming of the book is its relative lack of attention to policy and law. The United Nations Convention on the Law of the Sea surfaces, throughout, alongside a host of other sea-related legislative tools and policy guidance. With a few exceptions (e.g. Jay, this volume), the book only indirectly addresses how thinking spatially about the oceans and thinking of oceans as spaces informs, deforms, crosscuts and undercuts strategies and directives for the very futures of ocean spaces, their practical management for human use, and their more-than-human health. The United Nations Decade for Ocean Science (2021–2030) and emergent climate reports being conceived, researched, written and disseminated (e.g. the Intergovernmental Panel on Climate Change [IPCC] 2021) alongside measures to control migration and offshore asylum seekers and to deter pirates and prevent stowaways all point to the ways this book could stress, to a greater degree, what a spatial perspective on ocean issues could add to discourse and debate and to building more democratic ways of relating to the ocean in policy. Nonetheless, the book presents an attempt to take space seriously when working with the oceans, and to take seriously oceans as spaces of multiplicity, meaning, materiality, movement, and more. As Emma McKinley aptly writes in the foreword to this collection, "the lenses of geographical inquiry and their particular focus on the interconnectivity between space and spatiality can provide us with critiques and insight which will be invaluable to how we live, work and play in our ocean spaces, both now and in the future". This book starts this project. It is hoped others will continue it.

## References

Anderson J (2022) *Surfing Spaces*. Abingdon: Routledge.

Anderson J and Peters K (eds) (2014) *Waterworlds: Human Geographies of the Ocean*. Farnham: Ashgate.

Anim-Addo A (2011) 'A wretched and slave-like mode of labor': Slavery, emancipation, and the Royal Mail Steam Packet Company's coaling stations. *Historical Geography* 39: 65–84.

Aubert V (1982) *The Hidden Society*. London: Transaction Books.

Bear C (2017) Assembling ocean life: More-than-human entanglements in the blue economy. *Dialogues in Human Geography* 7(1): 27–31.

Blum H (2010) The prospect of oceanic studies. *PMLA* 125(3): 670–677.

Boochani B (2018) *No Friend But The Mountains*. Sydney: Picador.

Borovnik M (2007) Labor circulation and changes among seafarers' families and communities in Kiribati. *Asian and Pacific Migration Journal* 16(2): 225–249.

Borovnik M (2011) The mobilities, immobilities and moorings of work-life on cargo ships. *SITES* 9(1): 59–82.

Borovnik M (2017) Night-time navigating: Moving a container ship through darkness. *Transfers: Interdisciplinary Journal of Mobility Studies* 7(3): 38–55.

Brown M and Humberstone B (eds) (2015) *Seascapes: Shaped by the Sea*. Farnham: Ashgate.

Brown M and Peters K (eds) (2018) *Living with the Sea: Knowledge, Awareness and Action*. Abingdon: Routledge.

Campling L and Colás A (2018) Capitalism and the sea: Sovereignty, territory and appropriation in the global ocean. *Environment and Planning D: Society and Space* 36(4): 776–794.
Campling L and Colás A (2021) *Capitalism and the Sea: The Maritime Factor in the Making of the Modern World*. London: Verso.
Carson R (1989 [1951]) *The Sea Around Us*. Oxford: Oxford University Press.
Childs J (2020) Extraction in four dimensions: time, space and the emerging geo (-) politics of deep-sea mining. *Geopolitics* 25(1): 189–213.
Choi YR (2020) Slippery ontologies of tidal flats. *Environment and Planning E: Nature and Space* https://doi.org/10.1177/2514848620979312
Cowen D (2014) *The Deadly Life of Logistics: Mapping Violence in Global Trade*. Minneapolis, MN: University of Minnesota Press.
Cresswell T (1996) *In Place-Out of Place: Geography, Ideology, and Transgression*. Minneapolis, MN: University of Minnesota Press.
Cresswell T (2014) *Place: An Introduction* (second edition). Oxford: Wiley.
Davies A (2019) Transnational connections and anti-colonial radicalism in the Royal Indian Navy mutiny, 1946. *Global Networks* 19(4): 521–538.
DeLoughrey E (2019) Toward a critical ocean studies for the Anthropocene. *English Language Notes* 57(1): 21–36.
Dickson AJ (2021) The carceral wet: Hollowing out rights for migrants in maritime geographies. *Political Geography* 90 https://doi.org/10.1016/j.polgeo.2021.102475
Dittmer J (2018) The state, all at sea: Interoperability and the Global Network of Navies. *Environment and Planning C: Politics and Space* https://doi.org/10.1177/2399654418812469
Dodds K (2018) *Ice: Nature and Culture*. London: Reaktion.
Dunnavant JP (2021) Have confidence in the sea: Maritime maroons and fugitive geographies. *Antipode* 53(3): 884–905.
Evers C (2009) 'The Point': Surfing, geography and a sensual life of men and masculinity on the Gold Coast, Australia. *Social and Cultural Geography* 10(8): 893–908.
Fairbanks L, Boucquey N, Campbell LM and Wise S (2019) Remaking oceans governance: Critical perspectives on marine spatial planning. *Environment and Society* 10(1): 122–140.
Featherstone D (2005) Atlantic networks, antagonisms and the formation of subaltern political identities. *Social and Cultural Geography* 6(3): 387–404.
George R (2013) *Deep Sea and Foreign Going: Inside Shipping, the Invisible Industry that Brings you 90% of Everything*. London: Portobello Books.
George RY and Wiebe SM (2020) Fluid decolonial futures: Water as a life, ocean citizenship and seascape relationality. *New Political Science* 42(4): 498–520.
Gibbs L and Warren A (2014) Killing sharks: Cultures and politics of encounter and the sea. *Australian Geographer* 45(2): 101–107.
Glissant É (1997) *Poetics of Relation*. (trans. B Wing). Ann Arbor, MI: University of Michigan Press.
Gray NJ, Acton L and Campbell M (2020) Science, territory, and the geopolitics of high seas conservation. In: O'Lear S (ed) *A Research Agenda for Environmental Geopolitics*. Cheltenham: Edward Elgar Publishing, 30–43.
Hau'ofa E (1995). Our sea of islands. In: Wilson R and Dirlik A (eds) *Asia/Pacific as Space of Cultural Production*. Durham, NC: Duke University Press, 86–98.
Hau'ofa E (2008) *We are the Ocean: Selected Works*. Honolulu, HI: University of Hawaii Press.
Heins M (2016) *The Globalization of American Infrastructure: The Shipping Container and Freight Transportation*. New York, NY: Routledge.
Ingersoll KA (2016) *Waves of Knowing*. Durham, NC: Duke University Press.
Intergovernmental Panel on Climate Change (IPCC) (2021) Sixth assessment report. Available at: www.ipcc.ch/assessment-report/ar6/
Jay S (2018) The shifting sea: From soft space to lively space. *Journal of Environmental Policy and Planning* 20(4): 450–467.
Johnson ER (2015) Of lobsters, laboratories, and war: Animal studies and the temporality of more-than-human encounters. *Environment and Planning D: Society and Space* 33(2): 296–313.
Khalili L (2021) *Sinews of War and Trade: Shipping and Capitalism in the Arabian Peninsula*. London: Verso.
Lambert D, Martins L and Ogborn M (2006) Currents, visions and voyages: Historical geographies of the sea. *Journal of Historical Geography* 32(3): 479–493
Le N (2008) *The Boat*. Sydney: Penguin.

Legg S (2020) Political lives at sea: Working and socialising to and from the India Round Table Conference in London, 1930–1932. *Journal of Historical Geography* 68: 21–32.

Lehman J (2016) A sea of potential: The politics of global ocean observations. *Political Geography* 55: 113–123.

Lehman J (2018) From ships to robots: The social relations of sensing the world ocean. *Social Studies of Science* 48(1): 57–79.

Lehman J (2020) Making an anthropocene ocean: Synoptic geographies of the International Geophysical Year (1957–1958). *Annals of the American Association of Geographers* 110(3): 606–622.

Mack J (2013) *The Sea: A Cultural History*. London: Reaktion Books.

Martin C (2013) Shipping container mobilities, seamless compatibility, and the global surface of logistical integration. *Environment and Planning A* 45(5): 1021–1036.

Martin C (2016) *Shipping Container*. New York, NY: Bloomsbury Academic.

Massey D (2005) *For Space*. London: Sage.

Neimanis A, Åsberg C and Hayes S (2015) Post-humanist imaginaries. In: Bäckstrand K and Lövbrand E (eds) *Research Handbook on Climate Governance*. Cheltenham: Edward Elgar Publishing, 480–490.

Noxolo P (2017a) Introduction: Decolonising geographical knowledge in a colonised and re-colonising postcolonial world. *Area* 49(3): 317–319.

Noxolo P (2017b) Decolonial theory in a time of the re-colonisation of UK research. *Transactions of the Institute of British Geographers* 42(3): 342–344.

Olive R (2019) The trouble with newcomers: Women, localism and the politics of surfing. *Journal of Australian Studies* 43(1): 39–54.

Peters K (2010) Future promises for contemporary social and cultural geographies of the sea. *Geography Compass* 4(9): 1260–1272.

Peters K (2017) Oceans and seas: Physical geography. In: Richardson D (ed) *International Encyclopedia of Geography: People, the Earth, Environment and Technology*. New York, NY: Wiley.

Peters K (2020) The territories of governance: Unpacking the ontologies and geophilosophies of fixed to flexible ocean management, and beyond. *Philosophical Transactions of the Royal Society B* 375(1814) https://doi.org/10.1098/rstb.2019.0458

Peters K and Squire R (2019) Oceanic travels: Future voyages for moving deep and wide within the 'new mobilities paradigm'. *Transfers* 9(2): 101–111.

Peters K and Steinberg P (2019) The ocean in excess: Towards a more-than-wet ontology. *Dialogues in Human Geography* 9(3): 293–307.

Peters K, Steinberg PE and Stratford E (2018) (eds) *Territory Beyond Terra*. London: Rowman and Littlefield.

Peters K and Turner J (2015) Between crime and colony: Interrogating (im)mobilities aboard the convict ship. *Social and Cultural Geography* 16(7): 844–862.

Pugh J (2016) The relational turn in island geographies: Bringing together island, sea and ship relations and the case of the Landship. *Social and Cultural Geography* 17(8): 1040–1059.

Rediker M (1987) *Between the Devil and the Deep Blue Sea: Merchant Seamen, Pirates, and the Anglo-American Maritime World, 1700–1750*. Cambridge, MA: Cambridge University Press.

Satizábal P and Batterbury SP (2018) Fluid geographies: Marine territorialisation and the scaling up of local aquatic epistemologies on the Pacific coast of Colombia. *Transactions of the Institute of British Geographers* 43(1): 61–78.

Satizábal P and Dressler WH (2019) Geographies of the sea: Negotiating human-fish interactions in the waterscapes of Colombia's Pacific Coast. *Annals of the American Association of Geographers* 109(6): 1865–1884.

Savitzky S (2016) Icy futures: Carving the northern sea route. Unpublished doctoral dissertation, Lancaster University.

Schlüter A, Bavinck M, Hadjimichael M, Partelow S, Said A and Ertör I (2020) Broadening the perspective on ocean privatizations: An interdisciplinary social science enquiry. *Ecology and Society* 25(3) article 20 www.ecologyandsociety.org/vol25/iss3/art20/

Spence E (2014) Towards a more-than-sea geography: Exploring the relational geographies of superrich mobility between sea, superyacht and shore in the Cote d'Azur. *Area* 46(2): 203–209.

Squire R (2020) Companions, zappers, and invaders: The animal geopolitics of Sealab I, II, and III (1964–1969). *Political Geography* 82 https://doi.org/10.1016/j.polgeo.2020.102224

Squire R (2021) *Undersea Geopolitics*. London: Rowman and Littlefield.

Stafford J (2017). A sea view: Perceptions of maritime space and landscape in accounts of nineteenth-century colonial steamship travel. *Journal of Historical Geography* 55: 69–81.

Steinberg PE (2001) *The Social Construction of the Ocean*. Cambridge, MA: Cambridge University Press.
Steinberg PE (2009) Oceans. In: Kitchen R and Thrift N (eds) *International Encyclopedia of Human Geography*, Vol. 8. Oxford: Elsevier, 21–26.
Steinberg PE (2013) Oceans. In: Warf B (ed) *Oxford Bibliography of Geography*. New York, NY: Oxford University Press.
Steinberg PE (2017) Oceans and seas: Human geography. In: Richardson D (ed) *International Encyclopedia of Geography: People, the Earth, Environment and Technology*. New York, NY: Wiley.
Steinberg PE and Peters K (2015) Wet ontologies, fluid spaces: Giving depth to volume through oceanic thinking. *Environment and Planning D: Society and Space* 33(2): 247–264.
Stierl M (2021) The Mediterranean as a carceral seascape. *Political Geography* 88 https://doi.org/10.1016/j.polgeo.2021.102417
Underhill-Sem YTRROT (2020) The audacity of the ocean: Gendered politics of positionality in the Pacific. *Singapore Journal of Tropical Geography* 41(3): 314–328.
Waitt G (2008) 'Killing waves': Surfing, space and gender. *Social and Cultural Geography* 9(1): 75–94.
Walsh C and Döring M (2018) Cultural geographies of coastal change. *Area* 50(2): 146–149.
Wang CM and Chien KH (2020) Mapping the subaquatic animals in the Aquatocene: Offshore wind power, the materialities of the sea and animal soundscapes. *Political Geography* 83 https://doi.org/10.1016/j.polgeo.2020.102285
Warr B and Arias S (eds) (2009) *The Spatial Turn: Interdisciplinary Perspectives*. Abingdon: Routledge.

# SECTION II

# Ocean frameworks, ocean knowledges

# 2
# MAPPING
## Measuring, modelling and monitoring the oceans

*Jessica Lehman*

### Introduction

Mapping the ocean appears, at least superficially, to be a somewhat paradoxical notion. As Pezzani (2014: 156) writes, "if geography expresses, in its very etymology, the possibility to write and therefore read the surface of the earth, the sea seems to stand as the absolute opposite". Oceans have long appeared as the 'negative space' on many maps, and have frequently been described as resisting inscription and the imposition of boundaries or other markers of territory. Schmitt (2003: 43) describes the ocean as having "no character", a quality that forms the basis of his political economic theory, which poses a necessary binary between the land as a space of governance and the ocean as a space of freedom. Schmitt is not alone in this oceanic ontology; Barthes describes the ocean as a "non-signifying field [that] bears no message" (quoted in Pezzani, 2014: 156). And for Deleuze and Guattari (1987), the ocean is the most fundamental example of smooth space: uninscripted, ungriddable, ungovernable (see also Germond, this volume).

Yet these characterisations ignore the many efforts to territorialise and enclose the sea as a way to exert power over mobility and resources, as well as to pose an 'outside' to land-based territory; to govern freedom (see, for example, Campling and Colás, 2021; Steinberg, 2009). Moreover, against notions of the sea as unwritable, unknowable, and unmappable, there has emerged a relatively recent proliferation of marine geospatial data and related ways of mapping the sea in order to further a multitude of oft-conflicting interests. As Pezzani (2014: 158) writes, "scientifically, the possibility of understanding and controlling the dynamic forces that shape the ocean (whether human, oceanographic, or meteorological) hinge on the ability to draw (partial) lines and strategically exercise some sort of control over those parts". While mapping the sea – imposing lines of fixity on an unfixable entity – may appear to be contradictory and illusory, efforts to do so have real consequences in the world.

It is impossible to provide either an exhaustive index of contemporary ocean mapping efforts, or a complete history of these attempts. What I offer instead is a brief examination of some of the most influential ways that knowledge, power, and the materiality of the sea are folded together in attempts to map the ocean. In doing so, I examine attempts to create spatial

DOI: 10.4324/9781315111643-4

configurations of marine knowledge/power that go beyond mapping to include the closely related activities of measurement, modelling, and monitoring. In this chapter, mapping refers to efforts to exert territorial power across the ocean's surface, perhaps best exemplified during periods of exploration, marine mercantilism, and imperial expansion prior to the twentieth century. Measuring highlights efforts to quantify the ocean's properties, which has been a crucial part of governing resources and has gained importance with developing understandings of the ocean's role in climate. Modelling extends this attempt to know the ocean and produce marine geospatial data beyond what can be observed. Monitoring indicates mainly contemporary efforts to track the ocean as a changing, dynamic entity by measuring its characteristics in real time. While there is some sense of temporal progression through these categories, each instance that I explore shows that they operate in concert rather than as four distinct phases or sets of practices, and that they do not simply reveal the natural world but also configure dense nodes of power and knowledge. Through the joint activities of mapping, measuring, modelling, and monitoring, the contours of the close relationship between governance and knowledge of the ocean come into view (see Campbell et al., 2016).

Beyond revealing how these different logics and practices work together, this chapter shows that mapping the ocean exposes some of the paradoxes and tensions at the heart of knowing, governing, and living with the sea. With their adherence to fixed systems of latitude and longitude, and their divisions between land and water upon which modern systems of governance depend, maps "[fail] to communicate the complexity of the ocean as a mobile space whose very essence is constituted by its fluidity and that thereby is central to the flows of modern society" (Steinberg, 2013: 160). Nonetheless, maps play complex yet central roles in contested regimes of ocean governance and efforts to monitor the ocean's role in the global environment. Ultimately, attending to the practices of ocean mapping allows us to better understand the ocean's paradoxical nature – as frictionless space and abundant resource, as at once risky in its ability to threaten life on Earth, and as a fragile ecosystem at risk (see also Lehman, 2018). The practices of ocean mapping show us not only how these paradoxical understandings have developed, but also why they matter.

## Mapping oceans, building empires

By now it is commonly understood in the discipline of geography and beyond that mapping, along with other forms of representing geospatial data, is not a neutral, objective activity but is, in fact, laden with power. Moreover, mapping does not simply reflect the world but *produces* it; maps act as beings in the world, enabling certain possibilities and foreclosing others (see Kitchin and Dodge, 2007). Similarly, maps are inseparable from the conditions by which they are produced, and cartography itself is a socially determined way of creating knowledge about the world (see Pickles, 2012). We must understand mapping, then, as a contextual, productive, political, and unstable set of practices.

This understanding of mapping, as a contingent world-making practice, is immediately relevant to attempts to map the ocean. Maps are a central device for imposing regimes of territory on the ocean, a practice that is rife with difficulty given oceanic properties. As Phillips (2018: 60) writes, "efforts to construct territory in the deep ocean that build upon terrestrial ontological assumptions have been confounded by the movement of water and the human and non-human actors that move with it and through it". And yet, maps can also show how the ocean's lack of fixity has been vital to the very development of terrestrial ontologies of territory. For example, Steinberg (2009) argues that the ocean had to be mapped as a space of mobility in order to secure land-based territory within the rise of the nation-state system.

Tracing differing representations of oceans in influential world maps, Steinberg shows how the ocean was not only constructed as an 'outside' to the nation-state, but that this 'outside' must be understood as more than what was 'left over' when the territory had been defined. By attending to the features of world maps such as toponyms, rhumb lines, and grids, Steinberg (2009) traces understandings of the world ocean (in the Western world) from a "space of nature and society", to a "space of routes", to a "space of mathematics and memory". These shifts indicate not changes in understandings of what the ocean fundamentally *was*, but rather changes in its role in society. The ocean had to be constructed as a space outside of territorialisation, a space of mobility, in order for land-based governance to cohere. This construction was a social process that took centuries, and that both fundamentally involved and was reflected in cartographic practices.

Steinberg (2009) shows how maps participated in the creation of the ocean as a space outside of territorial control. This understanding of the ocean was perhaps most fully achieved in the eighteenth century with the application of elliptic lines, which suggest that "the ocean is a space of pure mathematical abstraction, a dematerialized arena of potentially limitless time–space compression, or idealized annihilation" (Steinberg, 2009: 485). This understanding was augmented by the inclusion of depictions of historic routes and voyages; yet these "avoided any implication that the ocean was a space for contemporary social activity or assertions of territorial power" (Steinberg, 2009: 485).

While we can trace, following Steinberg, how this understanding of the ocean developed through cartographic practice from the fifteenth to eighteenth centuries, true measurement of the ocean's properties did not inform graphic representations until about the middle of the nineteenth century (Reidy and Rozwadowski, 2014). Reidy and Rozwadowski (2014) tie the emergence of interest in systematically measuring and mapping the ocean's properties to the imperial development of the United States, through its shipping and whaling industries, and Britain, through its overseas territorial expansion. These attempts to scientifically map the ocean were financed and executed by imperial navies, initiating a relationship between oceanographic science and military might that continues to this day – a history and present evidenced not least in the geographies of underseas telecommunication cables (Starosielski, 2015). Moreover, Reidy and Rozwadowski (2014: 346) suggest that nineteenth-century oceanography has played a fundamental role in the development of many sciences precisely due to the shift in ocean mapping and measuring from "sporadic, experimental efforts into systematic, routine work". I will return to some of these themes in a following section.

The twentieth century saw ocean mapping take on a more internationalist character, though efforts were still certainly interwoven with nationalistic and imperial aims. Here the ocean came to more fully inhabit its paradoxical nature as both place and space. Maps play crucial roles in defining ocean space in the United Nations Convention on the Law of the Sea (UNCLOS), which was negotiated during the 1980s and ratified in 1994 (Hirst and Robertson, 2004). And yet producing and maintaining geospatial data for the purposes of defining ocean territory in accordance with UNCLOS is indicative of some of the tensions at the heart of ocean mapping. The low-water coastline provides a 'baseline' for the extension of sovereign territory into the sea. While "UNCLOS provides a guide as to how maritime boundaries should be determined, [it] is generally silent on how often the boundaries should be revised", calling into question the legal incorporation of contemporary practices of boundary monitoring (Hirst and Robertson, 2004: 3). Not only are more precise mapping technologies coming into use, but the low-water line may also be changing due to climate change and other geophysical processes, and various other factors may compel nations to review their baselines (Sammler, 2020). GIS now makes these constant revisions in measurements and maps possible. Although it seems perhaps intuitive that more accurate measurements are better, a certain stability is pragmatic for the enforcement of territorial claims, especially when it comes to issues such as offshore oil extraction, where infrastructures may not be especially flexible. Here we can see that relationships between sovereignty, mapping, and the sea are complex and continue to have numerous implications for both power and knowledge as a new agenda for ocean governance in an era of climate change and new ocean resources emerges (Campbell et al., 2016).

## Measuring oceans, making resources

As imperial expansion was linked to economic exploitation of the sea (at least among Western empires), the notion of the ocean as resource was born: it "transformed from highway to destination" (Reidy and Rozwadowski, 2014: 341). Reidy and Rozwadowski (2014) show that for territorial power to be extended over the ocean in the age of imperialism, representing it as a space of mobility would be insufficient, even though its construction as a space of freedom was still highly relevant. The important role of measurement in cohering power and representation becomes apparent here. As they write, "imperial practice and ideology led to the assumption that marine resources should be exploited – maximally – by people with the knowledge and power to identify and extract them" (2014: 350). It was also in this effort to exploit marine resources – and ocean space – for imperial gain that the ocean began to be studied as a volume as well as a surface, with the extension of ocean sounding surveys significantly further from the coast. If we are now compelled to think of the ocean as a space of volume (see Lehman, 2013; Steinberg and Peters, 2015), we can trace this ability to these measurements. Moreover, volume does not simply imply measurements in another dimension. Measuring the ocean volumetrically involves dealing with an entire set of dynamics, including "instability, force, resistance, depth, and matter alongside the simply vertical" (Elden, 2013: 45).

Thinking with the ocean and the forces that operate in its volumetric space appears to be conducive to thinking with a world in flux (Steinberg, 2013; Steinberg and Peters, 2015). Yet the movements of the ocean also exist in tension with other dynamics, other speeds and stillnesses. For example, writing of offshore oil mining, Phillips (2018: 51) proclaims that "the oil industry remains closely tied to place yet operates in environments where place is continually reformed by the movement of water and all that moves with it and through it", highlighting tensions between territory and different materialities in offshore oil fields. Thus, we can see that contending with the volumetric dynamics of the ocean introduces complications of time as well as space. In another context, Havice (2018) shows that following the governance of mobile ocean resources (such as migratory fish species) evidences the more-than-territorial dimensions of sovereignty.

Another facet of international boundary mapping, sovereignty, and marine materiality involves the continental shelf and the seabed. Mapping the boundaries of the continental shelf plays a key role in determining rights to seabed resources. As Phillips writes, "to define the limits of the continental shelf is to recodify a relationship between sovereignty and vertical and volumetric spaces" (2018: 55). Beyond the continental shelf, the seabed is defined in UNCLOS as the "common heritage of (hu)mankind". But Zalik (2018: 345) argues that when it comes to mineral resources in this area beyond national jurisdiction (called the Area in UNCLOS parlance), "a common property approach nominally prevails yet pre-existing investors claims are protected", and thus private companies are given advantage and transparent exchange of science and technology (an UNCLOS principle) is elided. In arguing that regimes of mineral exploitation in the Area are best understood through a geopolitical lens, Zalik (2018) shows not simply that politics penetrates into the ocean's most remote frontiers but also that the internationalism of the high seas is both a space where global South nations might make redistributional claims and where the power of corporate, national, and parastatal actors is reinscribed.

There is insufficient space here to undertake a complete history of ocean mapping, but by considering a contemporary form of organising and using marine geospatial data, we can see how understandings of the ocean as both a resource and a space continue to be held as variously complementary and in tension. Marine spatial planning (MSP) both indicates and advances a new form of enclosing the ocean, which

> has involved an unprecedented intensity of map-making that supports an emerging regime of ocean governance decisions where resources and their utilization are geocoded, multiple and disparate marine uses are weighed against each other, spatial tradeoffs are made, and exclusive rights to areas and resources are established.
>
> *Boucquey et al., 2019: 485*

While MSP is discussed in greater detail elsewhere (see Jay, this volume), it is relevant to mention here because of the degree to which it depends upon the production and representation of geospatial data. Map-making in MSP can be understood to support and enact the ongoing enclosure and neoliberalisation of ocean space, as it frequently advances the interests of capital within a managerial and technocratic framework (Flannery et al., 2018; Smith and Brennan, 2012). In MSP, maps become "obligatory passing points" for participation, though perhaps equally or more importantly they are the data portals that inform MSP decision-making (Smith and Brennan, 2012: 212; see also Boucquey et al., 2019). Maps are the "spatial representations into which actors are drawn", while data portals indicate both a governance outcome and a set of shifting relations between different actors (Smith and Brennan, 2012: 214). By attending to these practices of mapping and making geospatial data, we might understand not simply how ocean governance is changing, but also how different possibilities emerge in relation to hegemonic forces. Indeed, Fairbanks and colleagues (2018: 154) write, "even as [MSP] is a vehicle for state legibility of offshore environments and activities, the practice also provides communities and other actors with opportunities to subvert and reterritorialize the assemblage through data and intervene in governance and enclosure".

This section has explained how mapping ocean resources constructs the ocean as a particular kind of space. Yet, emphasising the challenges of mapping the sea and extending regimes of territory beyond land might miss key connections between these different spaces – connections with both spatial and temporal dimensions. The materialities of land, sea, ice, and air are unstable and indistinct (Steinberg and Kristoffersen, 2017). Moreover, "analysis of the practice of territory at sea shares conceptual ground with long-standing principles of terrestrial resource studies: that understanding relationships between enclosure, commodification, and struggle is central to understanding the transformation of landscapes" (Phillips, 2018: 67). If notions of the sea as unmappable and ungovernable elide certain realities of contemporary political power, they also miss opportunities for the analysis of economic and political processes that have both shaped the world, and that might suggest alternative outcomes.

## Monitoring the ocean, creating knowledge

Creating knowledge about the ocean is a technologically challenging exercise for the very same reasons that it is constructed as the 'outside' of land-based governance. The ocean is physically inhospitable to human bodies and technologies, long understood as "a forbidding and alien environment inaccessible to direct human observation", thus requiring constant technological mediation (Rozwadowski and van Keuren, 2004: xiii). Its size, in both extent and depth, challenges attempts at synoptic coverage. Its mobility and changeability, including perhaps especially its tightly coupled relationships with the atmosphere, pose challenges for time-series sampling. And regimes of ocean governance make much global ocean research both inherently international and bureaucratically complex. When it comes to the ocean, it is very difficult to "make *global* data", using Edward's (2010: xv) phrasing (my emphasis), or to gather systematic records of the ocean on a planetary scale. It is equally challenging to "make data *global*", or to create "coherent global data images [from] highly heterogeneous, time-varying observations" (Edwards, 2010: xv; emphasis added). These difficulties are greatly compounded in attempts not simply to map the ocean's surface or even its depths, but to understand it volumetrically, with all of the geophysical/chemical/biological dynamics and exchanges this implies.

Contemporary efforts to monitor the ocean take many different forms, and have many different ends, including widespread surveillance for border enforcement, global shipping, and marine conservation (see, for example, Davis et al., 2004; Pezzani and Heller, 2019). Here, I focus on the practices of ocean knowledge associated with the broad field of oceanography, which are increasingly applied to these different domains. Oceanography, whether undertaken by research institutes, government agencies (including military operations), or the private sector, is a largely data-driven science, particularly when it comes to attempts to map the ocean's physical characteristics at the global scale (see for example Cai et al., 2014). Ocean observations can be characterised by two conflicting tendencies. On one hand, ocean observations have been relatively scarce, especially when compared to analogous observations of the atmosphere, largely due to the challenges mentioned in the previous paragraph. On the other hand, ocean observations have undergone a tremendous technological revolution in a matter of decades, resulting in exponentially more ocean data circulating in the worlds of science and engineering than ever before, feeding equally revolutionary representations of the ocean that inform policy decisions and cultural conceptions of a dynamic sea (Conway, 2006; see also Lehman, 2018).

Ocean observations are collected using an ever-developing range of technologies, from ship-based sampling to satellite measurements. Without these observations, most of the representations of the ocean discussed in previous sections are impossible. Yet what does it mean to 'observe' the ocean, given its opacity and intractability? Ocean observations serve constant reminders of the limitations of the human body to make sense of the sea. And yet, the practice of making ocean observations constantly emphasises human relationships and embodied experiences with the sea, even as 'autonomous' technologies proliferate. Helmreich (2009) has argued that rather than remote sensing we should understand human/technical collaborations in the sea as a form of 'intimate sensing'. This sensing, Helmreich (2009; see also Helmreich, 2007) explains, involves transductions between humans and technologies as well as across human senses, where sound and hearing play an outsize role, revealing our overdependence on the visual as a way of apprehending the world (not least through mapping). As several scholars of ocean technology emphasise, observing technologies do more than extend human senses to the sea. The complex, varied relationship between humans and technology "transforms both our own bodies and the material world" (Camprubí, 2018). Modes of ocean sensing shape what it means to be a scientist, and what it means to know the natural world (Gabrys, 2016; Lehman, 2018).

A number of authors trace the development of observing technologies to show that technological shifts have resulted in changes to how we know the ocean as a particular kind of space. For example, Höhler (2002) follows a shift in methods of determining ocean depth from weight sounding to echo sounding. She shows that through this technical development of dense measurements, ocean depths became a space of knowledge, and thus "depth gained stability and validity" (2002: 128). Ocean depths went from something unknowable (and hence unmappable), to "scientifically coherent and convincing, tight and sound" (2002: 145). Significantly, it was not that scientific instrumentation revealed the hidden depths but rather "the questions to be asked and the hidden objects to be unveiled were gauged and defined in the process of depth measurement itself" (Höhler, 2002: 144). Other technologies for mapping and measuring the ocean, and their related data infrastructures, have generated further changes in what we understand the ocean to be, and what we consider to be appropriate objects of knowledge. Particularly relevant, perhaps, is the role these technologies have played in the transition from descriptive to dynamic oceanography that has occurred since the 1950s in western science (Hamblin, 2014; Mills, 2009). The latter can be characterised by a "preference for mathematical modeling, intensive data collection, integrated studies of several disciplines, and prediction" (Hamblin, 2014: 354).

This turn to dynamic oceanography is reliant on a set of networked technologies. Benson (2012) shows how Argos, a satellite-based observation infrastructure, both emerged from and advanced notions of the ocean as a global set of flows. In recent years, dynamic oceanography has found an apex in the re-organisation of ocean observations around principles of real-time monitoring and user-driven science, which have served to advance a notion of the ocean as a set of data streams (Lehman, 2016). This construction suggests an ocean that is constantly changing, even defined by potentiality (Lehman, 2016). At the same time, privileging one way of seeing the ocean means ignoring others. Hamblin (2014), for example, details an incident in the 1950s where dynamic oceanographers from hegemonic western oceanographic institutes were keen to use radioactive fallout as a tracer of ocean currents whereas descriptive oceanographers were more interested in understanding the cumulative effects of radiation on ocean life over time. Thus, the ocean that appears on maps and in models does not pre-exist the technologies and practices that construct it, even as it obscures both "its infrastructural history and conditions of possibility" and other forms of vision (Helmreich, 2011:1211; Hamblin, 2014).

## Modelling the ocean, making predictions

As Höhler (2002) suggests, despite the importance of observations, knowing the ocean is never solely a matter of direct observation (see also Goodwin, 1995). Nor is it solely a matter of visual observations at all. Gabrys (2016: 145) argues that "oceans have become sensor spaces with an extensive array of sensing nodes and drifting sensor points", and goes on to show how contemporary monitoring practices can bring dispersed and largely invisible phenomena like the pacific garbage patch into public consciousness. In contrast to visual observations taken in-situ or at a distance, as well as to abstracted mapping projects, contemporary ocean data is made by a heterogeneous assemblage of practices and technologies, including sensors that themselves "become environmental, [...] as drifting and circulating objects within enfolding gyres" (Gabrys, 2016: 140). It is this complex of ocean sensing that makes possible contemporary practices of dynamic oceanography and what are perhaps their most distinctive feature: predictive models.

Working in concert with sensors, remote imagery, and other observations, models are inescapable in contemporary understandings of the sea. They not only map the ocean's physical properties, but also create predictions of a changeable future (Hamblin, 2014; Lehman, 2016). More specifically, perhaps, models and simulations in oceanography provide important capacities to undertake activities such as testing hypotheses and exploring certain dynamics that might be otherwise impossible given the material and temporal constraints of doing science at sea (Lahsen, 2005). While computer models are a relatively recent development, numerical and physical forms of modelling have been vital to creating coherent images of the ocean as a global entity for decades if not centuries (see Camprubí, 2018; Edwards, 2010). Ocean modelling is a complex undertaking, and this has resulted in two different processes of modelling: one which attempts to model ocean dynamics in their three-dimensional complexity and one which seeks to construct " 'workable' oceans that can act as a boundary condition [to] atmospheric models but do not have the physical detail or response of the real ocean" (McGuffie and Henderson-Sellers, 2005: 188). As Lahsen writes,

> [i]n the coupled models [...] the ocean might tend to 'drift' away uncontrollably, a consequence of the linear rather than non-linear feedback structure of the model. In addition, large regions of modeled oceans have at times turned into solid ice.
>
> *Lahsen, 2005: 900*

Yet, the coupling between the ocean and the atmosphere is a key dimension of climate dynamics and thus considerable efforts have been made to improve coupled ocean–atmosphere models and increase the computing power necessary to make them utile (McGuffie and Henderson-Sellers, 2005).

Social sciences scholarship on computer modelling of both the ocean and the climate more broadly is still in its infancy (for exceptions see Edwards, 2010; Hastrup and Skrydstrup, 2013). However, scholars in the social sciences and humanities are increasingly interested in the digital representations of oceans that these models, along with related data visualisations, can produce (Gray, 2018; Helmreich, 2011). For example, Helmreich analyses Google Ocean as "a mottled mash of icons, indexes, and symbols of the marine and maritime world" (2011: 1211). On one hand, Helmreich argues, Google Ocean and related visualisations fulfil the longstanding dream of making the ocean depths transparent. Yet they also bring together "multiple representations, real and fictive, and multiple semiotic registers, iconic, indexical, symbolic, which can operate independently of one another (in different layers) while still forming part of a composite", opening the possibility for different interpretations for different publics (2011: 1232). Gray (2018) argues that new ocean representations and related technologies of calculation and visualisation play a key role in shifting ideas of the ocean from a resource frontier to a conservation and science frontier. In sum, the continuities and discontinuities between new digital representations and more static maps of the ocean seems a fruitful area for additional research.

Examining ocean models in the context of mapping also reveals certain important continuities in their epistemological and ontological functions. Lahsen (2005) finds that modellers frequently blur the lines between results from their models and observations in their thinking and discussion. Lahsen points out that in doing so they are more likely to leave the uncertainties of their models unacknowledged. Just as maps must be revealed not as mirroring reality but as both products and builders of worlds, so too must models. The nature of models as "truth machines" is perhaps even more powerful given the significant authority that models have come to inhabit (Lahsen 2005: 904; see also Hulme 2013). As Hume writes, "they need to be understood not merely as tools of scientific enquiry, but as powerful social objects" (2013: 41). Maps, measurements, and models, along with the monitoring practices on which they increasingly depend, materialise similar and related powers for explaining and making worlds.

## Conclusion

While the ocean may be the 'blank space' on many maps, throughout this exploration of mapping, measuring, and modelling, the power of ocean representations has been a constant refrain. Moreover, we have seen that ocean representations cannot be understood apart from the actual practices, the engagements with the ocean's physical properties, that both impede and facilitate such mapping (on this, see also Helmreich, 2014). Efforts to chart the ocean's various dimensions are inseparable from questions of its societal rule and geopolitical governance. Thus, the creation of ocean knowledge reveals the ways in which the ocean is variably and often simultaneously constructed as a space of risk, of resources, of opportunity, and of potential catastrophe on a planetary scale.

It is also worthwhile to note more broadly that while maps and related practices appear to emerge largely in the service of hegemons this is far from always the case. As Kitchin and Dodge write, following Pickles, "the power of maps as actants in the world (as entities that have effects) [is] diffuse, reliant on actors embedded in context to mobilize their *potential* effects" (2007: 334). As on land, counter-mapping at sea has the potential to contest oppressive state power. An example can be found in the work of Lorenzo Pezzani and Charles Heller, who founded a project called *Forensic Oceanography*. This project uses techniques of mapping and surveillance, namely drift modelling and remote sensing, to hold state actors accountable for migrant deaths in the Mediterranean (see Heller and Pezzani, 2020; Pezzani, 2014). Pezzani (2014: 159) describes *Forensic Oceanography* as one of a set of mapping projects that "attempt to make a certain political problem emerge by expanding the aesthetic and technological possibility in order to see and document the violations of the rights of migrants and transform the sea into an arena of conflict" (see also Heller et al., this volume).

Of course, this exploration of ocean mapping is far from exhaustive; I have provided a range of instances and developments that emerge from a variety of historical and geographic contexts. The purpose of these examples is to show some of the tensions of ocean mapping; tensions between land-based notions of territory and marine materiality, between internationalism and sovereignty, between freedom and fixity, between hegemonic and counter-hegemonic ocean politics, to name a few. The tensions, and the way they are borne out in the examples I have provided, underscore the necessity of understanding mapping as a contingent, political, world-making process.

## References

Benson E (2012) One infrastructure, many global visions: The commercialization and diversification of Argos, a satellite-based environmental surveillance system. *Social Studies of Science* 42(6): 843–868.

Boucquey N, St Martin K, Fairbanks L, Campbell LM and Wise S (2019) Ocean data portals: Performing a new infrastructure for ocean governance. *Environment and Planning D: Society and Space* 37(3): 484–503.

Cai W, Avery SK, Leinen M, Lee K, Lin X and Visbeck M (2014) Institutional coordination of global ocean observations. *Nature Climate Change* 5(1): 4–6.

Campbell LM, Gray NJ, Fairbanks L et al. (2016) Global oceans governance: New and emerging issues. *Annual Review of Environment and Resources* 41: 517–543.

Campling L and Colás A (2021) *Capitalism and the Sea.* London: Verso.

Camprubí L (2018) Experiencing deep and global currents at a 'Prototypical Strait', 1870s and 1980s. *Studies in History and Philosophy of Science Part A* 70: 6–17.

Conway EM (2006) Drowning in data: Satellite oceanography and information overload in the Earth sciences. *Historical Studies in the Physical and Biological Sciences* 37(1): 127–151.

Davis KLF, Russ GR, Williamson DH and Evans RD (2004) Surveillance and poaching on inshore reefs of the Great Barrier Reef Marine Park. *Coastal Management* 32(4): 373–387.

Deleuze G and Guattari F (1987) *A Thousand Plateaus: Capitalism and Schizophrenia.* (trans. B Massumi). Minneapolis, MN: University of Minnesota Press.

Edwards PN (2010) *A Vast Machine: Computer Models, Climate Data, and the Politics of Global Warming.* Cambridge, MA: MIT Press.

Elden S (2013) Secure the volume: Vertical geopolitics and the depth of power. *Political Geography* 34: 35–51.

Fairbanks L, Campbell LM, Boucquey N and St Martin K (2018) Assembling enclosure: Reading marine spatial planning for alternatives. *Annals of the American Association of Geographers* 108(1): 144–161.

Flannery W, Healy N and Luna M (2018) Exclusion and non-participation in Marine Spatial Planning. *Marine Policy* 88: 32–40.

Gabrys J (2016) *Program Earth: Environmental Sensing Technology and the Making of a Computational Planet*. Minneapolis, MN: University of Minnesota Press.

Goodwin C (1995) Seeing in depth. *Social Studies of Science* 25(2): 237–274.

Gray NJ (2018) Charted waters? Tracking the production of conservation territories on the high seas. *International Social Science Journal* 68(229–230): 257–272.

Hamblin JD (2014) Seeing the oceans in the shadow of Bergen values. *Isis* 105(2): 352–363.

Hastrup K and Skrydstrup M (eds) (2013) *The Social Life of Climate Change Models: Anticipating Nature*. New York, NY: Routledge.

Havice E (2018) Unsettled sovereignty and the sea: Mobilities and more-than-territorial configurations of state power. *Annals of the American Association of Geographers* 108(5): 1280–1297.

Heller C and Pezzani L (2020) Forensic oceanography: Tracing violence within and against the Mediterranean frontier's aesthetic regime. In: Lynes K, Morgenstern T and Paul IA (eds) *Moving Images: Mediating Migration as Crisis*. Bielefeld: transcript Verlag, 95–126.

Helmreich S (2007) An anthropologist underwater: Immersive soundscapes, submarine cyborgs, and transductive ethnography. *American Ethnologist* 34(4): 621–641.

Helmreich S (2009) Intimate sensing. In: Turkle S (ed) *Simulation and its Discontents*. Cambridge, MA: The MIT Press, 129–150.

Helmreich S (2011) From spaceship earth to Google ocean: Planetary icons, indexes, and infrastructures. *Social Research* 78(4): 1211–1242.

Helmreich S (2014) Waves: An anthropology of scientific things. *HAU: Journal of Ethnographic Theory* 4(3): 265–284.

Hirst B and Robertson D (2004) Geographic Information Systems, Charts and UNCLOS–Can they live together? *Maritime Studies* 136: 1–6.

Höhler S (2002) Depth records and ocean volumes: Ocean profiling by sounding technology, 1850–1930. *History and Technology* 18(2): 119–154.

Hume M (2013) How climate models gain and exercise authority. In: Hastrup K and Skrydstrup M (eds) *The Social Life of Climate Change Models: Anticipating Nature*. New York, NY: Routledge, 30–44.

Kitchin R and Dodge M (2007) Rethinking maps. *Progress in Human Geography* 31(3): 331–344.

Lahsen M (2005) Seductive simulations? Uncertainty distribution around climate models. *Social Studies of Science* 35(6): 895–922.

Lehman J (2013) Volumes beyond volumetrics: A response to Simon Dalby's 'The geopolitics of climate change'. *Political Geography* 37: 51–52.

Lehman J (2016) A sea of potential: The politics of global ocean observations. *Political Geography* 55: 113–123.

Lehman J (2018) From ships to robots: The social relations of sensing the world ocean. *Social Studies of Science* 48(1): 57–79.

McGuffie K and Henderson-Sellers A (2005) *A Climate Modelling Primer*. Chichester: Wiley.

Mills EL (2009) *The Fluid Envelope of our Planet: How the Study of Ocean Currents Became a Science*. Toronto: University of Toronto Press.

Pezzani L (2014) Mapping the sea: Thalassopolitics and disobedient spatial practices. In: Weizman I (ed) *Architecture and the Paradox of Dissidence*. Abingdon: Routledge, 151–162.

Pezzani L and Heller C (2019) AIS politics: The contested use of vessel tracking at the EU's maritime frontier. *Science, Technology, & Human Values* 44(5): 881–899.

Phillips J (2018) Order and the offshore: The territories of deep-water oil production. In: Peters K, Steinberg PE and Stratford E (eds) *Territory Beyond Terra*. London: Rowman and Littlefield, 51–67.

Pickles J (2012) *A History of Spaces: Cartographic Reason, Mapping and the Geo-Coded World*. Abingdon: Routledge.

Reidy MS and Rozwadowski HM (2014) The spaces in between: Science, ocean, empire. *Isis* 105(2): 338–351.

Rozwadowski HM and Van Keuren DK (2004) *The Machine in Neptune's Garden: Historical Perspectives on Technology and the Marine Environment*. Sagamore Beach, MA: Watson.

Sammler K (2020) The rising politics of sea level: Demarcating territory in a vertically relative world. *Territory, Politics, Governance* 8(5): 604–620.

Schmitt C (2003) *The Nomos of the Earth*. (trans. GL Ulmen). New York, NY: Telos Press.
Smith G and Brennan RE (2012) Losing our way with mapping: Thinking critically about marine spatial planning in Scotland. *Ocean & Coastal Management* 69: 210–216.
Starosielski N (2015) *The Undersea Network*. Durham, NC: Duke University Press.
Steinberg PE (2009) Sovereignty, territory, and the mapping of mobility: A view from the outside. *Annals of the Association of American Geographers* 99(3): 467–495.
Steinberg PE (2013) Of other seas: Metaphors and materialities in maritime regions. *Atlantic Studies* 10(2): 156–169.
Steinberg PE and Peters K (2015) Wet ontologies, fluid spaces: Giving depth to volume through oceanic thinking. *Environment and Planning D: Society and Space* 33(2): 247–264.
Steinberg PE and Kristoffersen B (2017) 'The ice edge is lost… nature moved it': Mapping ice as state practice in the Canadian and Norwegian North. *Transactions of the Institute of British Geographers* 42(4): 625–641.
Zalik A (2018) Mining the seabed, enclosing the Area: Proprietary knowledge and the geopolitics of the extractive frontier beyond national jurisdiction. *International Social Science Journal* 68(229–230): 343–359.

# 3
# SCIENCE
## Histories, imaginations, spaces

*Antony Adler*

### Introduction

Scholars working across various branches of the humanities and social sciences, from history to geography, sociology to media studies, are increasingly shedding their long-standing terrestrial bias and focusing their attention upon human activities at sea. Historians of science in particular, like other scholars of the maritime world, take as their starting point the understanding that the ocean is a place where humans have lived and sought to understand and influence the natural environment in diverse ways (Rozwadowski and Van Keuren, 2004: xi). Efforts by scientists to understand and influence aquatic, rather than terrestrial, environments have required the mediation of technologies as well as efforts of imagination that have not always paralleled developments in the natural sciences carried out on land. The ocean, in constant motion, and in many ways inhospitable to human life, long resisted the kind of scientific probing that has been possible on land. The area to be studied, covering two thirds of the globe's surface, enhanced the difficulties. "[T]he vastness of the areas to be considered", wrote American oceanographer Henry Bigelow (1931: 4), "determined the paths that the science of oceanography has followed". For scientists, piercing the shroud of the abyss has been as much an exercise in imagination as a technological feat. Attention to imagination is therefore critical when trying to understand the relationship between marine *science* and the ocean's spatial dimensions and features.

My goal in this chapter is not to provide a comprehensive overview of the development of the marine sciences – this has been accomplished by others (see, in particular, Deacon, 1971; Rozwadowski, 2005; Schlee, 1973) – but rather to explore how the relationship between marine science, marine space and the imagination has played out in different marine environments. With its focus on science, this chapter offers a complementary, yet distinctive, contribution to understanding relations between the art, the imagination and ocean space offered in this volume by Crawley, Critchley and Neudecker and to the "deep" offered by Squire, but it can be usefully read in connection. For this chapter, the oceans could easily be subdivided into a large number of specific geographical locals, and it would undoubtedly be profitable to take as case studies the history of scientific investigations of seamounts, submerged banks, specific currents, or any number of marine features. For the purpose of the task at hand, I will limit myself to the shallows, the open-ocean, and the seafloor. Before examining how imagination

 DOI: 10.4324/9781315111643-5

has played a role in the scientific exploration of these oceanic regions, let us begin by noting the distinctive role of imagination in marine science.

## Imagining vast expanses

Since the beginnings of modern marine science in the nineteenth century, and even earlier, understanding ocean spaces has required an interplay between the use of technology for remotely sensing the hidden regions of the ocean and imaginative inference about underwater realms. Although scientists have long sought to invent means of probing the depths, the harsh conditions of the marine environment have often resisted these attempts. Contemporary oceanographers have observed that we have better maps of the surface of the moon and Mars than of the seafloor. What is known about the geography of the ocean has traditionally been limited by what Anne-Flore Laloë (2014: 40–42) describes as a "shipped perspective" – a viewpoint constrained by being shipboard bound. The tracks of vessels on nautical charts and oceanographic survey maps can be read as representative of a "temporal relationship with the ocean, not a spatial one" since the track of a vessel on a chart marked the space occupied by a ship for only a brief period of time. Going to sea, oceanographers had to wrestle with extreme conditions hindering their investigations. As oceanographer William A. Nierenberg once remarked, "[t]o oceanographers the sea is an enormous and restless antagonist" (Fisher, 1969: 145). And marine geologist H. William Menard (1964: x) described the ocean as "regrettably unstable".

Thus, while oceanographers go to sea to study marine phenomena, a component of their work continues to rely on their ability to form mental images of a vast space beyond the reach of direct human perception. In a sense, the mental practices of marine scientists parallel those of astronomers who must often imagine vast, distant, and inaccessible terrains. It is not coincidental that exploration of the oceans and exploration of outer space are routinely rhetorically equated by both astronauts and ocean scientists. Two vessels of the American research fleet carry the names of former Astronauts (*R/V Sally Ride* and *R/V Neil Armstrong*) while NASA's space shuttles (with the exception of *Enterprise*, namesake of the fictional *Star Trek* spacecraft) all carried the names of famous ocean-going vessels.

In the twentieth and early-twenty-first centuries, the relationship between observation and the necessary efforts of imagination in marine science persists. Oceanographer Henry Stommel (1920–1992) described the process by which a marine scientist might gradually arrive at a better understanding of the physical characteristics of the ocean environment. "[W]e cannot see the currents in the real ocean", he tells us. "We do not deduce immediately the machinery of the ocean without the help of the insight which we obtain from a mental image that we invent ourselves". Observations, gathered in "the real ocean", he explains, are used to modify and correct this image. As this iterative process goes on, "it suggests new kinds of image and new kinds of data, and eventually we arrive at what we call a 'model ocean'". This process is repeated, but never finalised. "[A]t deeper levels the process always goes on and on: the final definitive model is never achieved" (Stommel, 1995: I8–I9). Stommel was describing the work of physical oceanographers – scientists who develop mathematical models to explain currents and wave movements (see also Helmreich, this volume; Lehman, this volume). However, the importance of a mental archetype for the interpretation of vast inaccessible oceanic space, and the interplay between imagination, observation, and theoretical model, carries across subdisciplines of oceanography. As Steinberg and Peters have argued, "any attempt to 'know' the ocean by separating it into its constituent parts serves only to reveal its unknowability as an idealized stable and singular object" (2015: 249–250). Attention to the role of imagination is thus imperative as geographers, environmental historians, and historians of science increasingly grapple with the ocean's three-dimensionality, volume, and ever-shifting nature.[1] The characteristics which differentiate oceans from the land shape oceanic politics in ways best understood using a 'wet' ontological framework. As Steinberg and Peters write, the ocean, "through its material reformation, mobile churning, and nonlinear temporality – creates the need for new understandings of mapping and representing; living and knowing; governing and resisting" (2015: 260–261). Similarly, ocean sciences are influenced by the material characteristics of the subject and context of their analysis. Taking the interplay between observation and imagination as an essential component of marine science, let us then consider what role it has played in scientific work conducted in different regions of oceanic space.

## The shoreline, shallows, and seabed

One of the first European naturalists to devote considerable attention to the study of the marine environment was the Italian polymath, military officer, and engineer, Luigi-Ferdinando Marsigli (1658–1730). Marsigli boasts in his writings that unlike other naturalists who merely gathered the testimony of fishermen and sailors, he had carried out his own direct experiments on the waters. His marine studies began in 1670, when he was sent on a diplomatic mission to Constantinople. There he heard from fishermen of the existence in the Bosporus Strait of a deep current flowing in the reverse direction of the surface current. Determined to observe this phenomenon first hand, he set out in a small boat and used a weighted line to sink a series of corks painted to enhance visibility. Watching the movement of the submerged corks through the water, he was able to determine the direction of the counter current (McConnell, 1982: 12). He then set about taking a series of measurements designed to yield information on current speed, water density, and tidal variation.

As a Northern Italian engaged in struggles against the Ottoman Turks, Marsigli's interest in military applications of hydraulic engineering prompted his studies of submarine topography and marine currents (Stoye, 2004: 27).[2] He travelled extensively, was fluent in Latin, Italian, and French, and corresponded with naturalists throughout Europe. When, in 1704, his military career was brought to an end after surrender of his troops at Breisach (he was second in command), he was banished to southern France. Yet exile opened further opportunities for scientific work (McConnell, 1993: 183). In France he carried out depth soundings in the Gulf de Gascoigne and he completed the work for which he is most remembered, his *Histoire Physique de la Mer*, published in 1725.[3]

Marsigli sought a unified theory for understanding the geological relationships between visible topography on land and hidden topography under the sea. He hypothesised that the highest mountains on land must be balanced by marine abysses of depth equal to the mountains' height. In unpublished papers, Marsigli referred to the "organic structure of the terraqueous globe", comparing his task to that of a human anatomist who must reveal inner workings of the human body not accessible to direct observation. He encouraged his readers to imagine the hidden world beneath the sea as a place of wonder, but also a place that could be understood through reason as well as by way of specially designed instruments. He dismissed the idea that some parts of the ocean are fathomless:

> The fishermen, venturing to this slope where they are accustomed to gather coral from 150 to 200 fathoms, and being unable to reach bottom with this much line, imagine that it is unattainable and say, […] with gross exaggeration, that the abyss has no bottom whatsoever and that there is no hope in finding it. This opinion, shared by

> persons interested in the sea [...] seems absurd to me, and rests only on their unwillingness to take the trouble and expense of preparing the necessities for this sounding.
>
> *Marsigli, 1725: 102*

Marsigli was nevertheless forced to admit that such means would be attainable only if "some Prince orders special ships and adequate instruments for the purpose" (Marsigli, 1725: 102). But in the eighteenth century the princes of Europe regarded ocean space as a byway, barrier, or battlefield, and the depths were of little practical interest; it would be almost two centuries before a vessel like the one Marsigli imagined was built.

Although true oceanographic expeditions were not launched until the end of the nineteenth century, many Victorian naturalists discovered the marine world, not in the open ocean, but at the shoreline as they collected seaweeds and marine creatures when the tide receded. Their entry into this liminal environment was governed by the rhythms of the sea, and collectors cast their sojourn below the tideline as a journey to an otherworldly realm, one which inevitably encouraged a different state of mind. "Only let there be sea, and plenty of low, dark rocks stretching out peninsular-like, into it; and only let the dinner-hour be fixed for high-water time, and the loving disciple asks no more of fate", wrote British seaweed collector Margaret Gatty (1809–1873).

> [A]ll the crowned heads of Europe may be shaken without his being able to feel that he cares. When the returning tide has [...] sent him home at last to dinner and things of the earth, earthy, the squabbles of nations may come in for a share of his attentions perhaps; but even then, only imperfectly.
>
> *Gatty, 1872: vii, see also Bryant et al., 2016 on the work of Margaret Gatty*

Thus framed, the shoreline became a distinct space for natural history and meditation as well as escape from worldly cares.

Amateur and professional naturalists alike, in the nineteenth century, flocked to the shores for health and recreation. Armed with guidebooks on natural history they scoured the rocks, tide pools, and beaches for marine plants and organisms, sometimes collected to stock home aquaria. For Victorian sensibilities, natural history was at once a pleasurable and self-improving pursuit. Guidebooks encouraged readers to contemplate the majesty of creation as evidenced by the complexity of the minute creatures discovered under the microscope. By the 1870s, dozens of marine biological stations had sprung up along the coasts of Europe, North America, and even as far afield as Japan. These laboratories, situated in favourable locations for the collection of living marine organisms, allowed naturalists to make systematic studies of coastal fauna. Instead of being lumped together as mysterious but homologous marine areas, shallow coastal waters became scientifically classified regions, distinguishable by naturalists for their distribution of marine life and the facility with which different organisms of particular interest for laboratory work could be collected.[4] In the Bay of Naples a naturalist could easily study sea urchins – useful for understanding embryological development – whereas another naturalist might travel to Woods Hole, Massachusetts, with the intention of studying horseshoe crabs or annelid worms.

## Discovering vertical dimensions of the open ocean

As amateur collectors and professional naturalists discovered the seashore, they also began adopting some of the collecting instruments of commercial fishermen. The most important of these was the oyster dredge, a net held open by a metal frame that could be dragged along the seabed collecting any animals that fell in its path. In 1849 British naturalist William H. Harvey wrote,

> [a]mong the amusements of the sea-shore there is, perhaps, none so capable of yielding a varied pleasure to a person whose taste for Natural History is awakened, as dredging, where it can be carried on under favourable circumstances. [...] When the water is clear and not very deep, the aspect of the bottom [...] often affords a charming submarine picture, as well as reveals the places where the dredge may be most profitably thrown down.
>
> *Harvey, 1849: 116–119*

Already a decade earlier, naturalist Edward Forbes, promoting the dredge as a scientific instrument, had persuaded the British Association for the Advancement of Science to form a "dredging committee" (Rehbock, 1979: 292–368). In an effort to aggregate data gathered by dredgers, the committee disseminated 'dredging papers', a form questionnaire with which naturalists could record the precise times and places of dredging, as well as the organisms recovered. By 1850 Forbes was urging that information gathered in this manner be "tabulated and reduced to an [sic] uniform language with advantage to science", declaring with satisfaction that, "no marine fauna in the world [had] been investigated with anything like the care devoted to that of the British seas" (Forbes, 1851: 193–194). Forbes established a system of zonation for the distribution of marine life: "the littoral zone", "laminarian zone", "region of corallines", and "region of deep-sea corals". He posited that these observed zones were "a representation in miniature of the entire bed of the ocean" (Forbes, 1844: 319).[5]

In the nineteenth century, European and North American interest in the open ocean was driven by the expansion of two industries: the fishery and deep-sea telegraphy. As the fishing fleets of northern Europe jockeyed for access to rich fishing grounds of the North Sea, this ocean region gained the attention of scientists as well. "The natural conditions of the bottom of this great North Sea is in a scientific sense less known than the deserts of the Sahara" lamented the British naturalist Frank Buckland (1881: viii). He argued that explorations of depths close to home were of greater importance than the study of "abyssal depths of far distant oceans" (Buckland, 1881: vii–viii).

Many nineteenth-century naturalists imagined the open ocean as a space full of exotic life. This was the popular view as well. "All voyagers on the wide Ocean concur in telling us that in their far wanderings they still and ever traverse living water", wrote French historian Jules Michelet (1861: 111). Though some, notably naturalist Edward Forbes, theorised that the depths might be devoid of life, the profligacy of life near the surface made others sceptical. Michelet, for example, noted that though

> it has been affirmed that, in the absence of solar light, life, also, must be absent; yet the darkest depths of the sea are studded with stars, living, moving, microscopic infusoriae and molluses [sic]. [...] [A] thousand strange and nameless creatures swarm in those uttermost depths [.] [...] The Sea! [G]lorious Sea, hath her own light, her own Sun, Moon, and Stars.
>
> *Michelet, 1861: 110–111*

Viewed in this way, the ocean presented an untapped field for biological discovery (see also Johnson, this volume).

As terrestrial regions were ever more thoroughly explored and mapped, ocean spaces offered the tantalising promise of 'virgin' terrain for important discoveries. Naturalist Charles Wyville Thomson wrote:

> I had long previously had a profound conviction that the land of promise for the naturalist, the only remaining region where there were endless novelties ready to the hand which had the means of gathering them, was the bottom of the deep sea.
>
> *Thomson, 1873: 49*

But Thomson and his collaborators also imagined in their search for deep-sea life that the importance of their discoveries lay in their usefulness for better determining the relationship of marine and terrestrial life.

Scientists in the early nineteenth century assigned to ocean spaces a temporal dimension distinct from that of terrestrial environments. In the depth scientists hoped to find living fossils like crinoids, primordial ooze (like the infamous *Bathybius haeckelii*), perhaps even prehistoric sea reptiles.[6] In a lecture in the late 1840s, Harvard university zoologist Louis Agassiz announced that he considered it "probable" a mariner would eventually find a living representative of the Ichthyosaurus or Plesiosaurus (Thomson, 1873: 434). In such imagination, probing the depths of the sea was made analogous to peering back in time. "[O]n the watch we were for missing links which might connect the present with the past", C. W. Thomson, future lead scientist of the British *Challenger* expedition, reported in 1873 (Thomson, 1873: 57). As marine scientists spent ever more time at sea and missing links failed to appear, this imagination of the ocean as a region untethered to terrestrial timescales waned.

## Ships, instruments, and ocean space

Although naturalists had long accompanied voyages of exploration, only in the late nineteenth century were expeditions specially organised for the study of the ocean itself. Thomson was already well acquainted with the difficulties of conducting science at sea when he sailed on the famous *H.M.S. Challenger* expedition in 1872. In the summer of 1868 he set out, along with the naturalist William Carpenter, in the surveying ship *Lightning* for a dredging expedition to the Faröe Banks. Thomson remembered the experience in this "cranky little vessel" without fondness. "We had not good times in the *Lightning*", he later wrote. "She kept out the water imperfectly, and as we had deplorable weather during nearly the whole of the six weeks we were afloat, we were in considerable discomfort" (Thomson: 1873, 57). Despite these hindrances the expedition served as proof of concept, showing that deep-sea dredging could produce valuable scientific results, and that scientists could collaborate to mutual advantage with the Admiralty.

Some naturalists described dredging, pulling an open net along the seafloor and bringing up whatever creatures happened to be drawn in, as an extension of the human senses. Léopold de Folin, who carried out oceanographic work on the French expedition of the *Travailleur* in 1881, described the dredge as "the hand of man applied to the bottoms of the abysses" (Folin, 1887: 1). But he also acknowledged the limitations of that technology; the snapshots the dredge provided of the seabed prompted him to consider all that could not be seen or sensed. "What we take from the depths is so little compared to the multitude of beings that live there and certainly the immense spaces upon which the efforts of exploration can barely touch", he lamented (Folin, 1882: 99). Thus, as they were carried across the ocean surface by ship, naturalists were prompted to imagine the hidden world passing beneath the keel. But ships designed for exploration also gave scientists new means for direct observation of marine spaces.

*HMS Challenger* was an experiment in and of itself – the vessel was modified to better serve the requirements of scientific collection and analysis (see Hardy, 2017: 86–93). Yet even with these preparations, the ship was enrolled in the scientific work in unexpected ways. Like other forms of equipment, the ship mediated scientists' experiences of ocean space. The most striking example of this is recorded by the expedition chemist, John Young Buchanan (1919: 125–126). He recalls that since *Challenger* could navigate by either steam or sail, propellers could be drawn up from the water so as to reduce drag when the ship was sailing:

> Looked into from the deck [...] sea-water appeared to be enclosed in it as the water is in a well, but with this difference, that the water, by day, was brilliantly illuminated from below. [...] The screw-well was, in effect, an artificial and perfected *Grotto di Capri*, which was carried round the world. [...] During the whole voyage the colour of the water was under observation in this very perfect apparatus.
>
> *Buchanan, 1919: 125–126*

The *Challenger* expedition proved that life was present everywhere in the ocean, both on the seafloor as well as distributed through the water column. Whereas seafloor creatures could be gathered using dredge nets (though not at extreme depths), determining the distribution of pelagic life in the water column proved technologically challenging. Oceanographic expeditions of the early twentieth century employed a variety of nets extended vertically on a single towed cable to simultaneously capture creatures at different depths. During the Atlantic cruise of the *Michael Sars* in 1910, two cables were deployed, drawing ten trawling nets of varying mesh size (Murray and Hjort, 1912: 48–49). Other methods were soon devised to tackle the problem of vertical sampling. German marine biologist, Carl Chun used a closeable net to study the distribution of marine life in the Gulf of Naples. He was able both to discern regions of pelagic life and also record the seasonal movements of animals in the water column. This led him to conclude that "surface fauna was apparently only the advance guard of the vast army below" (Field, 1892: 801). Using a variety of nets deployed at varying depths, scientists were able to visualise the vertical dimension of marine space and subdivide three-dimensional ocean space into different habitable zones of marine life.

Although earlier naturalists recorded the locations where sampling was conducted at sea, it was only in the late nineteenth century that scientists developed a standardised "observation station" system for coordinating sampling over the course of an oceanographic cruise (see Pinarldi et al., 2018). Wyville Thomson (1878: 258) wrote of the Challenger expedition observing stations (362 in total) that they "were fixed as nearly as possible in a straight line, if possible either meridional or on a parallel of latitude". Yet this collection method still amounted only to a series of snapshots of ocean conditions spread extremely far apart. Simultaneous observations over great distances long remained an impossibility, as were observations of change over time.

By the early twentieth century, developments in chemical and physical oceanography required further standardisation of sampling measurements and new instrumentation (water sampling bottles, specialised thermometers, new laboratory techniques). By 1936 the International Association of Physical Oceanography recommended the adoption of standard depths for measurement taking. These guidelines encouraged sampling at close depth intervals near the surface and larger intervals at greater depth.[7] Marine scientists now gave greater attention to the properties of the water column in its vertical dimension, and not only to the organisms found there.

## Domesticating the deep sea

In the mid-twentieth century a new regime of international maritime law allowed nations to extend exclusive economic rights 200 nautical miles outwards from their coasts. Some marine scientists viewed this extension of legal jurisdiction as an infringement on the freedom of scientific exploration at sea. Writing in 1984, American physical oceanographer Henry M. Stommel (1995: 171) described the 1982 Law of the Sea as "a disaster to those who would study the ocean". In his view, scientific work carried out at sea could benefit an international community and legal restrictions only hampered access to areas of study. By 1966, American oceanographer Roger Revelle (1967: 6), worried that, when compared to the costs of exploring outer space, the exploration of the oceans – "inner space" – was inexpensive, with the risk that soon there could be "hordes of moderately well-to-do amateurs" exploring the depths "perhaps getting in the way of the scientific submarines; perhaps making new discoveries on their own; certainly getting into trouble". Despite scientists' objections that they were losing primacy to ocean spaces, Cold War innovations in diving and submersible technology provided oceanographers with increased access to the ocean's vertical dimension and opened new areas of the deep sea for exploration (see Oreskes, 2003 and Squire, this volume). By 1966, oceanographers spoke of the "invasion of physical and biological oceanography by electronics" (Charlier and Dietz, 1966: 1421). Human-carrying submersibles could reach the deepest parts of the ocean, and the first crewless robotic vehicles were trialled.

In 1977, submarine investigations conducted with the *Alvin* submersible led to the discovery of hydrothermal vents off the coast of the Galapágos. The discovery of hydrothermal vents, and of the communities of marine organisms inhabiting these extreme environments, revolutionised the biological sciences by changing our understanding of how life makes energy (see Corliss et al., 1979). Soon thereafter, marine biologists hypothesised that these 'primeval' sites may have been where life first originated on earth (Baross and Hoffman, 1985: 327). Now scientists reimagined the seafloor as a vast desert interspersed by "oases of life" (Brazelton, 2017). Although oceanographic transect surveys remained important, the discovery of these sites gave specific locations in the deep sea newfound significance. Expeditions set out to visit specific known vent sites.

Were you to descend over 2000 metres in a submersible to visit a methane seep off the California coast, you might be surprised when looking out of the small porthole window to find a plastic pink flamingo looking in at you with 'MBARI' (acronym for Monterey Bay Aquarium Research Institute) scribbled on its side (Figure 3.1). Were you then to journey north across the seafloor to reach the Endeavour Hydrothermal vent field off the coast of Washington State, you might find a life-sized wooden human figure (the likeness of Alvin submersible pilot Dudley Foster), still recognisable, though eerily blackened by exposure to superheated vent fluid (see Delaney et al., 1992). These items are not haphazardly discarded garbage on the seabed. Rather, they were purposely installed as markers to indicate sampling locations or, in the case of the life-sized wooden mannequin, to serve as a scale reference for photography of deep-sea features. If, in your submersible, your fellow passenger is an oceanographer familiar with the Endeavour vent site, she might point to particular chimney features by name, such as: 'Hulk', 'Puffer', 'Peanut', or 'Dante'. These names will not be found on official government maps – they are not officially registered by the Federal government – but they do appear in scientific papers (Woods Hole Oceanographic Institution, 2021). The practice of naming specific underwater features and leaving markers on the seabed serves to transform the inhospitable and perplexing deep-sea environment into a familiar, comprehensible space, imbuing specific

*Figure 3.1* Pink flamingo seafloor marker. Screen capture.
*Source*: Courtesy of Dr Peter R Girguis.

regions and features of the seafloor with scientific importance. It also enlists these topographical features in the inside-jokes of deep-sea oceanographers long after an expedition has returned to land. Domesticating the deep sea by such practices helps to build a community of scientists with shared research focus, and marks the contributions of specific expeditions and individual scientists on the terrain. Increasingly, however, ocean spaces – even the most inhospitable – are no longer the exclusive realm of scientists.

## Conclusion: Toward a panoptic ocean space

In early 2013, a 14-year-old boy named Kirill Dudko, sitting in his bedroom in Donetsk, Hungary, witnessed something that had never before been seen by marine biologists. While watching a live video feed streamed from 900 metres beneath the surface off the coast of Vancouver Island, British Columbia, he observed a whiskered snout enter the frame and, in a single motion, slurp up a hagfish as though it were a strand of spaghetti. Hagfish produce large quantities of slime, a defence mechanism effective against predators like sharks who try to swallow them in several bites. Dudko's observation was the first time an elephant seal had been observed hunting hagfish. Though scientists knew about the depths that elephant seals could reach, and though hagfish had been found in their stomachs, it was the first direct observation of their hunting technique (Lavoie, 2013). The technology that allowed Dudko to make his observation was the NEPTUNE network (Northeast Pacific Time-Series Undersea Networked Experiments), a network of data recording instruments and video cameras placed on the seafloor and connected to land via fiberoptic cable. Ocean observatories allow real-time monitoring and observation of marine phenomena without the need for scientists to go to sea (see also Adler, 2014). In the twenty-first century, cabled observatories, remotely operated vehicles, and ship-to-shore satellite transmission have made remote ocean spaces ever more accessible (Duffy, 2017: 92). Advances in environmental genomics have given scientists the ability to detect the presence of marine creatures from trace amounts of DNA found in sea water samples (see McClenaghan et al., 2020). And innovations in robotics promise the ability to track biological movements in the open ocean even at the microscopic level (Zhang et al., 2021). These technologies extend scientists biological survey capabilities by allowing researchers to monitor species which had often eluded direct observation in the open ocean. At the same time, the growth of social media and spread of high-speed internet access are opening up the oceans to citizen scientists and lay interest.

Naturalists studying the oceanic spaces of our globe have long sought technological means of overcoming the limitations of our terrestrially evolved senses. Over time, technological innovations permitted them to engage the ocean in its full three dimensions. As historians of science have shown in other domains, the ongoing history of science at sea will require attention not only to developments in instrumentation and remote sensing technologies, but also to the mental heuristics that scientists use in order to structure and visualise a vast environment in constant flux. We can discern a back-and-forth relationship between how scientists *imagine* the marine environment and the technological inventions that facilitate interpretation of ocean spaces. Innovations in technology and instrumentation that mediate scientists' experience and understanding of ocean spaces, together with shifting political and economic interests, have reshaped scientific imaginations of oceanic space in the past and will undoubtedly continue to do so in the future.

## Notes

1 For more discussion of the role of verticality in the history of science, see Hardenberg and Mahony (2020).
2 For more on Marsigli, see Olson and Olson (1958) and McConnell (1999).
3 Many different spellings of Marsigli's name are found in the archival record (Marsigli, Marsilli, Marsili, or Marsilly). Ocean sciences historian Eric Mills suggests that the fact we associate Marsigli primarily with marine studies rather than with his work in geography, geology, limnology, or ethnology, is "an accident of history" (2001: 403).
4 The secondary literature on marine stations is extensive (see Alder, 2016).
5 In the Aegean Sea Forbes distinguished eight zones (see Mills, 1978: 513–514).
6 Bathybius haeckelii was identified in 1868 as a form of primordial slime. Subsequent studies determined it to be merely a chemical precipitate (see Rehbock, 1975: 504–533).
7 Temperature and salinity gradients are much more variable closer to the surface, thus these recommendations reflect improved understanding of the physical characteristics of the vertical water column (see Sverdrup et al., 1946: 356–357).

## References

Adler A (2016) The hybrid shore: The marine station movement and scientific uses of the littoral (1843–1910). In: Rozwadowski H and Anderson K (eds) *Soundings and Crossings: Doing Science at Sea 1800–1971*. Sagamore Beach MA: Science History Publications/Watson Publishing International, 157–191.

Adler A (2014) The ship as laboratory: Making space for field science at sea. *Journal of the History of Biology* 47(3): 333–362.

Baross JA and Hoffman SE (1985) Submarine hydrothermal vents and associated gradient environments as sites for the origin and evolution of life. *Origins of Life* 15: 327–345.

Bigelow HB (1931) *Oceanography: Its Scope, Problems, and Economic Importance*. Boston, MA: Houghton Mifflin Company.

Brazelton W (2017) Hydrothermal vents. *Current Biology* 27(11): R450–R452.

Bryant JA, Plaisier H, Irvine LM, McLean A, Jones M and Jones MES (2016) Life and work of Margaret Gatty (1809–1873), with particular reference to British sea-weeds (1863). *Archives of Natural History* 43(1): 131–147.

Buchanan JY (1919) *Accounts Rendered of Work Done and Things Seen.* Cambridge: Cambridge University Press.

Buckland F (1881) *Natural History of British Fishes.* London: Society for Promoting Christian Knowledge.

Charlier RH and Dietz RS (1966) Oceanography: Two reports on the recent international congress in Moscow. *Science* 153(3742): 1421–1428.

Corliss JB, Dymond J, Gordon LI, Edmond JM, von Herzen RP, Ballard RD, Green K, Williams D, Bainbridge A, Crane K and van Andel TH (1979) Submarine thermal springs on the Galápagos Rift. *Science* 203(4385): 1073–1083.

Deacon M (1971) *Scientists and the Sea, 1650–1900: A Study of Marine Science.* London: Academic Press.

Delaney JR, Robigou V, McDuff RE and Tivey MK (1992) Geology of a vigorous hydrothermal system on the endeavour segment, Juan de Fuca Ridge. *Journal of Geophysical Research* 97(B13): 19663–19682.

Duffy EJ (2017) Ocean 2.0. In: Kress WJ and Stine JK (eds) *Living in the Anthropocene: Earth in the Age of Humans.* Washington, DC: Smithsonian Books, 91–94.

Field GW (1892) The problem of marine biology. *The American Naturalist* 26(310): 799–808.

Fisher AC Jr. (1969) San Diego: California's Plymouth Rock. *National Geographic* 136(1): 114–146.

Folin L (1882) Les explorations sous-marines de l'aviso a vapeur Le Travailleur en 1880 et 1881. *Bulletin de la Société des Sciences, Lettres et Arts de Pau, 1881–1882* 11(11): 41–99.

Folin L (1887) *Sous les mers: Campagnes d'explorations du 'Travailleur' et du 'Talisman'.* Paris: Librairie J.-B. Ballière et Fils.

Forbes E (1844) On the light thrown on Geology by submarine researches; being the substance of a communication made to the Royal Institution of Great Britain, Friday evening, the 23rd February 1844. *Edinburgh New Philosophical Journal* 36: 318–327.

Forbes E (1851) Report on the Investigation of British Marine Zoology by means of the Dredge. *Report of the Twentieth Meeting of the British Association for the Advancement of Science; Held at Edinburgh in July and August 1850.* London: John Murray, Albemarle Street, 192–263.

Gatty M (1872) *British Sea-weeds. Drawn from Professor Harvey's 'Phycologia Britannica' (Vol. 1).* London: Bell and Daldy.

Hardenberg GV and Mahony M (2020) Introduction – Up, down, round and round: Verticalities in the history of science. *Centaurus* 62(4): 595–611.

Hardy P (2017) Where science meets the sea: Research vessels and the construction of knowledge in the nineteenth and twentieth centuries. Unpublished doctoral dissertation, Johns Hopkins University.

Harvey WH (1849) *The Sea Side Book: Being an Introduction to the Natural History of the British Coasts.* London: John Van Voorst.

Laloë AF (2014) 'Plenty of weeds & penguins': Charting oceanic knowledge. In: Anderson J and Peters K (eds) *Water Worlds: Human Geographies of the Ocean.* Farnham: Ashgate, 39–50.

Lavoie J (2013) Ukrainian teen solves deep-sea mystery off Vancouver Island. *Times Colonist,* 26 January.

Marsigli LF (1725[1999]) *Histoire Physique de la Mer* (trans. A McConnell). Bologna: Museo di Fisica dell'Università di Bologna.

McClenaghan B, Fahner N, Cote D, Chawarski J, McCarthy V, Rajabi H, Singer G and Hajibabaei M (2020) Harnessing the power of eDNA metabarcoding for the detection of deep-sea fishes. *PLoS ONE* 15(11) https://doi.org/10.1371/journal.pone.0236540

McConnell A (1993) L. F. Marsigli's visit to London in 1721. *Notes and Records of the Royal Society of London* 47(2): 179–204.

McConnell A (1999) Introduction. In: McConnell A (ed) *Histoire physique de la mer.* Bologna, Italy: Museo di Fisica dell Universita di Bologna, 6–28.

McConnell A (1982) *No Sea Too Deep: The History of Oceanographic Instruments.* Bristol: Hilger.

Menard HW (1964) *Marine Geology of the Pacific.* New York, NY: McGraw-Hill.

Michelet J (1861) *The Sea* (trans. Unknown). New York, NY: Rudd and Carleton.

Mills E (1978) Edward Forbes and John Gwyn Jeffreys. *Journal of the Society for the Bibliography of Natural History* 8(4): 507–536.

Mills E (2001) Essay review: Enlightened natural history of the beginnings of oceanic science? *Annals of Science* 58(4): 403–408.

Murray J and Hjort J (1912) *The Depths of the Ocean.* London: MacMillan.

Olson FCW and Olson MA (1958) Luigi Ferdinando Marsigli, the lost father of oceanography. *Quarterly Journal of the Florida Academy of Sciences* 21(3): 227–234.

Oreskes N (2003) A context of motivation: US Navy oceanographic research and the discovery of sea-floor hydrothermal vents. *Social Studies of Science* 33(5): 697–742.

Pinarldi N, Özsoy E, Latif MA, Moroni F, Grandi A, Manzella G, de Strobel F and Lyubartsev V (2018) Measuring the sea: Marsili's oceanographic cruise (1679–80) and the roots of oceanography. *Journal of Physical Oceanography* 48: 845–860.

Rehbock PF (1975) Huxley, Haeckel, and the oceanographers: The case of bathybius haeckelii. *Isis* 66(4): 504–533.

Rehbock PF (1979) The early dredgers: 'Naturalizing' in British Sea, 1830–1850. *Journal of the History of Biology* 12(2): 293–368.

Revelle R (1967) Unity and fission in oceanography: Presidential address to the International Association of Physical Oceanography. *International Association for the Physical Sciences of the Ocean, Procés-Verbaux* 10: 3–11.

Rozwadowski H (2005) *Fathoming the Ocean: The Discovery and Exploration of the Deep Sea*. Cambridge, MA: Harvard University Press.

Rozwadowski H and Van Keuren DK (2004) Foreword. In: Rozwadowski H and Van Keuren DK (eds) *The Machine in Neptune's Garden: Historical Perspectives on Technology and the Marine Environment*. Sagamore Beach, MA: Science History Publications, xi-xii.

Schlee S (1973) *The Edge of an Unfamiliar World: A History of Oceanography*. New York, NY: EP Dutton.

Steinberg PE and Peters K (2015) Wet ontologies, fluid spaces: Giving depth to volume through oceanic thinking. *Environment and Planning D: Society and Space* 33(2): 247–264.

Stommel HM (1995) Autobiography. In: Hoggand NG and Huang RX (eds) *Collected Works of Henry M. Stommel Vol. 1*. Boston, MA: American Meteorological Society, 5–112.

Stoye J (2004) *Marsigli's Europe 1680–1730: The Life and Times of Luigi Ferdinando Marsigli. Soldier and Virtuoso*. New Haven, CT: Yale University Press.

Sverdrup HU, Johnson MW and Fleming RH (1946) *The Oceans: Their Physics, Chemistry, and General Biology*. New York, NY: Prentice-Hall.

Thomson CW (1873) *The Depths of the Sea*. New York, NY: Macmillan and Co.

Thomson CW (1878) *The Voyage of the 'Challenger': The Atlantic*. New York, NY: Harper and Brothers.

Woods Hole Oceanographic Institution (2021) Godzilla, Sasquatch, and Homer Simpson: The curious names of deep-sea features. Dive and Discover: Expeditions to the Seafloor website. Available at: https://divediscover.whoi.edu/hot-topics/names/

Zhang Y, Ryan JP, Hobson BW, Kieft B, Romano A, Barone B, Preston CM, Roman B, Raanan B-Y, Pargett D, Dugenne M, White AE, Henderikx Freitas F, Poulos S, Wilson ST, DeLong EF, Karl DM, Birch JM, Bellingham JG and Scholin CA (2021) A system of coordinated autonomous robots for Lagrangian studies of microbes in oceanic deep chlorophyll maximum. *Science Robotics* 6(50) https://doi.org/10.1126/scirobotics.abb9138

# 4

# REPRESENTATION

## Seapower and the political construction of the ocean

*Basil Germond*

### Introduction

This chapter is concerned with oceanic representations. It unravels the implications, on the practice of seapower and maritime security, of the dominant discourse that consists in representing the sea as an 'empty space'. There is "substantial literature on marine representations", focused on art and literature and its social, cultural and geopolitical discourses (Steinberg, 2001: 33, and see, for example, Connery, 1995). However, this chapter explores how the exercise of seapower and maritime security rests on the sustained consensus – or construction – of the concept of *mare liberum*. Yet the recent resurgence of non-state threats at or from the sea have engendered a practice of maritime security that necessitates a move away from the 'empty sea' narrative towards a representation of the sea that emphasises control and governance not unlike on land. In this process, the sea is losing its particular, long-held, stabilised, discursive characteristics that have represented it as a 'free' and 'empty' space, in favour of a dominant discourse of security and control that is largely grounded in landed considerations.

Concentrating on the political, and mainly state, narratives and practices regarding the ocean, this chapter unpacks the way the political construction of the ocean interacts with states' power and security. It does so through a focus on the construction of 'sea power'. The 'political construction of the ocean' is an obvious nod to Philip Steinberg's *Social Construction of the Ocean* (2001), which, although framed within critical political economy and geography, still constitutes, to date, the main contribution to the study of the ocean from a power-knowledge perspective. Indeed, Steinberg's work is attentive to the power of representations in constructing an ocean for use, and for the expression of power (capital, military and resistant power) (see Steinberg, 2001: 32–38). Of course, the ocean is by no means only a representation – it is also a material and has a materiality that does political work (see Steinberg, 2013 building from Blum, 2010). Indeed, Steinberg has progressed a careful way of thinking with the ocean that does not reduce it to metaphor or abstracted representation, but also takes seriously its material complexity and our engagements with it (Steinberg, 2013 and also Steinberg and Peters, 2015). Nonetheless, thinking of how oceans are constructed is vital for interrogating their functioning – discursively and actually.

 DOI: 10.4324/9781315111643-6

Seapower (also written sea power) is a concept that has been popularised by the writings of US Captain Alfred Thayer Mahan at the end of the nineteenth century. Mahan's most important claim about seapower, based on his own experience in the US Navy as much as on his thorough knowledge of naval history, is that Nations' wealth strongly depends on a flourishing maritime commerce backed by a powerful navy (Mahan, 1890). However, beyond this relationship between naval strength and economic power, he never provided readers with a clear definition of the concept of seapower. The great majority of his writings do not discuss the concept of seapower at all but conclude or imply that states with a powerful navy and a thriving maritime commerce are 'powerful'. That said, Mahan included a short conceptual chapter in his 1890 *Influence of Sea Power* that proposes some elements, or constituents, of seapower (this, at the request of his publisher, c.f. Sumida, 1999: 46).

Seapower can be understood as a sum of geographical elements (such as the location of one's coasts on the global grid), material elements (such as demographical trends, economic and financial base, access to technology) and ideational elements (such as the existence of a maritime culture, maritime traditions, and the way elites and populations regard the sea) (Germond, 2015; Mahan, 1890; Steinberg, 2013). Since then, various scholars and practitioners have used the concept of seapower indiscriminately without paying much attention to its definition, except in the case of Admiral Richmond (1934) who usefully emphasised the importance of politicians' decisions as a cause of seapower. More recently, Geoffrey Till (2004), in his *Seapower: A Guide for the 21st Century*, defined seapower as inputs and outputs. Inputs relate to the geographical, material and ideational elements of seapower mentioned above. Outputs relate to the consequences of the enactment of seapower, such as exercising command of the sea (such as power and forces projection), the stabilisation of the liberal world order, or successful global ocean governance.

In this chapter, seapower is understood as the constitutive and symbiotic relationship between states' (and to some extent non-state actors') material and ideational power and the maritime domain, their use thereof, and the way it is politically constructed and practically exploited. Using a framework for analysis that draws from Foucault's discussion of the relationship between representations and practices (the knowledge-power matrix) and previous studies by Steinberg (notably 2001), this chapter discusses the way dominant *representations* of the sea have normalised particular *practices* of seapower over time. The representation of the sea in collective imaginaries has a strong political dimension in that it produces and normalises practices and governance structures. Discourse analysis is relevant only if it is put in relation with the practice that is normalised by said discourse and that at the same time frames the discourse (Fairclough, 1992). In other words, 'political constructions' are not only narratives about ocean space but are also the interactions between such narratives and the resulting practices. As O'Tuathail notes, "geography is not a natural given but a power-knowledge relationship" (1996: 10).

The discussion of the interlinkages between representations and practices of seapower can be framed within the dialectic of the smooth and the striated proposed by continental philosophers Deleuze and Guattari (1988) and also advocated by Steinberg (1999, 2011, 2018), Hannigan (2017) and Jones (2016), respectively in the fields of geography, sociology and law to account for two competing systems of spatial organisation. Deleuze and Guattari's point is that the sea is smooth by nature (in other words, not susceptible to political control in the same way as the land), but this smoothness generates a need for controlling ocean space, eventually leading to its striation. This dialectic of the smooth and the striated can be incorporated into a knowledge–power matrix framework that discusses the interlinkage between dominant discourse and

*Table 4.1* Analytical framework

| | *Representation of the sea* | *Practice of seapower* |
|---|---|---|
| **1) Smooth as positive** | Empty, void, *mare liberum* | Free trade, empire building, naval projection (modernity) |
| **2) Smooth as dangerous** | Unlawful, unregulated, marginal, endangered (environment) | Control, grid, surveillance, limited territorialisation (maritime security in a post-modern world) |
| **3) Smooth and Striated as a balance** | Dominant discourse of the free but controlled sea | All of the above; collective and non-military seapower |

*Source*: author.

normalised practices (see Table 4.1). Firstly, there is a 'positive' representation of the ocean as smooth that frames it as an 'empty space' and *mare liberum*, which is linked to a practice of free trade, empire building and naval projection. Secondly, there is a 'negative' representation of the ocean as smooth that frames it as an unlawful, unregulated, marginal and endangered space that is linked to a practice of control, striation and limited territorialisation. Thirdly, the dialectic attempts to resolve its opposites with a neomodern narrative of the ocean as striated that frames the ocean as in need of regulation, control, security in a bid to sustain the advantageous consequences of its smooth characteristics (seapower) and combat the negative consequences of *too* smooth an ocean (criminality, rogue actors). Eventually, a certain degree of striation is needed to maintain enough smoothness to secure the advantage of the empty, free sea. The exercise of seapower still rests on the sustained consensus around *mare liberum*, but the practice of maritime security necessitates a move away from representations of an 'empty sea' towards one that emphasises control and governance not unlike on land. This conception of seapower lies at the intersection between freedom and regulation and demonstrates that the sea is losing its specific discursive characteristics (notably the 'empty sea') in favour of a dominant discourse of security and control that is largely grounded in land considerations.

It is worth noting that this framework accounts for a dominant, mainly state-centric, and mainly western, discourse (see Fawcett et al., this volume; Waiti and Wheaton, this volume for alternatives). The concept of dominant discourse refers to the hegemonic status that some narratives have gained. Hegemonic discourses "will always be contested" (Fairclough, 2003: 207). This chapter analyses the evolution of the hegemonic discourse on the sea and seapower. However, within the same order of discourse, there are competing (sub)discourses that disagree with the representation of the sea as 'empty' while also criticising the move towards more striation in practice (for alternative sub-discourses, see Gillis, 2012; Hau'ofa, 1998; Jackson, 1995; Lehman, 2013; Smith, 2007; Steinberg et al., 2012; Tyrrell, 2006). It is also worth a reminder that the environmental protection/conservationist discourse might well not contradict the dominant discourse but can actually be aligned with it in its claim that the sea is in need of more control for the sake of marine environment protection (see also Gray, 2018).

To advance these discussions, the chapter is organised as follows: the next section discusses the 'traditional' representation of the ocean as 'empty' and the resulting modern practice of seapower. Then, the following section explores the impacts of post-modernity on ocean space and the subsequent desire to control the maritime domain in a way similar to the land. Finally, the synthesis of the smooth and the striated is discussed in relation to the current discourse and practice of collective seapower in a non-zero-sum space.

## The glorification of the 'smooth' sea: Empty space and *mare liberum*

The way in which the sea has traditionally been constructed in collective, western, imaginaries shall be put in relation with the interests and practice of dominant political and economic actors. The construction of the ocean as a 'void', or 'empty' space, has contributed to normalising the idea and practice of free flow of goods and capital across the ocean, which is thus a space that is at the same time exploited and supporting exploitation (Connery, 1995; Martin, 2013; Steinberg, 1999; Virilio, 1977). There is also a military/naval dimension of this representation. State actors can use the 'void' to 'freely' project military forces (Steinberg, 2001; see also Bueger, this volume; Depledge, this volume), which contributed to the creation of colonial empires in the seventeenth to nineteenth centuries (Mancke, 1999; Till, 2004: 16) and to the consolidation of the US-dominated western liberal world order in the twentieth century (Posen, 2003). Seapower has also been described as a condition for global leadership and is crucial in understanding hegemony (Cox, 1980) and explaining hegemonic cycles (Boswell and Sweat, 1991: 129–132; Modelski and Thompson, 1996). The particular nature of the sea has enabled a specific discourse to emerge and to become dominant (i.e. the void-empty space discourse), with all that this entails in terms of modern practices of empire building, stabilisation of the liberal world order, globalisation of the economy and the projection of power and norms.

The representation of the sea as a void relates to some of its natural, physical realities. Unlike the land, on which human beings can settle and polities are fixed, the sea is liquid, wet, in perpetual movement (Steinberg and Peters, 2015). For political actors, who have the "monopoly on the legitimate use of violence", it is, a priori, not possible to control the sea in the same manner as the land (for example, via occupation) (Corbett, 1911). The sea is a space beyond permanent habitation (see Steinberg, 1999). As a result, the sea, at least in political collective imaginaries, keeps the features of a 'final frontier', as conceived by Frederic Jackson Turner (1920), with characteristics ranging from an unlawful, dangerous space, to an empty space or even, in its extreme form – *mare nullius* – that is to say a space that legally, and by extension politically, belongs to nobody and thus can be 'claimed' by anyone powerful enough to back such claim (see also Fawcett et al., this volume). Thus, it becomes possible to 'roll away the frontier' or even, in Steinberg's words (2018) to somewhat 'striate' the frontier. Such a discourse (and the resulting practice) rests on the fact that the sea is free to use as long as nobody else claims otherwise. This has contributed to the development of a narrative that glorifies 'freedom of the seas'.

The concept of *mare liberum* originates in a legal claim made in 1609 by Hugo Grotius, who has been a major contributor to the development of positivist international law. However, international law must be understood as a set of rules adopted *by* someone *for* someone. In other words, there is a political dimension to it (Reus-Smit, 2004). In the case of Grotius, the aim was to make sure that the dominant maritime Powers of the time, such as Portugal and Spain (and to a lesser extent, England) would not be in a position to prevent the United Provinces of the Netherlands from fulfilling their overseas interests, notably in the Indian Ocean (Brito Viera, 2003). Therefore, the concept or representation of *mare liberum* has had a political dimension from its very inception. Since then, and throughout modern and contemporary history, it has been endorsed by dominant maritime Powers, for it has always represented a means to guarantee their right to operate and trade far away from home, in distant waters, without much legal constraint.

Dominant sea Powers possess the material and ideational leverage to shape the international maritime order by influencing the development of international law of the sea and making sure that any evolution of this law and the regimes related thereto proceeds to their advantage. In turn, then, maritime legal culture, norms and order play in favour of the dominant sea Powers of the time. In other words, there is a mutually reinforcing and beneficial relationship between being the main sea Power(s) and the shaping of international order at sea (Kraska, 2009: 117, 121; see also Depledge, this volume). UNCLOS and customary international law of the sea norms are certainly in dominant Powers' interest, since they guarantee freedom of commerce and limit to a strict minimum the constrains put on the military use of the sea (for example right of 'innocent passage' within territorial waters) while also allowing states to control flows of goods and people in the waters close to the shore (in territorial waters) and to some extent in the maritime space beyond their sovereign boundaries (Exclusive Economic Zones, global maritime surveillance and so on). Carl Schmitt, cited in Steinberg (2011) criticised the fact that the principle of *mare liberum* has been used to justify all sort of practices from an all-out freedom to do what you please at sea to the "[suppression of] those who would challenge the established rules" (Steinberg, 2011: 269). This fits with Modelski and Thompson's claim that seapower is a "necessary conditions for leadership", which is crucial not only for winning wars but also "in enforcing the new, postwar order; in policing sea lanes; and in deterring potential attacks on the world power and its allies and clients" (1996: 52) or, in other words, in stabilising the dominant (liberal) order via the production of norms and the enforcement thereof.

The *mare liberum* narrative complements, or is a corollary to, the 'void' narrative, since the 'emptier' the sea is the more prone it is to legal or political claims by competitors. For example, the Papal division of the world in 1493 between Spain and Portugal and the subsequent Treaty of Tordesillas (1494) were challenged by other European Powers. Indeed, at that time Europe was already divided between various polities within a zero-sum game, although the concept of 'fixed' borders would only crystalise in the seventeenth century with the Wesphalian system under the principles of sovereignty and non-interference in other states' domestic affairs. But the rest of the world's land space was still 'available' to conquer, exploit, and use by European states, as long as the sea remained free to use for economic, commercial and power and norms projection purposes. This led to a "partial negation of exclusive territoriality" over the seas (Anderson, 1996: 144), which was crystallised in the *mare liberum* principle. As a result, seapower within *mare liberum* contributed to overseas empire building (Mancke, 1999) and down the line to the consolidation of nation-states in Europe (Glete, 2000).

In addition to an 'empty' space, classic seapower writers and practitioners have also emphasised the idea that the sea is a lane of communication (Colomb, 1891; Mahan, 1890; Corbett, 1911). Here, the sea as *mare liberum* becomes a site for (unhindered) communication and movement on "a seemingly friction-free surface across which capital [but also, in this context, navy ships] can move without hindrance" (Steinberg, 1999: 416). In other words, the hypermobility noticed by Steinberg in reference to the flow of capital is also relevant in terms of naval power. The notion of sea lines of communication (SLOC) is crucial in maritime strategy. Indeed, the capacity for navies to exercise command of the sea rests on their capacity to control or secure relevant SLOCs. In addition, the protection of commerce requires securing control of relevant SLOCs, choke points and access to ports (Corbett, 1911).

In sum, the practice of seapower that consists in power projection, empire building, and the accumulation of capital has been normalised via a dominant representation and associated discourse of the 'empty sea' that is rather post-modern for it implies a non-territorial acceptation of the sea, which is constructed as a non-territorialized, smooth space. Not without irony, this dominant narrative has eventually been challenged by the reality of the post-modern world in the form of disruptive non-state actors operating at or from the sea.

## The limits of smoothness: The sea as a space in need of security

The post-Cold War era has witnessed the proliferation (or re-emergence) of criminal actors operating at or from the sea, such as pirates, terrorists, human traffickers and illegal fishers. This has highlighted the limits of the 'smooth sea' in that the free flow of goods, capital, and navy ships – as well as the sustainable exploitation of marine resources – are put in 'jeopardy' by non-state actors who benefit from the fact that the seas and maritime borders are hard to regulate, monitor and police. From 'empty' and 'free' the sea has become prone to the proliferation of harmful and/or undesirable non-state actors, which not only represent actual security threats but also disrupt the 'system', although processes of capital, and even state-sanctioned military endeavours can also, themselves, challenge the 'smooth seas' they are supposedly part of (Steinberg, 2001). Nonetheless, ever-growing uses of the sea as a space for crime and coercion have engendered the need to 'tackle' the threats and led to a process of securitisation of the maritime space.

Securitisation, as proposed by the Copenhagen School of International Relations, can be understood as a process by which a subject is constructed as a security issue (instead of a policy issue), thus justifying 'exceptional measures' instead of dealing with the issue in a 'business-as-usual' way (Buzan et al., 1998; Stritzel, 2007). The subject in question can be a phenomenon (such as migration) or any entity including a space (for example, the sea). The discourse by which the sea has been securitised emphasises the need to implement effective measures to deal with actors operating at or from the sea that the dominant discourse represents as 'disruptive'. The narrative stresses that something must be done to secure the sea, which has become "a disorderly geopolitical sphere in dire need of regulation, policing and management" in order to guarantee the "uninterrupted and unimpeded flow of commodities across the planet" (Campling and Colás, 2018). This implies a greater degree of control over the maritime domain, so as to be in a position to prevent, monitor, suppress, and repress illegal activities and disruptive agents. The securitisation of the sea has generated a need to extend states' sovereignty, or at least collective and functional sovereignty, beyond their external boundary to exercise a greater level of control over ocean space. Glück (2015) talks about the "production of security space" at sea that contributes to secure the free flow of goods. Ryan (2015) notes the resulting maritime spatial security practices via processes of zonation. Suarez de Vivero et al. (2009: 628) explain that states project "the rights of sovereignty [...] over seas and oceans", which results in "new patterns of territorial organisation" that contribute to "political and economic control".

In sum, the sea cannot be considered as a 'true' void where nothing but free movement happens. Rather, it is also represented as a space filled with threatening subjects and thus as a space in need of control and regulation; a space to be incorporated into states' jurisdictional, political and operational zone of control. Whereas traditional seapower has enabled power projection onto other states' territories (e.g. military invasions, foreign interventions), the securitisation of the sea rests on the projection of states' normative and policing power onto the maritime space so as to tackle the threats where they materialise.

This narrative shift is also framed within the broader discourse of sustainability. *Mare liberum* is a legal principle that "the high seas are open to all states, whether coastal or land-locked" (UN General Assembly, 1982: art. 87(1)), but it has a close connection to the more political concept of 'global commons'. Such a concept describes the sea (and perhaps more precisely its resources) as belonging to no one in particular but at the same time to everyone. Thus, it is free to use, as long as it is done in a way that is somewhat sustainable. The issue is one of the "struggle to govern the commons" (Dietz et al., 2003), which leads to technical ocean governance solutions to problems such as resource depletion and pollution. The 'empty sea' narrative is thus linked to a practice of managing ocean resources and protecting a fragile environment. This has led to the so-called 'stewardship discourse', deconstructed by Steinberg, which notes that the ocean is recognised as fragile and in need of governance, management, and spatial planning in a way that tends to justify the non-territorial exercise of power at sea (1999: 419). The securing of the 'global commons' (that now also includes airspace, outer space and perhaps even cyberspace) has, then, a security or even military dimension (Jasper, 2010). In other words, both stewardship of marine resources and more generally ocean governance are processes that are guided by states' interest in controlling, securing and eventually exploiting/using the sea.

To account for this, the 'empty sea' narrative is complemented by a 'frontier' narrative that emphasises the hybridity of the maritime domain in terms of freedom *versus* control. As shown by Deleuze and Guattari (1988) and Steinberg (2018), the dialectic of smoothness/striation and open/closed spatiality plays a very important role in shaping the sea as a frontier space, where the clash between political and administrative freedom (notably freedom of movement) and initiatives to regulate and order the sea are still very present. The sea has traditionally been considered as "a space that was best governed by an absence of enclosure" and thus that "should not be constructed as a frontier" (Steinberg, 2018: 23). However, the current frontier characteristics of the sea result from economic and governance actors' initiatives which aim to close the frontier, or in other words to transform the sea into a space that is as similar to the land as possible (with precise jurisdictions and borders, spatial planning mechanisms) whilst keeping the advantages of the freedom of the sea discussed above.

That said, the frontier characteristics of the sea, and especially its security dimension, are also related to the sea as both a gateway to the rest of the world and an entryway to one's territory (Germond, 2010). Classical writers have emphasised the importance of the sea as a way to extend one's own territory up to the enemy's coast (for example, Colomb, 1896: 20; Mackinder, 1904: 428–429). As a corollary, however, the sea is also an entryway into one's own territory for hostile forces or criminals and terrorists. So, the ocean as a 'void' has always been double-sided. States can make the most of the sea as a free space to use at their convenience and to their advantage; they can 'roll back' the frontier as far away as other states' coasts, but on the other hand they need to protect themselves from other actors doing the same 'against' or 'towards' them. This is true in wartime with military expeditions (power and forces projection) but also in peacetime when it comes to controlling the incoming (and to a certain extent outgoing) flows of goods and people, so as to tackle issues such as illegal fishing, piracy, terrorism, arms and drug smuggling, human trafficking and other forms of maritime criminality. In practice, this has led states to operate as far away from their coast as possible to push back the security frontier, so as to benefit from enough strategic depth (Germond, 2010). Whereas within the frontier "the 'inside' gradually becomes an 'outside'" (Steinberg, 2018: 1), when states perform authority within their maritime margins, the outside gradually becomes the inside following a process of closure and striation.

The 'empty sea' narrative responds to the need to limit the process of territorialisation of the sea, or in other words to prevent states from 'closing' the sea, in a bid to make sure that the sea remains free for all to use in the spirit of free trade (see also Steinberg, 2018). But the need to steward marine resources and to secure the maritime domain justifies practices consisting in controlling human activities and flows at and from the sea. The *mare liberum* narrative has contributed to cement the sea as a post-modern, 'free' space where sovereignty does not apply as strictly as on land. However, the maritime security and frontier narratives, which represent the sea as a dangerous, unlawful and unregulated space calls for a territorialisation of the sea, a process that consists in limiting freedom of the sea to certain actors and certain activities while controlling flows, structures and agents in the same way as on land. This has resulted in various layers of legal titles over areas and 'multiple' or 'multi-layered' sovereignty over the high seas, leading to "a new sea-based territoriality" (Suarez de Vivero and Rodriguez Mateos, 2014: 62). States attempt to control the sea by applying forms of sovereignty and experience "drawn from land", resulting in a hybrid and functional forms of "terraqueous territoriality" (Campling and Colás, 2018: 776; see also Peters et al., 2018; Peters, 2020).

In sum, the 'smoothness' of the sea has allowed modern nation-states to thrive (including via empire building), but it has also allowed post-modern disruptive actors to operate at the margin of, or against, 'the system'. This has induced the need to regulate, control and striate the sea more than before (see also Hannigan, 2017). Consequently, the twenty-first-century dominant narrative on ocean space is one of securitisation and limited territorialisation at least as much as freedom.

## Smooth and striated: The sea as a non-zero-sum space and collective seapower

States have projected their sovereignty over the oceans, but contrary to the land, territorialities are not mutually exclusive at sea. International relations scholars have used the concept of the 'zero-sum game' to refer to the nature and outcomes of the relationships between states on the world's stage. Scholars from the realist tradition argue that world politics is a zero-sum game – for one state's power gain results in other states' power loss – which encourages the pursuance of relative gains over absolute ones (Waltz, 1979). Scholars from the liberal tradition argue that world politics is not a zero-sum game, for absolute gains are more important than relative gains, which explains states cooperation. This framework for analysis can be applied to geographical spaces (for example, land or sea) rather than processes (such as international relations). Whereas the myth of exclusive sovereignty has been critically deconstructed (Agnew, 1994), it is important to stress that the land is more prone to exclusive sovereignty than the sea. From a political perspective, the land is constructed as a zero-sum space in that sovereignties and territorialities are represented as mutually exclusive and static in the Westphalian narrative. This dominant discourse is part of the "largely successful strategy for establishing the exclusive jurisdiction implied by state sovereignty" (Agnew, 2005: 437) and for putting "statehood outside of time" (Agnew, 1998: 50).

Sea space is different. Firstly, the sea is not prone to the mutual exclusion of sovereignties due its fluid nature and to the fact that it is difficult to 'occupy' (Corbett, 1911; Steinberg, 1999; Steinberg and Peters, 2015). Secondly, as discussed above, it is not in states' interest to 'close' the sea as much as the land, since fixed territorialities are not compatible with, or as 'useful' as, 'smoothness' and the free sea. This results in a hybrid, limited territorialisation of the sea that combines forms of striation (spatial planning, zonation, UNCLOS areas, surveillance and control) with the inherent/fluid and legal characteristics of the sea, that is conducted in a collective, non-zero-sum way (for example, shared ownership).

This narrative is linked to the reconfiguration of the concept of seapower in the twenty-first century. Seapower has traditionally been associated with Mahanian, navalist policies, and generally with navies as the instruments of state power. Tackling transnational, criminal threats as well as protecting the marine environment and resources has contributed to the development of a post-Mahanian, post-modern form of seapower, which is not only less state-centric and less naval but also more collective (Germond, 2019; Pugh, 1996; Till, 2007). Maintaining 'good order' at sea, stewarding marine resources and regulating human activities in the maritime domain is dependent on the enactment of a non-military and collective form of seapower in support of global ocean governance and maritime security. Thus, ordering the maritime domain requires state and non-state actors, security and economic stakeholders to *cooperate* at various scales. Structures, policies and objectives are collective; expected gains/benefits are absolute, shared between actors and not relative. For example, the 1,000-ship navy, an initiative launched by the US in 2005 then rebranded Global Maritime Partnership, accounts for the fact that regulating the maritime domain is not possible without the participation of all relevant states and private stakeholders, which share similar objectives (Mullen, 2006). Even though such an initiative has received some negative feedback (Till, 2008), it demonstrates that seapower can also be understood and enacted collectively.

Collective seapower fits with the description of the sea as a non-zero-sum space. Whereas, as Geoffrey Till claims, seapower can be *relative* "since some countries have more than others" (Till, 2004: 2), when it comes to global ocean governance and maritime security (that is to say 'ordering' ocean space) then seapower can also be *absolute*, in that it is enacted jointly; it depends on the involvement of state *and* non-state stakeholders and the benefits of its enactment are shared among stakeholders in a non-mutually exclusive way. Agents of collective seapower do not need to share more than the common desire to maintain both freedom of the seas and security and stability in the maritime domain. This fits with the English School of International Relations' concept of a pluralistic international society of sovereign states, which share the common desire to maintain a certain degree of order, stability and certainty within the international system without systematically sharing similar values and identities (Bull, 1977, discussed in Germond, 2019).

Representing the sea as a non-zero-sum, shared space – as smooth and striated – accounts for the fact that states' control over portions of the seas remains limited and that the ultimate goal is not to transform the sea into a land space where sovereignties and territorialities are mutually exclusive but to secure the maritime domain so as to guarantee that state and non-state actors can fully benefit from the advantage of the free sea. Whereas naval power remains a prerogative of the nation states, seapower in its post-modern, collective acceptation, transcends the boundaries of the Westphalian order and this process is facilitated by the physical nature and legal characteristics of the sea. In the post-2022 Ukraine invasion era, upholding freedom of navigation is likely to become a core objective of the West's opposition to authoritarianism and state violence, thus reinforcing the *mare liberum* dimension of collective seapower.

## Conclusion: Towards neo-modern seapower?

The post-modern characteristics and characterisation of the sea as a fluid, empty space, has helped produce and cement what is known as 'modernity' (via colonial empires, bureaucratisation and nation-state building). These developments, forged and formed through oceanic representations of an empty sea for the taking, are by no means positives from which post-modernity can be contrasted. As noted earlier, although the two representations of the ocean diverge (an ocean of freedom, to one whose freedom then demands control), the ocean of 'emptiness' is also one that is characterised by crime, by unsustainable exploitation and by violence. Yet the point remains, a representation exists that splits the two, creating practices for how the ocean is then materially enacted. Indeed, the 'need' to control the sea (striation) becomes apparent. On the one hand, *mare liberum*, which has supported modernity, rests on the endorsement of a non-territorialized vision of the sea, which has thus been associated with post-modern attributes (e.g. 'empty sea', void 'belonging to nobody', porous borders) even at the height of the modern era (nineteenth century). On the other hand, the securitization of the sea has been accompanied by a narrative emphasising the need to control, regulate and order ocean space, justifying a practice of limited territorialisation. The dialectic of smoothness

*Table 4.2* The evolving narrative of the smooth and striated sea

| *Narrative of the smooth sea* | *Narrative of the striated sea* | *Narrative of the smooth* and *striated sea* |
|---|---|---|
| Void/empty | Grid | Shared ownership |
| Non-territorialized | Territorialized | Limited territorialisation |
| Free/unregulated | Controlled/regulated | Secured |
| Open | Closed | Governed |
| Gateway | Entryway | Frontier space |

*Source*: author.

(empty sea, *mare liberum*, non-territorialisation) and striation (regulation, control and limited territorialisation) has led to the transformation of the sea into a hybrid, non-zero-sum space that combines elements of unboundedness and territoriality both from a legal/jurisdictional and a social/political interactions perspective. This has resulted in a form of *neo-modernity* in that the dichotomy between smooth and striated has been transcended in a dialectical way: 1) the sea must be smooth enough to freely trade, build empires, and is thus represented as 'empty' and free; 2) the sea must be striated: regulated enough to make sure that state and non-state actors can benefit from the advantages of its smooth characteristics; and 3) smooth and striated are merged into a neo-modern form of narrative of freedom and security and a practice of limited territorialisation, both reinforcing each other.

Table 4.2 shows this evolving (somewhat dialectical) representation, from smooth to striated to smooth *and* striated. The narrative on ocean space does not follow either/or binaries, since the sea can be both smooth and striated at the same time. Neither empty and free, nor fully controlled and regulated, the sea has to be secured; neither open nor closed the sea has to be governed. At the time, each of these categories and uses is tension-filled and complex.

States have security and economic interests in governing the sea in the same was as they govern the land: as a regulated, monitored, striated space. Assuring security and 'prosperity' on land – which has always been the main objective of seapower, for human beings live on land and not at sea (Gray, 1994: 3–4) – calls for a practice of security, regulation and striation at sea that bear land characteristics and reflect underlying power relations. The sea has thus gradually been transformed into a land-like space in representation and practice.

However, the sea is not the land because of immutable physical attributes that cannot be denied. The sea cannot be occupied, striated or controlled like the land can be. In addition, (dominant) states agree that it is in their interest (and in the system's interest) to adhere to the principle of *mare liberum*. Thus, the sea has not become the land, but an order at sea has developed, which combines all of the above in a neo-modern form of governance of a non-territorializable, non-zero-sum space, compromising between *mare liberum* and total security. Collective seapower epitomises neo-modernity in that it mixes modern forms of limited territorialisation with collective, post-modern forms of control over ocean space.

## References

Agnew J (1994) The territorial trap: The geographical assumptions of international relations theory. *Review of International Political Economy* 1(1): 53–80.

Agnew J (1998) *Geopolitics: Re-envisioning World Politics*. London and New York, NY: Routledge.

Agnew J (2005) Sovereignty regimes: Territoriality and state authority in contemporary world politics. *Annals of the Association of American Geographers* 95(2): 437–461.

Anderson J (1996) The shifting stage of politics: New medieval and postmodern territorialities? *Environment and Planning D: Society and Space* 14(2): 133–153.

Blum H (2010) The prospect of oceanic studies. *PMLA* 125(3): 670–677.

Boswell T and Sweat M (1991) Hegemony, long waves, and major wars: A time series analysis of systemic dynamics, 1496–1967. *International Studies Quarterly* 35(2): 123–149.

Brito Vieira M (2003) Mare liberum vs. mare clausum: Grotius, Freitas, and Selden's debate on dominion over the seas. *Journal of the History of Ideas* 64(3): 361–377.

Bull H (1977[1995]) *The Anarchical Society: A Study of Order in World Politics*. London: Macmillan.

Buzan B, De Wilde J and Weave O (1998) *Security: A New Framework for Analysis*. Boulder, CO and London: Lynne Rienner.

Campling L and Colás A (2018) Capitalism and the sea: Sovereignty, territory and appropriation in the global ocean. *Environment and Planning D: Society and Space* 36(4): 776–794.

Colomb PH (1896) Imperial defence. In: Colomb PH (ed) *Essays on Naval Defence*. London: Allen, 1–30.

Colomb PH (1891) *Naval Warfare: Its Ruling Principles and Practice Historically Treated*. London: Allen.

Connery CL (1995) Pacific rim discourse: The US global imaginary in the late Cold War years. In: Wilson R and Dirlik A (eds) *Asia/Pacific as Space of Cultural Production*, Durham, NC: Duke University Press, 30–56.

Corbett J (1911 [1988]) *Some Principles of Maritime Strategy*. Annapolis, MD: United States Naval Institute.

Cox RW (1980) The crisis of world order and the problem of international organization in the 1980s. *International Journal* 35(2): 370–395.

Deleuze G and Guattari F (1988) *A Thousand Plateaus: Capitalism and Schizophrenia*. London: Athlone (trans. B Massumi).

Dietz T, Ostrom E and Stern PC (2003) The struggle to govern the Commons. *Science* 302(5652): 1907–1912.

Fairclough N (1992) *Discourse and Social Change*. Cambridge: Polity Press.

Fairclough N (2003) *Analysing Discourse: Textual Analysis for Social Research*. Abingdon: Routledge

Germond B (2010) From frontier to boundary and back again: The European Union's maritime margins. *European Foreign Affairs Review* 15(1): 39–55.

Germond B (2015) *The Maritime Dimension of European Security: Seapower and the European Union*. London and New York, NY: Palgrave Macmillan.

Germond B (2019) Seapower and small navies: A collective and post-modern outlook. In: Speller I, Sanders D and McCabe R (eds) *Europe, Small Navies and Maritime Security*. Abingdon: Routledge, 26–35.

Gillis JR (2012) *The Human Shore*. Chicago, IL: University of Chicago Press.

Glete J (2000) *Warfare at Sea, 1500–1650: Maritime Conflicts and the Transformation of Europe*. London: Routledge.

Glück Z (2015) Piracy and the production of security space. *Environment and Planning D: Society and Space* 33(4): 642–659.

Gray CS (1994) *The Navy in the post-Cold War World: The Uses and Value of Strategic Sea Power*. University Park, PA: The Pennsylvania State University Press.

Gray N (2018) Charted waters? Tracking the production of conservation territories on the high seas. *International Social Science Journal* 68(229–230): 213–368.

Hannigan J (2017) Toward a sociology of oceans. *Canadian Review of Sociology/Revue canadienne de sociologie* 54(1): 8–27.

Hau'ofa E (1998) The ocean in us. *The Contemporary Pacific* 10(2): 392–410.

Jackson SE (1995) The water is not empty: Cross-cultural issues in conceptualising sea space. *Australian Geographer* 26(1): 87–96.

Jasper S (ed) (2010) *Securing Freedom in the Global Commons*. Stanford, CA: Stanford University Press.

Jones H (2016) Lines in the ocean: Thinking with the sea about territory and international law. *London Review of International Law* 4(2): 307–343.

Kraska J (2009) Grasping 'the Influence of Law on Sea Power'. *Naval War College Review* 62(3): 113–35.

Lehman JS (2013) Relating to the sea: Enlivening the ocean as an actor in Eastern Sri Lanka. *Environment and Planning D: Society and Space* 31(3): 485–501.

Mackinder HJ (1904) The geographical pivot of history. *The Geographical Journal* 23(4): 421–437.

Mahan AT (1890 [2007]) *The Influence of Sea Power Upon History, 1660–1783*. New York, NY: Cosimo Classics.

Mancke E (1999) Early modern expansion and the politicization of oceanic space. *Geographical Review* 89(2): 225–236.

Martin C (2013) Shipping container mobilities, seamless compatibility, and the global surface of logistical integration. *Environment and Planning A* 45(5): 1021–1036.

Modelski G and Thompson WR (1996) *Leading Sectors and World Powers: The Coevolution of Global Economics and Politics*. Columbia, SC: University of South Carolina Press.

Mullen M (2006) What I believe: Eight tenets that guide my vision for the 21st century navy. *US Naval Institute Proceedings* 132(1) www.terrorfreetomorrow.org/upimagestft/Mullen%20Jan%2017%20 2006.pdf

Ó'Tuathail G (1996) *Critical Geopolitics*. London: Routledge.

Peters K, Steinberg PE and Stratford E (eds) (2018) *Territory beyond Terra*. London: Rowman and Littlefield.

Peters K (2020) The territories of governance: Unpacking the ontologies and geophilosophies of fixed to flexible ocean management, and beyond. *Philosophical Transactions of the Royal Society B* 375(1814) https://doi.org/10.1098/rstb.2019.0458

Posen BR (2003) Command of the Commons. *International Security* 28(1): 5–53.

Pugh M (1996) Is Mahan still alive? State naval power in the international system. *The Journal of Conflict Studies* 16(2): 109–123.

Reus-Smit C (ed) (2004) *The Politics of International Law*. Cambridge: Cambridge University Press.

Richmond H (1934) *Sea Power in the Modern World*. London: G. Bell and Sons.

Ryan BJ (2015) Security spheres: A phenomenology of maritime spatial practices. *Security Dialogue* 46(6): 568–584.

Smith DP (2007) The 'buoyancy' of 'other' geographies of gentrification: Going 'back-to-the water' and the commodification of marginality. *Tijdschrift voor economische en sociale geografie* 98(1): 53–67.

Steinberg PE (1999) The maritime mystique: Sustainable development, capital mobility, and nostalgia in the world ocean. *Environment & Planning D: Society and Space* 17(4): 403–426.

Steinberg PE (2001) *The Social Construction of the Ocean*. Cambridge: Cambridge University Press.

Steinberg PE (2011) Free sea. In: Legg S (ed) *Spatiality, Sovereignty and Carl Schmitt: Geographies of the Nomos*. Abingdon: Routledge, 268–275.

Steinberg PE (2013) Of other seas: Metaphors and materialities in maritime regions. *Atlantic Studies* 10(2): 156–169.

Steinberg PE (2018) Editorial: The ocean as frontier. *International Social Science Journal* 68(229–230): 237–240.

Steinberg PE, Nyman E and Caraccioli MJ (2012) Atlas swam: Freedom, capital, and floating sovereignties in the seasteading vision. *Antipode* 44(4): 1532–1550.

Steinberg PE and Peters K (2015) Wet ontologies, fluid spaces: Giving depth to volume through oceanic thinking. *Environment & Planning D: Society and Space* 33(2): 247–264.

Stritzel H (2007) Towards a theory of securitization: Copenhagen and beyond. *European Journal of International Relations* 13: 357–383.

Suarez de Vivero JL, Rodriguez Mateos JC and Florido del Corral D (2009) Geopolitical factors of maritime policies and marine spatial planning: State, regions, and geographical planning scope. *Marine Policy* 33(4): 624–634.

Suarez de Vivero JL and Rodriguez Mateos JC (2014) Changing maritime scenarios. The geopolitical dimension of the EU Atlantic Strategy. *Marine Policy* 48: 59–72.

Sumida J (1999) Alfred Thayer Mahan, geopolitician. *Journal of Strategic Studies* 22(2–3): 39–62.

Till G (2004) *Seapower: A Guide for the 21st Century*. London: Frank Cass.

Till G (2007) Maritime strategy in a globalizing world. *Orbis* 51(4): 569–575.

Till G (2008) 'A cooperative strategy for 21st century seapower': A view from outside. *Naval War College Review* 61(2): 25–38.

Turner FJ (1920 [1996]) *The Frontier in American History*. New York, NY: Henry Holt.

Tyrrell M (2006) From placelessness to place: An ethnographer's experience of growing to know places at sea. *Worldviews: Global Religions, Culture, and Ecology* 10(2): 220–238.

UN General Assembly (1982) *Convention on the Law of the Sea*, 10 December 1982, Available at: www.refworld.org/docid/3dd8fd1b4.html.

Virilio P (1977) *Vitesse et politique*. Paris: Gallilée.

Waltz K (1979) *Theory of international Politics*. Reading, MA: Addison-Wesley.

# 5

# EMPIRE

## Towards errant and interlocking maritime spaces of power

*Andrew Davies*

### Introduction

Empires are complex and multifaceted political formations which seek to re-organise space to their own advantage, usually geopolitically and economically, but which frequently incorporate social and cultural and more diverse processes to facilitate these aims. On the one hand, they are about the extension of sovereignty beyond the territorial or other political boundaries of a particular nation-state, or the formal establishment of a political empire. But, in addition to this relatively formal political process, the reality of imperialism has considerable links to a broader range of practices such as: expropriation and extractivism; cultural dominance and the imposition of social hierarchies; colonialism; settler colonialism; and racialised forms of capitalism. Empires, and particularly the modern European empires emerging from the fifteenth century, were engines of time-space compression and emergent forms of globalisation, fundamentally bringing seemingly distant places into close relation, and encouraging the movement of peoples across ocean spaces in numbers never before seen. The extension of these modern imperial spatial forms (and concomitant resistance to them) into and through oceanic or maritime spaces were and are essential components to many of the practices of empire/imperialism, particularly as imperial powers sought to codify and organise ocean spaces (c.f. Steinberg, 2001). These attempts to variously limit, order and control ocean spaces were also bound together with diverse activities which challenged and resisted them and thus there are considerable links here to many of the other chapters contained within this Handbook.

Putting empire at the centre of our analysis of ocean space for this chapter means an analysis of the practices of imperialism in both the past and the present. Whilst it is impossible to cover every way in which imperialism interacts with maritime spaces there have been a number of emerging trends in the maritime-focused studies of empire and imperialism. Thus, the first section of this chapter explores how the spatial framings used to understand ocean spaces have shifted as a result of the two interrelated trends. Firstly, the turn towards global, trans-imperial and -national histories which seek to understand relationships across ocean spaces, and, secondly increasingly decolonial readings of ocean spaces, which emphasise the limitations of academic categorisations of space and society which are rooted in imperial or Eurocentric thought. Broadly, these trends have emphasised that, whilst imperialism (and its associated processes) has


DOI: 10.4324/9781315111643-7

always produced new spatial relations, they are increasingly read in ways which emphasise the plurality and diversity of these processes. A second section explores how these broader conceptual trends actually 'touch down' in particular places and contexts, showing how experiences of imperialism were and are in port-cities, ship-, military- and carceral-spaces. Discussion of imperialism must also necessarily cover how imperialism was and is contested. A final two sections explore these latter issues in relation to both explorations of anticolonialism in the past, but also the contested maritime spaces of imperialism in the present.

## Spatial metaphors and challenging imperial limits

The study of imperialism and empires requires, by its nature, a certain amount of thinking across the spatial boundaries of the nation-state. However, oceanic studies of empire have been at the forefront of understanding how limited a nation-centric or single-empire lens can be. There has been a huge increase in studies of imperialism which place transnational or transimperial oceanic spaces at their centre. This imperative is well established within maritime studies, often linked to Braudel's 'Mediterranean World' studies, but whilst much of this is well known, there are important aspects of this work which continue to challenge some of the core spatial scales and metaphors which we rely upon to organise our geographical conceptions of the world. The most obvious is the turn towards oceanically-scaled radicalism which was spurred by Linebaugh and Rediker's *The Many Headed Hydra* (2000), which still provides a focal point in understanding the Atlantic World of the seventeenth century 'from below'. The book spurred a range of similar projects thinking at an oceanic scale (c.f. Pearson, 2003; and also, Igler, 2013), and which often worked alongside similar trends towards global histories (c.f. Osterhammel, 2014). Such works obviously emphasise interconnection and a range of social, political and economic processes not necessarily linked to imperialism, but the role of imperial expansion in reshaping existing connections (such as the effects of European forms of capitalism upon pre-existing trade networks) was always central.

However, at the same time, a range of studies emerged complicating and challenging the reification of certain scales – particularly the oceanic – as given. Whilst interconnection was important, it was clear that drawing political or conceptual limits or boundaries around ocean spaces often obscured as much as it revealed – if one of the purposes of studying maritime spaces was to emphasise their spatially extensive relations, why should we rely on the arbitrary lines provided by our conceptions of where an ocean or sea 'ends' in order to structure our studies of them? As a result, a range of studies have emerged which either focus on different, sub-oceanic, scales (Amrith, 2013; Subramanian, 2016) or which deliberately seek to challenge the boundedness of these oceanic 'territories' and to hybridise our sense of them (Armitage, 2018; Hofmeyr, 2007, 2010; Lowe, 2015; North, 2018). Here, it is also useful to think about the intellectual challenges presented by these works, and the imperatives which they in turn create. By drawing on heterogenous and spatially extensive social, political, economic and material processes, such studies call into question both the geographical but also intellectual limits which such spatial metaphors, often unwittingly, impose. Drawing on Hofmeyr's work on the Indian Ocean, Sharad Chari (2019) has argued, the intellectual 'errantry' (after Glissant, 1997) which this endeavour involves should not be an attempt to either impose yet more Euro-American concepts onto maritime spaces, nor should it be an attempt to discover or rescue a forgotten historical or contemporary 'other' in the form of the subaltern. Rather, this should require a broadening of the terms by which we explore maritime spaces (for a related argument about the links between the subaltern and the conceptual, see Jazeel, 2014). This requires understanding maritime spaces alongside, in Chari's words, the 'material, political-economic, ecological, multi-species and infrastructural' (2019: 203–204). More than just 'adding' additional categories to our analysis, such errantry requires deliberate attempts to think across and beyond normal conceptions of social, cultural and historical work. This is similar in approach to calls across the academy for decolonisation, and thus intellectual engagements with the sea and empire are at the forefront of efforts to think beyond and across established conceptual boundaries.

This form of scholarship is important to this chapter's context for two reasons. Firstly, ocean spaces with their inherent interdisciplinarity and materiality, are crucial in evidencing the complex ontological realities (Peters and Steinberg, 2019; Steinberg and Peters, 2015) which require such boundary crossings. Secondly, whilst the focus of much of this research is often anti-imperialist, or at least critical in nature (and more on this later), such scholarship is often implicitly decolonial in orientation – resisting and challenging the orthodox boundaries which were often produced in and through imperialist practices in the past and present. As Sivasundaram (2020) has argued, historical research on ocean spaces, in his case, the Indian and Pacific Oceans during the so-called Age of Revolutions, has recently begun to challenge orthodox historical trajectories which were established by imperialist historiographies in the nineteenth and twentieth centuries. Sivasundaram excavates alternate, often subaltern, histories which show how groups like indigenous peoples were unevenly enrolled into imperial jurisdictions as they expanded into these 'new' oceanic territories. Such work is not searching for a 'pure' oceanic subaltern, but rather continues the trends of Linebaugh and Rediker's *Hydra* mentioned above towards more hybrid and heterogenous readings of imperial space. As well as expanding the *geographical* scope away from the Atlantic to different oceanic spaces, it necessarily expands the intellectual, political, spiritual, material, economic, and more, contexts which ensured imperialism was not a monolithic or universalistic process (despite the intentions in this regard of many imperial projects, and indeed many imperialist historians and geographers in their attempts to impose 'order' on knowledge).

As well as exposing or recovering such subaltern and hybrid histories, a core imperative noted above of 'errantry' (after Édouard Glissant) necessitates thinking beyond disciplinary boundaries. Exemplar here in challenging the terracentricity of much analysis, but also blurring the shoreline between sea/ocean is Sunil Amrith's work which challenges the intellectual boundaries of history, science, migration studies, environmental studies and more. Whilst his *Crossing the Bay of Bengal* (2013) places a distinct maritime space – the eponymous Bay – at its centre, exploring the histories and cultures which emerged from imperialism and which interconnect across its waters, it challenges the spatial limits of the Bay of Bengal as a discrete territorial unit – showing how it has always exceeded the limits of its supposed borders. To Amrith, the looming climate catastrophe and its effects on the Bay expose how inadequate territorially- and disciplinary-limited studies are in understanding the transnational and intersecting challenges such regions face. More recently, Amrith (2020) has taken this transdisciplinary imperative further in his water-driven history of efforts to understand and regulate the monsoon in his brilliant *Unruly Waters* – which whilst less explicitly 'maritime' in some respects, makes the boundaries between land, sea and atmosphere much more porous. For example, to eighteenth- and nineteenth-century scientists in South Asia seeking to understand the monsoon, it swiftly became clear that an understanding of the complex meteorological systems which spanned the Indian Ocean region was required, and that such systems were often exacerbated by the failures of imperial governance, which often overstated the potential of technological changes and led to disastrous human results (see also Davis, 2000).

However, in addition to such work which integrates the social and environmental categories, further strands of work have worried away at the maritime limits of imperial subjecthood. As Abraham (2015) has argued, much resistance to imperialism has proved hard to categorise as there has been a tendency to *a priori* assume that the nation-state, and the subjectivity provided by it, was and is the only possible outcome of struggles against colonialism – i.e. that national forms of independence, and national identity, are all that was ever desired by colonised subjects. In Abraham's case, the amorphous and unclear political subjectivities generated by the Singapore Mutiny of 1915 indicate the limitations of imperial categorisations of their subjects in the early twentieth century, and, similar to Chari's arguments about 'subaltern' seas, examining events such as this highlights the fissiparous nature of colonial subjectivity, rather than placing such events as 'anomalies' or exceptions which do not fit into the imperial order of things. This debate is important as it highlights the continued terminological challenges which are presented by spatial and sociological framings which emerged from Eurocentric knowledge. Abraham's paper uses the term 'international' to frame the diverse (counter)currents of empire, and whilst resisting such simple scalar characterisations, the spatial terminologies of empire, nation and ocean remain open for some debate. It is with these conceptual challenges in mind that the next section begins to show how these debates have played out in particular contexts.

## Interlocking imperial contexts and mobilities: Ports, ships, militaries, carceralities

Whilst the work in the previous section highlighted the ways in which large-scale imperial processes have been understood, an equally important trend within work on empire has continued to unpick the varied and heterogenous ways empires and imperialism helped to reshape spaces in a range of more specific contexts. For example, in his globally scaled primer on colonialism, Kris Manjapra (2020) places the colonial port city as a key space where what he terms the "interlocking histories" of the European colonial past were shaped. To Manjapra, the rapacious qualities of European capitalism expanded through the trading spaces created by colonial ports, and as such, ports provided enclave spaces by which European capitalism could establish "islands of Europe" (2020: 113) across the world. Whilst ports and port cities have long been recognised as key spaces of encounter and as nodal points in the trade links of empire, work on mobility and materiality has increasingly shown how port cities became core spaces where the contradictions of empire came into sharp relief. Jon Hyslop's work has made numerous contributions here, from exploring the attempts (and failures) of the Government of India to control smuggling into its ports (Hyslop, 2009), the racialised politics of disembarkation in South Africa (2018), to the role of ports in diverse forms of left politics in the interwar period (2019). Ports were places where the management of imperial difference were seen to be especially important (Fischer-Tiné, 2009), and often involved the very micropolitical practices of governing bodies and their positions became important. Thus, a variety of research has explored the roles that particular buildings and spaces within port cities played in governing the diverse populations who moved through them. For instance, Minca and Ong (2016) explore the roles hotels played in acting as both spaces of care and control in the transhipment of migrants overseas. Whilst not explicitly 'imperial' their study of the Royal Dutch Lloyd hotel brings into conversation the political geographies of the camp and incarceration with maritime governance routines – particularly in the ways bodies were ordered in spaces like hotels.

Other port city organisations and institutions, such as sailors' homes or seamen's missions, provided particular micro-political spaces of imperial organisation and regulation. For example, Justine Atkinson (2020) has shown how seaman's missions exemplify the tensions of imperial space. Envisioned as spaces for evangelical proselytisation during the encounter between British and Chinese empires in Guangdong. Seamen's Missions here are bound together with the 'civilising mission' of European empire as vehicles for delivering Christian universalism and steamship networks became a connecting locus binding groups together, with the port city-based mission providing a way to advance such ideologies. However, such missions were also spaces where attempts to manage the disreputable 'sailor class' were undertaken – thus missions fulfilled a dual purpose of spreading the civilising mission amongst both the colonial 'other' but also to regulate and instil suitable morals amongst the sailor whilst ashore. Similar tensions about how to govern the white 'sailor class' in colonial Calcutta are also central to Manikarnika Dutta's (2021) study of sailor's homes in Calcutta. Elsewhere, Hannah Martin's (2021) work exploring the non-exceptional histories of race and racism in the interwar North-East of England provides an insight into the ways in which so-called 'Arab Boarding Houses' were often microcosms of the British empire's gender, race and class politics. Such work tells more complex histories of how sailors, or in the latter case, the often white, British, wives/partners of the male 'Arab' owners of boarding houses, occupied intermediate subjectivities in the imagination of colonial authorities – neither being colonised subjects, nor able to be fully trusted as members of a subordinate or subaltern class themselves.

However, whilst ports provided important nodal spaces for maritime empires, ships and their crews formed the sutures that bound the different interlocking geographies of empire together. Various work has explored the ways in which imperial steamship mobilities helped to reconfigure existing geographies by creating rhythms and regularity to imperial networks (Anim-Addo, 2014; Steel, 2015) but also created imperial anxieties as populations of people began to move across space in potentially undesirable ways (Ballantyne, 2014). As such, the lived experiences of travel and movement aboard ships are increasingly relevant. On the one hand, this is about the recognition of how oceanic mobility was regulated and ordered as the examples above showed, but also how such spaces were affectively felt and experienced. Significantly, whilst oceanic travel was constitutive of new economic and political geographies, it was also productive of new emotional and affective geographies. For instance, Jeffrey Auerbach's *Imperial Boredom* devotes a significant chapter to the mundane and monotonous nature of shipboard travel in the British Empire. As Auerbach puts it,

> the ocean was not just a scenic backdrop to human events, but actively shaped the human experience, helping to produce feelings of boredom that had never been felt before, as well as the time and space to write about them.
>
> *Auerbach, 2018: 21*

As empires and travel expanded individual's horizons, so too did this change individuals' understandings of and relationship to space and place. As Auerbach reminds us, these changes were often mundane, but are nonetheless vital to understand how empires changed the way individuals made sense of the world they encountered. Importantly, the nature and scope of this shipboard monotony varied according to social categories, where race, class and gender all shaped the experience of shipboard life, but also were impacted by technological change – for example, the transition from sailing to steaming created seemingly fewer opportunities for (manual) work and/or distraction, which thus increased the banality and monotony of shipboard routines, even as they increased the speed of transit across the sea. Likewise, the emergence of the telegraph and radio communication meant that the seemingly isolated space of the ship became more likely to be connected to 'landed' spaces, which altered the means of finding distractions from the monotony of shipboard life.

Mobility, monotony and empire have also intersected with maritime forms of carcerality. The prominence of the slave ship as a central technology of Atlantic world slavery is well known (Glissant, 1997; Rediker, 2007). Understanding elsewhere, the broader global systems of transportation and punishment established under colonial rule have been central to work by Clare Anderson (2009, 2012). Anderson emphasises the relational ways in which empires shifted undesirable populations around the globe, and marks another addition to work which develops subaltern approaches to understanding oceanic space. Similarly, Peters and Turner (2015) have shown how the internal (im)mobilities of convict ship-space, along with its material infrastructure in the form of shackles, chains and cells were key ways in which prisoners' bodies were governed and controlled on long sea journeys.

The controversy over the SS *Komagata Maru* has proved an important area for work for those exploring the legal and migrational frameworks by which empires sought to govern maritime space (Mawani, 2018; Roy and Sahoo, 2016). The *Komagata Maru* incident was an attempt to allow South Asians to migrate into Canada through the port of Vancouver, which was ultimately unsuccessful after a long standoff between the ship's crew and passengers and the port and government authorities in Vancouver and Ottawa. Whilst there is not the space to go into detail here (see also Featherstone, this volume), the *Komagata Maru* exposed the racial inequalities of the British–Canadian colonial system for all to see, showing how movement across imperial jurisdictions was inherently racialised, but this also provided an impetus for transnational forms of anticolonialism to emerge, particularly in inspiring the Ghadar movement (c.f. Ramnath, 2011).

Lastly in this section, there is an emerging field of research that has explored the social, cultural and political geographies of military ship-space, often driven through the turns towards assemblage thinking which have emerged in geography from the late 2000s. Whilst some of this work does not make explicit links to empire or imperialism, these links are often present. For example, Dittmer and Waterton (2018) note only briefly in their affective and embodied exploration of the museum ship *HMS Belfast* that such military ships "materialised the global ambition of the British Empire" (2018: 706). Whilst *HMS Belfast* is a particularly apt choice for this statement as a cruiser, a class of ship which was often relied upon for trade protection and providing 'power-projection' across the British Empire, this recognition that such militarised forms of heritage are bound together with empire is important, as such links are often occluded through a focus on the technological aspects of such ships, or the tendency towards patriotic accounts of the exploits of the ship and its crew during wartime. Similarly, whilst the turn towards volumetric accounts of territory and geopolitical control have been prominent in both maritime and geopolitical geographies, Williams (2017) has explored the links between these trends by examining the naval aircraft carrier as a tool of geopolitical power projection for the US fleet in the interwar Pacific Ocean. The aircraft carrier literally mobilised US naval (and, by extension, imperial) power – to Williams, aircraft carriers and the aircraft upon them, acted as a form of 'mobile island' which, by being at sea project a different form of geopolitical power, which is more spatially expansive and multi-dimensional than the previous forms of battleship or gunboat diplomacy which were limited by the range of a ship's guns, or the visibility of the ship itself. Crucially, the technical developments of shipboard aviation in the interwar years (such as developing more robust ship and aircraft designs) allowed the US Navy to both better manage the material instabilities of the ocean (e.g. being able to launch aircraft in rougher seas) but additionally to begin establishing a dominant military presence across the Pacific which it has not relinquished since defeating the Empire of Japan in World War Two – and more on this below. More explicitly (anti)imperial, my own work (Davies, 2014) has explored how naval doctrines, ship-spaces and oceanic mobilities in the British-ruled Government of India's Royal Indian Navy (RIN) created a distinct series of assemblages which sought to both govern and civilise the colonial sailor. However, the limits to colonial practices of discipline were exposed as the crude translations of Royal Naval doctrines to colonial settings intersected with imperial racisms, and ultimately led to the Mutiny of the RIN in 1946. Such work all suggests that the intersections between maritime/naval militaries, imperialism and current debates within human geography are an important but under-explored issue.

## Maritime anticolonialisms

As well as the particular spaces of imperial ordering, governing and mobilities discussed above, and as already hinted earlier, resistance to these modes, and the very real limits of control which the imperial state was and is able to impose are the subject of the final two sections of this chapter. Whilst not wishing to separate the varied experiences of empire into a binary of domination and resistance, and indeed, much of the discussion in this chapter so far has shown how entangled these experiences were, it is worth drawing out the role which maritime spaces played in facilitating anticolonialism. On the one hand, there is a burgeoning literature which, again taking its cue from global or transnational studies, has explored the ways in which empire opened up a worldwide network of opportunities, particularly for anticolonial revolutionaries. As Sujit Sivasundaram (2020: 3) puts it "[t]hose who took passage on European ships, or who worked on grand projects as labourers and technicians, could use this moment of opportunity to contemplate their selfhood and futures in radically new ways". Harper (2020) is also explicit about the ways in which port cities and their environs provided a great 'village abroad' where those inclined towards anticolonial activities could meet, even as they were exiled or fugitives from their homes. Steamship networks formed a crucial locus of connection, allowing political ideas to be smuggled as much as more material contraband (Hyslop, 2009). These networks of revolution and resistance were deeply imbricated with colonial capital and labour (Ahuja, 2009; Balachandran, 2012), and were often closely aligned with the particular maritime networks established by imperial nation states, such as the shipping lines of the Netherlands' Empire (Alexanderson, 2019).

A central contribution to the importance of maritime labour to colonial resistance has come through the work of David Featherstone. Featherstone has consistently argued (2015, 2019, 2020, and this volume) that seafarers were central in developing subaltern forms of cosmopolitanism which contested and reshaped the imperial order. These cosmopolitanisms necessarily involved building of solidarity across colonial lines of difference, recognising the often systemic and trans-imperial hierarchies which position colonial subjects negatively compared to their (often European or white) others. This again continues to show the errantry of Chari above, and marks more research exploring how political agency was heterogenous and contested across maritime imperial spaces.

Once again, specific material spaces, such as the ship, are important. One important aspect here is the temporality and space afforded to anticolonialists whilst travelling at sea. Mohandas Gandhi's *Hind Swaraj*, the anticolonial text he wrote where his ideas of both *swaraj* (freedom) and *satyagraha* (truth-force, or truth-struggle) were first fully expressed for a South African and Indian audience, was written whilst travelling from London to South Africa on the *SS Kildonan Castle*.[1] However, far from spaces of writerly solitude, ships were also important political spaces as Stephen Legg (2020) has most recently shown in his examination of the experience and politics of transit to and from India to the UK for the Round Table Conferences in the 1930s. Legg argues that whilst transit on board ships could be an example of Auerbach's (2018) term 'Imperial Boredom' as individuals were sequestered on board ships for long periods of time, they also proved important spaces for political negotiations amongst different groups heading towards the conference(s), but also provided important spaces for newspapers to generate commentary on interesting characters, like Gandhi – with stories appearing about his supposedly remarkable habits and lifestyle. The importance here is that the ship becomes not a technological mechanism by which the vectors of anticolonial organisation were simply 'transported' from place to place, but rather a specific place where anticolonial politics was shaped by the material and atmospheric nature of the ship. Lastly, as well as the RIN mutiny mentioned above, my own work (Davies, 2019) explored how the Swadeshi Steam Navigation Company, an Indian nationalist steamship company set up to contest British shipping monopolies between South India and Ceylon (Sri Lanka) in the early twentieth century altered the land-based anticolonial activities in the far south of India. Here the ships and offices of the shipping line became powerful nationalist symbols which mobilised anticolonial activity in the urban spaces of South India, but which fundamentally imagined an Indian Ocean-wide form of resistance to European imperialism and capitalism. Thus, again, the interlocking and entangled nature of maritime imperialisms defies simple categorisations. The next section of this chapter continues this trend by exploring research which covers more recent contexts.

## Contemporary imperialisms and the oceanic

The largely historical focus of the chapter so far is not to suggest that practices of imperialism and maritime space do not intersect in the present. As maritime geographers have been at pains to point out over the past decade and more, the seeming invisibility of ship- and ocean-spaces from the land has been an important lacuna for geography which is only now being adequately addressed, and this arguably extends even more so to studies of the colonial present. As such, this final substantive section of this chapter outlines a few important trends in the literature.

Whilst contemporary infrastructural geographies have often explored, more or less explicitly, the role large-scale megaprojects and/or logistics play in facilitating imperial geopolitical forms (c.f. Cowan, 2014), the interlocking geographies emphasised above by Manjapra in relation to port cities are clearly visible in the present-day logics of ports and shipping. Particularly useful here is Laleh Khalili's (2020) *Sinews of War and Trade*, where the geopolitical present of the Middle East, particularly the Arabian Gulf and its role in global trade, is framed through the intervention of past (largely British) imperialisms, but continues to be shaped by the USA's imperial and extractive interests in the region. What Khalili's study shows with often brutal effectiveness is the genuine interlocking nature of the imperial present, shaped as it is by the reworking of geopolitical and legal sovereignties to suit imperial needs, but how this plays out in dynamic ways – from the dredging of harbours and the destruction of ecologies, through to the reform of labour rights on board ships and in ports. This again shows how the historical and theoretical work highlighted earlier in this chapter which destabilises and challenges imperial categorisations is not only ever a matter of history or theory. Khalili's work provides an important and substantive work on the multifaceted nature of imperialistic processes in oceanic spaces – and provides a caustic reminder of the fact that ships, shipping and trade remain essential components of geopolitical control. Similar work, not explicitly about 'imperialism' but instead about how capitalist forms of austerity, which I would argue could be read as indicative of imperialist processes, become realised in maritime and estuarine/riverine spaces, is Laura Bear's (2015) ethnographic work on the Hooghly river in West Bengal. This explores how sovereign indebtedness operates to create interlocking socio-economic, environmental, and political results in postcolonial India. In Khalili and Bear's work the importance of both past colonial and postcolonial history are fundamental to understand the complex realities of the twenty-first-century maritime economy.

In addition to these economy-centred readings of global shipping, there are also more militaristic readings of twenty-first-century maritime imperialism. Sasha Davis, combining interests in archipelagic and assemblage literatures, has undertaken a range of work which explores how the USA extends and maintains its geopolitical supremacy across the Pacific. This intersects with Williams' work on aircraft carriers and the development of US imperial control mentioned above, but Davis shows how military island bases and their environs remain an important part of the colonial present of the twenty-first century. Whilst this clearly involves a process of colonial control by which the US seeks to both claim island territories and maintain hegemonic dominance (Davis, 2011, 2015), this exists alongside processes of resistance and organisation amongst islanders and related social movements which seek to contest such imperialistic impulses (Davis, 2017). Davis' work forms part of the wider island studies/archipelagic turn within maritime studies which has contested stereotypical imaginings of islands but also helps to expose the continuing coloniality of the seas.

Importantly here, alongside other works on islands and colonialism, this work provides avenues for research which is more collaborative and decolonial in scope, promoting diverse epistemologies in further efforts to resist colonial framings of these spaces (c.f. Grydehøj et al., 2021; Pugh, 2016), but also drawing on linkages to social movement and participatory geographies. Research on these island spaces and their relations chimes closely with other scholarship, most notably Alison Mountz's (2011, 2020) longstanding and vital mapping of the asylum and enforcement archipelago, as well as the carceral works outlined in the previous section. Elsewhere, imperialism's long legacies are visible in other island spaces, from the ongoing struggle of indigenous Chagos islanders with British and US geopolitical interests (Zondi, 2020) or in the particular forms of creolisation formed through the geographies of indentured labour (Durgahee, 2017). Lastly, of course, islands have proved to be one of many spaces where China's expansionist geopolitical aims have become more visible, in both the South China Sea and in the Maritime Silk Road Initiative (Blanchard and Flint, 2017). Whilst it is likely still too much to suggest that China is in possession of a formal geopolitical empire, the imperialist impulse in contesting US hegemony and in expanding Chinese geopolitical interests means that there are certainly important correlatives which can be drawn here.

Thus, this last section shows that, whilst imperialism and anti-imperialism may not be as visibly bound to particular steamship mobility, labour practices, or, to port city spaces as it was in the early twentieth century, maritime spaces and the wider circuits of geopolitical and capitalist relations which manifest as port, maritime, naval infrastructures and more, continue to be spaces where imperial power in its various twenty-first-century forms are made manifest.

## Conclusion

Maritime experiences and understandings of empire are necessarily diverse and contested, and the range and scope of research undertaken clearly exceeds the space a short overview such as this chapter allows. However, across the various contexts, scales and spaces highlighted above is a tendency towards thinking across and beyond a singular ocean space or a single academic category/concept towards understandings which emphasise two trends. The first of these, drawing upon Manjapra, is the increased recognition of the ways in which empire and imperialism reworked existing, or created new, interlocking relationships between diverse spaces and places. These are more than political, economic or cultural relationships, but also increasingly recognised as material, ecological, ontological and more. The geographies of empire in the past and the present bring diverse, sometimes contested, processes into close relation, and so research is increasingly inter- or trans-disciplinary in orientation as scholars try to engage across or beyond traditional disciplinary boundaries. Relatedly, the second trend is towards intellectual 'errantry' (after Glissant, 1997), drawing upon anti- and decolonial thought which seeks to challenge dominant categorisations and worldviews which are rooted in European and colonial imperialism. These two trends have significant overlaps and complement each other, but the overall trend is towards more nuanced understandings of how maritime spaces were and are important to imperialism's ability to reshape, relations across space. These relations often exceed simple categorisations, and, despite the variety and breadth of scholarship covered above, it is clear that such work is only the beginning of exciting trends which will continue to expand and challenge our knowledge of maritime empires and imperialisms.

## Note

1 As Hyslop (2011) notes, an important moment prior to Gandhi writing *Hind Swaraj* was also the defeat of the Russian Navy by the Empire of Japan at the Battle of Tsushima in 1905 which gave impetus to dreams of an Asia free from European imperialism, but also stoked racist fears of a 'yellow peril'.

## References

Abraham I (2015) 'Germany has become Mohammedan': Insurgency, long-distance travel, and the Singapore mutiny, 1915. *Globalizations* 12(6): 913–927.

Ahuja R (2009) Networks of subordination – networks of the subordinated: The ordered spaces of South Asian maritime labour in the age of imperialism (c. 1890–1946). In: Tambe A and Fischer-Tine H (eds) *The Limits of British Colonial Control in South Asia: Spaces of Disorder in the Indian Ocean Region.* Abingdon: Routledge, 13–48.

Alexanderson K (2019) *Subversive Seas: Anticolonial Networks across the Twentieth Century Dutch Empire.* Cambridge: Cambridge University Press.

Amrith S (2013) *Crossing the Bay of Bengal* Cambridge, MA: Harvard University Press.

Amrith S (2020) *Unruly Waters* London: Penguin.

Anderson C (2009) Discourses of exclusion and the 'convict stain' in the Indian Ocean (c. 1800–1850). In: Tambe A and Fischer-Tiné H (eds) *The Limits of British Colonial Control in South Asia: Spaces of Disorder in the Indian Ocean Region.* Abingdon: Routledge, 105–120.

Anderson C (2012) *Subaltern Lives: Biographies of Colonialism in the Indian Ocean World* Cambridge: Cambridge University Press.

Anim-Addo A (2014) 'The great event of the fortnight': Steamship rhythms and colonial communication. *Mobilities* 9(3): 369–383.

Armitage D (2018) World history as oceanic history: Beyond Braudel. *The Historical Review-La Revue Historique* 15(1): 341–361.

Atkinson J (2020) 'On their own element': Nineteenth-century seamen's missions and merchant seamen's mobility. In: Lambert D and Merriman P (eds) *Empire and Mobility in the Long Nineteenth Century.* Manchester: Manchester University Press, 92–111

Auerbach JA (2018) *Imperial Boredom: Monotony and the British Empire* Oxford: Oxford University Press.

Balachandran G (2012) *Globalizing Labour: Indian Seafarers and World Shipping, c. 1870–1945.* Oxford: Oxford University Press.

Ballantyne T (2014) Mobility, empire, colonisation. *History Australia* 11(2): 7–37

Bear L (2015) *Navigating Austerity: Currents of Debt Along a South Asian River.* Stanford, CA: Stanford University Press.

Blanchard J-M and Flint C (2017) The geopolitics of China's maritime Silk Road initiative. *Geopolitics* 22(2): 223–245.

Chari S (2019) Subaltern Sea? Indian Ocean errantry against subalternization. In: Jazeel T and Legg S (eds) *Subaltern Geographies.* Athens, GA: University of Georgia Press, 191–209.

Cowan D (2014) *The Deadly Life of Logistics: Mapping Violence in Global Trade.* Minneapolis, MN: University of Minnesota Press.

Davies AD (2014) Learning 'large ideas' overseas: Discipline, (im)mobility and political lives in the Royal Indian Navy mutiny. *Mobilities* 9(3): 384–400.

Davies AD (2019) *Geographies of Anticolonialism: Political Networks Across and Beyond South India, c. 1900–1930.* Chichester: Wiley-Blackwell.

Davis M (2000) *Late Victorian Holocausts: El Nino Famines and the Making of the Third World.* London: Verso.

Davis S (2011) The US military base network and contemporary colonialism: Power projection, resistance and the quest for operational unilateralism *Political Geography* 30(4): 215–224.

Davis S (2015) *The Empires' Edge: Militarization, Resistance, and Transcending Hegemony in the Pacific*. Athens, GA: University of Georgia Press.

Davis S (2017) Apparatuses of occupation: Translocal social movements, states and the archipelagic spatialities of power. *Transactions of the Institute of British Geographers* 42(1): 110–122.

Dittmer J and Waterton E (2018) "You'll go home with bruises": Affect, embodiment and heritage on board HMS Belfast. *Area* 51: 706–718.

Durgahee R (2017) *The Indentured Archipelago: Experiences of Indian Indentured Labour in Mauritius and Fiji, 1871–1916*. Doctoral dissertation, University of Nottingham.

Dutta M (2021) The sailors' home and moral regulation of white European seamen in nineteenth-century India. *Cultural and Social History* 18(2): 201–220.

Featherstone D (2015) Maritime labour and subaltern geographies of internationalism: Black internationalist seafarers' organising in the interwar period. *Political Geography* 49: 7–16.

Featherstone D (2019) Maritime labour, transnational political trajectories and decolonization from below: The opposition to the 1935 British Shipping Assistance Act. *Global Networks* 19(4): 539–562.

Featherstone D (2020) Anti-colonialism, subaltern anti-fascism and the contested spaces of maritime organising. In: Brasken K, Copsey N and Featherstone D (eds) *Anti-Fascism in a Global Perspective: Transnational Networks, Exile Communities and Radical Internationalism*. Abingdon: Routledge, 155–175.

Fischer-Tiné H (2009) Flotsam and jetsam of empire? European seamen and spaces of disease and disorder in Mid-Nineteenth Century Calcutta. In: Tambe A and Fischer-Tiné H (eds) *The Limits of British Colonial Control in South Asia: Spaces of Disorder in the Indian Ocean Region*. Abingdon: Routledge, 121–154.

Glissant É (1997) *Poetics of Relation* (trans. B Wing) Ann Arbor, MI: University of Michigan Press.

Grydehøj A, Bevacqua ML, Chibana M, Nadarajah Y, Simonsen A, Su P, Wright R and Davis S (2021) Practicing decolonial political geography: Island perspectives on neocolonialism and the China threat discourse. *Political Geography* https://doi.org/10.1016/j.polgeo.2020.102330

Harper T (2020) *Underground Asia: Global Revolutionaries and the Assault on Empire*. London: Allen Lane.

Hofmeyr I (2007) The Black Atlantic meets the Indian Ocean: Forging new paradigms of transnationalism for the global south – literary and cultural perspectives. *Social Dynamics* 33(2): 3–32.

Hofmeyr I (2010) Universalizing the Indian Ocean. *Proceedings of the Modern Language Association of America* 125(3): 721–729.

Hyslop J (2009) Guns, drugs and revolutionary propaganda: Indian sailors and smuggling in the 1920s. *South African Historical Journal* 61(4): 838–846.

Hyslop J (2011) An "eventful" history of *hind swaraj*: Gandhi between the battle of Tsushima and the union of South Africa. *Public Culture* 23(2): 299–319.

Hyslop J (2018) The politics of disembarkation: Empire, shipping and labor in the port of Durban, 1897–1947. *International Labor and Working Class History* 93: 176–200.

Hyslop J (2019) German seafarers, anti-fascism and the anti-Stalinist left: The 'Antwerp Group' and Edo Fimmen's International Transport Workers' Federation, 1933–40. *Global Networks* 19(4): 499–520.

Igler D (2013) *The Great Ocean: Pacific Worlds from Captain Cook to the Gold Rush*. Oxford: Oxford University Press.

Jazeel T (2014) Subaltern geographies: Geographical knowledge and postcolonial strategy. *Singapore Journal of Tropical Geography* 35(1): 88–103.

Khalili L (2020) *Sinews of War and Trade: Shipping and Capitalism in the Arabian Peninsula*. London: Verso.

Legg S (2020) Political lives at sea: Working and socialising to and from the India Round Table Conference in London, 1930–1932. *Journal of Historical Geography* 68: 21–32

Linebaugh P and Rediker M (2000) *The Many Headed Hydra*. London: Verso.

Lowe L (2015) *The Intimacies of Four Continents* Durham, NC: Duke University Press.

Manjapra K (2020) *Colonialism in Global Perspective*. Cambridge: Cambridge University Press.

Martin H (2021) *The Intersection of Race, Class and Politics in the North East of England, 1919–1939*. Unpublished doctoral dissertation, Northumbria University.

Mawani R (2018) *Across Oceans of Law: The Komagata Maru and Jurisdiction in the Time of Empire*. Durham, NC: Duke University Press.

Minca C and Ong C-E (2016) The power of space: The biopolitics of custody and care at the Lloyd Hotel, Amsterdam. *Political Geography* 52: 34–46.

Mountz A (2011) The enforcement archipelago: Detention, haunting and asylum on islands. *Political Geography* 30(3): 118–128.

Mountz A (2020) *The Death of Asylum: Hidden Geographies of the Enforcement Archipelago* Minneapolis, MN: University of Minnesota Press.

North M (2018) Connected Seas I. *History Compass* 16(12) https://doi.org/10.1111/hic3.12503

Osterhammel J (2014) *The Transformation of the World.* Woodstock: Princeton University Press.

Pearson M (2003) *The Indian Ocean.* Abingdon: Routledge.

Peters K and Steinberg PE (2019) The ocean in excess: Towards a more-than-wet ontology. *Dialogues in Human Geography* 9(3): 293–307.

Peters K and Turner J (2015) Between crime and colony: Interrogating (im)mobilities aboard the convict ship. *Social and Cultural Geography* 16(7): 844–862.

Pugh J (2016) The relational turn in island studies: Bringing together island, sea and ship relations and the case of the Landship. *Social and Cultural Geographies* 17(8): 1040–1059.

Ramnath M (2011) *Haj to Utopia.* Berkeley, CA: University of California Press.

Rediker M (2007) *The Slave Ship: A Human History*. London: Penguin.

Roy AG and Sahoo AK (2016) The journey of the Komagata Maru: National, transnational, diasporic. *South Asian Diaspora* 8(2): 85–97.

Sivasundaram S (2020) *Waves across the South: A New History of Revolution and Empire.* London: William Collins.

Steinberg PE (2001) *The Social Construction of the Ocean.* Cambridge, Cambridge University Press.

Steinberg PE and Peters K (2015) Wet ontologies, fluid spaces: Giving depth to volume through oceanic thinking. *Environment and Planning D: Society and Space* 33(2): 247–264.

Steel F (2015) The "missing link": Space, race and transoceanic ties in the settler-colonial pacific. *Transfers* 5(3): 49–67.

Subramanian L (2016) *The Sovereign and the Pirate: Ordering Maritime Subjects in India's Western Littoral.* Oxford: Oxford University Press.

Williams AJ (2017) Aircraft carriers and the capacity of mobilise US power across the pacific, 1919–1929. *Journal of Historical Geography* 58: 71–81.

Zondi S (2020) The post-colonial is neocolonial in the Indian Ocean region: The case of Chagos seen through the African-centred decolonial theoretical lens. *Africa Review* 12(2): 119–132.

# 6

# FRONTIERS

## Ocean epistemologies – privatise, democratise, decolonise

*Leesa Fawcett, Elizabeth Havice and Anna Zalik*

### Introduction

The oceans are often typified as riddled with 'knowledge gaps' and as under-researched relative to terrestrial space (see also Alder, this volume; Waiti and Wheaton, this volume). Only 0.04 to 4 per cent of total research dollars worldwide goes to ocean science, a pattern that has led humanity to know more about Mars than about Earth's oceans (Intergovernmental Oceanographic Commission, 2019). Knowledge gaps in dominant scientific[1] understanding of the oceans are frequently implicated in the so-called 'ocean crisis', based on the assumption that what 'we' know about oceans and how 'we' govern them are inextricably linked. For instance, in describing the rationale for the United Nations (UN) Decade of Ocean Science, Visbeck (2018: 1) argues that the vast volume of oceans is "neither fully observed, nor adequately understood", and that enhancing understanding is critical to ocean governance.

In this chapter, we conceptualise knowledge as part of an *epistemological frontier* defining contemporary attempts to harness the ocean and its resources (see also Havice and Zalik, 2018). We think about how current knowledge and knowledge gaps in human understanding of the oceans are co-constituted with governance of spaces beyond direct human observation. We use the notion of a 'frontier' in this chapter in two ways. First, the oceans, particularly the high seas and deep marine zones, are often described as the last planetary frontier; a metaphor which uncritically divides the human from 'wilderness' and leaves 'the frontier' as a boundary that remains intact and out of human reach. In western historiography, the frontier develops and is conquered through imperialism and resource-fuelled global capitalism, processes that are informed and made through epistemological tools such as cartography and surveying (see Lehman, this volume). The application of these tools helps to make 'terrain' ripe for civilising and human dominance (e.g., Elden, 2010). Frontiers combine commodity-formation with cultural and territorial control to make a range of natural and social processes available for appropriation (see Gustavsson and Allison, this volume; Thomas et al., this volume). Second, frontiers are zones where a range of interest groups and agents, with varying degrees of power, seek to implement scientific and technological knowledge to reshape political, economic, social and ecological relationships in their interests (Peluso and Lund, 2011; Vandergeest, 2018). At resource frontiers, no single institution or actor exercises political authority, though the already


DOI: 10.4324/9781315111643-8

powerful are clearly at an advantage in shaping political-economic, social and environmental dynamics.

Below, we focus on *epistemology* as a frontier in and of itself, and as a foundational element of a broader notion of ocean frontiers: marine zones that are presently beyond full incorporation into capitalist circuits, but increasingly in their sights (e.g., Silver and Campbell, 2018). We are interested in the role that knowledge plays in shaping ongoing tensions between attempts to enhance extraction and conservation of ocean spaces through governance processes that might involve enclosure or commoning (Tladi, 2011). Governance, we define broadly as combinations of actors, institutions, legal processes, political economy relations and knowledges involved – directly or indirectly – in environmental decision-making (Bridge and Perrault, 2009). We recognise the emergence of *governance* in relation to the rise of neoliberal 'flexible regulation' where corporate and NGO activity is to complement more traditional state- and interstate environmental decision making, and which progressive forces hoped would remain open to de-colonial possibilities. We find resonance in these tensions with a guiding principle of maritime law that emerged through the United Nations Convention on the Law of the Sea (UNCLOS): that of the common heritage of (hu)mankind, which is among the key equity principles in international law (Okereke, 2008; *International Social Science Journal*, 2018).

The slow socio-ecological violence (Nixon, 2011) arising from anthropocentric and terrestrial approaches to oceans governance in part arises from thinking of oceans as 'external' to human and terrestrial worlds. Yet thinking about human attempts to harness ecologies in the deepest parts of the oceans and the seabed requires attention to the knowledges being developed, deployed, or overlooked in shaping the contemporary oceans. In the contemporary era, new data technologies are revealing ocean space, more than human natures and human activity in the oceans in forms previously impossible. These hold promises for future conservation, and potentially transformative socio-ecological systems that are based on democratised knowledge and commoning in oceans governance (Havice et al., 2018). However, there is also the risk and potential that new forms of knowledge are produced around and in relation to human extractivist agendas and are privatised and utilised to enclose the oceans as resource frontiers (Coumans, 2018; Zalik, 2018). Furthermore, some knowledges and ways of knowing are completely excluded from or outside of dominant oceans governance relations. Thus, we use this chapter to consider whether knowledge generation at epistemological frontiers reinvigorate, ignore or negate the principle of 'common heritage' upon which States – led by the Global South – sought to base twentieth-century ocean jurisprudence via UNCLOS. This requires attention to the historical origins, legacies and power relations that form the epistemological foundations of human approaches to governing the oceans.

In what follows we offer examples of three frontier epistemologies prosecuting, or essential to understanding, dynamics in oceans governance. First, we discuss how Cold War-era proprietary knowledge concerning the oceans, held in part by firms affiliated with state militaries, has shaped contemporary resource extraction and regulatory debates over mining and conservation of the seabed. These are playing out at the International Seabed Authority (ISA), the UNCLOS-established entity that oversees seabed mining in areas beyond national jurisdiction. We then turn to how organisations that develop 'new data technologies' employed in contemporary dominant science make information on the oceans available in novel forms. We examine new data technologies as a new frontier for ocean knowledge that is presumed to lead to better governance by rendering oceans and human activity in them visible *to all*, and thus subject to human governance. These hold potential to democratise oceanic knowledge, and in turn governance, while also reifying knowledge produced by specific scientific entities. Finally, following from this, we consider the vital place of decolonised knowledges. We explore how

Indigenous epistemologies and ontologies, as well as the knowledge of more-than-humans in the oceans, sit external to human-imposed governance, yet within ideas of the common heritage of humankind. Drawing on the case of sound pollution, specifically how the more-than-human knowledge possessed and transferred is disrupted through industrial activity, we examine what is at stake when such knowledges are overlooked.

## Privatising knowledge: Proprietary data and the deep seabed

The ISA is a UN-agency created and mandated under UNCLOS to manage the seabed beyond national jurisdiction, a zone known in international law as the 'Area'. The ISA oversees seabed mineral mining, and in principle hydrocarbon and other seabed resource collection, beyond national jurisdiction. In the 1970s as UNCLOS negotiations unfolded, interest in deep sea minerals mining was considerable, but with the global economic downturn of the 1980s, interest waned. However, at the turn of the millennium and with mineral commodity booms, a rush toward mining the deep seabed once again has been reinvigorated, in part now to gain access to minerals and rare earths as inputs for emerging and purportedly less carbon-intensive, technologies. As of the end of 2019, the ISA had granted 30 contracts for exploration to firms sponsored by ISA member states. These are located in seven regions of the Area where seabed mineral mining in polymetallic nodules, polymetallic sulphides, or cobalt rich crusts is considered viable – most significantly the Clarion Clipperton Zone between Hawaii and Mexico to which 18 exploration contracts have been allocated.[2] However, the ISA has not granted any *extraction* concessions in the Area and the body is currently developing a code to guide exploitation. Knowledge about minerals and other dimensions of the deep oceans are central to the future of extractive practices on the seafloor.

Informed by the principle that emerged through UNCLOS negotiations asserting that the international seabed is the 'common heritage of (hu)mankind', the Global South states those calling for a New International Economic Order (NIEO), advocated for information and technology transfer on the Area's ecology and potentially minable resources. The objective here was to promote redistribution of knowledge and power from military and capital centres in the Global North/First and Second World. Information sharing was also to include the technologies developed to harness those resources, in particular seabed minerals. By the time UNCLOS was ultimately ratified in 1994, the same year the World Trade Organization came into effect, the common heritage principle that shaped the ISA was overshadowed by the neoliberal agenda associated with the 'Washington Consensus'. Thus, the terms of the UNCLOS implementing agreement on Part XI of the Convention pertaining to the creation of the ISA, specifically favoured 'market principles' in the governance of the Area (United Nations, 1994). The Implementing Agreement, as such, underscored the protection of 'pioneer investors', states and their firms (parastatal or private) which had undertaken research in the Area during the Cold War, prior to UNCLOS coming into effect. The negotiations over pioneer investor protections continue to favour the activities that certain states and firms conducted prior to UNCLOS ratification; these were permitted under the 'reciprocating agreements' negotiated largely in secret (Hayashi, 1989, 2005). Among the most important pioneer investors was the conglomerate OMCO (Oceans Mineral which today holds exploration concessions under the ISA [Zalik, 2015, 2018]) whose rights were ultimately transferred to global weapons manufacturer Lockheed Martin. Given US non-ratification of UNCLOS, Lockheed's interests as holder of ISA exploratory contracts are held by its wholly-owned British subsidiary, UK Seabed Resources Ltd (UKSRL). While the ISA unveiled a new data sharing platform in 2019, to date it does not appear that proprietary information held by the pioneer investors is made available there.

Accordingly, among the key points of contention in the emerging mining/exploration code at the ISA relate to what fiscal mechanism will be employed to redistribute profits from mining the zone of 'common heritage', and what ecological requirements and environmental impact assessment (EIA) will be required of mining contractors. The ecology of the deep seabed remains a *knowledge frontier*, but the research entailed by EIA processes, and the capacity of such research to promote ecological conservation, is shaped by the conditions of investment that makes it possible, which we discuss below. Crucial also is that proponent firms rely on proprietary data held by firms active in the Area prior to UNCLOS ratification (Zalik 2018, 2021). Indeed, at today's ISA, considerable quantities of the limited – but growing – knowledge concerning the deep seabed is held by pioneer investor firms, notably UKSRL/Lockheed. A competitor deep sea mining firm, Nautilus Minerals, explicitly documents its dependence – and by extension that of various other firms – on data held today by Lockheed, collected by a former OMCO staff member in the 1970s and 1980s. Similarly, Ocean Minerals Singapore holds rights to an area adjacent to a UKSRL exploration area under the 'parallel system' and explicitly acknowledges its partnership with Lockheed as providing it the ability to employ the deep sea data that the firm controls. The parallel system was intended to promote redistribution and technology transfer to the Global South but, as per the key intervention of this chapter, also underlines how proprietary knowledge shapes industrial partnerships.

Patent books document the proprietary deep ocean technology created in the 1970s which offered prototypes to develop contemporary technologies under competing ISA contracting firms. The principle of 'common heritage' was consequently compromised by the protection of intellectual property that the Part IX Implementation Agreement upholds (UNCLOS Implementation Agreement 5: 17). Critically, the baseline data required for the protection of the marine environment is formally mandated as the responsibility of the contractor. Thus, while in principle there should be access to ecological information, there is no mechanism to ensure full disclosure from the contracting firm. The firm may itself provide the transportation and resources necessary for marine biologists to undertake exceptionally costly ultra-deep marine research for which it is dependent on finance capital. Yet the financial capital that makes such research possible is extended upon the basis of subsequent returns to the investing firm from seabed mineral extraction, thus militating against the ability to implement a precautionary principle which open environmental impact assessments should entail (Zalik, 2018, 2021).

The above conditions point at the key role mining capital plays in carrying out ecological research – and the ultra-deep ecologies placed at risk – due to extension of financing for the very processes and assessments intended to make them knowable. This finance requires subsequent mining returns and thus impels extraction. Financing for ecological research on the premise of ultimate returns from extraction poses the risk of encouraging extraction terms which would restrict potential redistribution of the profits that may accrue. A fiscal regime presented to the ISA by MIT consultants in 2018 stresses the need for favourable investor terms in order to make high-risk mining of the Area attractive.[3] The production, here, of information presented as expert knowledge is designed to advance a governance regime favourable to capital.

Although Global South states frequently raise the 'common heritage of (hu)mankind' as a fundamental UNCLOS principle at ISA Council and Assembly meetings, the balance of power at the ISA rests with powerful states, including the US and UK. The current Secretary General of the ISA, elected in 2017, is a former ISA staff person and British lawyer, Michael Lodge. Lodge is the first Secretary General of the Authority who does not hail from the Global South but who multiple interviewees suggested was supported by the UK, the US as observer state, and various allies, including Canada. Recent publications co-authored by Lodge and US-based policy specialists and researchers (Lodge and Verlaan, 2018; Lodge et al., 2017) emphasise the importance of fiscal conditions that favour contractors. Thus, the NIEO's pursuit of a commons-oriented approach to information and technology, is subsumed under proprietary considerations and a proprietary approach to revenue distribution. Nevertheless, attempts at democratisation and decolonisation of the seabed persist. A range of NGO and science community observers to the ISA, as well as the advocacy coalition the Deep Sea Mining Campaign, have made crucial contributions to the debate over exploration regulations with the DSMCC explicitly calling for a moratorium on deep-sea mining.[4] In its submission on 2019 draft exploitation regulations, the scientist organisation Deep Ocean Stewardship Initiative (DOSI) attends to the controversy over 'confidential information' in their multiple submissions on the draft. In their 2019 submission, they write, "Our scientists agree that no environmental data should be withheld from public scrutiny for any time period. Such practice of withholding environmental data amounts to the privatisation of information obtained from an area that belongs to all humankind".[5] Scientists are resisting capital's enclosure for extractive purposes, for example by seeking memorialisation of the transatlantic trade to commemorate enslaved peoples who died in the Middle Passage (Turner et al., 2020). To date, however, the ability of these critics to shape governance at this ocean frontier, and to promote a precautionary approach to the use of nature beyond state jurisdiction, is prefigured by the privately-held state and military knowledges (Zalik, 2018) used to exploit it. Despite the fanfare associated with the 2019 launch of ISA's Deep Data initiative, to date it does not redress the protection of privately held proprietary information.

## Democratising knowledge, democratising oceans governance?

We have seen above that proprietary ocean knowledges appear to predominate in contestation between the interests of extraction by private firms and those who wish to privilege the seabed as common heritage. Concurrently, however, efforts to develop new data technologies to fill knowledge gaps about the ocean – and to make the results widely available – are also emerging, some of which may support the common heritage camp. Remote sensing, ocean observing systems, and satellite tracking illuminate ocean spaces and species from above, while remote underwater vehicles do so from below (Campbell et al., 2016: 57–58; Lehman, 2016, this volume). Satellite tags on animals and vessels turn mobile 'things' into sources of data collection as they move through the oceans (Blair, 2019). These new data technologies – which encapsulate methods of data collection and generation, the data themselves, and the platforms, analytical techniques and infrastructures developed to interpret or make sense of these data for governance purposes (see Havice et al., in press) – and the organisations that mobilise them are providing insights into the oceans' physical, chemical, ecological and biological materiality (e.g., Boustany et al., 2002; Halpin et al., 2006; Sayre et al., 2017), human 'impacts' on them (e.g., Halpern et al., 2008; Kroodsma et al., 2018) and rendering ocean spaces and resources legible and governable in new ways (Havice et al., 2018). Those gathering and processing these kinds of data aim to account for the fluid and voluminous mobilities that constitute the oceans, in part out of recognition that the material features of oceanic processes shape what is institutionally possible for their management (Acton et al., 2019; Havice et al., 2018; Peters and Squire, 2019).

Unfolding new data technologies in the oceans are fuelled by NASA-type satellite and remote sensing hardware and centralised private data collected by militaries, as well as 'start-ups' that are, for instance, releasing small satellites and gathering data at higher resolution and lower cost than previously possible. In contrast to the case of proprietary knowledge about the seafloor outlined above, the wide range of actors that are developing, deploying and utilising novel data sources present the potential to *democratise* knowledge about the oceans, with an express aim of intervening in and potentially destabilising existing oceans governance practices and power relations (see Drakopulous et al., forthcoming). In doing so, they present potential to expand the scope of actors armed with information necessary to intervene in oceans governance processes typified by opaque inter-state politics and tasked with governing ocean objects (e.g., fish) and processes (e.g., fishing) that are out of sight and difficult to monitor (e.g., Campbell et al., 2016; Havice and Campling, 2010). Some such initiatives purport to fill knowledge gaps and offer a contrast to state and inter-state bodies such as the International Maritime Organization (Psaraftis and Kontovas, 2020) and the International Seabed Authority (Ardon, 2018) governance processes that do not require full disclosure of data or decision-making processes and operate behind partially closed doors.

To take one example of this type of 'commoning' epistemological frontier, we look to Global Fishing Watch's (GFW) recent efforts to render 'illegal, unregulated and unreported' (IUU) fishing activity visible and thus, governable. IUU fishing is a top concern in fisheries management because it is estimated to affect one in every five fish caught with an annual cost of up to US$23 billion, to threaten sustainability, and to be embroiled in equity and human rights concerns (Food and Agricultural Organization, 2018). Despite a host of bureaucratic and market-based tools to tackle IUU fishing, it remains a vexing management challenge in part because of knowledge gaps: it is notoriously difficult to monitor vessels in time and space, and to assess if a vessel is abiding by law (see e.g., Urbina, 2015; Vandergeest, 2018). Global Fishing Watch relies on data from multiple satellite technologies, combined with machine learning innovations and visualisation tools to reveal fishing vessel activity. The overall aim is to 'fight' IUU fishing (among other oceans governance challenges associated with mobility and knowledge gaps), and more broadly to advance ocean sustainability and stewardship by increasing transparency.

GFW is a collaboration that typifies the shift from government to governance in the oceans. It was founded by three non-governmental partners: Oceania (international ocean conservation organisation); Skytruth (a non-profit organisation specialising in using satellite technology to protect the environment that carries the tag-line, 'If you can see it, you can change it', [Skytruth, 2020]); and Google (a private firm with a wide range of tech-based products that in the partnership provided tools for processing 'big data'). GFW is now an independent, international non-profit organisation with a core team of employees and specified relationships with academic and research labs around the globe that gather and analyse data to conduct novel analyses of vessel activity. Global Fishing Watch research partners and technicians develop algorithms to learn and look for patterns in large data sets to determine type of ship, kind of fishing gear, and based on movement patterns, where and when it is fishing. The range of organisations involved in its founding and the collaborative format through which the organisation works exemplifies how knowledge is an object with potential to broaden the range of participants involved in oceans governance.

Global Fishing Watch offers access to data and its analysis and near real-time tracking of global commercial fishing activity and other shipping vessel activity, aiming to track all large-scale fishing to cover all 300,000 boats responsible for upwards of 75 per cent of global marine catch and as much as 80 per cent of fishing on the high seas (Global Fishing Watch, 2020). GFW's data set began with vessel tracking systems, most commonly, the automatic identification system (AIS), a GPS-like device that large ships use to broadcast their position to avoid collision. The International Maritime Organization (and many national governments) require large vessels to use AIS, reflecting one way that the non-state GFW effort is intimately intertwined with the history of state- and inter-state-based oceans governance. More recently, it has begun to develop methods to use multiple satellite technologies to identify fishing vessels that do not broadcast their positions while fishing illegally (Park et al., 2020).

GFW makes this information available through downloadable data, interactive online maps, promising that "anyone with an internet connection can trace the movements of about 60,000 commercial fishing boats, along with their name and flag state, in near real time" (Global Fishing Watch, 2020). The outcome has provided an ontological opening in the oceans by replacing a blank and empty map of the sea, with a concrete and actionable vision of an 'ocean on fire with fishing activity' and in need of national and international, state and non-state conservation actions (Gray, 2018). In making data and visualisation tools available, GFW aims to garner insight and action on illegal fishing, inform economic and policy in government and private spheres, and to capture the imaginations of a concerned public (Global Fishing Watch, 2020).

More broadly, Global Fishing Watch presents an opportunity to examine the intersection among epistemological frontiers of new data technologies, representations of the oceans and the future of oceans governance. It reflects a knowledge-based theory of change that suggests that there is a 'need' to know the oceans from a techno-scientific perspective because better knowledge, available to all, will increase accountability and in turn drive better and more informed governance. This knowledge-driven theory of change is premised on the underlying assumption that more comprehensive and higher quality data will lead to more effective environmental governance; a premise based on the ideal that science–policy relations are linear and technical, rather than co-produced (Gabrys, 2016; Goldman et al., 2011; Jasanoff, 2004). Review of GFW papers published in peer-reviewed journals reveals that while GFW researchers frequently couple their technical results with a call for stronger governance, pathways for governance improvements or for using the GFW to 'take action' are generally underspecified.

GFW is perhaps the highest profile example of new data technologies in the oceans, but it is hardly alone among rapidly proliferating organisations generating novel forms of data about the oceans and accompanying promises to use such knowledge to improve governance (see Drakopulous et al., 2022). Here, it is useful to return to a theorisation of the frontier as a site where new knowledge and technologies come into contact with and hold potential to reinforce, challenge and reformulate existing authorities, hegemonies and sovereignties. New data technologies, and attendant open access data-based tools and cartographic techniques reveal the oceans in time and space. In doing so, they present what Rankin (2016) describes as a new geo-epistemology, or way of knowing and using the earth with profound implications and possibilities for governance: in this case, they present the potential for commoning knowledge about the oceans that is otherwise invisible or enclosed by private firms and states (e.g., navies) to inform a more just oceans governance in which capitalist extraction can be monitored for compliance with regulations. That is, democratised knowledge might be used as a basis to 'strengthen' existing state- and inter-state-based regulation, and/or lead to alternative approaches for regulation by broadening the field of governance to include more kinds of knowledge, new regulatory tools, and more voices.

However, empirical examinations of the application of new data technologies in oceans governance reveal that new data technologies cannot provide singular and clear 'solutions' to governance problems and instead, continue to be deeply entangled with existing marine (or maritime) politics, and reformulate, rather than resolve, governance challenges and politics (Havice et al., 2018, in press). That is, it remains unclear if and how new data technologies have a role to play in unsettling historical capitalist power in oceans as they are embodied in and through existing political and governance frameworks, as outlined in the case of the ISA above. The *epistemological frontier* is unfolding around if and how new data technologies developed and deployed by a wide range of stakeholders, including those from outside of the traditional firms and states involved in oceans governance, present new governance possibilities for the oceans.

## Decolonising knowledge: Unknowable and unheard more-than-human knowledges

As our previous section demonstrates, contemporary orthodox scientific knowledge production, whether privatised or democratised, predominates in the generation of data intended to shape oceans governance. But what knowledges and ways of knowing are absent from this dominant formulation? And how do we approach ocean knowledge frontiers given that different epistemological starting points lead to distinct governance responses and approaches? Seeds of an alternative approach lie in the philosophical premise that *ethics precedes knowledge-making*: we encounter others (human and more-than human) first ethically and then we may or may not be able to come to 'know' them (Cheney and Weston, 1999). Ethics before knowledge is a radical critique of mainstream approaches to epistemological action where first one seeks knowledge about a subject and then constructs how to behave ethically (see also Plumwood, 2002). Following Cheney and Weston (1999), an ethics-based epistemological approach to oceans governance means that: a) oceans are neither easily nor simply knowable; b) ethics is not extensionist and incremental, but pluralistic and dissonant; and c) because hidden possibilities surround us, the task of ethics is to elucidate and improve the world (see Fawcett, 2005, 2013). To seriously consider the oceans as a common heritage of humankind (Tladi, 2011) requires an ethics-based epistemological approach that champions sharing knowledge and technologies over privatisation – an epistemological stance for democratising knowledge as our previous sections attest.

Ethical distinctions undergird the variation between the private knowledge of the ISA mining regime and the democratised knowledge of Global Fishing Watch: divergent transnational approaches to oceans governance. They point at the tensions around which humans are able to benefit from the oceans, which principles should drive governance regimes, and which institutions and interests determine these benefits and principles. The tension between knowledge gathered, technology deployed, and the supposed unknowability of the oceans raises the question of whether part of the problem in oceans governance is thinking about the high seas as "Earth's last conservation frontier" (Gjerde et al., 2016: 56).

The Common Heritage of Humankind, brought forth by legal scholars from the Southern hemisphere, is a counterpoint to this proprietary assumption. South African legal scholar, Dire Tladi (2016) argues for a "paradigm shift towards solidarity and the conservation of good in the oceans for all our benefits" and has championed inter-generational equity in the distribution of ocean resources. Indigenous justice scholar, Anishinaabeg Deborah McGregor (2014; 2016) reiterates the responsibilities for knowledge and care of our planetary waters as evidenced in the knowledge systems of Indigenous peoples. Furthermore, Dene scholar Coulthard (2014: 171) asserts settler colonialism's misrecognition calls for critically revised Indigenous politics and astutely asks: "[w]hat forms might an Indigenous political-economic alternative to the intensification of capitalism on and within our territories take?" Could we (re)learn how to live with the oceans as commons, indeed rethink the meaning of 'commons' in a form that could resist enclosure while continuing to create knowledge for decolonising, democratising, intergenerational justice?

Historically, the seas have been the site of the violence of the slave trade, world wars, and the overall bolstering of imperial and colonial structures (McKittrick and Woods, 2007). The prevailing knowledges from these historical processes have largely ignored Indigenous relationships to the ocean (see Waiti and Wheaton, this volume). But what knowledges and epistemologies, Indigenous sciences and research autonomous from proprietary, extractive purposes, have been lost or gone unheard in these processes? And what possibilities, lived relations and risks are present when there is attention to these knowledges? Indigenous peoples have made knowledge with and about oceans and successfully migrated across them for thousands of years (Atleo, 2004). Polynesian epistemology considers humans as inhabitants of ocean, not land (Hau'ofa, 1994, 1998; Lopez, 2019). Hawaiian scholar, Ingersoll (2016) calls for seascape epistemologies that in their fluidity resist rigid land-locked notions usually applied to oceans. Building upon Ingersoll (2016), in their quest for decolonial ocean futures, George and Wiebe (2020: 4) analyse from epistemologies across archipelagos in Kanaka Maoli (Native Hawaiian) and Coast Salish (First Nations) how to "challenge the foundational underpinnings of extractivist, property-centric settler-colonial liberal governmentality by turning away from land-locked property-centric territorial geographies and engage with more embodied, fluid, storied, and vibrant ways of being, knowing and sensing the world". The Consortium for Ocean Leadership (2020) recent workshop to identify national ocean exploration priorities in the Pacific reports the vital necessity of "sustained interactions with Indigenous communities" – community relationships that "must be continuously 'relational' rather than 'transactional'".

Anishinaabeg scholar, Leanne Betasamosake Simpson (2008: 33) emphasises that "animal clans were highly respected and were seen as self-determining, political 'nations'". How could these multi-species political nations function in oceanic spaces defined by nation states, international institutions and global capital? Gray (2018) raises these questions in detailing how Indigenous participants at the 2016 World Conservation Congress rejected the idea of modern control of ocean spaces by nation-states. Cree scholar Billy Ray Belcourt (2015: 4) argues that "anthropocentrism, is the anchor of speciesism, capitalism, *and* settler colonialism".

Building from these insights, how do we collectively imagine a different approach to knowing the ocean commons that is neither anthropocentric, nor based on nation states and territoriality? The case of whales and anthropogenic sound pollution helps address this question. Sound travels farther and faster in the oceans than in air. Whales traverse the oceans from surface waters to darker depths and have minimal control over the soundscapes in which they find themselves, yet they depend on being vocal (Clark et al., 2009; Weilgart, 2007). Toothed whales (odontocetes) and baleen whales (mysticetes) rely on echolocation, singing and calling – forms of aural communication that embody material perceptual information between whales, their environments and each other, thus helping to sustain their lives. Some baleen whale vocalisations can be heard thousands of miles across an ocean (Clark, 2019). Endangered North Atlantic, female right whales are known to 'whisper' to their offspring to avoid being heard by predators – a form of acoustic hiding (Parks et al., 2019). Some cetacean biologists believe whales are social learners passing on intergenerational knowledge and forming distinct cultures (Whitehead and Rendell, 2015). In a distinct intercultural way, Nuu-chah-nulth scholar, Richard Atleo (2004) discusses the harmony between his whaling peoples and whales as part of the theory of Tsawalk, whereby 'everything is one'.

Anthropogenic noise pollution originates from vessel traffic, seismic surveys and explosives, military and naval exercises (including anti-submarine training), dredging and coastal construction (Simmonds et al., 2004), as well as industrial products of oceans that have long provided an open route for the "smooth movement of capital, resources and militaries" (Havice and Zalik, 2018: 219). Most commonly used in oil and gas exploration, seismic testing involves the operation of exceedingly loud airguns that reflect sound off the ocean floor. Unfortunately, the frequency band of seismic airguns coincides with the sound band used by baleen whales (Clark,

2019). The sounds could cause marine mammals disorientation, stranding, communication disruption, and changes in vocalisations, feeding or behaviour patterns, contributing to other factors that appear as the cause of death/injury. Even the constant drone of low-level shipping traffic could distract whales leading to increased collisions between whales and ships and causing increased whale mortalities (Nowacek et al., 2007).

Clark (2019) argues that regulations to protect whales have a set standard for what constitutes harm, but that this does not thoroughly consider the cumulative harm of noise pollution over time. It is not only about what the sound does to marine mammals in the moment, but how it affects their ability to survive, communicate and thrive with *constant* noise pollution. A team of researchers found a decreased concentration of cortical steroid (which indicates stress) in right whale faeces around the time of 9/11 when ship and airplane traffic – and noise – stopped all together (Rolland et al., 2012).

Currently, dominant human epistemological frames lack data about the deleterious effects of sound pollution on the social fabric of marine lives (McCarthy, 2004). Still, citizen scientists and activists rally to fight the acoustic war against marine mammals with the line "[a] deaf whale is a dead whale" (Horwitz, 2014: 165), questioning the potentially deadly outcomes of transboundary sound pollution. Given the constant, ungoverned amount of noise pollution in the oceans there is an epistemological need to "complicate conventional assumptions about violence as a highly visible act that is newsworthy because it is event focused, time bound, and body bound" (Nixon, 2011: 3). Sound pollution created by and for humans is a form of slow, unending violence in the oceans, and humans' inability to 'know' it and its impacts on a more-than-human world under contemporary governance approaches enable it. Meanwhile, Steinberg (2018) rightly questions the safeguards in place to protect ocean inhabitants and environments; sound pollution is not a priority, except perhaps for the ocean's actual inhabitants. If oceans governance was to draw upon Indigenous knowledge systems, it would recognise a reciprocal responsibility to know and to act on behalf of other animal nations (see e.g., Simpson, 2011).

In contrast to industrial opportunism and its closure to equity principles in knowledge generation, intergenerational justice informs the epistemological horizon of an oceans' commons. Multilateral Marine Protected Areas (MPAs) are one oft-proposed solution to underwater noise pollution (McCarthy, 2004), but MPAs are critiqued as less about conservation and more about western legal procedures and territory-making in the oceans (Gray, 2018), and also for incongruities between the ecological scale of migratory animals (such as, turtles or whales) and the various jurisdictional and governance scales in place (or not) for MPAs and other spatial conservation tools (Havice et al., 2018). A western-science based epistemology is generally privileged in such tools. In drawing attention to the underwater cultural heritage of the oceans as a type of archaeological frontier, Lehman (2018) reminds us to pay attention to who is at the table, and who is telling the stories that affect ocean governance.

Taking animal lives seriously requires decolonising knowledge and accountability to, and inclusiveness of, Indigenous politics (Belcourt, 2015). How do we seriously question the idea that only humans can be political subjects and form political communities and enlarge our vision to include the more-than humans (Nadasdy, 2016: 2–3, following Anishnaabe scholar John Borrows; see also Johnson, this volume)? The overwhelming narratives of oceans 'in crisis' demand new epistemologies about governance in commoning, enclosing or envisioning the oceans otherwise. Moving away from proprietary data towards democratising and decolonising ocean knowledge creates space for collaboration, ingenuity and transformative changes for the ocean and its inhabitants addressing Lubchenco and Gaines' (2019) call for new ocean narratives. Returning to an ethics-based epistemology is a radical turn towards the unknowable ocean world as pluralistic yet full of co-existing possibilities; the task of ethics is to elucidate these possibilities (Fawcett and Johnson, 2019) and work towards inclusive governance models. These normative questions are completely imbricated with questions of epistemological control and attendant power relations concerning science and governance on ocean frontiers.

## Conclusion

This chapter draws out epistemological frontiers as a vital site of contestation and generative possibility in oceans governance. Frontiers are frequently dominated by the already powerful. This is apparent in the contemporary debates and dynamics shaping the exploitation regime to govern deep seabed mining. The ISA example reveals the central role of privatised knowledge in the geopolitical relations and ongoing practices shaping contemporary seabed regime formation. In contrast, the GFW case explores how emergent knowledge brokers may use new data technologies to reveal exploitative practices at sea, with hopes that such 'transparency' will foment more democratic governance of ocean space; though pathways for change remain in-the-making. By contrast, the more-than human natures and political ecologies that constitute the oceans and seabed, and Indigenous ocean knowledge and worlds are little understood and appreciated by dominant science, much less incorporated or legible in formal global oceans governance. The cases that we have reviewed here illustrate the role of knowledge in ongoing contestation over the definition and control of resources emerging from both historic and new patterns of exploration, extraction, conservation and commodification.

The colonial and anthropocentric underpinnings of dominant ocean epistemologies have shaped understandings of past and contemporary ocean spaces, and are also enrolled in its possible future articulations. That is, while frontiers are sites where new hegemonies are being constituted, they are also sites where pre-existing power relations and hegemonies seek to reassert themselves. Here we have shown that knowledges – whether privatised, democratised and/or decolonised – are central to the ways in which the ocean is being enrolled in socio-spatial projects. As such, the oceanic epistemological frontier is "spatial and material as well as conceptual, political and procedural; it is simultaneously about limits and edges and the promise and peril of transcending those limits" (Havice and Zalik, 2018: 221). A radical transformation of these relations would need to ally those who consider oceans to resist/escape and/or remain outside colonialism and capitalism (e.g., George and Wiebe, 2020; Gilroy, 1993; Linebaugh and Rediker, 2013; McKittrick and Woods, 2007). Oceanic epistemologies would need to recentre the apparent 'blank space' of the frontier as centre (Lehman, 2018), rather than limit. Falling back on practices of knowledge creation and use that continue to inform the colonial and anthropocentric institutions and relations that have shaped oceans is to miss potential openings to foreground an ethics of commoning and common heritage in the oceans.

## Acknowledgements

The authors wish to thank Kimberley Peters for encouragement and generous feedback on an earlier draft. Support for research informing this chapter was provided by a Social Science and Humanities Research Council of Canada Connections Grant involving all three authors. Havice was also supported by the National Science Foundation Geography and Spatial Science Division (Award #2026345 and Award #1539817).

## Notes

1 Following Liboiron (2021), the term 'dominant science', rather than 'Western science', signals the power relations inscribed in scientific practice and that not all Western science is dominant.
2 For polymetallic nodules, the ISA has granted exploration contracts in the CCZ (18), and the Central Indian Ocean Basin (1); six for polymetallic sulphides in the South West Indian Ridge, Central Indian Ridge and the Mid-Atlantic Ridge; and four for cobalt-rich crusts in the Western Pacific Ocean.
3 See www.isa.org.jm/document/mit-presentation-council-july
4 See www.eureporter.co/frontpage/2019/05/29/fisheries-and-environmental-organizations-issue-joint-call-for-moratorium-on-deepseamining/
5 See www.dosi-project.org/wp-content/uploads/DOSI-Comment-on-ISA-Draft-Exploitation-Regulations-October-2019.pdf, part IX DR 89, 3, p 13.

## References

Atleo E Richard (Umeek) (2004) *Tsawalk: A Nuu-chah-nulth Worldview*. Vancouver: UBC Press.

Acton L, Campbell LM, Cleary J, Gray NJ and Halpin PN (2019) What is the Sargasso Sea? The problem of fixing space in a fluid ocean. *Political Geography* 68: 86–100.

Ardron J (2018) Transparency in the operations of the International Seabed Authority: An initial assessment. *Marine Policy* 95: 324–331.

Belcourt BR (2015) Animal bodies, colonial subjects: (Re)locating animality in decolonial thought. *Societies* 5(1): 1–11.

Blair JJ (2019) Tracking penguins, sensing petroleum: "Data gaps" and the politics of marine ecology in the South Atlantic. *Environment and Planning E: Nature and Space* https://doi.org/10.1177/2514848619882938

Boustany AM, Davis SF, Pyle P, Anderson SD, Le Boeuf BJ and Block BA (2002) Expanded niche for white sharks. *Nature* 415: 35–36.

Bridge F and Perreault T (2009) Environmental governance. In: Castree N, Demeritt D, Liverman DM and Rhoads B (eds) *A Companion to Environmental Geography*. Oxford: Wiley-Blackwell, 475–497.

Campbell LM, Gray NJ, Fairbanks L, Silver JJ, Gruby RL, Dubik BA and Basurto X (2016) Global oceans governance: New and emerging issues. *Annual Review of Environment and Resources* 41: 517–543.

Cheney J and Weston A (1999) Environmental ethics as environmental etiquette: Toward an ethics-based epistemology. *Environmental Ethics* 21(2): 115–134.

Clark C (2019) Written Testimony of Dr CW Clark before the House Natural Resources Committee, Subcommittee on Water, Oceans and Wildlife. 7 March 2019. Washington, DC.

Clark C, Ellison WT, Southhall BL, Hatch L, Van Parijs SM, Frankel A and Ponirakis D (2009) Acoustic masking in marine ecosystems: Intuitions, analysis, and implication. *Marine Ecology Progress Series* 395: 201–222.

Consortium for Ocean Leadership (2020) Ocean exploration: Report on the workshop to identify national ocean exploration priorities in the Pacific. 10 July–22 September 2020. Available at: https://oceanleadership.org/wpcontent/uploads/2020/11/OceanExploration_PacificPriorities_WorkshopReport_NOV2020.pdf

Coulthard GS (2014) *Red Skin, White Masks: Rejecting the Colonial Politics of Recognition*. Minneapolis, MN: University of Minnesota Press.

Coumans C (2018) Into the deep: Science, politics and law in conflicts over marine dumping of mine waste. *International Social Science Journal* 68 (229–230): 303–324.

Drakopulous L, Havice E and Campbell LM (forthcoming) Architecture, agency and ocean data science initiatives: Data-driven transformation of oceans governance. *Earth Systems Governance*.

Drakopulos L, Havice E, Crisp K, Zurita Posas A and Campbell LM (2022) Catalog of Ocean Data Science Initiatives. *Qualitative Data Repository*. Available at: https://doi.org/10.5064/F6ZQWQJS. QDR Main Collection. V1

Elden S (2010) Land, terrain, territory. *Progress in Human Geography* 34(6): 799–817.

Fawcett L (2005) Bioregional Teaching: How to climb, eat, fall and learn from porcupines. In: Tripp P and Muzzin L (eds) *Teaching as Activism: Equity Meets Environmentalism*. Montreal: McGill University Press, 269–280.

Fawcett L (2013) Three degrees of separation. In: Stevenson RB, Brody M, Dillon J and Wals REJ (eds) *International Handbook of Research on Environmental Education*. Abingdon: Routledge, 409–417.

Fawcett L and Johnson M (2019) Coexisting entities in multispecies worlds: Arts-based methodologies for decolonial pedagogies. In: Lloro-Bidart T and Valerie SB (eds) *Animals in Environmental Education*. Cham: Palgrave Macmillan, 175–193.

Food and Agricultural Organization of the United Nations (2018) *Growing Momentum to Close the Net on Illegal Fishing*. Available at: www.fao.org/news/story/en/item/1137863/icode/

Gabrys J (2016) *Program Earth: Environmental Sensing Technology and the Making of a Computational Planet*. Minneapolis, MN: University of Minnesota Press.

George YR and Wiebe SM (2020) Fluid decolonial futures: Water as a life, ocean citizenship and seascape relationality. *New Political Science* 42(4): 498–520.

Gilroy P (1993) *The Black Atlantic: Modernity and Double Consciousness*. Cambridge, MA: Harvard University Press.

Gjerde KM, Reeve LLN, Harden-Davies H, Ardron J, Dolan R, Durussel C, Earle S, Jimenez JA, Kalas P, Loffoley D, Oral N, Page R, Ribeiro MC, Rochette J, Spadone A, Thiele T, Thomas HL, Wagner D, Warner RM, Wilhelm A and Wright G (2016) Protecting Earth's last conservation frontier: Scientific, management and legal priorities for MPAs beyond national boundaries. *Aquatic Conservation: Marine and Freshwater Ecosystems* 26(S1): 45–60.

Global Fishing Watch (2020) Available at: https://globalfishingwatch.org/

Goldman MJ, Nadasdy P and Turner MD (2011) *Knowing Nature: Conversations at the Intersection of Political Ecology and Science Studies*. Chicago, IL: University of Chicago Press.

Gray NJ (2018) Charted waters? Tracking the production of conservation territories on the high seas. *International Social Science Journal* 68(229–230): 257–272.

Halpern BS, Walbridge S, Selkoe KA, Kappel CV, Micheli F, D'Agrosa C, Bruno JF, Casey KS, Ebert C, Fox HE, Fujita R, Heinemann D, Lenihan HS, Madin EMP, Perry MT, Selig ER, Spalding M, Steneck R and Watson R (2008) A global map of human impact on marine ecosystems. *Science* 319(5869): 948–952.

Halpin P, Read AJ, Best BD, Hyrenbach KD, Fukioka E, Coyne MS, Crowder LB, Freeman SA and Spoerri C (2006) OBIS-SEAMAP: Developing a biogeographic research data commons for the ecological studies of marine mammals, seabirds, and sea turtles. *Marine Ecology Progress Series* 316: 239–246.

Hau'ofa E (1994) Our sea of islands. *The Contemporary Pacific* 6(1): 147–161.

Hau'ofa E (1998) The ocean in us. *The Contemporary Pacific* 10(2): 392–410.

Havice E, Campbell LM and Boustany AM (in press) New data technologies and the politics of scale in environmental management: Tracking Atlantic bluefin tuna. *Annals of the American Association of Geographers*. https://doi.org/10.1080/24694452.2022.2054766.

Havice E, Campbell LM and Braun A (2018) Science, scale and the frontier of governing mobile marine species. *International Social Science Journal* 68(229–230): 273–290.

Havice E and Campling L (2010) Shifting tides in the Western and Central Pacific Ocean tuna fishery: The political economy of regulation and industry responses. *Global Environmental Politics* 10(1): 89–114.

Havice E and Zalik A (2018) Ocean frontiers: Epistemologies, jurisdictions, commodifications. *International Social Science Journal* 68(229–230): 219–235.

Hayashi M (1989) Registration of the first group of pioneer investors by the preparatory commission for the international sea-bed authority and for the International Tribunal for the Law of the Sea. *Ocean Development & International Law* 20(1): 1–33.

Hayashi M (2005) Military and intelligence gathering activities in the EEZ: Definition of key terms. *Marine Policy* 29(2): 123–137.

Horwitz J (2014) *War of the Whales: A True Story*. New York, NY: Simon and Schuster.

Ingersoll KA (2016) *Waves of Knowing*. Durham, NC: Duke University Press.

Intergovernmental Oceanographic Commission (2019) *UNESCO*. Available at: www.ioc-unesco.org/index.php?option=com_oe&task=viewEventRecord&eventID=2366

*International Social Science Journal* (2018) Our Common Heritage: International seas threatened by mining. *International Social Science Journal* 68(229–230): 361–363

Jasanoff S (2004) *States of Knowledge: The Co-production of Science and Social Order*. New York, NY: Routledge.

Kroodsma DA, Mayorga J, Hochberg T, Miller NA, Boerder K, Ferretti F, Wilson A, Bergman B, White TD, Block BA, Woods P, Sullivan B, Costello C and Worm B (2018) Tracking the global footprint of fisheries. *Science* 359(6378): 904–908.

Lehman J (2016) A sea of potential: The politics of global ocean observations. *Political Geography* 55: 113–123.

Lehman J (2018) Marine cultural heritage: Frontier or centre? *International Social Science Journal* 68(229–230): 291–302.

Libriron M (2021) *Pollution is Colonialism*. Durham: Duke University Press.

Linebaugh P and Rediker M (2013) *The Many-Headed Hydra: Sailors, Slaves, Commoners, and the Hidden History of the Revolutionary Atlantic*. Boston, MA: Beacon Press.

Lodge MW and Verlaan PA (2018) Deep-sea mining: International regulatory challenges and responses. *Elements: An International Magazine of Mineralogy, Geochemistry, and Petrology* 14(5): 331–336.

Lodge MW, Segerson K and Squires D (2017) Sharing and preserving the resources in the deep sea: Challenges for the International Seabed Authority. *International Journal of Marine and Coastal Law* 32(3): 427–457.

Lopez B (2019) *Horizon*. Toronto: Random House Canada.

Lubchenco J and Gaines S (2019) A new narrative for the ocean. *Science* 364(6444): 911.

McCarthy E (2004) *International Regulations of Underwater Sound: Establishing Rules and Standards to Address Ocean Noise Pollution*. Boston, MA: Kluwer Academic Publishers.

McGregor D (2014) Traditional knowledge and water governance: The ethic of responsibility. *AlterNATIVE: An International Journal of Indigenous Peoples* 10(5): 493–507.

McGregor D (2016) Living well with the Earth: Indigenous rights and environment. In: Lennox C and Short D (eds) *Handbook of Indigenous Peoples' Rights*. New York, NY: Routledge, Taylor and Francis, 167–180.

McKittrick K and Woods C (2007) *Black Geographies and the Politics of Place*. Toronto: Between the Lines.

Nadasdy P (2016) First nations, citizenship and animals, or why northern indigenous people might not want to live in zoopolis. *Canadian Journal of Political Science/Revue canadienne de science politique* 49(1): 1–20.

Nixon R (2011) *Slow Violence and the Environmentalism of the Poor*. Cambridge, MA: Harvard University Press.

Nowacek DP, Thorne LH, Johnston DW and Tyack PL (2007) Responses of cetaceans to anthropogenic noise. *Mammal Review* 37(2): 81–115.

Okereke C (2008) Equity norms in global environmental governance. *Global Environmental Politics* 8(3): 25–50.

Park J, Lee J, Seto K, Hochberg T, Wong BA, Miller NA, Takasaki K, Kubota H, Oozeki Y, Doshi S, Midzik M, Hanich Q, Sullivan B, Woods P and Kroodsma DA (2020) Illuminating dark fishing fleets in North Korea. *Science Advances* 6(30): eabb1197.

Parks SE, Cusano DA, Van Parijs SM and Nowacek DP (2019) Acoustic crypsis in communication by North Atlantic right whale mother–calf pairs on the calving grounds. *Biology Letters* 15(10): https://doi.org/10.1098/rsbl.2019.0485

Peluso NL and Lund C (2011) New frontiers of land control: Introduction. *Journal of Peasant Studies* 38(4): 667–681.

Peters K and Squire R (2019) Oceanic travels: Future voyages for moving deep and wide within the "new mobilities paradigm". *Transfers* 9(2): 101–111.

Plumwood V (2002) *Environmental Culture: The Ecological Crisis of Reason*. New York, NY: Routledge.

Psaraftis H and Kontovas C (2020) Influence and transparency at the IMO: The name of the game. *Marit Econ Logist 22*: 151–172.

Rankin W (2016) *After the Map: Cartography, Navigation, and the Transformation of Territory in the Twentieth Century*. Chicago, IL: University of Chicago Press.

Rolland RM, Parks SE, Hunt KE, Castellote M, Corkeron PJ and Nowacek DP (2012) Evidence that ship noise increases stress in right whales. *Proceedings of the Royal Society B: Biological Sciences* 279(1737): 2363–2368.

Sayre RG, Wright DJ, Breyer SP, Butler KA, Van Graafeiland K, Costello MJ, Harris PT, Goodin KL, Guinotte JM, Basher Z, Kavanaugh MT, Halpin PN, Monaco ME, Cressie N, Aniello P, Frye CE and Stephens D (2017) A three-dimensional mapping of the ocean based on environmental data. *Oceanography* 30(1): 90–103.

Silver J and Campbell L (2018) Conservation, development and the blue frontier: The Republic of Seychelles Debt Restructuring for Marine Conservation and Climate Adaptation Program. *International Social Science Journal* 68(229–230): 241–256.

Simmonds M, Dolman S and Weilgart L (2004) Oceans of noise: A WDCS science report. Whale and Dolphin Conservation Society. Available at: https://uk.whales.org/wp-content/uploads/sites/6/2018/08/Oceans-of-Noise.pdf

Simpson L (2008) Looking after Gdoo-naaganinaa: Precolonial Nishnaabeg diplomatic and treaty relationships. *Wicazo Sa Review* 23(2): 29–42.

Simpson L (2011) *Dancing on Our Turtle's Back: Stories of Nishnaabeg Re-creation, Resurgence and a New Emergence*. Winnipeg: ARP Books.

Skytruth (2020) Skytruth website. Available at: https://skytruth.org/

Steinberg PE (2018) Commentary: The ocean as frontier. *International Social Science Journal* 68(229–230): 237–240.

Tladi D (2011) Ocean governance: A fragmented regulatory framework. In: Jacquet P, Pachauri R and Tubiana L (eds). *Oceans: The New Frontier*. New Delhi, TERI Press: 99–111.

Tladi D (2016) 26th Meeting of State Parties to the 1982 United Nations Convention on the Law of the Sea. 20–24 June. United Nations, New York, NY.

Turner PJ, Cannon S, Deland S, Delgado JP, Eltis D, Halpin PN, Kanu MI, Sussman CS, Varmer O and Van Dover CL (2020) Memorializing the Middle Passage on the Atlantic seabed in Areas Beyond National Jurisdiction. *Marine Policy* https://doi.org/10.1016/j.marpol.2020.104254

United Nations (1994) *UNCLOS Implementation Agreement on Part XI of the Convention*. Available at: www.un.org/Depts/los/convention_agreements/convention_overview_part_xi.htm

Urbina I (2015) The Outlaw Ocean. *The New York Times*. 25 July.

Vandergeest P (2018) Law and lawlessness in industrial fishing: Frontiers in regulating labour relations in Asia. *International Social Science Journal* 68(229–230): 325–342.

Visbeck M (2018) Ocean science research is key for a sustainable future. *Nature Communications* 9(1): 1–4.

Weilgart LS (2007) The impacts of anthropogenic ocean noise on cetaceans and implications for management. *Canadian Journal of Zoology* 85(11): 1091–1116.

Whitehead H and Rendell L (2015) *The Cultural Lives of Whales and Dolphins*. Chicago, IL: University of Chicago Press.

Zalik A (2015) Trading on the Offshore: Territorialization and the ocean grab in the international seabed. In: Ervine K and Fridell G (eds) *Beyond Free Trade*. London: Palgrave Macmillan, 173–190.

Zalik A (2018) Mining the seabed, enclosing the Area: Ocean grabbing, proprietary knowledge and the geopolitics of the extractive frontier beyond national jurisdiction. *International Social Science Journal* 68(229–230): 343–360.

Zalik A (2021) World making and the deep seabed: Mining the Area Beyond National Jurisdiction. In: Havice E, Himley M and Validiva G (eds) *The Routledge Handbook of Critical Resource Geography*. New York: Routledge, 412–424.

# 7
# CULTURE
## Indigenous Māori knowledges of the ocean and leisure practices

*Jordan Waiti and Belinda Wheaton*

### Introduction

Aotearoa New Zealand (NZ) is a small island nation situated in the South Pacific Ocean. The Māori (Indigenous peoples of the land) arrived in Aotearoa NZ on waka hourua (double-hulled voyaging canoes) over several planned journeys between 800–1350AD (Buck, 1950; Walker, 2004). The 3000-km journey from tropical East Polynesia (Hawaiki) to the temperate and sub-Antarctic waters of Aotearoa NZ was one of the longest ocean voyages of the pre-industrial age (Walter et al., 2017).

Initial settlement involved adapting to the new environment for food and shelter. Indeed, the development of horticulture became necessary for survival (Walker, 2004), which often meant relocating near waterways such as lakes, rivers and the ocean where additional food sources such as seafood (fish, shellfish, seaweed) could be harvested alongside crops (Selby et al., 2010). For the next 5–700 years of relative isolation, Māori would continue to successfully adapt to life in Aotearoa NZ as population stability and social balance was achieved (Durie, 2003). Ultimately, Māori would develop social, political and economic systems which would serve them sufficiently until the arrival of the first Europeans and subsequent colonisation.

Ocean space has played a significant role in the development of Māori society and culture, providing nutritional sustenance, a space for recreational activities, and a space to re-connect and re-affirm relationships with ancestors and the Ātua Māori (Māori deity's). Waterways including rivers, lakes and the ocean are regarded by Māori as a taonga (treasure), reflecting the concept of whakapapa (genealogy) and connection to place (Jackson et al., 2017; Tapsell, 1997). Yet, in Aotearoa New Zealand's contemporary (post)colonial political context, water, seashore and oceans, are also sites of cultural and political contestation (Strang, 2014; see also Thomas et al., this volume).

In this chapter we discuss ocean cultures – how the ocean spaces are known, understood and then practiced – and connection between culture, knowledge and lived experience through indigenous Māori past and present relationships with the ocean. We do this to situate the ocean and oceanic culture beyond the dominant western gaze, essential to decolonialising understandings of ocean space. We focus particularly on Māori engagements through three recreational, ocean-based cultural practices – waka hourua (double-hulled voyaging canoe),

DOI: 10.4324/9781315111643-9

waka ama (outrigger canoe) and heke ngaru (surfing) – that were in existence in Aotearoa NZ since the first arrival of Māori, and that imbue Māori cultural customs. We outline how during the process of colonisation their existence and influence on Māori life diminished, but more recently, they have been experiencing a resurgence in popularity within Māoridom. Our discussion then explores how these water-based cultural practices are important spaces for the active and ongoing promotion of Māori cultural traditions and self-determination.

To help understand the content and context of this chapter, and how it provides a lens for thinking about ocean space, we first introduce briefly the connections between culture and the ocean, before exploring ocean cultures through a Māori worldview and the Indigenous knowledge system (Mātauranga Māori) that underpins it.

## Oceanic cultures: A Māori worldview

Within the raft of work that has sought to bring the sea to scholarly attention in geography, and the wider social sciences, culture has been an important frame of understanding. As geographers and sociologists have asked: How does the ocean become a space of cultural meaning and practice through leisure pursuits such as surfing? (see Anderson, 2012, 2013, 2014; Evers, 2009; Ford and Brown, 2005; Olive, 2015). As Olive and Wheaton (2021) discuss in introducing a special issue focused on sport and recreation in oceanic blue spaces, these leisure pursuits are central ways in which people access, experience and give cultural meaning to oceanic spaces. They illustrate increasing academic interest across social science and humanities disciplines, and attention to the cultures of diverse physical cultural practices including (but certainly not limited to); surf sports, fishing (e.g. Eden and Bear, 2011), sailing (Couper, 2018), paddling (Liu, 2021), ocean and wild swimming (Britton and Foley, 2020; Gould, et.al, 2021; Moles, 2021), beach-combing, surf rescues, tourism, diving (Squire, 2017; see also chapters in Brown and Humberstone, 2015).

Yet, as Peters has noted in a paper arguing for social and cultural geographies of the seas,

> [a]though the maritime world is beginning to feature in contemporary social and cultural geographies, there are further ways in which oceans… may be harnessed as spaces and places of study and thus contribute towards more consistent contemporary research examining the seas.
>
> *Peters, 2010: 1265*

In spite of this call, arguably there remains much to be done to progress understandings of cultures connected to the oceans, especially those that are non-western (Olive and Wheaton, 2021; Nemani, 2015). Indeed, much existing work on cultures at sea, or borne from relations with the sea, remain centred on the Global North through historical and contemporary analysis (Peters, 2010; see also, Anderson and Peters, 2014; Wheaton et al., 2020, 2021).

Yet there are many ways to view the world, and for Māori, like many other Indigenous people, this is viewed through a different lens to many western cultures. In pre-colonial times, Māori society was based on "decentralised tribal autonomy" and the "organic solidarity of kinship" (Ministerial Advisory Committee, 1988: no page). This structure centred on four organisational levels (as shown in Figure 7.1), with all levels linked by kinship to a common ancestor. This structure provided for the foundations of one's identity as it emphasised genealogical connections that stretched back to Hawaiki.

Firstly, waka (canoe) consisted of a group of tribes who were all descendants of a common ancestor who had voyaged to Aotearoa NZ from Hawaiki. As a social group waka were made

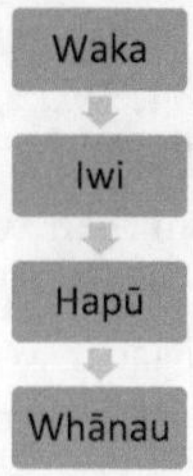


*Figure 7.1* Structure of Māori society.
*Source*: authors.

up of a cluster of iwi (tribe) who all descended from one of the original waka crew members (Walker, 2004). Under the leadership of the ariki (paramount chief), these tribes often formed their own independent, self-sufficient and self-governing set of rules surrounding economic, social and customary practices (Ministerial Advisory Committee, 1988; Te Awekotuku, 1991).

Tribes were also divided into smaller organisational units called hapū (sub-tribe). Hapū were similarly organised around a common ancestor whom they were named after. The hapū provided the means through which tribal structures and activities could be more easily managed (Ministerial Advisory Committee, 1988). Whānau (the biological, extended and joint family) would collectively constitute a hapū and was typically the smallest collective entity that lived together or within close proximity to each other (Buck, 1950). This four-level organisational system typified Māori societal, structural and political systems through the early periods of European contact, as well as the 1800s and early 1900s. Moreover, this structure provided a sense of identity for individuals as it exemplified an unbroken genealogical link that stretches back to Hawaiki.

While this organisational system (see Figure 7.1) allowed for Māori society as a whole to flourish, day-to-day activities and interactions were guided by a variety of kawa and tikanga (Māori customs and concepts). These kawa and tikanga are numerous and defining or outlining each of them is beyond this chapter, however those that are applicable to the context are explained below.

Embodying the kinship structure is the spiritual and physical connection to Te Taiāo (the environment). Māori relationships with the environment include the spiritual, genealogical, affective, cognitive and behavioural ties to a physical location as a result of the meanings and history within it. The language and practices of Māori ancestors are interwoven into the landscapes and histories, which emphasise the links to Māori origins (Ka'ai and Higgins, 2004). Māori see the environment as an interconnected whole – all parts connected by whakapapa, with no dichotomy between humans and the natural world (Durie, 1998a, 1998b).

Indeed, the significance of whakapapa within a Māori worldview is embodied in the relationships with tribal lands, the reverence for tipuna (ancestors), and the determination to exercise kaitiakitanga (the exercise of customary custodianship) and rangatiratanga (chiefly authority) (Panelli and Tipa, 2007). Whakapapa is often personified in the natural world, and extends beyond people to include environmental features such as mountains, rivers, lakes and oceans (Roberts et al., 1995). For example, pepeha is the act of reciting one's genealogy through historical linkages with environmental features such as an awa (river) or moana (beach or ocean) or a landform such as maunga (mountain). Declaring one's pepeha defines their identity and locates that person within a geographical area. While whakapapa provides the genealogical knowledge, pepeha is able to locate people in time (Graham, 2009). Moreover, pepeha not only connotes the origin of the Māori people, but also reflects the mana (authority), tribal identity and tribal boundaries (Walker, 2004; Wikaire and Newman, 2013). Underlying this connection between the environment and kinship is a strong sense of identity. Māori continue to gain a sense of identity and belonging from their whakapapa and connection with the natural environment (Smith, 2004; Waiti and Awatere, 2019).

Ka'ai and Higgins (2004) explain that a Māori worldview is comprised of both spiritual and physical realms, and the interaction between these two realms are practiced through various kawa and tikanga. These kawa and tikanga pay homage to tīpuna, the Māori gods, and encompasses cosmology, flora and fauna, the environment and landscape. For example: Tangaroa is the deity that governs the ocean and the creatures within; Tanemahuta is the deity that governs the bush and the creatures within; while Tawhirimatea governs the winds and all other meteorological aspects (Reed, 2004; Roberts et al., 1995). Interacting within these environments and acknowledging the pertinent kawa and tikanga provides the opportunity to acknowledge and connect with these various Ātua.

The custom of kaitiakitanga is the practice of guardianship and environmental management (Selby et al., 2010). This commitment is underpinned by an "inherent obligation we have to our tūpuna and to our mokopuna [descendants]; an obligation to safeguard and care for the environment for future generations" (Selby et al., 2010: 1). Kaitiakitanga ensures that Māori obligations to coastal spaces remains a taonga tuku iho (treasures handed down by our ancestors) so that whānau can continue to utilise these spaces for generations to come (Raureti, 2018).

## Mātauranga Māori

Integral to a Māori worldview is Māori thought and knowledge, termed Mātauranga Māori. Like other forms of Indigenous knowledge (IK), Māori knowledge systems come in many forms and can encompass the spiritual beliefs or esoteric messages that set the foundations upon which that particular society is based (Kerr, 2007). In essence, it is the knowledge and thoughts that allows a society to survive and interact.

Mead (2003) suggests that mātauranga Māori is a recently revived construct which incorporates Māori knowledge from the past, the present, and is still developing. It is a term that has been utilised by many to describe "Māori systems of knowledge" (Durie, 1998a: 76), an "epistemology of Māori" (Tau, 1999: 15), a "Theory of Māori Knowledge" (Royal, 1998: 2), and "traditional Māori knowledge forms" (Doherty, 2009: 18). Indeed, mātauranga Māori is embedded in Māori epistemology, and is used to differentiate between Māori knowledge and other forms of knowledge (Edwards, 2009).

Mātauranga Maōri was traditionally transferred through kawa, tikanga, leisure activities and cultural activities such as song, story-telling, arts and crafts. This ensured the proper transmission of knowledge throughout generations. When this transmission ceases to exist, kinship structures are fractured, connections with the environment are diminished, and one's sense of identity suffers.

## The impact of European colonisation

The arrival of Europeans throughout Aotearoa NZ began in the late 1700s. In most regards, initial contact with Europeans was welcomed by Māori as trade was established and new technologies were introduced (O'Malley, 2019; Walker, 2004). Whilst Māori enjoyed the new technologies, the large influx of land hungry settlers soon put a strain on the relationship. In an attempt to improve this relationship, the Treaty of Waitangi was signed between the

British Crown and some Māori chiefs in 1840. As has been widely recognised, the Māori signatories to the Treaty believed that it would provide a shared agreement that upheld three principles – partnership (in the running of the country), participation (for Māori in all matters concerning society) and the protection (of Māori beliefs and customs) (Durie, 1998a; Walker, 2004). However, soon after the signing it became apparent that the British Crown and settlers had a different understanding.

The arrival of European settlers and missionaries and the subsequent signing of the Treaty of Waitangi brought about significant change for Māori. The combination of the New Zealand Land Wars, introduced diseases, land confiscations, missionary influence and settler government legislation all lead to negative social change for Māori (Durie, 1998a; O'Malley, 2019; Walker, 2004). This resulted in cultural alienation, economic hardship and illness, which all contributed to the rapid depopulation of Māori (Durie, 1998b; Walker, 2004). It has been estimated that the Māori population dropped from 150,000 in 1800 (Durie, 1998b), to only 42,000 by 1896 (when an actual census was taken) (Pool, 1977). By then the European population was 15 times that of Māori (Lange, 1999). This depopulation, permeated with racial ideologies seeing Māori as an 'inferior race', prompted a number of unsympathetic responses from Pākehā (New Zealanders primarily of European descent), who believed that Māori dying away was both an inevitability and unproblematic (Pool, 1967).

This negative outlook on Māori well-being was perpetuated by the desire of the British to substantiate their imperialistic and colonising behaviour. As such, Mātauranga Māori, like many other IK systems, have suffered from the effects of colonising cultures that undervalue and undermine these systems (Kerr, 2007). Indeed, Māori values, customs and concepts (see Mātauranga Māori) were lost as a result of these undermining systems of the colonial government.

In contemporary times, bi-culturalism dominates contemporary political discourse based on the partnership between Māori and the Crown (through the Treaty of Waitangi), and the state's recognition of the languages, cultures and traditions of both Pākehā and Māori. Yet, Māori continue to be adversely affected by colonial structures and practices (Bell, 2014) including undermining Mātauranga Māori and the Treaty of Waitangi. Ongoing inequality and disadvantage is highlighted by a range of negative socio-economic, education and health outcomes experienced by Māori (Durie, 1998a; Robson and Harris, 2007; Walker, 2004).

The theft of Māori land remains central to ongoing political tension (Salmond, 2014). A succession of governments have passed laws to commandeer, and then privatise resources previously held as 'commons', including sea beds and fishing grounds (Salmond, 2014), actions which Māori have contested, most successfully through the Waitangi Tribunal.[1] In the early 2000s, conflict over the ownership of the country's foreshore and seabed came to the fore. A major Māori claim was instigated aimed at regaining control of the foreshore, which the government resisted by introducing the 2004 Foreshore and Seabed Act (Strang, 2014). The June 2003 Court of Appeal ruling, was regarded by many New Zealanders as a defining moment in the evolution of the country's 'post-colonial' relationship between the Crown and Māori. In recent years, local Iwi protests to deep sea oil drilling off of the East and West coasts of Aotearoa New Zealand have again tested the Crown–Māori relationship (see also Thomas et al., this volume).

Having outlined Aotearoa's (post) colonial context, the next sections will outline three important ocean-based cultural practices that imbue Māori cultural customs. We discuss how they were impacted through colonisation, their subsequent demise and recent resurgence within Māoridom. We do this to demonstrate how cultural systems matter to engagements with the ocean today.

## Māori associations with the ocean

As we have outlined above, Māori connections to the ocean extend back in time to East Polynesia (Buck, 1950; Tuaupiki, 2017a, 2017b), and the ocean has played a significant role in the development of Māori society, including food, recreation, and a space to re-connect and re-affirm relationships with ancestors and the Ātua Māori.

Despite the detrimental effects of colonisation, Māori continue to enact and revive their associations with the ocean through the revival of traditional concepts, customs and knowledge. This section will highlight activities that represent three different domains of the 'ocean': waka hourua (double-hulled voyaging canoes) predominantly in the open ocean; waka ama (outrigger canoes) in coastal waters; and heke ngaru (surfing) at the sea shore. Waka is better understood as a type of canoe, which have symbolic and spiritual meanings in Māori genealogy and represent "tribal identity, mana and territory" (Walker, 2004: 38). Waka are more than a functional vessel (i.e., transport, food gathering, fishing and recreation), they are a symbolic, cultural, genealogical and spiritual feature of Māori society which is directly related to Māori through whakapapa (Wikaire and Newman, 2013).

## Waka hourua

It has been suggested that the wider prehistoric Polynesian expansion was the most dramatic burst of overwater exploration in human prehistory (Crowe, 2018; Kerr, 2007; Walter et al., 2017). Polynesians lacked sailing technologies such as compasses and charts, yet were masters of navigational arts and of sailing canoe technology (Crowe, 2018; Kerr, 2007; Tuaupiki, 2017a). Waka hourua were the vehicles that allowed the discovery of Aotearoa NZ, and provided the communication links to the South Pacific Ocean (Crowe, 2018; Kerr, 2007; Tuaupiki, 2017a, 2017b). Drawing on Mātauranga whakatere waka (Māori navigation knowledge systems) (Tuaupiki, 2017a), this journey required knowledge on the current, tides, moons, birdlife and cloud formations. As Kerr outlines (2007: 27–28), "waka carried the cultural, material and spiritual treasures of the people. They were the space ships of the ancestors, seeking out these new lands and opportunities".

Waka hourua and Mātauranga whakatere waka has seen an increased interest from Māori (and non-Māori) and other Indigenous researchers and practitioners over the past couple of decades (see Crowe, 2018; Diamond, 2007; Kerr, 2007; Mita, 2014; Tuaupiki, 2017a, 2017b). As Kerr (2007: 27) notes, "this revival in waka Māori coincides with a general recovery of the indigenous knowledge specific to waka throughout the Pacific", and in particular the ways it illustrates "the area of Māori science and technology".

Whilst colonisation resulted in the loss of various Māori customs, concepts and beliefs (Durie, 1998b; O'Malley, 2019; Walker, 2004), waka hourua traditions were mostly likely lost before the arrival of Europeans (Adds, 2012; Kerr, 2007). As Kerr (2007) notes, this loss may well be due to the notion that the Māori knowledge based moved from small island living and ocean-based existence to a land-based existence (including seaside) during the settlement in Aotearoa NZ. So, as time went on, there was no need for trans-ocean voyagers.

The germinal (contemporary) resurgence of double-hulled voyaging within Polynesia occurred in the 1970s when Hokule'a sailed from Hawaii to Tahiti in 1976. Following on from this, Matahi Brightwell sailed upon the waka Hawaikinui from Tahiti to Aotearoa NZ (Nelson, 1998). A major reason for the recent resurgence in waka hourua was to disapprove a common myth – a western cultural knowledge – perpetuated by non-Māori scholars (e.g. Sharp, 1957) that Māori simply 'drifted' to Aotearoa NZ (Kerr, 2007; Tuaupiki, 2017b).

Many iwi and Māori organisations throughout Aotearoa NZ have now acquired or built their own waka hourua to help promote Mātauranga whakatere waka and cultural development. A range of voyaging societies have flourished over the past 15–20 years. Some are utilising their waka hourua as 'floating classrooms', whereby the waka and associated Mātauranga whakatere waka (Maori navigation knowledge) are utilised to disseminate cultural knowledges and practices. In 2019, many of these waka hourua have been on show during the Tuia-Encounters 250 commemoration, a celebration of Aotearoa New Zealand's Pacific voyaging heritage and acknowledges the first onshore encounters between Māori and Pākehā in 1769–70.[2]

## Waka ama

Waka ama were rarely used by Māori in traditional times. Rather, waka taua (war canoes), waka tīwai (river canoes) and waka tētē (sea-fishing canoes) were more common (Barclay-Kerr, 2006). However, the genesis for the resurgence of Waka ama in Aotearoa NZ began in the early 1980s when Matahi Whakataka-Brightwell was living in (Pacific neighbour) Tahiti, and noticed the strong presence amongst the Kanaka Maohi (Indigenous peoples of Tahiti). He pioneered competitive Waka ama in Aotearoa NZ when he established the first club in Gisborne in 1985 (Mita, 2014). Whakataka-Brightwell noted that:

> Māori people haven't had the opportunity for nearly seven generations to enjoy the world of Tangaroa. Wind, sea, canoe, air – its massaging the whole being of a person. It's giving the person a completely new feeling of what nature is all about on the sea and that's Tangaroa.
>
> *Nelson, 1998: 60*

Waka ama has since experienced a swell in popularity, especially among the Māori community (Mita, 2014, 2016; Waka Ama New Zealand, 2018; Wikaire and Newman, 2013), and as a form of recreation has attracted most research interest to date (Mita, 2014, 2016; Mita et al., 2016; Wikaire and Newman, 2013), especially amongst Māori academics. As Wikaire and Newman (2013) discuss, the cultural practice has been gradually transformed from an affordable indigenous grassroots activity to a profitable highly competitive and commercial sport. According to the NGB Waka Ama NZ (Nga Kaihoe o Aotearoa Inc), there are 83 clubs with over 5,419 members (Waka Ama New Zealand, 2017). Internationally, between 1992 and 2012, the New Zealand team often ranked in the top three positions among over 20 competing countries and regions in the *Va'a World Sprint Championships* (Waka Ama New Zealand, 2013, cited in Liu, 2018). Waka ama racing events (from long distances to sprint) are coordinated by the NGB, Nga Kaihoe o Aotearoa Inc. Many clubs have crews that compete in regional and national competitions. These include the *Annual ActivePost Waka Ama National Sprint Championships*, which had its 25th anniversary in 2015. The event involved teams from over 55 clubs represented over 2,500 competitors aged seven to 70, with participation numbers continuing to increase.[3] Likewise, the *NZ National Waka Ama Secondary School Championship* has been running for over 16 years, with over 80 schools competing from all over NZ.[4] Liu (2018, 2021) also illustrates that waka ama has been particularly popular amongst Maori women. Membership statistics of waka ama clubs in New Zealand show that registered female members consistently outnumber their male counterparts, particularly in the Open, Masters (over 40 years old) and Senior (over 50 years old) divisions (Waka Ama New Zealand, 2017, 2018). This pattern was also evident in the waka ama club Liu participated in, where female paddlers and coaches dominated (Liu, 2018), most of whom started paddling because of their love of water and water sports.

For Māori, however, 'sport' often takes different meaning than maximising the performance, with, spiritual development considered as important as the physical pursuit (Hokowhitu, 2007). As Wikaire and Newman (2013: 61) have argued, "waka ama can be seen as a cultural site in which long-standing Euro-centric norms of leadership can be challenged and a uniquely indigenous sport reinstated within indigenous communities in Aotearoa/New Zealand". Liu's research interrogates the physical culture of waka ama paddling based on ethnographic research in one club in Aotearoa (Liu, 2018, 2021). She illustrates that although waka ama has become a worldwide competitive sport and participants in the club she studied trained purposefully for competitions, while paddling, these participants still considered it as a way to connect to the sea, their kaitiaki (guardian) and tīpuna. Through waka ama, Māori participants' experience embodied, emotional and spiritual connections with their ancestry, colonial history, nature and other people (Jackson et al, 2016; Liu, 2018, 2021; Liu and Bruce, 2019). Jackson et al. (2016) describe paddling the waka as a way for Māori to reawaken connections with water as part of their role as kaitiaki. Likewise, Wikaire and Newman (2013: 60) suggest that waka ama "provides a unique physical cultural space through which tino-rangatiratanga (self-determination), Māori identity and Māori culture can be shared and promoted". For example, the Hauteruruku ki Puketeraki club utilise waka ama as a tool for the transmission of water safety skills and knowledge, and Māori cultural values (Jackson et al., 2016).

Liu's (2021) research participants often made a distinction between waka ama as a competitive sport and waka ama as a culturally informed physical activity, and made sure both types of paddling were practiced in the club. For example, Māori cultural protocols, such as karakia (incantations) and rāhui (restrictions/prohibitions), were regularly practiced. She also notes that sustaining "healthy mutual relations with the ocean" was a goal pursued by the club, and that paddlers' environmental responsibilities, such as care for the ocean and the marine life, co-existed with training. For example, club members along with members of a local hapū (sub-tribe) helped in a mussel reef restoration project. As one of her interviewees explained, pollution had killed off natural food resources from the sea: "We should be willing to give back to it [the bay]. Not just take from it all the time. You know it's our playground and we take care of it" (cited in Liu, 2021: 150). Liu argues that this co-existence of the Māori cultural protocols alongside the norms of 'western' sport have shaped waka ama in New Zealand. On one hand, waka ama stays as a cultural site for Māori people's self-identification and self-determination in post-colonial New Zealand society and, on the other hand, it becomes a sport in the neo-liberalist economic climate and driven by corporations' commercial interest (Wikaire and Newman, 2013). The multiple functions of waka, as sport, recreation, cultural practice, and specifically indigenous space exists throughout the country, although these different aspects differ across communities.

## Heke ngaru

Heke ngaru is another traditional pastime that imbues Māori connections to, and knowledge of, the ocean and is undergoing a revitalisation amongst Māori. 'Contemporary' surfing is said to have arrived in Aotearoa NZ in 1915 when legendary Hawaiian surfer Duke Kahanamoku gave surfing exhibitions at various beaches around the two islands (Warshaw, 2005). However, Māori had been riding waves before their first arrival in Aotearoa NZ, and descendants of the Aotea waka speak of their ancestors surfing back in Hawaiki. Indeed, the written and observed evidence shows that Māori had been riding waves since their time in Hawaiki, and right up until the effects of colonisation came into effect. As Best (1924: 40) notes, Māori "practised surf riding, with and without boards, as also in small canoes, as did the Hawaiians and others". Māori surf riders used wooden boards (kopapa), logs, waka and pōhā (bags of kelp filled with air) (Beattie, 1919; Best, 1924).

Skinner (1923) observed a number of Māori males and females riding waves on a two-man waka. He took particular note of the 60-year-old Rangatira (chief) of the local hapū, named Te Rangi Tuataka Takere, being amazed at his skill in catching and riding the waves. He noted that,

> The most lasting impression made on my mind in this surfing incident, was that of the poise and skill of Te Rangi Tuataka Takere, the high-born rangatira, as he sat statue like, steering-paddle firmly grasped, his fine muscular figure and clean cut tattooed features, reproducing, with the general surroundings, a grand picture of pure Maoridom as it had been for centuries prior to A.D. 1884.
>
> *Skinner, 1923: 37*

Contrary to much non-Indigenous knowledge and understanding of surfing being a male-dominated activity, both Best (1924) and Skinner (1923) above note that it was a pastime enjoyed by all ages and genders. Furthermore, Best (1924: 43) notes "this sport was indulged in by both youth and adults, including females". Masterson (2018a, 2018b) and Walker (2011) also highlight the strong presence of females within the pre-European history of Hawaiian surfing culture.

Waiti's informal discussions with Māori surfers have highlighted various reasons why the practice of heke ngaru may have diminished after the arrival of Europeans in Aotearoa NZ. Cooler water temperatures in Aotearoa NZ compared to Eastern Polynesian may have meant that it was only comfortable to ride waves within the summer months. With the arrival of colonisation, factors such as a lack of time due to competing priorities (i.e., warfare, retaining land), land confiscations and relocations away from coast, and the transfer of knowledge (i.e., customs and beliefs) pertaining to heke ngaru, would have also contributed to this demise.

Participation figures for informal and nomadic recreational practices like surfing tends to be unreliable (Gilchrist and Wheaton, 2017). Surfing New Zealand have estimated that less than 5 per cent of the population surf (cited in Wilkinson, 2017), whereas according to Sport NZ data, 8 per cent of adults had participated in surfing/bodyboarding in the past 12 months (2015: 53). Sport NZ's 2013/4 data gives a more detailed analysis of surfing participation, suggesting that 6.2 per cent of men and 2.5 per cent of women had surfed in the past year, further suggesting that NZ European were the highest group of participants, with Māori only constituting 5.3 per cent of the total (Sport NZ, 2015). It is likely that like other sport and leisure pursuits, socio-demographic factors as well as geography, and parental and community support, play a role in understanding lack of participation. Nonetheless, despite being a minority, Māori surfers have featured strongly in Aotearoa's competitive surfing both nationally and internationally. During the 1980s, Māori surfers from the Waitara Bar Boardriders Club in Taranaki dominated the national contests. A new cohort of Māori surfers emerged from Whaingaroa/Raglan and Gisborne, and began to dominate national contests in the 1990s. In the 2000s Māori surfers from around Aotearoa NZ featured strongly nationally. During these decades, the Aotearoa Māori National Surfing Titles have been an important event for Māori surfers, including various age divisions and surf craft divisions (i.e., bodyboarding, stand-up paddleboarding, longboarding) across male and female divisions. This event allows Māori wave riders to reaffirm and connect with whakapapa whānau and kaupapa whānau (i.e., other Māori surfers). It demonstrates also the connection between knowledge and practice in contemporary surfing activities.

Research exploring the experiences of contemporary Māori surfers is limited (see, however, Nemani, 2015; Waiti and Awatere, 2019; Wheaton et al., 2017). Nemani's (2015) research with body-boarders suggest that surfing often has different cultural meanings for Māori and Pacific Island participants. She noted that a unique form of cultural capital existed amongst the Māori and Pacific Island participants she interviewed, in which respect, courtesy and fairness were given more value than the demonstrations of physical capital dominant amongst Pākehā participants.

Waiti and Awatere (2019) investigated the 'sense of place' among Māori wave riders (or Kaihekengaru). The results found that these Kaihekengaru experience a sense of place reflecting both the ocean and landmarks nearby, which is underpinned by mātauranga Māori. Drawing on their whakapapa, surfing enables these Kaihekengaru to connect with Iwi-specific environmental features, their ancestors and the various Ātua as they are imbued in the environment and immersed in the ocean. As shown by Table 7.1, the Māori atua influence surfing and surfing conditions. As one respondent noted,

> I feel a connection with the Ātua when I'm surfing. Particularly with Ranginui, Papatuanuku, Tawhirimatea and Tangaroa. All of them combine to create the conditions ideal for surfing. They all have a large presence physically and spiritually when in the water.
>
> *Waiti and Awatere, 2019: 39*

These findings have many similarities with research on wave riding in Hawaii (Ingersoll, 2016; Masterson, 2010, 2018a, 2018b; Walker, 2011), and in other indigenous communities (e.g. McGloin, 2007), in the ways that their indigenous knowledge systems embrace different practices and assumptions about what water means and how relationships with it are made. Ingersoll's (2016) research in Hawaii illustrates how indigenous Kanaka Maoli surfers view the sea bed as "part of their genealogy" (Ingersoll, 2016: 47), immersing them in very different relationships to the ocean than non-indigenous surfers. She developed the concept of 'seascape epistemology' to articulate an indigenous Hawaiian way of knowing and ontology, founded on a sensorial, intellectual and embodied literacy of the ocean.

*Table 7.1* Ātuatanga and their influence on surf conditions

| *Ātua* | *Domain* | *Influence on surf conditions* |
|---|---|---|
| Tāne Mahuta | Forest, birds, and insects | Protection against wind and sand deposits |
| Papatūānuku | Earth | Bathymetry, Seabed (reef, sand, boulders) |
| Tāwhirimātea | Wind and Weather conditions | Low pressure systems and wave face conditions |
| Tamanui-te-Rā | Sun and Solar energy | Causes sea breezes and glare |
| Rūaumoko | Earthquakes & Volcanoes | Influences the seabed and bathymetry |
| Tangaroa/Hinemoana | Ocean and sea life (Male and Female element) | Affective experiences, safety, and harm minimisation |

*Source*: Waiti and Awatere (2019: 39).

There are there also diverse community initiatives emerging around Aotearoa New Zealand that use surfing, or surfing related activities for engaging Māori communities and knowledge systems. These include various surfing for development type programmes (see for example Pretorius, 2019; Wheaton et al., 2017). Aotearoa Water Patrol is a newly incorporated society that is made up of Māori watermen with the purpose of sharing and promoting water safety knowledge and skills to their whānau and broader communities through heke ngaru. The Hauteruruku Ki Puketeraki club takes a similar approach to the delivery of water safety skills and knowledge, albeit through the use of waka ama and waka hourua (Jackson et al., 2016).

## Discussion

In this last brief discussion, we highlight how these water-based cultural practices are important sites for active and ongoing promotion of Māori cultural tradition and self-determination. As Wikaire and Newman argue (2013), research that illustrates the role of indigenous sports like waka ama as a vehicle of social change within indigenous communities is limited. Yet these indigenous sports play an important role in cultural transmission, cultural revitalisation and self-determination.

As we have shown, waka hourua, waka ama and heke ngaru can be seen as Māori attempts to reclaim traditional knowledge, to reclaim Māori 'space' within the ocean, and through which tino rangatiratanga (self-determination), leadership, Māori identity and Māori culture can be shared and promoted. Tino Rangatiratanga is highlighted by the reclamation of Matauranga Maori customs and concepts, and also by the leadership shown by the Māori organisations who have played an important role in the revitalisation process. Examples of key Māori lead societies include, in voyaging, the various Aotearoa NZ voyaging organisations (e.g. *Te Toki Voyaging Trust, Tairawhiti Voyaging Trust, Raukawa Moana Voyaging Trust*), in waka ama the national organisation (*Waka Ama NZ*) and numerous local waka ama clubs (e.g. *Mareikura Waka Ama Club, Matangirua, Horouta Waka Ama Club*). In surfing too, despite the informality of the sport, there are several vibrant kaupapa Māori-focused Boardriders clubs (e.g. *Waitara Boardriders Club, Ahipara Boardriders Club, Aotearoa Water Patrol*). Importantly, through leading these organisations, and their decision making, Māori have been able to secure financial resources.

Leadership is also exhibited within these kaupapa Māori organisations through advocating for the need for Māori to have the necessary knowledge and skills to ensure the safety of everyone who undertakes these activities. They have promoted and facilitated the accreditation of their members with qualification and knowledge to ensure they are able to take a lead in water safety amongst Māori communities. For example, Coastguard Education qualifications (e.g. Day Skippers, VHF Radio, Boat Master qualifications) are emphasised in the waka communities. This focus on water safety and drowning prevention reflects a sense of kaitiakitanga, which includes protecting other human beings. The value of this indigenous-led approach to the promotion and delivery of culturally relevant water safety programmes is emphasised by Gollob, Giles and Rich (2013).

## Conclusions and future considerations

In this chapter we have attended to themes of culture and ocean space, focusing on indigenous Māori past and present relationships with the ocean particularly through engagements in recreational-based cultural practices. Maori knowledge systems certainly share some values with other IK systems, particularly those in the Pacific, and help to understand the agencies of oceanic spaces, and people's physical, spiritual and genealogical connections with them (Waiti and Awatere, 2019; Wheaton et al., 2020). Yet, there are also important differences in their colonial histories, and present cultural and political status, which as we have illustrated impact contemporary social relations, including those more-than-human.

We have illustrated the ways in which indigenous Māori culture is deeply connected with nature, and particularly the sea. Māori not only perceive the surrounding environment, tribal history and ancestry *as parts of* themselves, but *as* themselves (Hoskins and Jones, 2017). These relationships are commonly found in Māori people's experiences of sport and recreation in the ocean, and are illustrated by participants' different understandings of, the meanings given to, and identity constructions around these practices. We suggest that the examination of Matauranga Māori helps us to move beyond understandings of oceanic bluespaces as merely a passive medium for human activities (Liu, 2021), and of ocean spaces as western constructions, where persistent cultural knowledge create dominant ways of knowing the oceans that are not representative of all peoples, all cultures who live with the seas (c.f. Olive and Wheaton, 2021; Wheaton et al., 2020).

While recreational-based cultural practices are clearly gaining visibility, academic research remains limited. There is a need for further research in this area, particularly by indigenous researchers to explore important questions about how culture and practice connect, in order to develop community initiatives and facilitate equitable social change. As others have recognised, this dearth of research reflects the politics of academic work in Aotearoa more widely; that indigenous researchers are under-represented, and that indigenous knowledges are undervalued (Tuhiwai Smith, 2012). Furthermore, this has an important impact on the ways in which resources are distributed. For example, despite the popularity of waka ama it is vastly underfunded in relation to activities like rowing and canoeing that remain popular with the socio-economically privileged Pākehā population. This is despite similar participation levels (see Waka Ama NZ, 2018; Rowing NZ, 2018). Perhaps then, in light of this, it is time for both policy makers and the media to take note of these practices and support them more equitably.

In Aotearoa New Zealand's 'post'-colonial political context, contestation over water is not just concerned with issues such as property rights or 'managerial responsibilities', but are also "deeply felt affective concerns about human-environmental relations" (Strang, 2014: 124). These are issues of cultural awareness, support and respect for different knowledge systems. Ongoing contestation over rights and responsibilities of and for water and land are deeply embedded in Māori ongoing battle for self-determination. The desire for Māori self-determination through these ocean-based sporting activities can be seen as examples of this ongoing struggle.

## Notes

1 The Waitangi Tribunal is a permanent commission of inquiry that makes recommendations on claims brought by Māori relating to Crown actions which breach the promises made in the Treaty of Waitangi.
2 See, for example, https://mch.govt.nz/tuia250
3 See, www.wakaama.co.nz/
4 Likewise, see www.wakaama.co.nz

## References

Adds P (2012) Long-distance prehistoric two-way voyaging: The case for Aotearoa and Hawaiki. *Journal of the Royal Society of New Zealand* 42(2): 99–103.

Anderson J (2012) Relational places: The surfed wave as assemblage and convergence. *Environment and Planning D: Society and Space* 30(4): 570–587.

Anderson J (2013) Cathedrals of the surf zone: Regulating access to a space of spirituality. *Social and Cultural Geography* 14(8): 954–972.

Anderson J (2014) Exploring the space between words and meaning: Understanding the relational sensibility of surf spaces. *Emotion, Space and Society* 10: 27–34.

Anderson J and Peters K (eds) (2014) *Waterworlds: Human Geographies of the Ocean*. Farnham: Ashgate.

Barclay-Kerr H (2006). Waka – canoes - Other types of waka. *Te Ara – The Encyclopedia of New Zealand*. Available at: www.TeAra.govt.nz/en/waka-canoes/page-5

Bell A (2014) *Relating Indigenous and Settler Identities: Beyond Domination*. London: Palgrave Macmillan.

Beattie H (1919) Traditions and legends collected from the natives of Murihiku (Southland, New Zealand). *The Journal of the Polynesian Society* 28(XI): 212–225.

Best E (1924) *Games and Pastimes of the Maori*. Wellington: AR Shearer.

Britton E and Foley R (2020) Sensing water: Uncovering health and well-being in the sea and surf. *Journal of Sport and Social Issues* 45(1): 60–87.

Brown M and Humberstone B (eds) (2015) *Seascapes: Shaped by the Sea*. Farnham: Ashgate.

Buck P (1950) *The Coming of the Maori* (2nd edition). Wellington: Whitcombe and Tombs Ltd.

Couper PR (2018) The embodied spatialities of being in nature: Encountering the nature/culture binary in green/blue space. *Cultural Geographies* 25(2): 285–299.

Crowe A (2018) *Pathway of the Birds: The Voyaging Achievements of Māori and their Polynesian Ancestors*. Honolulu, HI: University of Hawai'i Press.

Diamond JM (2007) *Collapse: How Societies Choose to Fail or Succeed*. New York, NY: Viking.

Doherty W (2009) Mātauranga Tūhoe: The centrality of mātauranga-ā-iwi to Māori education. Doctoral dissertation, Auckland University. Available at: https://researchspace.auckland.ac.nz/

Durie MH (1998a) *Te mana, te kāwanantanga: The Politics of Māori Self-determination*. Auckland: Oxford University Press.

Durie MH (1998b) *Whaiora: Maori Health Development* (2nd edition). Auckland: Oxford University Press.

Durie MH (2003) *Ngā kāhui pou: Launching Māori Futures*. Wellington: Huia Publishers.

Eden S and Bear C (2011) Reading the river through 'watercraft': Environmental engagement through knowledge and practice in freshwater angling. *Cultural Geographies* 18(3): 297–314.

Edwards S (2009) Titiro whakamuri kia marama ai te wao nei: Whakapapa epistemologies and Maniapoto Maori cultural identities. Doctoral dissertation, Massey University. Available at: http://mro.massey.ac.nz/handle/10179/1252

Evers C (2009) 'The Point': Surfing, geography and a sensual life of men and masculinity on the Gold Coast, Australia. *Social and Cultural Geography* 10(8): 893–908.

Ford N and Brown D (2005) *Surfing and Social Theory: Experience, Embodiment and Narrative of the Dream Glide*. London: Routledge.

Gilchrist P and Wheaton B (2017) The social benefits of informal and lifestyle sports: A research agenda. *International Journal of Sport Policy and Politics* 9(1): 1–10.

Golob M, Giles A and Rich K (2013) Enhancing the relevance and effectiveness of water safety education for ethnic and racial minorities. *International Journal of Aquatic Research and Education* 7(1): 39–55.

Gould S, McLachlan F and McDonald B (2021) Swimming eith the Bicheno "Coffee Club": The textured world of wild swimming. *Journal of Sport and Social Issues* 45(1): 39–59.

Graham JPH (2009) Whakatangata kia kaha: toitū te whakapapa, toitū te tuakiri, toitū te mana: An examination of the contribution of Te Aute College to Māori advancement. Unpublished doctoral dissertation, Massey University.

Hokowhitu B (2007) Māori sport: Pre-colonisation to today. In: Collins C and Jackson S (eds) *Sport in Aotearoa/New Zealand Society*. Palmerston North: Dunmore Press, 78–95.

Hoskins TK and Jones A (2017) *Critical Conversations in kaupapa Māori*. Wellington: Huia Publishers.

Ingersoll KA (2016) *Waves of Knowing: A Seascape Epistemology*. Durham, NC: Duke University Press.

Jackson AM, Mita N and Hakopa H (2017) *Hui-te-ana-nui: Understanding kaitiakitanga in our Marine Environment*. Wellington: National Science Challenge Sustainable Seas.

Jackson AM, Puketeraki k, Hauteruruku, Mita N, Kerr H, Jackson S and Phillips C (2016) One day a waka for every marae: A southern approach to Maori water safety. *New Zealand Physical Educator* 49(1): 26–28.

Ka'ai T and Higgins R (2004) Te ao Māori: Māori world-view. In: Ka'ai TM and Moorfield JC, Reilly MJP and Mosely S (eds) *Ki te Whaiao: An Introduction to Māori Culture and Society*. Auckland: Pearson Education, 13–25.

Kerr H (2007) Indigenous navigational knowledge and recovery. *Toroa-te-nukuroa* 2: 26–35.

Lange R (1999) *May the People Live: A History of Māori Health Development 1900–1920*. Auckland: Auckland University Press.

Liu L (2018) Female Chinese and Māori sport participants' embodied experiences of risk, pain and injury. Doctoral thesis, University of Auckland. Available at: https://researchspace.auckland.ac.nz/handle/2292/37549

Liu L (2021) Paddling through bluespaces: Understanding waka ama as a post-sport through indigenous Māori perspectives. *Journal of Sport and Social Issues* 45(2): 138–160.

Liu L and Bruce T (2019) Extending understandings of risk in organised sport. *International Review for the Sociology of Sport* 55(6): 726–746.

McGloin C (2007) Aboriginal surfing: Reinstating culture and country. *International Journal of Humanities* 4(1): 93–110.

Masterson IA (2010) Hua Ka Nalu: Hawaiian surf literature. Unpublished Master of Arts dissertation, University of Hawaii, HI.

Masterson IA (2018a) E hua e: Surfing and sexuality in Hawaiian society. In: lisahunter (ed) *Surfing, Sex, Genders and Sexualities*. London: Routledge, 48–67.

Masterson IA (2018b) Surfing ali 'i, kahuna, kupua and akua: Female presence in surfing's past. In: lisahunter (ed) *Surfing, Sex, Genders and Sexualities*. London: Routledge, 191–207.

Mead HM (2003) *Tikanga Māori: Living by Māori values*. Wellington: Huia Publishers.

Ministerial Advisory Committee (1988) *Puao te atatu: The Report of the Ministerial Advisory Committee on a Maori perspective for the Department of Social Welfare*. Wellington: Government Printer.

Mita N (2014) Ko Au te Waka, Ko te Waka Ko Au: Examining the Link Between Waka and Hauora. Unpublished Honours dissertation, University of Otago.

Mita N (2016) Hauteruruku ki Puketeraki – Connecting to Te Ao Takaroa. Unpublished Masters dissertation, University of Otago.

Mita N, Flack B, Flack S, Ferall-Heath H, Jackson A and Taiapa-Parata W (2016) Application of Maori worldview, connection with Te Ao o Takaroa. *Out and About* (32): 12–16.

Moles K (2021) The social world of outdoor swimming: Cultural practices, shared meanings, and bodily encounters. *Journal of Sport and Social Issues* 45(1): 20–38.

Nelson A (1998) *Nga waka Māori: Māori Canoes*. Wellington: IPL.

Nemani M (2015) Being a brown bodyboarder. In: Humberstone B and Brown M (eds) *Seascapes: Shaped by the Sea*. Farnham: Ashgate, 92–108.

O'Malley V (2019) *The New Zealand Wars – Ngā Pakanga o Aotearoa*. Wellington: Bridget Williams Books.

Olive R and Wheaton B (2021) Understanding blue spaces: Sport, bodies, wellbeing, and the sea. *Journal of Sport and Social Issues* 45(1): 3–19.

Olive R (2015) Surfing, localims, place-based pedagogies and ecologcial sensibilities in Australia. In: Humberstone B, Prince H and Henderson KA (eds) *Routledge International Handbook of Outdoor Studies*. London: Routledge, 501–510.

Panelli R and Tipa G (2007) Placing well-being: A Maori case study of cultural and environmental specificity. *EcoHealth* 4: 445–460.

Peters K (2010) Future promises for contemporary social and cultural geographies of the sea. *Geography Compass* 4(9): 1260–1272.

Pool DI (1967) Post-War trends in Maori population growth. *Population Studies* 21(2): 87–98.

Pool DI (1977) *The Māori population of New Zealand 1769–1971*. Auckland: Auckland University Press.

Raureti T (2018) Connection between swimming and a Māori identity. Masters dissertation. University of Otago.

Reed AW (2004) *Reed Book of Māori Mythology*. Wellington: Reed Publishing (NZ) Ltd.

Roberts M, Norman W, Minhinnick N, Wihongi D and Kirkwood C (1995) Kaitiakitanga: Māori perspectives on conservation. *Pacific Conservation Biology* 2: 7–20.

Robson B and Harris R (eds) (2007) *Hauora: Māori Standards of Health IV. A Study of the Years 2000–2005*. Wellington: Te Rōpū Rangahau Hauora a Eru Pōmare.

Rowing NZ (2018) *Annual Report 2018*. Available at: www.rowingnz.kiwi/Category?Action=View&Category_id=517.

Royal TAC (1998) Mātauranga Māori: Paradigms and politics. Paper presented to the Ministry for Research, Science and Technology. Available at: www.charles-royal.com/assets/mm,paradigms%20politics.pdf

Salmond A (2014) Tears of Rangi: Water, power, and people in New Zealand. *HAU: Journal of Ethnographic Theory* 4(3): 285–309.

Selby R, Mulholland, M and Moore P (2010) *Māori and the Environment*. Wellington: Huia Publishers Ltd.
Sharp A (1957) *Ancient Voyagers in the Pacific*. London: Penguin Books.
Skinner WH (1923) Surf-riding by canoe. *Journal of the Polynesian Society*: 32: 35–37.
Smith A (2004) A Maori sense of place? Taranaki waiata tangi and feelings for place. *New Zealand Geographer* 60: 12–16.
Sport NZ (2015) *Sport and Active Recreation Profile: Surfing – Findings from the 2013/14 Active New Zealand Survey*. Wellington: Sport New Zealand.
Squire R (2017) "Do you dive?": Methodological considerations for engaging with "volume". *Geography Compass* 11(7) https://doi.org/10.1111/gec3.12319
Strang V (2014) The Taniwha and the Crown: Defending water rights in Aotearoa/New Zealand. *Wiley Interdisciplinary Reviews: Water* 1(1): 121–131.
Tapsell P (1997) The flight of Pareraututu: An investigation of taonga from a tribal perspective. *The Journal of the Polynesian Society* 106(4): 323–374.
Tau RTM (1999) Mātauranga Māori as an epistemology. *Te Pouhere Korero* 1(1): 10–23.
Te Awekotuku N (1991) *He tikanga whakaaro: Research Ethics in the Māori Community - A Discussion Paper*. Wellington: Ministry of Māori Affairs.
Tuaupiki JW (2017a). Ngā waka o Tāwhiti. *MAI Journal* 6(3): 305–321.
Tuaupiki JW (2017b) Te mātauranga whakatere waka. *Te Kōtihitihi: Ngā Tuhinga Reo Māori* 4: 36–57.
Tuhiwai Smith L (2012) *Decolonizing Methodologies: Research and Indigenous Peoples*. London and New York, NY: Zed.
Waiti J and Awatere S (2019) Kaihekengaru: Māori surfers' and a sense of place. *Journal of Coastal Research* 87: 35–43.
Waka Ama New Zealand (2017) *Waka Ama New Zealand Annual Report 2017*. Available at: www.wakaama.co.nz/content/files/5a5e85eac3673/Waka ama NZ annual report 2017.pdf
Waka Ama NZ (2018) *Waka Ama New Zealand Annual Report 2018*. Available at: www.wakaama.co.nz/content/files/5c5343a791417/2018%20Waka%20Ama%20NZ%20Annual%20Report.pdf
Walker IH (2011) *Waves of Resistance: Surfing and History in Twentieth-Century Hawai'i*. Honolulu, HI: University of Hawai'i Press.
Walker R (2004) *Ka whawhai tonu matou: Struggle without End*. Auckland: Penguin.
Walter R, Buckley H, Jacomb C and Matisoo-Smith E (2017) Mass migration and the Polynesian settlement of New Zealand. *Journal of World Prehistory* 30(4): 351–376.
Warshaw M (2005) *The Encyclopedia of Surfing*. Florida, FL: Houghton Mifflin Harcourt.
Wikaire RK and Newman JI (2013) Neoliberalism as neocolonialism?: Considerations on the marketisation of Waka Ama in Aotearoa/New Zealand. In: Hallinan C and Judd B (eds) *Native Games: Indigenous Peoples and Sports in the Post-Colonial World*. Bingley: Emerald, 59–83
Wheaton B, Roy G and Olive R (2017) Exploring critical alternatives for youth development through lifestyle sport: Surfing and community development in Aotearoa/New Zealand. *Sustainability* 9(12) https://doi.org/10.3390/su9122298
Wheaton B, Waiti J, Cosgrove M and Burrows L (2020) Coastal blue space and well-being research: Looking beyond western tides. *Leisure Studies* 39(1): 83–95.
Wheaton B, Waiti J, Olive R and Kearns R (2021) Coastal communities, leisure and wellbeing: advancing a trans-disciplinary agenda for understanding ocean-human relationships in Aotearoa New Zealand. *International Journal of Environmental Research and Public Health* 18(2) www.mdpi.com/1660-4601/18/2/450
Wilkinson J (2017) Largest surfing survey in New Zealand shows massive rise in paddleboarders over 10 years. *Stuff*. Available at: www.stuff.co.nz/national/89628631/largest-surfing-survey-in-new-zealand-shows-massive-rise-in-paddleboarders-over-10-years

## Glossary

| | |
|---|---|
| Ātua | Māori deity's akin to Mātauranga Māori |
| Awa | River |
| Iwi | Tribe |
| Hinemoana | The female deity of the ocean |
| Kaihekengaru | Māori wave riders |
| Karakia | incantations |

| | |
|---|---|
| Kawa | Māori customs and concepts |
| Kaupapa whanau | a group of people who share a common bond, other than descent, it may be geographical location or a shared purpose. |
| Maunga | mountain |
| Mana | authority |
| Mātauranga Māori | a Māori knowledge system |
| Moana | ocean or beach |
| Mokopuna | descendants |
| Pepeha | the features of the landscape that are determined by one's whakapapa |
| Rāhui | restrictions, prohibitions |
| Rangatira | Chief |
| Taonga | treasure |
| Taonga tuku iho | treasures handed down by ancestors |
| Tangaroa | the male deity of the ocean |
| Te Taiāo | the environment |
| Tikanga | Māori customs and concepts |
| Tino Rangatiratanga | self-determination |
| Tipuna | ancestor(s) |
| Waka | canoe |
| Waka ama | outrigger canoe |
| Waka hourua | double-hulled voyaging canoe |
| Hapū | sub-tribe |
| Hawaiki | the traditional homeland of Māori, situated in East Polynesia |
| Heke ngaru | surfing, wave riding |
| Kaitiakitanga | the practice of guardianship and environmental management |
| Whakapapa | genealogy, genealogical connections, lineage |
| Whakapapa whanau | a family or grouping of people who share the same genealogy |

# SECTION III

# Ocean economies, ocean labour

# 8

# FISHING

## Livelihoods and territorialisation of ocean space

*Madeleine Gustavsson and Edward H. Allison*

### Introduction

Ocean space is increasingly being mapped and designated to different maritime activities – such as renewable energy, aquaculture, fisheries and seabed mining (see also Fawcett et al., this volume; Hadjimichael, this volume; Jay, this volume; Lehman, this volume). As discussions around what the Blue Economy may offer become more frequent (European Commission, 2012; Silver et al., 2015; Voyer et al., 2018), there is uncertainty around the place of traditional actors – such as (often small-scale) fisheries – who have used the ocean for centuries (Gustavsson and Morrissey, 2019). Some authors have suggested that the most immediate concern (and existential threat) to many small-scale fisheries is coastal 'grabbing' or 'squeezing' (Bavinck et al., 2017; Cohen et al., 2019; Said et al., 2017). Coastal grabbing, it is argued, is caused by a displacement of fishing communities from the use of coastal space which has been seen to occur when, for example: i) fishers are displaced from traditional landing beaches to make space for tourism; ii) coastal fishing villages become gentrified; and iii) conservation displaces fishing activities (Bavinck et al., 2017; Said et al., 2017; Cohen et al., 2019). This chapter will extend these discussions by examining the potential impacts of *territorialisation* of ocean space on fishing livelihoods.

First, we discuss the emergence of a 'spatial turn' in ocean and coastal management (St. Martin and Hall-Arber, 2008), alongside the development of the recent discourse around the Blue Economy (Silver et al., 2015; Winder and Le Heron, 2017; Voyer et al., 2018). Within the Blue Economy, narrative ocean space is seen as having economic potential beyond fisheries into also accommodating often high-tech industries such as renewable energy, seafloor mining and intensified aquaculture (Morrissey, 2017). Second, we discuss what is commonly meant by 'fishing livelihoods', how it has been conceptualised and applied in different ways – beyond viewing fishing as an economic practice, followed by a discussion of what the territorialisation of ocean space might mean to fisherfolk. In light of the development of *The Voluntary Guidelines for Securing Small-Scale Fisheries in the Context of Food Security and Poverty Eradication* (FAO, 2015: no page), the chapter discusses how practices of territorialisation need to consider the pre-existing, and particular, place of fisherfolk – and their livelihoods – within management of ocean space. The chapter ends with a discussion around emerging research areas in the

DOI: 10.4324/9781315111643-11

context of understanding potential impacts on fishing livelihoods from the territorialisation of ocean spaces.

## Territorialisation of ocean space

The underlying discourses of ocean management and governance have recently undergone a 'spatial turn' (St. Martin and Hall-Arber, 2008) from a focus on sectoral management, such as fisheries, aquaculture, tourism or shipping, to spatial and 'holistic' approaches, including both the social and natural environment (such as Marine Spatial Planning [MSP]) (Douvere, 2008; Jay, this volume). It has been argued that this shift has been motivated by an ambition to reduce conflict between different users alongside an increased demand for ocean space by multiple actors (Kidd et al., 2011). As a predecessor to MSP, Integrated Coastal Zone Management (ICZM) – which focused on the spatial management of the terrestrial coastal zone – has been on the policy agenda since the early 1970s (Douvere, 2008). Following on from this, Merrie and Olsson (2014) trace the emergence of the MSP policy discourse and argue it has emerged from the use of spatial ocean management within the Great Barrier Reef, Australia. Merrie and Olsson (2014) suggest that several events and national experiments in spatial management, as well as the introduction of GIS technologies, enabled and shaped the upscaling of MSP to the international level.

Alongside the discourse and practice of MSP, the idea of the Blue Economy has become widespread at regional, national and international levels (European Commission, 2012; Silver et al., 2015). The basic idea is that the ocean offers new and previously underexplored avenues for economic capital development from industries such as offshore energy and bio-technology (see Morrissey, 2017). Given the proliferation of ocean uses, there is concern that conflicts between current uses and users of the ocean (such as small-scale fisheries) will emerge when new actors and uses are introduced. As such, to ensure the balance between users, MSP and territorialisation of ocean space are considered as complementary to facilitating growth of the Blue Economy (Kidd et al., 2011).

Drawing on political economy work and writings concerned with the 'enclosure of the commons' – that is, the privatisation of previously open access resources, such as the ocean (see Fawcett et al., this volume; Hadjimichael, this volume) – Fairbanks et al. (2018: 144) argue that new forms of enclosures, such as MSP, hold "the potential to both close and open the seas for ocean communities, environments and other actors". Boucquey et al. (2016) further highlight the role of GIS technologies, and how data availability can shape the material outcomes of MSPs. Furthermore, Satizábal and Batterbury (2018: 68) cite the work of Steinberg and Peters (2015) in arguing that "[t]he land-water binary has been enforced by the modern state, in which solid land is a social space, while the liquid sea is a place to compete for resources and territorial sovereignty, disregarding marine social processes". Echoing this perspective, St Martin and Hall-Arber (2008: 780) suggest that "mining, shipping, energy development, recreational fisheries, tourism, etc., to the degree they are mapped, are represented as occurring in locations at-sea but those locations and activities are only rarely linked to onshore locations or dependent communities". As such, they argue that GIS approaches to MSP have a 'missing layer' in that it does not fully capture the 'human dimension' of marine space (St Martin and Hall-Arber, 2008) – such as the social, cultural and economic relations between fishing communities and ocean space. Extending this argument, Boucquey et al. (2016: 1) pose the question: "[i]n the evolving outcome of [MSP] there is much at stake – who and what counts as citizens of the ocean?" Following this line of thought, Flannery and Ellis (2016: 121) write: "[w]hile MSP is quickly becoming the dominant marine management paradigm, there has been comparatively little assessment of the potential negative impacts and possible distributive impacts that may arise from its adoption". Here we have to consider that when ocean space is designated to new actors within the Blue Economy, some actors will potentially 'lose' that space. Indeed, Jay (2013: 520) suggests "the distribution of resources achieved through zoning will be contested and perceived by some as inequitable". Jay (2013) further suggests that 'zoning' of the seas with its 'inherent dynamism', fluidity and mobility and the desire to order that which cannot be 'fixed' stems from ideas and methods developed in the natural sciences – highlighting the need for social science perspectives on spatial management of the ocean.

Recent studies have tried to understand the views of multiple stakeholders with regards to the uneven distributions of economic impacts within the Blue Economy across different geographical scales (Gustavsson and Morrissey, 2019). As will be discussed below in relation to fishing livelihoods, social processes are fluid along the ocean–land continuum and changes in one area could have impacts in another. Having provided a context to debates in territorial ocean management, the next section will discuss what is commonly meant by fishing livelihoods, followed by a discussion of how ocean space and fishing livelihoods relate.

## (More-than-economic) fishing livelihoods

Fishing has commonly been understood as a 'way of life' with significance to fisherfolk beyond the economic dimension of making a living (Urquhart et al., 2011). Fishers' adherence to this fishing way of life often serves as the 'cultural explanation' for why fishers tend to fish despite decreasing economic viability (McGoodwin, 2001). However, important for many authors is the idea that fishers derive non-economic benefits from fishing and from being fishers. An example is fishers' sense of job satisfaction which is linked to their sense of independence, freedom, pride and a fascination with risks (Ross, 2013; Van Ginkel, 2001). Furthermore, by interviewing men in the small boat fisheries of Newfoundland, Power (2005) found that fishers do not distinguish their work from their sense of 'self' – instead they define their sense of self in relation to their occupational identity. Such findings have been echoed by, for example, LiPuma (1992) who has studied the Galician fishery in Spain and argues that there is a strong identity tied to being a fisher. Extending this individual perspective, researchers like Gustavsson and Riley (2018) and Van Ginkel (2014) has stressed the importance of family and intergenerational dimensions to what it means to be a fisher. Yet the more-than-economic importance of fishing seldomly finds its way into discussions and formulations of marine policy (Symes and Phillipson, 2009) – such as MSP and ideas around the Blue Economy.

A number of authors have focused on fishing communities and have taken various theoretical and methodological approaches to explore their livelihoods (Angerbrandt et al., 2011). From an economic perspective, the fishing community includes not only fishers but also 'interlinked industries' such as fish processing facilities (Morrissey and O'Donoghue, 2012). Yet, the importance of fishing has been recognised as greater than its economic value (Urquhart and Acott, 2014) and work on fishing 'dependency' has moved from a focus on economic dependency, such as employment and income, to recognise the socio-cultural dependencies of individuals and local communities (Urquhart and Acott, 2014). In particular, Ross (2013) explores how the working culture of fishing in the Scottish fishery is dependent on strong interpersonal relations and reciprocity amongst fishers and other local people. As such, the social identity of the fishing community is suggested to lie in the social relations within it – as underpinned by informal labour structures (Symes and Phillipson, 2009). Within this context, Munro (2000) explores the ways the 'self' and the 'community' in a northeastern Scottish fishing village are interrelated. Drawing on Foucault, Munro (2000) provides a critique of individualism and argues that social relationships such as family and community are important for the choices that individuals make in relation to their work and family life. She explores the themes of marriage, childcare, kinship and social participation, and how 'appropriate behaviour' in relation to these positions is socially constructed in time and place, thus shaping the choices and behaviours of fishers.

In addition to exploring what it means to be a fisher or be living in a fishing community there are two dominant frameworks used by research and policy in trying to capture the meaning of fishing lives and livelihoods in more holistic ways. One example is the Sustainable Livelihoods Approach (SLA), which was first applied to fisheries in a paper by Allison and Ellis (2001) in which they sought to understand the strategies fisherfolk use when confronted with fluctuating resources of fish – particularly in the context of poverty. The SLA was developed to understand the different capabilities rural families had to cope with crises – building on concepts of vulnerability, sustainability and resilience. A livelihood, according to Ellis (2000: 10) can be described by five assets: i) physical; financial; natural; human; and social capital (see also Allison, 2003). The framework has been used to understand the role fisheries play in rural communities in developing countries (see Béné, 2003), and how livelihood diversification can promote a resilience amongst fishing families and coastal communities to cope with external threats (Brugère et al., 2008). The SLA has also been used in considering how gender modifies access and opportunities to the five forms of assets enabling women and men to pursuit diverse livelihood strategies (Fröcklin, 2014). The SLA has been widely adopted in development policies and practices and can be seen as a 'boundary object'. That is, it allowed natural scientists and technical experts in agronomy, forestry, water resources, engineering (amongst others) to start a dialogue with social scientists and human development specialists, and it encouraged economists to think more broadly about utility functions and how people navigate amid imperfect information and markets. In essence, the SLA draws attention to the often place-based economic relationships (mediated by institutions, including social relations) in which fishing activities are embedded, highlighting the interconnectedness of fishing and other coastal economic activities, whether rural or urban. Whilst the SLA has mainly been applied in the context of developing countries, others have similarly explored how *pluri-activity*, such as combining fishing with wage work, has been used as a coping strategy by fishing households in the Global North to maintain their culture and livelihoods (see Salmi, 2005 for the case of Finland).

Another common approach to studying fisheries from a social science perspective has been using the well-being lens (Gough and McGregor, 2007). This is an approach which draws on a three-dimensional framework which links the 'material', 'subjective' and 'relational' well-being of the lives of fishers and their communities (Britton and Coulthard, 2013; McGregor, 2009) and through this narrative it tries to incorporate the social and the 'natural' world (Coulthard et al., 2011). The well-being framework (for a comprehensive review, see Weeratunge et al., 2014) is particularly used to understand how individuals adapt to change depending on their material resources (what an individual has), their relational resources (interactions individuals engage in via social relationships) and subjective resources (feelings about what one does and has) (Coulthard, 2012). It has been used in studies of fisheries in the Global South (see Coulthard et al., 2011) as well as the Global North (see White, 2017) and studies have used both less structured (qualitative) and more structured (both qualitative and quantitative) approaches in documenting different well-being dimensions (see White, 2014 for a discussion on methods). An edited volume entitled *Social Wellbeing and the Values of Small-Scale Fisheries* (Johnson et al., 2017) presents several well-being studies: linking well-being to ideas of cultural ecosystem service and place (Acott and Urquhart, 2017); using a constructivist approach to well-being examining symbolic dimensions of resilience in Norfolk, UK (White, 2017); examining how well-being can help us understand how small-scale fishers can adapt to biophysical change in Brazil (Idrobo, 2017), amongst many other case studies. In a study of Northern Ireland (UK), Britton and Coulthard (2013) examine how fishers, their families and their communities derive well-being from the fishery and highlighted the important role(s) of women as 'well-being agents' in their support of the well-being of fishing families and communities (Britton, 2012). Extending this argument, Urquhart and Acott (2013) have, through field observations of the physical place and semi-structured interviews, studied how the Southeastern English fishing town of Hastings is socially constructed as a fishing place and, most importantly, how the cultural landscape of fishing contributes to well-being in coastal communities.

Another emerging and associated way to think about livelihoods is through the lens of 'values' (see for example Johnson, 2017; Song, 2017) in considering the benefits society derives from fisheries and the relationships fisherfolks have to the bio-physical environment and each other (see Chan et al., 2016 on 'relational values'). All these approaches enable a deeper examination of the social and economic implications of the emergent interest in MSP and arguably ought to be better understood in relation to the Blue Economy.

## Ocean space and fishing livelihoods

There is an emerging discussion around the importance of ocean and coastal space to fishing livelihoods (de la Torre-Castro et al., 2017; Jentoft and Knol, 2014). We highlight two dimensions in particular that illustrate how the relationship between fisheries and ocean space can be complex and presents a challenge to the discourse in which oceans space could be mapped, controlled and designated to different users (as implied in MSP and the expansion of the Blue Economy). First, we discuss the importance specific ocean spaces and places can have for fishing livelihoods followed by a discussion of mobilities in fishing livelihoods.

Drawing on Bourdieu's ideas of capitals – particularly paying attention to social capital (stemming from, and reaffirmed by, social contacts), cultural capital (skills, knowledge and dispositions which may be gained by education and socialisation) and economic capital – Gustavsson et al. (2017) have noted that knowledge and skills take on symbolic meaning in fishing communities. Gustavsson et al. (2017) found that through displaying place-specific knowledge and skills – by, for example, moving fishing gear in seasonally appropriate ways – fishers became known to others in their communities as 'good fishers'. Being a 'good fisher' allowed them access to social capital in the form of help from others, in, for example, situations of need whilst at (a sometimes dangerous) sea. The wider significance of this, in the context of territorialisation of ocean space and spatial management techniques, is that spatially specific fishing activities are embedded in – or even constitute (as part of an assemblage) – pre-existing marine social processes. For example, in their study of a regulator's attempt at implementing highly protected Marine Conservation Zones (MCZ) in North Wales (UK), Gustavsson et al. (2017) found that fishers fiercely opposed the disruption to the pre-existing informal ownership system around fishing territories, which would have been the consequence if fishers had been displaced. Luckily for them, the highly protected MCZs were never implemented. Studies such as these reveal how knowledge, identity and place are connected and reinforce each other in the case of fishing livelihoods and, perhaps more significantly, that the use of ocean space is key to these relations.

In additional to place-specificity on land (such as the location of harbours and landing beaches as well as the communities in which fishers live) and at sea (for example, specific fishing grounds and informally claimed territories), fishing activities can be spatially variable and transgress national and jurisdictional boundaries – sometimes leading to severe conflicts as seen in the case of Pal Bay in Sri Lanka involving Indian trawlers (Scholtens, Bavinck and Soosai,

2012) as well as in the more recent conflicts between English and French scallop fishers in the English Channel (Sommerlad, 2018). Fishing livelihoods are also often mobile (see Nunan, 2010 for a review) with Allison and Seeley (2004: 220) arguing "many fisherfolk – even small-scale fisheries doing day-trips are geographically mobile over their lifetimes and can often be classified as seasonal or long-term migrant". There are also nomadic fishing communities who live mostly at sea on boats – most famously the Sama-Bajau ('sea nomads') of Southeast Asia (Stacey and Allison, 2019). On a global scale, fishers often migrate in response to fish availability or to access spatially dispersed livelihood opportunities in fisheries, agriculture and the urban economy (Nunan, 2010; Overå, 2001). Whilst migrating fishing communities have existed for a long time, Nunan (2010) argues migration is often viewed negatively in policy terms because migrants are presumed to lack interest in stewardship of resources that are not 'local' to them. Much migration is, however, either long term or repeated seasonally; migrants do not wander at random and can have long-term relationships in other places. These livelihood patterns and place-based connections are observed not only in the context of the Global South but also, for example, in northern Norway where fishers move with seasonal patterns of fish migration (Gerrard, 2013; Jentoft and Knol, 2014).

The policy challenge, in the context of territorialisation then, is that sustaining small-scale fisheries requires both a recognition of place (such as a long-standing fishing community, landing beach and traditional fishing grounds) *and* the facilitation of mobility and flexibility to pursue mobile fish stocks and spatially dispersed livelihood opportunities. In this sense, some fisheries can be likened to transhumant pastoralist systems (Jones, 2005; Turner et al., 2016) and require similar policy measures to sustain them. There is therefore much to be learned by examining the policies that have revived transhumance in Europe (Kerven and Behnke, 2011).

The complex position of fisherfolk in relation to ocean space has recently been recognised by the Food and Agricultural Organisation of the United Nations in their document *The Voluntary Guidelines for Securing Small-Scale Fisheries in the Context of Food Security and Poverty Eradication* (FAO, 2015; hereafter referred to as the SSF Guidelines). The document draws on a human rights-based approach which recognises that fisheries policy has to go beyond the allocation of fishing opportunities to address underlying causes of social and economic marginalisation in fishing communities (see Allison et al., 2012; Ratner et al., 2014). The SSF guidelines highlight duty-bearers' responsibility to ensure that principles of equity, accountability and justice are upheld when considering the governance of coastal and riparian regions, so that fisherfolk are not discriminated against or excluded from the benefits of citizenship that accrue to others. These principles apply across issues ranging from women's rights, the rights of migrants, to rights to food, health and decent work. In particular, the Guiding principle 2: *Respect of cultures* "recognizing and respecting existing forms of organization, traditional and local knowledge and practices of small-scale fishing communities" (Food and Agricultural Organisation of the United Nations, 2015: 2) is of importance in relation to the move towards territorialisation of ocean space. The document discusses the importance of tenure rights in stating that "small-scale fishing communities need to have secure tenure rights to the resources that form the basis for their social and cultural well-being, their livelihoods and their sustainable development" (Food and Agricultural Organisation of the United Nations, 2015: 5). The wider significance of these principles is the recognition that small-scale fishing communities and fishers have a right to fish in their traditional territories (be those 'local' or 'mobile'), using traditional forms of fishing practices, to pursue their own culture, livelihood and well-being goals.

Whilst governments are not obliged to adopt these voluntary guiding principles, their formulation could be seen as a strong argument that small-scale fisheries need to be considered sensitively in the context of potential impacts that fishing livelihoods and communities could experience from the increased territorialisation of ocean space in management strategies (such as MSP). Whilst the SSF Guidelines focus on eradicating poverty in the Global South, the guiding principles have been applied in diverse geographical contexts and it has been suggested that the principles are relevant and would benefit small-scale fisheries in the Global North as well (Jentoft et al., 2017). In light of attempts at implementing the SSF there is a need to ask: how can the specific rights formulated in the SSF Guidelines be integrated in (or potentially conflict with) Blue Economy initiatives? We will return to this question below.

## Discussion and future directions

This chapter has highlighted tensions between the dual objectives of enabling the growth of the Blue Economy through ocean territorialisation and securing small-scale fisheries and fishing livelihoods more broadly. Whilst both Blue Economy thinking and SSF guidelines are international in scope, it could be argued that the relations between ocean space and fishing livelihoods are local or regional, and context-specific. As such, a fruitful way for future research to untangle some of the dynamics and potential impacts of ocean territorialisation could be to explore how these issues play out in varying geographical and local contexts. By understanding these (often) pre-existing marine social processes (such as mobile and place-dependent fishing practices), we can begin to think about how to minimise potential impacts of territorialisation of ocean space on fishing livelihoods. Efforts are already being undertaken in this regard with scholars and practitioners starting to think about how the FAO's SSF Guidelines can be implemented at national levels (Jentoft et al., 2017) and how the expanding 'Blue Economy' and the associated territorialisation of inshore waters can accommodate and support existing oceans rights-holders such as small-scale fisherfolk (Allison et al., 2020). Examples of how to – in practice – secure the 'special status' of fisherfolk in the Blue Economy and territorialisation of the ocean include the Puruvesi winter seining fishery in Finland which, as Mustonen explains, has

> received a special distinction in the EU. The EU has provided a Protected Geographical Indication (PGI)[1] to the vendace fish (*Coregonus albula*) from Puruvesi, recognizing both the biological qualities of the fish, as well as how it is harvested, i.e. traditional seining.
>
> *Mustonen, 2018: 103*

Furthermore, in 2017 this fishery became included in the 'National Inventory of Living Heritage in Finland', which covers a wide range of fields of intangible culture in Finland (Snowchange Cooperative, 2017). Later on in 2017 the same fishery was shortlisted for UNESCO's list of intangible cultural heritage (Mustonen, 2018). Whilst securing the position of certain fisheries is important, there are arguably issues around how this approach could be upscaled and transferred to other geographical localities where fishers are using more standardised fishing methods that may not so easily go under the label of 'cultural heritage'. More research and policy work are needed to solve these issues.

Ideas around colonisation of the sea and ocean grabbing (Bennett et al., 2015; Foley and Mather, 2018) are also pertinent, as they address concerns such as displacement of fishing communities by the expansion of marine protected areas, the granting of oil and gas exploration and exploitation licenses in inshore and offshore waters and, lately, the spread of large-scale transnational aquaculture and its financialisation (Knott and Neis, 2017). Partly stemming from these discussions, there is an emerging debate within the small-scale fisheries research and policy communities arguing we need to shift the discourse from the Blue Economy to 'Blue Justice'

(Too Big To Ignore, 2019). This could potentially be a productive avenue for future research on fishing livelihoods and ocean space if it can be effective in arguing for socio-cultural as well as economic impacts stemming from territorialisation of ocean space.

A further avenue for future research on fishing livelihoods and ocean space follows Porter's (2012: 60) argument that "in very few [studies on fisheries] do we find recognition that behind the hand that fishes, there lies a fully social person (usually male), and behind him [sic], a family and a community (which includes women and children as well as men)". Whilst livelihood research often focuses on households, using ocean space is often seen as being something that men do. As scholars have argued (see Frangoudes and Gerrard, 2018), men are often (but far from always) the main actor on the sea capturing fish. However, fishing activities and livelihoods do not only rely on ocean space but also on coastal and land spaces and the connections between them. Whilst some authors have attempted to understand women's activities as part of fishing livelihoods (de la Torre-Castro et al., 2017; Fröcklin, 2014) it is still largely undocumented how impacts from the territorialisation of ocean space can be gendered. In addition to this, Neis et al. (2013) have argued that gender – and intergenerationally – blind fishing policies can have unintended consequences for the future sustainability and resilience of fishing families, communities and livelihoods.

The social science research community is just beginning to explore the diverse and complex impacts on individuals, families, communities and local economies that emerge from the increasing territorialisation of ocean space. Both the material and subjective dimensions of these impacts – whether experienced in ocean or terrestrial spaces – are important areas for study by 'marine social scientists' (see Gustavsson et al., 2021), be they economists, sociologists, anthropologists or human geographers. Policy-engaged critical and transdisciplinary scholarship can provide important support to coastal communities who seek 'Blue Justice' in a territorialised ocean.

## Note

1 For more info on EU's Geographical indications, see: https://ec.europa.eu/trade/policy/accessing-markets/intellectual-property/geographical-indications/

## References

Acott T and Urquhart J (2017e) Co-constructed cultural ecosystem services and wellbeing through a place-based approach. In: Johnson DS, Acott T, Stacey N, Urquhart J (eds) *Social Wellbeing and the Values of Small-scale fisheries*. Amsterdam: Springer, 23–44.

Allison EH (2003) Potential applications of a 'Sustainable Livelihoods Approach' to management and policy development for European inshore fisheries. In: Hart P and Johnson M (eds) *Who Owns the Sea?* Hull: The University of Hull, 23–43.

Allison EH and Ellis F (2001) The livelihoods approach and management of small-scale fisheries. *Marine Policy* 25(5): 377–388.

Allison EH, Kurien J and Ota Y (2020) *The Human Relationship with Our Ocean Planet*. Washington, DC: World Resources Institute. Available at: https://oceanpanel.org/blue-papers/HumanRelationshipwithOurOceanPlanet

Allison EH and Seeley JA (2004) HIV and AIDS among fisherfolk: A threat to 'responsible fisheries'? *Fish and Fisheries* 5: 215–234.

Allison EH, Ratner BD, Åsgård B, Willmann R, Pomeroy R and Kurien J (2012) Rights-based fisheries governance: From fishing rights to human rights. *Fish and Fisheries* 13(1): 14–29.

Angerbrandt H, Lindström L and de la Torre-Castro M (2011) What is this thing called 'community' good for? In: Chuenpagdee R (ed) *World Small-Scale Fisheries Contemporary Visions*. Delft: Eburon, 353–365.

Bavinck M, Berkes F, Charles A, Esteves Dias AC, Doubleday N, Nayak P and Sowman M (2017) The impact of coastal grabbing on community conservation – A global reconnaissance. *Maritime Studies* 16(1) https://doi.org/10.1186/s40152-017-0062-8

Béné C (2003) When fishery rhymes with poverty: A first step beyond the old paradigm on poverty. *World Development* 31(6): 949–975.

Bennett NJ, Govan H and Satterfield T (2015) Ocean grabbing. *Marine Policy* 57: 61–68.

Boucquey N, Fairbanks L, St. Martin K, Campbell LM and McCay B (2016) The ontological politics of marine spatial planning: Assembling the ocean and shaping the capacities of 'Community' and 'Environment'. *Geoforum* 75: 1–11

Britton E (2012) Women as agents of wellbeing in Northern Ireland's fishing households. *Maritime Studies* 11(16) https://doi.org/10.1186/2212-9790-11-16

Britton E and Coulthard S (2013) Assessing the social wellbeing of Northern Ireland's fishing society using a three-dimensional approach. *Marine Policy* 37: 28–36

Brugère C, Holvoet K and Allison EH (2008) Livelihood diversification in coastal and inland fishing communities: misconceptions, evidence and implications for fisheries management. In *Working paper, Sustainable Fisheries Livelihoods Programme (SFLP)*. Rome: FAO/DFID, 1–39.

Chan KMA, Balvanera P, Benessaiah K, Chapman M, Díaz S, Gómez-Baggethun E, Gould R, Hannahs N, Jax K, Klain S, Luck GW, Martín-López B, Muraca B, Norton B, Ott K, Pascual U, Satterfield T, Tadaki M, Taggart J and Turner N (2016) Opinion: Why protect nature? Rethinking values and the environment. *Proceedings of the National Academy of Sciences* 113(6): 1462–1465.

Cohen PJ, Allison EH, Andrew NL, Cinner J, Evans LS, Fabinyi M, Garces LR, Hall SJ, Hicks CC, Hughes TP, Jentoft S, Mills DJ, Masu R, Mbaru EK, and Ratner BD (2019) Securing a just space for small-scale fisheries in the Blue Economy. *Frontiers in Marine Science* 6 (171): 1–8.

Coulthard S (2012) What does the debate around social wellbeing have to offer sustainable fisheries? *Current Opinion in Environmental Sustainability* 4(3): 358–363.

Coulthard S, Johnson D and McGregor JA (2011) Poverty, sustainability and human wellbeing: A social wellbeing approach to the global fisheries crisis. *Global Environmental Change* 21(2): 453–463.

de la Torre-Castro M, Fröcklin S, Börjesson S, Okupnik J and Jiddawi NS (2017) Gender analysis for better coastal management: Increasing our understanding of social-ecological seascapes. *Marine Policy* 83: 62–74.

Douvere F (2008) The importance of marine spatial planning in advancing ecosystem-based sea use management. *Marine Policy* 32(5): 762–771.

Ellis F (2000) *Rural Livelihoods and Diversity in Developing Countries*. Oxford: Oxford University Press.

European Commission. (2012) *Blue Growth Opportunities for Marine and Maritime Sustainable Growth*. Available at: https://eur-lex.europa.eu/legal-content/EN/TXT/PDF/?uri=CELEX:52012DC0494&from=EN

Fairbanks L, Campbell LM, Boucquey N and St. Martin K (2018) Assembling enclosure: Reading Marine Spatial Planning for alternatives. *Annals of the American Association of Geographers* 108(1): 144–161.

Flannery W and Ellis G (2016) Exploring the winners and losers of marine environmental governance. *Planning Theory & Practice* 17(1): 121–151.

Foley P and Mather C (2018) Ocean grabbing, terraqueous territoriality and social development. *Territory, Politics, Governance* 7(3): 1–19.

Food and Agricultural Organisation of the United Nations (FAO). (2015) *Voluntary Guidelines for Securing Sustainable Small-Scale Fisheries*. Rome: FAO.

Frangoudes K and Gerrard S (2018) (En)gendering change in small-scale fisheries and fishing communities in a globalized world. *Maritime Studies* 17: 117–124.

Fröcklin S (2014) Women in the seascape: Gender, livelihoods and management of coastal and marine resources. Unpublished doctoral dissertation, Stockholm University.

Gerrard S (2013) Mobilities, materialities, and masculinities: Interconnected mobility practices in Norwegian coastal fisheries. *Norsk Geografisk Tidsskrift - Norwegian Journal of Geography* 67(5): 312–319.

Gough I and McGregor JA (2007) *Wellbeing in Developing Countries: From Theory to Research*. Cambridge: Cambridge University Press.

Gustavsson M, Riley M, Morrissey K, and Plater A (2017) Exploring the socio-cultural contexts of fishers and fishing: Developing the concept of the 'good fisher'. *Journal of Rural Studies* 50: 104–116.

Gustavsson M and Riley M (2018) The fishing lifecourse: Exploring the importance of social contexts, capitals and (more than) fishing identities. *Sociologia Ruralis* 58(3): 562–582.

Gustavsson M and Morrissey K (2019) A typology of different perspectives on the spatial economic impacts of Marine Spatial Planning. *Journal of Environmental Policy & Planning*. 21(6): 841–853.

Gustavsson M, White C, Phillipson, J and Ounanian K (eds) (2021) *Researching People and the Sea: Methodologies and Traditions*. London: Palgrave.

Idrobo CJ (2017) Adapting to environmental change through the lens of social wellbeing: Improvements and trade-offs associated with a small-scale fishery on the Atlantic forest coast of Brazil. In: Johnson DS, Acott T, Stacey N, Urquhart J (eds) *Social Wellbeing and the Values of Small-scale fisheries*. Amsterdam: Springer, 75–96.

Jay S (2013) From disunited sectors to disjointed segments? Questioning the functional zoning of the sea. *Planning Theory & Practice* 14(4): 509–525.

Jentoft S, Chuenpagdee R, Barragán-Paladines MJ and Franz N (eds) (2017) *The Small-Scale Fisheries Guidelines: Global Implementation*. Amsterdam: Springer.

Jentoft S and Knol M (2014) Marine spatial planning: Risk or opportunity for fisheries in the North Sea? *Maritime Studies* 13(1) https://doi.org/10.1186/2212-9790-13-1

Johnson DS (2017) The values of small-scale fisheries. In Johnson DS, Acott T, Stacey N, Urquhart J (eds) *Social Wellbeing and the Values of Small-scale fisheries*. Amsterdam: Springer, 1–22.

Johnson DS, Acott T, Stacey N and Urquhart J (eds) (2017) *Social Wellbeing and the Values of Small-scale fisheries*. Amsterdam: Springer.

Jones S (2005) Transhumance re-examined. *The Journal of the Royal Anthropological Institute* 11(2): 357–359.

Kerven C and Behnke R (2011) Policies and practices of pastoralism in Europe. *Pastoralism: Research, Policy and Practice* 1(28): 1–5.

Kidd S, Plater A and Frid C (eds) (2011) *The Ecosystem Approach to Marine Planning and Management*. London: Earthscan.

Knott C and Neis B (2017) Privatization, financialization and ocean grabbing in New Brunswick herring fisheries and salmon aquaculture. *Marine Policy* 80: 10–18.

LiPuma E (1992) Social identity and the European community: An Iberian example. *Maritime Studies* 5(2): 46–73.

McGoodwin JR (2001) Understanding the cultures of fishing communities: A key to fisheries management and food security. *Fisheries Technical Paper 401*. Rome: FAO.

McGregor JA (2009) Human wellbeing in fishing communities. *ESPA Workshop 1: Building Capacity for Sustainable Governance in South Asian Fisheries: Poverty, Wellbeing and Deliberative Policy Networks*. Chennai, India: Institute of Development Studies

Merrie A and Olsson P (2014) An innovation and agency perspective on the emergence and spread of Marine Spatial Planning. *Marine Policy* 44: 366–374.

Morrissey K (2017) *Economics of the Marine: Modelling Natural Resources*. London: Rowman and Littlefield.

Morrissey K and O'Donoghue C (2012) The Irish marine economy and regional development. *Marine Policy* 36(2): 358–364.

Munro G (2000) *How do Individuals Relate to their Local Communities Through Work and Family Life? Some fieldwork evidence*. Aberdeen: University of Aberdeen.

Mustonen T (2018) Lake Puruvesi, North Kareliaand South Savo, Finland – Representing the Ecoregion: Saimaa. In Tunón T (ed.) *Biodiversity and Ecosystem Services in Nordic Coastal Ecosystems: An IPBES-Like Assessment Volume 2 The Geographical Case Studies*. Copenhagen: Rosendahls, 201–212.

Neis B, Gerrard S and Power NG (2013) Women and children first: The gendered and generational social-ecology of smaller-scale fisheries in Newfoundland and Labrador and Northern Norway. *Ecology and Society* 18(4) https://doi.org/http://dx.doi.org/10.5751/ES-06010-180464

Nunan F (2010) Mobility and fisherfolk livelihoods on Lake Victoria: Implications for vulnerability and risk. *Geoforum* 41(5): 776–785.

Overå R (2001) Institutions, mobility and resilience in the Fante migratory fisheries of West Africa. Bergen: Chr. Michelsen Institute (CMI Working Paper WP 2001:2). Available at: www.cmi.no/publications/900-institutions-mobility-and-resilience-in-the-fante

Porter M (2012) Why the coast matters for women: A feminist approach to research on fishing communities. *Asian Fisheries Science* 25: 59–73.

Power NG (2005) *What Do They Call a Fisherman? Men, Gender, and Restructuring in the Newfoundland Fishery*. St. John's: Institue of social and economic research, Memorial University.

Ratner BD, Åsgård B and Allison EH (2014) Fishing for justice: Human rights, development, and fisheries sector reform. *Global Environmental Change* 27(1): 120–130.

Ross N (2013) Exploring concepts of fisheries 'dependency' and 'community' in Scotland. *Marine Policy* 37: 55–61.

Said A, MacMillan D, Schembri M and Tzanopoulos J (2017) Fishing in a congested sea: What do marine protected areas imply for the future of the Maltese artisanal fleet? *Applied Geography* 87: 245–255.

Salmi P (2005) Rural pluriactivity as a coping strategy in small-scale fisheries. *Sociologia Ruralis* 45(1–2): 22–36.

Satizábal P and Batterbury SPJ (2018) Fluid geographies: Marine territorialisation and the scaling up of local aquatic epistemologies on the Pacific coast of Colombia. *Transactions of the Institute of British Geographers* 43(1): 61–78.

Scholtens J, Bavinck M and Soosai AS (2012) Fishing in Dire Straits: Trans-boundary incursions in the Palk Bay. *Economic and Political Weekly* 47(25): 87–96.

Silver JJ, Gray NJ, Campbell LM, Fairbanks LW and Gruby RL (2015) Blue Economy and competing discourses in international oceans governance. *Journal of Environment and Development* 24(2): 135–160.

Snowchange Cooperative (2017) *Puruvesi Winter Seining and Snowchange Receive National Recognition for Cultural Heritage*. Available at: www.snowchange.org/2017/11/puruvesi-winter-seining-and-snowchange- receive-national-recognition-for-cultural-heritage/

Sommerlad J (2018) *Scallop wars: A Brief History of British and French Fishermen Musselling in on Each Other's Catch*. Available at: www.independent.co.uk/news/uk/home-news/scallop-wars-britain-france-fishing-rights-english-channel-history-a8512871.html

Song AM (2017) How to capture small-scale fisheries' many contributions to society? - Introducing the 'Value-Contribution Matrix' and applying it to the case of a swimming crab fishery in South Korea. In: Johnson DS, Acott T, Stacey N, Urquhart J (eds) *Social Wellbeing and the Values of Small-scale fisheries*. Amsterdam: Springer, 125–146.

Stacey N and Allison EH (2019) Sea nomads: Sama-Bajau mobility, livelihoods and marine conservation in Southeast Asia. In: King, TJ and Robinson G (eds) *At Home on the Waves: Human Habitation of the Sea from the Mesolithic to Today*. New York, NY: Berghahn Books, 309–331.

Steinberg PE and Peters K (2015) Wet ontologies, fluid spaces: Giving depth to volume through oceanic thinking. *Environment and Planning D: Society and Space* 33(2): 247–264.

St Martin K and Hall-Arber M (2008) The missing layer: Geo-technologies, communities, and implications for marine spatial planning. *Marine Policy* 32(5): 779–786.

Symes D and Phillipson J (2009) Whatever became of social objectives in fisheries policy? *Fisheries Research* 95(1): 1–5.

Too Big To Ignore (2019) *Blue Justice: Small-Scale Fisheries are Too Important to Fail!* Available at: http://toobigtoignore.net/blue-justice-small-scale-fisheries-are-too-important-to-fail/

Turner MD, McPeak JG, Gillin K, Kitchell E and Kimambo N (2016) Reconciling flexibility and tenure security for pastoral resources: The geography of transhumance networks in eastern Senegal. *Human Ecology* 44: 199–215.

Urquhart J, Acott T, Reed M and Courtney P (2011) Setting an agenda for social science research in fisheries policy in Northern Europe. *Fisheries Research* 108: 240–247.

Urquhart J and Acott T (2013) Constructing 'the Stade': Fishers' and non-fishers' identity and place attachment in Hastings, south-east England. *Marine Policy* 37: 45–54.

Urquhart J and Acott T (2014) A sense of place in cultural ecosystem services: The case of Cornish fishing communities. *Society & Natural Resources* 27(1): 3–19.

Van Ginkel R (2001) Inshore fishermen: Cultural dimensions of a maritime occupation. In: Symes D and Phillipson J (eds) *Inshore Fisheries Management*. Dordrecht: Kluwer academic publishers, 177–194.

Van Ginkel R (2014) A Texel fishing lineage: The social dynamic and economic logic of family firms. *Maritime Studies* 13(10) https://doi.org/10.1186/s40152-014-0010-9

Voyer M, Quirk G, McIlgorm A and Azmi K (2018) Shades of blue: What do competing interpretations of the Blue Economy mean for oceans governance? *Journal of Environmental Policy and Planning* 20(5): 595–616.

Weeratunge N, Béné C, Siriwardane R, Charles A, Johnson D, Allison EH, Nayak PK and Badjeck M-C (2014) Small-scale fisheries through the wellbeing lens. *Fish and Fisheries* 15(2): 255–279.

White CS (2014) Structured interview tools: insights and issues from assessing wellbeing of fishermen adapting to change using scoring and ranking questions. In: *SAGE Research Methods Cases*. London: SAGE. Available at: https://ueaeprints.uea.ac.uk/id/eprint/50049

White CS (2017) Symbols of resilience and contested place identity in the coastal fishing towns of Cromer and Sheringham, Norfolk, UK: Implications for social wellbeing. In: Johnson DS, Acott T, Stacey N, Urquhart J (eds) *Social Wellbeing and the Values of Small-Scale Fisheries*. Amsterdam: Springer, 45–74.

Winder GM and Le Heron R (2017) Assembling a Blue Economy moment? Geographic engagement with globalizing biological-economic relations in multi-use marine environments. *Dialogues in Human Geography* 7(1): 3–26.

# 9

# PLANNING

## Seeking to coordinate the use of marine space

*Stephen Jay*

### Introduction

Amongst the recent developments in understanding human interaction with the coasts, seas and oceans, one of the approaches giving most explicit attention to the spatial dimensions of this interaction has been marine spatial planning (MSP). In fact, MSP seeks not so much to *understand* as to *govern* marine space, as MSP is practiced by planners seeking to address problems such as rising pressures on resources, potential user conflicts and resulting environmental damage, connected to major global challenges. It does this by introducing to the sea processes of spatial planning that have long been established on land; this is usually undertaken by government bodies with planning or marine responsibilities working closely with relevant stakeholders.

The rise of MSP raises questions about the kind of spatiality that it is bringing to bear. To what extent is 'MSP spatiality' being shaped by the spatialities of terrestrial planning, for instance? And how much adaptation is taking place in response to the characteristics of the marine environment and maritime activities? What influence are other disciplines and professions engaged in this exercise having on the ways that MSP conceives of and handles marine space? This chapter explores these questions, firstly, by presenting an overview of the emergence of MSP, its international uptake and broad principles and practices. Secondly, a number of MSP processes are described, particularly with reference to their underlying spatiality, which range from physically deterministic to conditional and exploratory understandings. Thirdly, the potential for MSP to demonstrate greater adaptation to marine characteristics is discussed.

### The emergence of marine spatial planning

Marine spatial planning (also known as maritime spatial planning, ocean planning and similar terms) is a relatively new approach to marine management. It is intended to help coastal nations manage more sustainably their internal and territorial waters and, in many cases, their extensive Exclusive Economic Zones (EEZ) and continental shelf areas (Ehler et al., 2019). As its name suggests, the idea is that the practice of terrestrial, or land-use, planning should be extended to the sea. This is based on the assumption that the uses of the seas are increasing to the point that they are coming into conflict with each other and leading to further environmental damage. Moreover, MSP is driven by a recognition that there is insufficient coordination of sea uses by



DOI: 10.4324/9781315111643-12

government. Proponents of MSP argue that these problems can be addressed by a system of planning that guides the arrangement of activities and introduces better inter-sectoral regulation. MSP thus represents a 'spatial turn' in marine management that could optimise the sustainable use of the seas (Douvere, 2008; Gilliland and Laffoley, 2008).

The idea and practice of MSP have spread internationally over recent years. Its origins lie in a zoning exercise for the Great Barrier Reef Marine Park, Australia, in the early 1980s (Day, 2002). China also played an innovating role, implementing a system of marine functional zoning in some Chinese waters from the late 1980s (Fang et al., 2011). Other early experience was gained in North America, with environmentally led initiatives. However, most progress has been made since the early 2000s, largely driven by an MSP programme of UNESCO's Intergovernmental Oceanographic Commission and other scientific and international policy initiatives. Take-up has been greatest in Europe, with individual nations and the European Union promoting the concept and setting up mechanisms for implementing MSP (CEC, 2008; Douvere and Ehler, 2009). This led to the adoption in 2014 of the EU's 'Maritime Spatial Planning Directive', which required all coastal Member States to prepare cross-sectoral maritime spatial plans by 2021 (EPC, 2014). Many European nations now have official plans in place or in preparation, focusing on their key maritime activities, including both traditional uses such as shipping and fishing, and new uses, such as offshore renewable energy and aquaculture. In the USA, some states have made plans for their coastal waters, with some federal support (Bates, 2017). MSP is also being rolled out in other parts of the world, including in the Global South, although these initiatives tend to be at an earlier stage of development; they are generally focused on environmental concerns. The EU and UNESCO are now collaborating in a global MSP initiative (UNESCO, no date).

As part of this international momentum, broad, guiding principles for MSP have gained consensus. These include such things as: taking an ecosystem-based approach, effective stakeholder engagement, cross-border cooperation, using good quality data and adaptive management (Kidd et al., 2011; Long et al., 2015). MSP processes have also been developed, including one recommended by UNESCO (Ehler and Douvere, 2009). However, despite some efforts at standardisation, MSP is being carried out in a wide variety of ways, reflecting different national and sub-national contexts. At sea, just as on land, planning is subject to different planning traditions and legal and administrative frameworks. Practice is also shaped by the particular characteristics of the environment and human activities of the area in question, and by the social and political priorities of the day. Moreover, not all MSP processes are officially recognised; many, especially in a start-up phase, or in less developed countries, are voluntary initiatives, sometimes funded by research bodies or NGOs. Increasingly, however, MSP is in the hands of authorities producing plans with statutory weight. The variety of MSP practices and outputs is now coming to light through comparative studies of MSP processes around the world (Blau and Green, 2015; Rodriguez, 2017).

Nonetheless, there are common features in MSP processes (Foley et al., 2010). Typically, there are preparatory steps, such as deciding on the geographical range of the plan and defining its overall objectives. Following this, spatial data is collected, relating to existing conditions and activities; this includes information about many aspects of the natural environment and the main maritime activities, according to the availability of information. This data is generally managed in a geodatabase and exported to a geographic information system (GIS), which provides the key visual resource for developing the plan. This may be made publicly available via a web portal (Campbell et al., 2020). This visualisation leads to an analysis of key issues and pressures, including possible areas of conflict, and an exploration of preferred future spatial arrangements. In some MSP systems, this culminates in formal allocation of discrete areas for specified uses, possibly in a comprehensive zoning exercise. In other systems, the emphasis is more upon setting criteria for future use, taking into account what is now known about the area. Other measures may also be set out in the plan to control activities, such as environmental management measures. Most MSP processes involve a greater or lesser degree of public and stakeholder engagement, and require political approval before a plan comes into force.

## The spatialities of MSP in practice

MSP discourse and practice emphasise the spatial distribution of maritime activities and their inter-relation with the marine environment, not least through the prominence of GIS mapping in MSP processes and the resources dedicated to this (see also Lehman, this volume). This invites reflection on the understandings of spatiality that are at work. At first sight, a fairly conventional, physically absolutist understanding of space predominates (Jay, 2012). This is expressed in references to 'the allocation of space'. A mosaic model is presented, whereby the sea is subdivided into bounded units for different uses, such as conservation, wind energy and shipping (Douvere, 2008). In MSP undertaken from this perspective, the language is particularly stark in the arguments for ocean zoning, with comparisons drawn with the seemingly fixed boundaries and zones of terrestrial planning. This passage from Doherty's *Ocean Zoning* exemplifies this perspective:

> Ocean zoning is similar to land-use zoning wherein specific areas are designated for particular uses. On land, for example, we separate residential and commercial areas and separate incompatible uses, so that playgrounds are not located next to city dumps… Similarly, we wouldn't want dragging to occur in areas with sensitive benthic habitats.
>
> *Doherty, 2003: 2*

GIS comes to the fore here, as the tool that can enable a satisfactory spatial geometry to be drawn (St. Martin and Hall-Arber, 2008). A range of related, and sometimes elaborate, decision-support tools are also promoted in order to optimise the spatial design needed to achieve selected objectives (Stelzenmüller et al., 2013).

This logic is not unreasonable. Terrestrial planning does have its origins in the physical space of traditional architecture, extended into the public realm and writ city-large. And some planning systems have resulted in comprehensive zoning, whereby clearly defined areas are exclusively designated for certain types of development, with strict criteria regarding things like the size and spacing of buildings (Stach, 1987). Interestingly, the zoning exercise for the Great Barrier Reef Marine Park (GBRMP) referred to above emulates one classic example of urban zoning, that for the village of Euclid, Ohio in the USA. In the 1920s, the municipality established a zoning system restricting the type of development that could take place within its territory, in an attempt to protect itself from the industrial expansion of nearby Cleveland. A series of zones was defined, with varying degrees of restriction. This can be described as a pyramidical system, with the type of zone with the greatest number of restrictions at the top, slightly fewer restrictions in the type below, and so on. Euclid's system rose to fame by surviving a legal challenge, thus establishing the right for other municipalities to follow suit (Cullingworth and Caves, 2003). There are notable parallels with the GBRMP. Firstly, zoning in the GBRMP was also a defence against an external threat; here, it was the spectre of oil exploration and limestone mining moving in, as well as growing pressures from tourism and fishing (Kenchington and Day, 2011). Likewise, zoning was promoted in an effort to preserve environmental quality (although natural, in the case of the Great Barrier Reef, rather than residential, in the case of

| ACTIVITIES (see *Zoning Plan* for full details) | General Use Zone | Habitat Protection Zone | Conservation Park Zone | Buffer Zone | National Park Zone | Preservation Zone |
|---|---|---|---|---|---|---|
| Boating, diving | Yes | Yes | Yes | Yes | Yes | No |
| Collecting (e.g. bêche-de-mer, shells, coral, aquarium fish) | Permit | Permit | No | No | No | No |
| Line fishing | Yes | Yes | Yes | No | No | No |
| Mesh netting | Yes | Yes | No | No | No | No |
| Bait netting | Yes | Yes | Yes | Yes | No | No |
| Trolling (for pelagic species) | Yes | Yes | Yes | Yes | No | No |
| Spearfishing | Yes | Yes | No | No | No | No |
| Pole and line tuna fishing | Permit | Permit | No | No | No | No |
| Trawling | Yes | No | No | No | No | No |
| Traditional fishing and collecting | Yes | Yes | Yes | Yes | Yes | No |
| Traditional hunting | Permit | Permit | Permit | Permit | Permit | No |
| Cruise ships | Yes | Permit | Permit | Permit | Permit | No |
| General shipping (other than shipping area) | Yes | No | No | No | No | No |
| Crayfishing | Yes | Yes | No | No | No | No |
| Mariculture | Permit | Permit | No | No | No | No |

Islands: All Commonwealth owned islands in the Far Northern Section are zoned "Commonwealth Island Zone". See Zoning Plan for full details.

Zoning: This map does not purport to show zoning for areas outside the Great Barrier Reef Marine Park Far Northern Section.

**Emergencies: Access to all zones is allowed in emergencies.**

*Figure 9.1* Former zoning matrix for the Great Barrier Reef Marine Park.
*Source*: Day (2002: 142).

Euclid). Secondly, as in Euclid, a pyramidical system was introduced for the GBRMP; different types of fishing and shipping were increasingly prohibited as one moved up the sequence of six types of colour-coded zone, from 'general use zone' to 'preservation zone' (see Figure 9.1) (Australian Government, no date; Day, 2002). Thirdly, the GBRMP has also achieved a landmark status, by pioneering this form of control at sea; like Euclid, the GBRMP now holds a hallowed place in the history of planning, or at least that of MSP.

In some of the more recent examples of MSP, conservation-oriented zoning still predominates. For example, a series of (non-statutory) plans produced for Canada's North Pacific coast includes maps in which the whole of the plan area is divided into coloured zones (MaPP, no date). There are General Management, Special Management and Protection Management Zones, with increasing emphasis on conservation through the range of zones. The latter are further divided into low, medium and high levels of protection, following internationally recognised conservation categories. However, overlaying this contouring of environmental importance are areas of economic potential; the Special Management Zones are so designated because of their potential for activities such as shellfish aquaculture, marine renewable energy and tourism. Whereas the GBRMP zoning system permits certain uses in some locations, the North Pacific plans actively promote them in selected areas. And overlaying the whole system are two high-level discourses. Firstly, an ecosystem-based framework is repeatedly referred to, such that all uses are conditional on good environmental management. Secondly, partnership between the provincial government (British Colombia) and First Nations is stressed, recognising First Nations' values and territorial rights in the region (Jones et al., 2010). The plans set out detailed proposals for economic opportunities, such as for community-based fishing and marine renewable energy, with constant regard to these principles. The spatiality of the plans is expressed not just in mapping, but also in finely-worded reasoning that is not simply related to the zones; extended sections of text introduce more expansive understandings of spatiality, with notions of environment and society, and all their political underpinning, infusing the GIS-dominated representation of knowledge and connecting explicitly to wider rationales. This is generally true of MSP as a whole. Even the GBRMP system extends beyond strict zoning, by including wider management initiatives, such as public education and engagement, best practice codes and industry partnerships (Kenchington and Day, 2011).

In other examples, the concept of zoning has been taken up, but in a more partial and selective way. This can be seen, for instance, in the plans for Germany's federally governed EEZ in the North and Baltic Seas. These are amongst the earliest national European plans to be completed (BSH, no date). Here, clearly defined areas are designated for just a few key uses: shipping, offshore wind energy, cables and pipelines, and research (see Figure 9.2). Zoning does not cover the entire EEZ, but just those expanses considered necessary to protect or facilitate these activities. In fact, the designated areas fall into two categories: priority areas, in which the specified use has priority over other uses, and a lesser class of reservation areas, in which special consideration is given to the specified use. In this MSP system, the act of mapping, along with government approval, grants these designations legal status (the document as a whole is a legal ordinance). In addition to these special areas, other areas are shown, for nature conservation, mineral and aggregate extraction and military training, but these are by way of information only; they are not a product of the plans, but were already established through other official processes. There is a great deal of overlap between these various areas; for instance, some shipping areas are superimposed on conservation areas. However, there are also segments of the EEZ that are completely void. This is very different to the zoning examples described above, in which there is a complete mosaic, with no overlapping and no gaps.

The German EEZ plans set out a legal spatial hierarchy, in which uses that are reckoned to need the plans' attention gain legal protection from any other uses that are deemed incompatible. Notably, these are mostly resource-centred activities, exploiting and importing the sea's energy resources and transporting goods to and from Germany's ports. Moreover, these uses, especially shipping, are granted vast portions of space. These plans are far removed from the conservation-led plans of the previous two examples; they are driven by other political and economic priorities, arguably reflecting the pre-existing shipping and offshore renewable energy responsibilities of the authority charged with producing the plans. Other uses have no legal status accorded by the plans, but are shown in the background on the maps, simply by way of information (and this was only after lobbying by their representatives during the plan-making process). Moreover, one of the longest-standing uses, fishing, is notably absent from the maps, in part because fishers refused to identify the areas of the EEZ that they considered to be most important. The absence of fisheries on the maps may also reflect the declining economic significance of fisheries. In short, the German EEZ plans present a very different spatiality to that of the zoning ideal; national policy priorities for certain sectors find favour and are given choice

*Figure 9.2* Map for the North Sea section of the German EEZ.
*Source:* Courtesy of Bundesamt für Seeschiffahrt und Hydrographie (BSH) (Federal Maritime and Hydrographic Agency of Germany) (no date).

areas, with no attempt to comprehensively allocate all of the plans' territory to the range of competing interests (Jay et al., 2012).

Even further away from the ideal of comprehensive zoning are those MSP processes that are policy-centred, intended simply to guide the development of future activities without prescribing exact locations. This approach is illustrated by the United Kingdom's marine spatial plans. Although these are quite diverse, reflecting the devolution of MSP to sub-national authorities, they have a broadly common approach (Defra, 2009). They do not contain maps indicating strict allocation of uses, but instead establish a decision-making framework that is intended to help proponents to bring forward acceptable proposals. The plans contain extended text setting out development criteria, expressed in policy statements with supporting justification and information. They do include maps that provide information about existing conditions and activities, and possibly indicate broad areas where certain uses may be appropriate, but this is far from the designation of specific areas for particular uses. For example, the East plans, which cover part of the North Sea, are like the German EEZ plans in that they support the development of offshore wind energy. However, they do so simply through statements such as: "Policy WIND2: Proposals for Offshore Wind Farms inside Round 3 zones, including relevant supporting projects and infrastructure, should be supported" (HM Government, 2014: 121). The 'Round 3 zones' referred to here are large, strategic areas considered suitable for the next major expansion of the industry (drawn up through an exercise separate to that of the MSP process itself).

This reflects the UK approach to planning on land, in which plans do not define the exact details of development but instead set out preferred terms of development. This is often characterised as a 'discretionary' system, in which planners have far more freedom when considering development proposals than in more 'regulatory' systems, where plans are more decisive and there is less room for manoeuvre (Booth, 1999). This distinction is being played out at sea as well as on land. The German EEZ plans, following national tradition (Booth, 1996), clearly belong in the regulatory camp, such that a developer who proposes a wind farm in one of the offshore wind priority areas can be confident of success; the space already bears the legal stamp for this use. But a developer proposing a wind farm in the territory of the UK's East plans will need to negotiate more fully with authorities to arrive at a location and specifications within a broadly acceptable area; here, marine space remains more open to exploration and deliberation.

The variety of approach expressed in these examples from Australia, Canada, Germany and the UK illustrates the very different forms of expression that MSP is finding as it is being put into practice, despite the rather monolithic model that was presented in the founding arguments for MSP. This calls into question the assumption that the end point of an MSP process is the allocation of space. This is no more universally the case at sea than on land; diverse traditions of planning, and their associated spatialities, are carried offshore too (Kidd and Ellis, 2012). The physically deterministic model of space only partially holds sway, in MSP as in wider planning practice. The UK presents a particularly striking example of a different approach. In this MSP system, it is the text, rather than maps, that plays the main role in shaping possible spatial outcomes. Moreover, the text does not describe in detail intended outcomes, but provides material to work with when it comes to developing and implementing specific initiatives. The plans therefore present a more provisional and conditional spatiality than that expressed in designated areas or zones. The spatial reality is yet to be determined, and the plans are yet to play their role in helping to determine that reality (Beauregard, 2015). Conditionality is taken even further in this system by the allowance that the plan is unlikely to be the only source upon which decisions are based; other 'material considerations' may come into play, and may even override the carefully thought-out position of the plan (Defra, 2007: 4.86).

This conditionality is in fact an underlying feature of MSP in general, though often in a less obvious manner. Even when MSP processes result in maps with clear allocation of marine uses, many of the mapped activities still await realisation; developers still need to come forward with their investment decisions, and conservation measures still need to be implemented. Maps indicate a desired reality, and the production of maps may help to bring about this reality, but much will still depend on actors and mechanisms beyond the control of the MSP process (see also Lehman, this volume). Moreover, this will be an approximate and continuing process, in which spatial realities may gradually converge with the mapped intentions, or may take a different turn and disrupt the plan's intentions. Maps in MSP therefore tend to represent a desired end, the achievement of which is far from certain, and so run the risk of presenting an illusory spatiality. It is perhaps a strength of textual, policy-centred approaches to MSP that they avoid the over-confidence that may come with drawing polygons for different uses. The sea and all its unknowns and dynamisms does not lend itself to such human certainty; a more openly provisional approach to planning may match more closely the characteristics of the sea and our tentative relationship with it (Boelens and de Roo, 2015).

## Edging towards new approaches

It is a fitting coincidence that the village of Euclid, referred to above, was named after the classical Greek mathematician who is associated with the geometric spatiality of which zoning is an example *par excellence*. Euclidean space is now frequently critiqued as an overly limiting material with which to work in the spatial disciplines and professions. It presents a physical, 'container' view of space that fails to allow, on the one hand, for social dimensions in the production of space, and on the other hand, for relational dynamics that transcend juxtaposed units (Davoudi and Strange, 2009). Indeed, the practice of urban zoning has received repeated criticism for its simplistic approach, leading to monotonous, exclusionary neighbourhoods and failing to address wider regional issues (Haar and Kayden, 1989; Jacobs, 1961). Moreover, it failed to protect Euclid in the way intended, as the village was eventually overrun by the expansion of Cleveland. The Great Barrier Reef Marine Park mirrors this experience too; zoning has proved powerless in the face of large-scale forces such as deteriorating water quality (partly due to coastal development), and increasing water temperature and acidification, which have wreaked havoc on the coral ecosystem (Hassan and Alam, 2019).

However, as shown above, MSP has not confined itself to the narrowness of zoning, nor even to less rigorous variants of spatial allocation. More generalised policy has also taken central place, sometimes alongside a spatial allocation system of one type or another. This is being expressed through background information, planning criteria, formal policy statements, proposed management measures and so on. Most marine spatial plans are in fact dominated by text, not maps (though the maps may grab the most attention). Typically, the text is the result of lengthy fact-finding, consultation, cross-sectoral and wider policy framework considerations, alignment with objectives and development of management options. Through the text, nuanced arguments are made in which different, frequently competing, interests are balanced and judgements made about their relative importance. Actual spatial solutions are often deferred to an 'implementation' phase of planning, when decisions will be reached on the basis of a plan, but will also take into account continuing accumulation of knowledge, strengthening or changing policy agendas and emerging opportunities. But even then, planning never reaches the point of completion; progress is always incremental, decisions are open to reconsideration and revision, players are subject to shifting priorities, and new initiatives are always on the horizon.

Arguably, understanding and going about planning in this way is more attuned to the uncertain trajectories of human interaction with the sea than the pretended certainties of zoning (Jay, 2013; Retzlaff and LeBleu, 2018). And even though this approach has much in common with terrestrial planning, the marine realm calls for greater contingency of this kind. Indeed, there is plenty of recognition in the MSP literature that the sea is not the land, and that planning needs to adapt to its distinctiveness, especially its less controllable characteristics (Jay, 2018).

The first and most obvious difference between planning at sea and at land is the sea's physical and biological character, including its three-dimensional nature, its complex temporal variations and fluid materiality. These characteristics are reflected in the range of activities that exploit different parts of the water column, the seabed and atmosphere, often with seasonal and tidal variations, and they create a complex spatial and temporal pattern of human use and interaction with the ocean's surroundings. MSP generally recognises this complexity, but struggles to accommodate it. For instance, the static, two-dimensional representations of sea use, as offered by GIS, and which are part and parcel of zoning, have serious shortcomings here. To date, there has been little uptake in MSP practice of more mobile and three-dimensional graphical representations of sea use, or of models for exploring more dynamically interacting arrangements of activities. One of the obstacles to this is the tradition, often buttressed by regulatory requirements, of producing a paper document that is 'the plan'. There have been calls for more adaptable representations, such as vertical zoning, but there is little evidence of take-up within MSP processes. At present, it is the text of plans that offers more opportunity to grapple with these complexities. So, for example, although the German maps present two-dimensional representations of the German EEZ, the accompanying text refers to the interrelations of seabed pipelines and cables and shipping lanes in any one (vertical) place.

Secondly, and linked to these natural characteristics, is the much weaker human 'hold' on the sea. Compared to the land, the sea has a short history of state control, and what jurisdiction has been gained is limited, receding in effect from the coastline. Similarly, there are weak forms of ownership of offshore territory and resources, especially by private interests. There are also greater difficulties when it comes to marking out definitive areas, monitoring human activities and enforcing rules about those activities. This changes the context of planning radically, as planning on land is typically implemented in an environment of clearly delineated units of land and strong frameworks of land and property ownership, development rights and policing of activities. Although the marine environment is seeing a trend in which governance is being strengthened, and resource, if not territorial, rights are being allocated to certain players, this remains far from the much more settled situation on land, and MSP needs more supple ways of engaging with this administratively weak context. This supports the case for MSP being less ambitious and more indicative, with plans restricted to setting out possibilities for future use rather than trying to emulate the clear-cut parameters of land-use planning. There is a strong argument for MSP to incorporate visioning and scenario-building exercises, in which alternatives for the long-term future of the seas can be explored, relating to such things as sustainable energy production, ecosystem recovery and health, and increasing international cooperation (McGowan et al., 2019). In fact, the relative weakness of human control presents an opportunity for wider-reaching preferences to be set before options are foreclosed.

Thirdly, and connected to all these features, human presence is much less dominant at sea than on land. The sea is not well populated by people; human 'settlement' is mostly fleeting and marginal to society as a whole, and the sea remains distant and risk-laden for most people.

Vessel-based activities are mobile and transitory, and fixed installations are thinly dispersed, generally remote and temporarily occupied. This is in no way equivalent to the 'built environment' on land. This presents challenges for MSP, not least because public representation is a key feature of most planning systems. In MSP, stakeholder interests tend to be well-represented (though not equally) through industry and government and NGOs. But wider public voices are largely absent, except perhaps coastal communities expressing concerns about local issues. There is potential for more open forms of public engagement, using, for example, online participation in assembling views about possible futures for the seas and exploring what is desirable in terms of resource use, conservation and cultural dimensions of the sea. A related issue is that marine activities are relatively limited in their scope, restricted largely to functional or industrial uses such as transport and resource exploitation, and these sit uneasily alongside growing concerns for the sea's ecological well-being and the cultural assets that it holds. MSP finds itself at the heart of the growing tension between these competing sets of interest, and has the task of mediating a wide range of societal views, heightened by current global-scale concerns about climate change and pollution, and their impact upon the oceans and their role in helping to regulate natural systems (Ntona and Morgera, 2018).

## Conclusions

MSP introduces notions of spatiality to our understanding of the seas and oceans that are drawn from the wider practice of spatial planning. Initially, this has been through the rather conventional contribution of perceiving space in physically deterministic ways, expressed through zoning and other forms of spatial allocation, as represented by two-dimensional mapping. However, it has also introduced more agile spatial thinking through the more discursive tools of planning, such as policy development, by which varying spatial claims are considered and prioritised, and complex conditions for development are set out. This can be traced back to the traditions of planning as practiced on land, which are generally being inherited by MSP along the lines of varying national planning cultures and systems.

At the same time, MSP is introducing complementary ways of understanding the specific spatiality of the seas and oceans. It spurs planners to be cognisant of the differences between land and sea, highlighting the stark differences in the manner and extent of human occupancy and development and the particular ways that planning could adapt to the physical and social characteristics of the sea. But this remains at a very preliminary stage; MSP is generally aware of these challenges, but is not yet developing particularly innovative means of dealing with marine realities. To do so may mean the more radical insertion into MSP thinking and practice of spatial ontologies that attempt to embody the relationality and liveliness of the sea and human interaction with it (Boucquey et al., 2016). This may involve, for example, re-conceptualising the space-being-planned as bio-physically flexing, teeming with actants, constituted by relations, and having a tendency to forming complex interactions that are themselves continuously reforming. Moreover, MSP must find ways of sharing such dynamics in its own engagement with this space (Jay, 2018, 2020).

## References

Australian Government (no date) *Great Barrier Reef Marine Park Authority*. Available at: www.gbrmpa.gov.au/access-and-use/zoning

Bates A (2017) Revisiting approaches to marine spatial planning: Perspectives on and implications for the United States. *Agricultural and Resource Economics Review* 46(2): 206–223.

Beauregard R (2015) *Planning Matter: Acting with Things*. Chicago, IL: University of Chicago Press.

Blau J and Green L (2015) Assessing the impact of a new approach to ocean management: Evidence to date from five ocean plans. *Marine Policy* 56: 1–8.

Boelens L and de Roo G (2015) Planning of undefined becoming: First encounters of planners beyond the plan. *Planning Theory* 15(1): 42–67.

Booth P (1996) *Controlling Development: Certainty and Discretion in Europe, the USA and Hong Kong*. Abingdon: Routledge.

Booth P (1999) Discretion in planning versus zoning. In: Cullingworth B (ed) *British Planning: 50 Years of Urban and Regional Policy*. London: Athlone Press, 31–44.

Boucquey N, Fairbanks L, St. Martin K, Campbell L and McCay B (2016) The ontological politics of marine spatial planning: Assembling the ocean and shaping the capacities of 'Community' and 'Environment'. *Geoforum* 75: 1–11.

Bundesamt für Seeschiffahrt und Hydrographie (BSH) (no date) *Maritime Spatial Planning*. Available at: www.bsh.de/EN/TOPICS/Offshore/Maritime-spatial-planning/Maritime-spatial-planning_node.html

Campbell L, St. Martin K, Fairbanks K, Boucquey N and Wise S (2020) The portal is the plan: Governing US oceans in regional assemblages. *Maritime Studies* 19: 285–297.

Commission of the European Communities (CEC) (2008) *Roadmap for Maritime Spatial Planning: Achieving Common Principles in the EU*. Brussels: COM 791, CEC.

Cullingworth J and Caves R (2003) *Planning in the USA: Policies, Issues and Processes*. Abingdon: Routledge.

Davoudi S and Strange I (eds) (2009) *Conceptions of Space and Place in Strategic Spatial Planning*. Abingdon: Routledge.

Day J (2002) Zoning: Lessons from the Great Barrier Reef Marine Park. *Ocean and Coastal Management* 45(2/3): 139–156.

Department for Environment, Food and Rural Affairs (Defra) (2007) *A Sea Change: A Marine Bill White Paper*. London: Cm 7047, Defra.

Department for Environment, Food and Rural Affairs (Defra) (2009) *Implementing Marine Planning*. London: Defra.

Doherty P (2003) *Ocean Zoning: Perspectives on a New Vision for the Scotian Shelf and Gulf of Maine*. Halifax, Nova Scotia: Ecology Action Centre.

Douvere F (2008) The importance of marine spatial planning in advancing ecosystem-based sea use management. *Marine Policy* 32(5): 762–771.

Douvere F and Ehler C (2009) New perspectives on sea use management: Initial findings from European experience with marine spatial planning, *Journal of Environmental Management* 90(1): 77–88.

Ehler C and Douvere F (2009) Marine Spatial Planning: A step-by-step approach toward ecosystem-based management. Intergovernmental Oceanographic Commission and Man and the Biosphere Programme, UNESCO, Paris.

Ehler C, Zaucha J and Gee K (2019) Maritime/marine spatial planning at the interface of research and practice. In: Zaucha J and Gee K (eds) (2019) *Maritime Spatial Planning: Past, Present and Future*. Cham: Palgrave Macmillan, 1–21.

European Parliament and Council (EPC) (2014) *Directive 2014/89/EU of the European Parliament and of the Council of 23 July 2014 Establishing a Framework for Maritime Spatial Planning*, Official Journal of the European Union, L 257, 135–145.

Fang Q, Zhang R, Zhang L and Hong H (2011) Marine functional zoning in China: Experience and prospects. *Coastal Management* 39(6): 656–667.

Foley M, Halpern B, Micheli F, Armsby MH, Caldwell MR, Crain CM, Prahler E, Rohr B, Sivas D, Beck MW, Carr MH, Crowder LB, Duff JE, Hacker SD, McLeod KL, Palumbi SR, Peterson CH, Regan HM, Ruckelshaus MR, Sandifer PA and Steneck RS (2010) Guiding ecological principles for marine spatial planning. *Marine Policy* 34: 955–966.

Gilliland P and Laffoley D (2008) Key elements and steps in the process of developing ecosystem-based marine spatial planning. *Marine Policy* 32(5): 787–796.

Haar C and Kayden J (eds) (1989) *Zoning and the American Dream: Promises Still to Keep*. Chicago, IL: Planners Press.

Hassan D and Alam A (2019) Marine spatial planning and the Great Barrier Reef Marine Park Act 1975: An evaluation. *Ocean & Coastal Management* 167: 188–196.

HM Government (2014) *East Inshore and East Offshore Marine Plans*. London: Defra.

Jay S (2012) Marine space: Manoeuvring towards a relational understanding. *Journal of Environmental Policy and Planning* 14(1): 81–96.

Jay S (2013) From disunited sectors to disjointed segments? Questioning the functional zoning of the seas. *Planning Theory and Practice* 14(4): 509–525.

Jay S (2018) The shifting sea: From soft space to lively space. *Journal of Environmental Policy and Planning* 20(4): 450–467.

Jay S (2020) Measured as the water flows: The striated and smooth in marine spatial planning. *Maritime Studies* 19: 255–268.

Jay S, Klenke T, Ahlhorn F and Ritchie H (2012) Early European experience in marine spatial planning: planning the German exclusive economic zone. *European Planning Studies* 20(12): 2013–2031.

Jacobs J (1961) *The Death and Life of Great American Cities*. New York, NY: Random House.

Jones R, Rigg R and Lee L (2010) Haida marine planning: First Nations as a partner in marine conservation. *Ecology and Society* 15(1): 1–17.

Kenchington R and Day J (2011) Zoning, a fundamental cornerstone of effective Marine Spatial Planning: Lessons learnt from the Great Barrier Reef, Australia. *Journal of Coastal Conservation* 15: 271–278.

Kidd S and Ellis G (2012) From the land to sea and back again? Using terrestrial planning to understand the process of marine spatial planning. *Journal of Environmental Policy and Planning* 14(1): 49–66.

Kidd S, Plater A and Frid C (eds) (2011) *The Ecosystem Approach to Marine Planning and Management*. London: Earthscan.

Long R, Charles A and Stephenson R (2015) Key principles of marine ecosystem-based management. *Marine Policy* 57: 53–60.

Marine Plan Partnership for the North Pacific Coast (MaPP) (no date) *Marine Plan Partnership for the North Pacific Coast (MaPP)*. Available at: http://mappocean.org/

McGowan L, Jay S and Kidd S (2019) Scenario-building for Marine Spatial Planning. In: Zaucha Y and Gee K (eds) *Maritime Spatial Planning: Past, Present & Future*. Cham: Palgrave, 327–351.

Ntona M and Morgera E (2018) Connecting SDG 14 with the other sustainable development goals through marine spatial planning. *Marine Policy* 93: 214–222.

Retzlaff R and LeBleu C (2018) Marine spatial planning: Exploring the role of planning practice and research. *Journal of Planning Literature* 33(4): 466–491.

Rodriguez N (2017) A comparative analysis of holistic marine management regimes and ecosystem approach in marine spatial planning in developed countries. *Ocean & Coastal Management* 137: 185–197.

St. Martin K and Hall-Arber M (2008) The missing layer: Geo-technologies, communities, and implications for marine spatial planning. *Marine Policy* 32(5): 779–786.

Stach P (1987) Zoning: To plan or to protect? *Journal of Planning Literature* 2(4): 472–481.

Stelzenmüller V, Lee J, South A, Foden J and Rogers S (2013) Practical tools to support marine spatial planning: A review and some prototype tools. *Marine Policy* 38: 214–227.

UNESCO (no date) Balancing sustainable use and conservation through Marine Spatial Planning. Available at: http://msp.ioc-unesco.org/

# 10
# DOCKING
## Maritime ports in the making of the global economy

*Charmaine Chua*

### Introduction

As Alan Sekula (1996) notes in his photographic essay *Fish Story*, Friedrich Engels' classic *The Conditions of the Working Class in England* (1845) somewhat unexpectedly opens its investigation of the English industrial working class not from the alleyways and factories of industrial Britain, but from the maritime standpoint of the deck of a ship coming into harbour:

> I know of nothing more imposing than the view which the Thames offers during the ascent from the sea to London Bridge. The masses of buildings, the wharves on both sides... the countless ships along both shores, crowding ever closer and closer together, until, at last, only a narrow passage remains in the middle of the river, a passage through which hundreds of steamers shoot by one another; all this is so vast, so impressive, that a man cannot collect himself, but is lost in the marvel of England's greatness before he sets foot upon English soil.
>
> *Engels, 1968: 30*

For Engels, the harbour's busy movements afford at first sight an admiration and optimism about the economic life of the city, giving rise to the imagination of seaport towns as centres of "commerce, wealth and grandeur" (1968: 68). But this is quickly replaced by a sober realisation of the immiseration at the heart of these developments. It is through the spatial move from the maritime scene to the street that Engels dramatises how "capital, the direct or indirect control of the means of subsistence and production", carries out a "social warfare" on "the poor man" (1968, 69). "The sacrifices which all this has cost become apparent later", Engels writes.

> After roaming the streets of the capital a day or two... one realizes for the first time that these Londoners have been forced to sacrifice the best qualities of their human nature, to bring to pass all the marvels of civilization which crowd their city...
>
> *Engels, 1968: 68*


DOI: 10.4324/9781315111643-13

The arteries of maritime trade that bring commodities to shore and beyond them are, for Engels, simultaneously conduits of wealth and poverty, providing the basis for a crucial political economic insight: that a fundamental feature of capitalism's development is the inequality and exploitation that underlie the social relations of production.

Docking, known by many other names – to berth, to moor, to 'come alongside', to land, to anchor – refers to the process by which a vessel arrives along a port's berth, comes to a stop, and is secured to land by a rope. A seemingly innocuous process, perhaps. Yet docking is also the moment at which the fluidity and expansiveness of the open sea meets the channel, the shore, and the port terminal – and all the juridical, infrastructural, and corporate weight that comes with the hinterland of territorial space. While docking is a process that acts as an interface between land and sea, this chapter does not presuppose a strict binary between the two but rather highlights how port governance processes on land are both implicated by and in turn structure the maritime circulation of commodities. If the docks provided Engels with a narrative entry point into his analysis of nineteenth-century working-class conditions, today few scholars of political economy would begin an analysis of contemporary capitalist relations with one's arrival from the sea. The harbour rarely features in the public eye as a place central to economic and social life, and ships carry containers, bulk goods, or cars, but very few humans to shore. Even though maritime supply chains bring almost all of the commodities we own and consume to the marketplace, these spaces of circulation remain relatively invisible to large sectors of the population, hidden behind walled districts and industrial zones on the outskirts of urban life. As ports have moved out of urban centres and into spaces hidden from sight to most people, the world market has become more familiar to most people as an abstract and immaterial sphere, articulated through stock prices and data flows rather than through the concrete materialities that move the world's trade through a vast infrastructural system of ships, warehouses, and other transport infrastructure.

This chapter will demonstrate how docking – as the interface between oceanic circulation and land – is a crucial force in the making of no less than the global economy. Maritime transport moves at least 80 per cent of internationally traded cargoes, making ports not only crucial spaces of capital accumulation, but also spaces of state power that reproduce global economic circulation in communities around the waterfront and in the hinterland. Variations in the relationship between state and capital also reveal uneven dynamics and distributional tensions between competing ports. Beginning with a historical overview of ports as sites of long-distance trade and exchange, I trace how ports have played crucial roles in the social, political and economic development of cities, countries, and regions from well before the colonial era and into the present (Hoyle and Pinder, 1992; Wang and Ducruet, 2013). Today, the port has rapidly transformed from a bustling place of labour and city life into an increasingly securitised and corporate site, shaped by changing technologies of transportation and communication that revolutionised port systems in the 1960s (see also Heins, this volume). Structured by broader economic trends such as neoliberal corporatisation and the rise of logistics, the port today acts as a dynamic node in a complex web of maritime systems, and is a crucial component in facilitating commercial transactions around the world (Ng et al., 2014). Port actors have correspondingly sought to reassess the governance and operational structures of ports, seeking to facilitate efficient port services in order to gain a competitive edge over surrounding ports. In particular, ports play an essential role in the development of national economies: as commercial trade has become increasingly important for national accumulation strategies, the state–capital nexus seeks to promote the internationalisation of capital through the expansion of port functions. In this chapter, I follow these shifts in port governance, focusing on how the mixture of public and private partnerships in port governance results in the rapid expansion of large-scale port infrastructures on the one hand, and increasing corporatisation of port services on the other. Importantly, the rapid expansion of port capacities in some places and increasing competition to become a logistics hub means that the benefits of port expansion are unevenly distributed depending on ports' capacities to develop their infrastructures and services. As a result, it is important to consider the social and political economic consequences of port governance, as ports increasingly adapt their legal and infrastructural apparatuses to compete for increasingly rapid and large quantities of trade.

## A brief history of ports

As Alan Sekula writes, "ports are fulcrums of history, the levers many, and the results unpredictable" (1996: 32). Ports are first and foremost the meeting place of long-distance trade. As Karl Polanyi traces, maritime trade was in fact the true starting point of the market. Refuting the nineteenth-century orthodox myth that trade first began from an individual propensity to barter, out of which came the necessity of local markets, the division of labour, and eventually the necessity of trade, Polanyi reverses the sequence of that argument. The starting point of the market was in fact long-distance trade; the geographical location of different goods and the division of labour determined by location provided individuals the possibility of bargaining, leading to the propensity to exchange, for instance, English woollens for Portuguese wine. Rather than developing from man's essentially competitive spirit, then, the world market emerges because local markets were confined only to the goods they could produce in a given region, giving rise to the development of the port as "the meeting place of long distance trade" (Polanyi, 2001: 62). Fernand Braudel (1981) points out that long-distance oceanic travel enabled earlier forms of proto-capitalism: even in medieval times, merchants from Egypt to Japan ploughed the seas in Cairo, Aden, and the Persian Gulf ports.

The harbour that Engels imagined to contain romance and wonder – a site that inspired poetry, paintings, and imaginations of freedom (Casarino, 2002; Taussig, 2002) – has actually been a site of brutal work, impoverished conditions, and the slave trade (see, for example, Ahuja, 2006; Linebaugh and Rediker, 2013). In the nineteenth century, architects of British empire aligned a conception of merchant imperialism with the exploitation of faraway spaces not only through the domination of land – by setting up extractive institutions of settlement and plantation – but also by seizing critical gateways to the world's oceans, which was then used to control access to crucial shipping lanes (Subrahmanyam, 2006; Tracy, 1990; see also Davies, this volume).

As Ince argues, colonial networks were "central as social spaces providing the concrete conditions for imagining and experimenting with new ways of organizing social production for profit" (2014: 112). Between the fifteenth and eighteenth centuries, as Britain's overseas empire grew, and with it the national debt that funded colonial wars, the country needed a system of trucking from which long-distance markets could develop. Where carriers had to halt at fords, seaports, and riverheads, ports developed at the places of transshipment. The long-distance monopolies of trade afforded by the British empire's control of sea-borne trade routes prompted innovation in shipping technologies. As Braudel notes, despite the significance of the luxury trade to the Spice Islands and elsewhere in the initial accumulation of wealth by European trading companies, it was only the upsurge of the industrial revolution that prompted the need to exchange heavy goods in enormous proportions. As overland routes proved too costly, the economic growth of the time made large-scale investments in oceangoing transport "both 'profitable' and necessary" (Braudel, 1981: 430). It was this "struggle against distance", "guaranteed by de facto or de jure monopolies", that finally "made the luxury of large tonnage ships possible" (Braudel, 1981: 430; 423). The port's evolution as a site for securing the conditions of global circulation is thus deeply rooted in imperial history. As contemporary supply chains intensify processes of maritime commercial exchange, the transformation of modern ports echoes these histories, underscoring that the separation Engels sought between an open sea and an exploitative urban centre is instead a circulating space of long-distance exploitation between metropole and colony, and core, periphery, and semi-periphery.

Today, the port continues to serve as the bridge between oceans and land, but it has transformed under new demands. With the worldwide adoption of the 'intermodal' shipping container in the 1970s – a steel box that transports freight between multiple modes of transportation (from ships to rail and truck) – port actors have reassessed their operational and governance structures, while integrating themselves more intricately into global supply chains, and have become crucial in facilitating transactions around the world (see Danyluk, 2019; Heins, this volume; Nam and Song, 2011; Ng and Liu, 2014). As shipping containerisation created a global system of regularised compatibility, bringing previously disaggregated sectors of sea and land transport into an integrated network, it also reconfigured the cultural and regional geographies of port cities.

The transformation of the London harbour that so enlivened Engel's narrative provides an example of these shifts. In 1961, the British Cabinet commissioned an inquiry into the efficiency of British ports, which found that port operators should prepare their facilities for containerised vessels in order to keep the port commercially viable (Great Britain House of Commons, 1962). The Rochdale Report (1962) found that Britain would risk losing crucial container traffic to other continental ports such as Rotterdam if it did not make adaptations to port infrastructure. Following these findings, the Port of London Authority (PLA) decided that to remain competitive, it would have to move its main operations out of the Pool of London to Tilbury, on the Essex Coast (Martin, 2012: 147). The PLA invested heavily in deep water berthing at the Tilbury Docks, constructing seven container berths by 1967. Within a year, Tilbury was handling seven-eighths of London's entire tonnage, leaving the London harbour virtually empty of cargo ships in a short time. Were Engels to enter London through the Docklands today, he would witness an entirely transformed urban geography: under subsequent 'regeneration' that occurred under the Thatcher government in the 1980s (Martin, 2012; Smith, 1989), the area has become a major financial centre, and the docks serve no commercial purpose except as a tourist attraction.

The containerisation of goods revolutionised the maritime supply chain by providing an intermodal technology that has allowed shippers to efficiently combine shipping with storage and logistics at the port, warehousing, and transportation to other modes. In the process, it has necessitated a reconfiguration of not only docking facilities, but also the other infrastructural linkages that produce an integrated transport system across different modes of sea, land, and rail. Containerisation has drastically reduced the amount of labour required to handle cargo by almost 90 per cent, leading not only to significant cost savings for shipping companies, but also providing an economically viable way to offshore manufacturing to underdeveloped nations where labour costs were cheaper.

As such, Deborah Cowen argues:

> without the rapid and reliable movement of stuff through space – from factories in China to US big box stores, for instance – cheap labour in the global South cannot be 'efficiently' exploited, and globalized production systems become as inefficient economically as they are environmentally.
>
> *Cowen, 2010: 601*

While literatures in economic and transportation geography often address how containerisation and logistics are positive technological breakthroughs for ports and shipping companies (Hayuth, 1987; Hilling and Hoyle, 1984; Olivier and Slack, 2006), much less remarked upon are the social, political, and environmental consequences of logistics innovations on local populations and the politics of port governance. As an emerging body of literature critically studying the effects of logistics suggests, however, these transformations in the supply chain are also formations of power, producing forms of daily life and violence throughout the maritime supply chain (Campling and Colás, 2021; Chua et al., 2018; Coe, 2020; Cowen, 2014; Khalili, 2020; Stenmanns, 2019). As such, it is important to understand the broad trends of internationalisation that have shifted the operations of ports, as well as the effects these transformations have prompted in the dynamics of accumulation in local, national, and international settings.

## The politics of port governance

Since the logistics revolution of the 1960s and 70s, scholars have argued that the advent of just-in-time delivery has made transportation an integrated component of production systems (Coe, 2020; Mezzadra and Neilson, 2019; Sheppard, 2016; Wang and Olivier, 2006). When logistics operations emerged in the 1960s and 70s, they took Fordist principles as the basis of efforts to achieve economies of scale, efficiency gains, and standardisation. New developments in just-in-time delivery, however, have shifted the grounds on which shippers compete, from economies of scale to "economies of scope", based on the integrated services companies can provide such as logistics tracking, IT-based supply chain design, and other value-added services (Ng et al., 2014). As logistics providers have integrated and consolidated their operations, they have been able to take control of larger segments of the supply chain, where companies such as Kuehne+Nagel, DB Schenker and Nippon Express have evolved from road haulage or freight forwarding companies to full logistics providers that coordinate various elements of the supply chain from production and purchasing to transportation, warehousing, and supply chain design. This growth of logistics companies means that, as Theo Notteboom (2006: 46) argues, "ports are increasingly competing, not as individual places that handle ships but within transport chains or supply chains".

In this way, ports insert themselves into global intermodal networks defined by the imperatives of global capital accumulation, rather than by local or public interests. As Slack (1993) notes, "ports are becoming pawns in a game of commerce that is global in scale, and on a board where the major players are private corporations whose interests rarely coincide with the local concerns of the port administrations" (in Olivier and Slack, 2006: 1414). Logistics thus changes the raison d'être of seaports. Although seaports were historically owned by the public sector, they are now not only subsumed by the dictates of the world market, but also by corporate agendas that promote the commercialisation of the docks, the internationalisation of capital, and the devolution of public port authorities to corporate entities (Brooks and Cullinane, 2006; Parola et al., 2013). Studies in maritime transportation have shown that ports have become relatively substitutable within a given region, where container shippers are "port blind, leaving the choice of port and touring to the carrier they have chosen" (Brooks, in Olivier and Slack, 2006: 1414). As Notteboom (2009) has shown, the ports of Antwerp and Rotterdam, similar in scale and located along the Rhine Delta, are increasingly acting as substitutes for each other, where carriers centralise their vessel calls in a single port of call and rely on a hinterland distribution system once they have docked.

Whereas shippers used to select ports on the basis of their strategic geographical location (as was the case in the establishment of the ports of Singapore, Malta, and other colonial entrepôts at key points in colonial trade routes), ports are selected today on the basis of the entire network in which they are embedded (Danyluk, 2019; Khalili, 2020). Ports are chosen not in terms of where they are located in absolute distance from the warehouse or customer, but where the sum total cost of sea, port, and hinterland transportation will be cheapest and most efficient. In order to accommodate the rapid increases in the size of mega-vessels, ports have had to move farther away from city centres and into sites that can be dredged, terraformed, and shaped to accommodate deep water berths and adequate landside access for ever-growing mega-ships to dock.

For example, the busiest port in the world by cargo tonnage is the Port of Ningbo-Zhoushan, which handled 1.17 billion tons of cargo in 2020.[1] Ningbo-Zhoushan is 400 kilometres from the major commercial and retail destination of Shanghai, a transport distance that would have been prohibitive for just-in-time delivery two decades ago. As ships grow larger to capture economies of scale, however, considerations of geographical location become subordinate to assessments of the port's docking capacity, where the chief consideration is that ports have enough space and equipment to unload and berth megaships in a timely manner, combined with its ability to transport goods quickly to the city centre. To overcome the spatial distance between Shanghai and Ningbo, which possessed these spatial qualities, the Chinese state built a $1.5 billion, 448 m (1470 feet) long trans-oceanic bridge, connecting the two cities across a bay and cutting the travel time between them from four to two hours. For a seaport to attract mega-carriers and large logistics companies, it has to position itself not only as a convenient geographical location, but as an intermodal distribution service centre that is connected to extensive transport and communications networks. Ports in this way must be understood as only one node within a wider supply chain, and their ability to compete is determined by their networked efficiency, and ability to create connections and synergies with other transportation nodes in the wide supply and value chains of which they are a part (Dua, 2018; Notteboom, 2006; Robinson, 2002).

The increasing influence of transnational capital in port operations has meant that state functions in the port have receded while private entities have grown in influence. Worldwide, the dominant model of port governance is the 'landlord port', in which the public sector owns the land and infrastructure, and leases these to private operators as a concession, with equipment and operations managed by the private sector. The World Bank's taxonomy of ports consists of four administrative forms, distinguished by their levels of public and private ownership and operation: at one end of the spectrum, service ports, which are predominantly publicly run and owned, and whose numbers are declining; tool ports, in which the port authority owns and develops port infrastructure, equipment, and land; landlord ports; and fully privatised ports (World Bank, 2007: 81–83). The landlord port model has been promoted heavily by the World Bank (2007) as an optimal model, given that it strikes a balance between public and private interests, squaring with the World Bank's long-standing promotion of neoliberal economic policy and market-friendly intervention.

The worldwide support for this public–private partnership model can be understood both historically and ideologically. The landlord port developed in the 1980s in the context of a reorganisation of public power around neoliberal reforms and the marketisation of public goods. In the context of port reforms, the shift to corporatisation was also heavily shaped by dockworker militancy and organisation around the world, but especially in Europe and North America. The new transport geographies enabled by containerisation and the rise of global supply chains aided the political aims of the neoliberal state in pushing back against the post-war gains made by organised labour, resulting in the automation of the docks that restructured labour contracts and reduced union power, allowing the Thatcherite and Reaganite suppression of labour struggles.[2] Debrie et al. (2013) argue that port reforms developed not only because of institutional convergences, but because of a merging ideological consensus. The privatisation of port services, the sharing of responsibility between private operators and public regulators, and the corporatisation of port authorities to "encourage a more entrepreneurial culture" were all institutional reforms that developed in complementary step with a growing ideological agreement about the purported benefits of privatisation and neoliberal reform (Debrie et al., 2013: 58). The landlord port model is thus far from an innocuous mode of governance, but rather enabled a series of ideological and political shifts that created a system "necessarily hostile to all forms of social solidarity that put restraints on capital accumulation" (Harvey, 2007: 73).

In line with growing needs for private investment, public port authorities have thus turned to public–private partnership (PPP) arrangements in order to overcome budgetary constraints (Hammami et al., 2006), supply operational expertise in the management and administration of equipment (Parola et al., 2013), and attract foreign direct investment to supplement public funding of port infrastructure (Mu et al., 2011). Yet, this does not explain the purpose of maintaining a public stake in logistics and port infrastructure. While the reasons for maintaining a public role vary across contexts, Fawcett has argued that while, in general, "the port is responsible for facilitating economic development via private enterprise", it is "also often a public agency responsible for its actions as it manages the port in the public interest" (2007: 217). Port managers and authorities often seek to balance neoliberal efforts to promote private sector activities with public interests in community economic development, urban rejuvenation, and the reduction of environmental pollution, goals that are not always in sync with the encroachment of the private sector. As logistics systems increase the spatial distance of trade and delivery, the work attached to commodity trade also moves away from local communities and towards distant locations, producing a contradiction between private profit accumulation and local community interests.

How the consequences of PPP arrangements unevenly play out in different contexts depends in large part on state capacity and the extent of state capture that relate to the financial structure of port funding. The most important determinant for the persistence of the public sector has to do with the massive legal and physical infrastructure needed to move containers and goods – from container cranes, rail infrastructure, and deep water berths to the jurisdictional zoning of special economic zones required for foreign direct investment – that require financial costs beyond the realm of feasibility for most private terminal and shipping companies to amortise over the duration of their leases at the port. As James Fawcett explains, in the context of the United States and other developed countries in the global North, the public financing capability of port authorities gives them unique access to capital markets and allows them in some cases to float bond issues tax-free to the buyer (2007: 218). "One of the more compelling reasons for public ownership of these vital facilities", then, "is in the ports' capability to borrow sufficient funds for their own capital expansion" (Fawcett, 2007: 218). In contrast, in developing nations in the global South, public authorities lack the capacity for public financing that might include grants, interest rate and tax subsidies, loan guarantees and insurance premiums to manage the risks of port investments. As such, governments in developing countries are commonly incentivised to gain private access to capital by opening seaports to corporate investment, with few conditionalities on private investment.

The differential tensions between the mixed outcomes of public–private partnerships underscores the fundamentally uneven dynamics of capitalist development. In expanding the search for relative surplus value, capital is "driven to convert" spaces across the globe that are seen as "external" and "relatively undeveloped" into places of production and accumulation. Through such processes of primitive accumulation, even "external" space is internalised and produced "*within and as part of the global geography of capitalism*" (Smith, 2008: 187). This global process of integration is inherently uneven. As inherited disparities in levels and conditions of development produce regions with different determinations of the value of labour power, the cost of materials, and other elements of production, a "powerful centripetal force is felt as uneven geographical investments in transport systems feed further uneven geographical developments" (Harvey, 2005: 101). In this way, the hierarchical networks that transport systems help to circulate ensure that capitalist development sustains itself "not through absolute expansion in a given space but through the internal differentiation of global space, that is, through the production of differentiated spaces" (Smith, 2008: 120).

Thus, pressure from certain states, firms, and social forces directed toward corporatising market forces inevitably deliver unequal dividends across the world. Ports in many nations in the global South have to make capital intensive adaptations to their docking technology to gain access to the economic opportunities containerised traffic provides, but lack of access to capital often means that developing countries have to open their ports to corporatisation and the dictates of neoliberal capitalism. One example of these uneven dynamics can be seen in Buenaventura, a port city in Colombia. Austin Zeiderman (2016) traces how both national and local governments have envisioned the future of the city of Buenaventura as a "world-class" port, tied to projections of booming trade expansion with East Asia. In order to attract carriers and port terminal operators to Buenaventura, public and private capital from the Colombian government and investors from Europe, Asia and the Middle East have funnelled money into large-scale infrastructure projects, constructing a set of connected ports, container terminals, cargo trains, dredging projects, a logistics operations centre, and a highway into the hinterland.

This economic development plan has, however, privatised the port through concessions, abolished labour unions in an effort to lower costs, and favoured relocation programs that displace the city's inhabitants. The port only generates a limited amount of employment in Buenaventura, with many workers brought in from foreign locations, while the city is Colombia's poorest and least developed, with a 30 per cent unemployment rate and 80 per cent of its inhabitants living below the poverty line (Zeiderman, 2016: 810). Such corporatised strategies of port expansion are not limited to Colombia; multinational companies have also applied to establish operations in Cuba's new Mariel Special Development Zone, a free-trade zone being established next to the Mariel container port (Danyluk, 2017), while in 2016, Chinese-owned COSCO Group Ltd acquired 67 per cent of the shares of Piraeus Port Authority near Athens, and has been granted the management of all port services there through a concession agreement that expires in 2052 (Hatzopoulos and Kambouri, 2018: 155–174). These examples of the increasing corporatisation of ports and the corresponding increase in the precarity of life for their workers and city's inhabitants are thus reflective of the late liberal formations of power that Zeiderman argues are "precarious forms of political life", which ensconce populations within neoliberal market interventions, and that force them to "navigate formations of liberal governance and their logics of vulnerability and protection" (2016: 182).

As these examples illustrate, the changing relationship between public and private entities at the port heavily shape outcomes of port development. State reforms have constructed the port as a space that has, over time, responded to capital's demands for place-specific regulatory, institutional, and infrastructural arrangements that can enlarge capital's space of operation (Colás, 2017; Easterling, 2016). As competitiveness amongst regions, coastal cities and nations are dependent on their ability to facilitate maritime circulation, states compete by creating zones of exclusion – variously called special economic zones or special development zones – that facilitate export and import processing without the heavy burdens of surveillance and taxation. From export-processing zones to free-port areas and logistics corridors (Grappi, 2018), spaces of "extrastatecraft" (Easterling, 2016) aim to enhance market competition. States thus play a key role in pursuing top-down political strategies of standardisation and fragmentation, integrating policy frameworks, and creating institutionalised frameworks that facilitate flexible accumulation regimes (Grappi, 2018; Harvey, 2007).

It is important in pointing to the uneven development of port governance, however, that neither the public nor the private sector is a homogenous entity. The political and social consequences of port governance should not be simplified in terms of neat models that distinguish between private, networked entities on the one hand, and public, territorial ones on the other (Hesse, 2004). Rather, variations in port governance models, and in particular their impacts on local communities, should be understood to result not simply from economically determined structural factors such as neoliberal marketisation alone, but also from the strategic choices made by port actors themselves within a set of available possibilities structured by the global capitalist system as well as by national socio-political constraints. States make varied choices with regard to the degree to which they allow private interests to trump those of waterfront communities; so too with the enforcement of labour law and environmental standards. In the context of the bargaining power that dockworkers have attained to negotiate their working conditions against private companies and port authorities, for example, the degree to which states exert control over labour has important effects on the outcomes of port governance over labour on the docks, which in turn affects the degree to which the public sectors protect community interests over private interests. For example, as Katy Fox-Hodess (2019) has shown in her work comparing Colombian and Chilean dockworkers, Colombian workers' bargaining power was weakened by a broader context of state-sanctioned violence and absence of labour law enforcement. In contrast, in the Chilean case, a relatively normalised context for trade unionism meant that Chilean dockworkers "had available to them a variety of strategic pathways" and ultimately adopted a "class struggle unionism" strategy (Fox-Hodess, 2019: 50). This class struggle unionism approach was successful because of a high degree of shop-floor power, rather than because of particular models of port governance or workers' inherent structural positions in the economic system.

Recognising the variations within state approaches to port governance, then, is an important step towards understanding how encroaching forms of neoliberalisation of ports can be contested. Although some might claim that the logistics revolution and increasing port privatisation has weakened unions, increased contingency, and lowered labour standards, the networked effect of the dynamic port sector has also created vulnerabilities within the global capitalist system that workers have increasingly sought to contest. Kim Moody argues for instance that logistics is in fact at the forefront of labour struggles, since industries continue to be reliant on "the ties that bind" the production of goods and services, making it possible that the concentration of workers in key "nodes" and linkages create clusters that renew the potential for mass labour struggles (2017: 59–70). Writing of the strategic importance of logistics workers, Bonacich and Wilson likewise argue that "the global transportation and warehousing sector is absolutely vital to the success of global capitalism as we know it" (2008: 249). Because just-in-time systems depend on uninterrupted commodity flows across long distances, port workers have the potential power to coordinate nationally and internationally across transnational supply chains. Indeed, from Durban, South Africa to the San Francisco Bay Area, dockworkers have done precisely so, as the rich history of militant dockworker internationalism attests to workers' ability to forge resistance across global supply chains in solidarity with struggles across oceans and in far-flung locations (Cole, 2018; Fox-Hodess, 2019; Alimahomed-Wilson and Ness, 2018; see also Featherstone, this volume).

If port governance models seem to push waterfronts further toward the consolidation of corporate power and private interests, it is the workers whose labour has been crucial to the functioning of the docks that have actively contested these trends, and sought to make ports into places of internationalist solidarity. This reminder of the importance of workers' power might thus return us to the opening image of this chapter. If, for Engels, the heart of immiseration and inequality lay in the streets and factories that evinced capital's warfare on the working class in nineteenth-century London, it may well be that in the twenty-first century the key to contesting the violence of global capitalism lies in mobilising logistics and supply chain workers' immense power on the docks.

## Notes

1 While the ports of Shanghai and Singapore are ranked above Ningbo-Zhoushan as the busiest ports in the world because of the speed and number of container vessels passing through them, the port of Ningbo-Zhoushan has handled the most cargo by volume for twelve consecutive years since 2008, increasing its throughput by 4.7 per cent year on year (Si, 2021). This is due in large part to its role in enhancing the circulation of goods and materials to and from the Yangtze economic zone and Belt and Road regions.

2 At the same time, dockworkers have been a powerful force of labour internationalism and transnational solidarity across the world. See Cole, 2018, and Fox-Hodess, 2017, 2019, 2020.

## References

Ahuja R (2006) Mobility and containment: The voyages of South Asian seamen, c. 1900–1960. *International Review of Social History* 51(2006): 111–141.

Alimohamed-Wilson J and Ness I (2018) *Choke Points: Logistics Workers Disrupting the Global Supply Chain.* London: Pluto Press.

Bonacich E and Wilson J (2008) *Getting the Goods: Ports, Labor, and the Logistics Revolution.* Ithaca, NY: Cornell University Press.

Braudel F (1981) *Civilization and Capitalism, 15th–18th Century: Volume I, The Structures of Everyday Life* (trans. S Reynolds). New York, NY: Harper and Row.

Brooks M and Cullinane K (eds) (2006) Devolution, Port Governance and Port Performance special issue. *Research in Transportation Economics* 17: 1–686.

Campling L and Colás A (2021) *Capitalism and the Sea: The Maritime Factor in the Making of the Modern World.* London: Verso.

Casarino C (2002) *Modernity at Sea: Melville, Marx, Conrad in Crisis.* Minneapolis, MN: University of Minnesota Press.

Chua C, Danyluk M, Cowen D and Khalili L (2018) Introduction: Turbulent circulation: Building a critical engagement with logistics. *Environment and Planning D: Society and Space* 36(4): 617–629.

Coe NM (2020) Logistical geographies. *Geography Compass* 14(10) https://doi.org/10.1111/gec3.12506

Colás A (2017) Infrastructures of the global economy. In: McCarthy DR (ed) *Technology and World Politics: An Introduction.* Abingdon: Routledge, 146–164.

Cole P (2018) *Dockworker Power: Race and Activism in Durban and the San Francisco Bay Area.* Chicago, IL: University of Illinois Press.

Cowen D (2014) *The Deadly Life of Logistics: Mapping Violence in Global Trade.* Minneapolis, MN: University of Minnesota Press.

Cowen D (2010) A geography of logistics: Market authority and the security of supply chains. *Annals of the Association of American Geographers* 100(3): 600–620.

Danyluk M (2017) Capital's logistical fix: Accumulation, globalization, and the survival of capitalism. *Environment and Planning D: Society and Space* 36(4): 630–647.

Danyluk M (2019) Fungible space: Competition and volatility in the global logistics network. *International Journal of Urban and Regional Research* 43(1): 94–111.

Debrie J, Lavaud-Letilleul V and Parola F (2013) Shaping port governance: The territorial trajectories of reform. *Journal of Transport Geography* 27(2013): 56–65.

Dua J (2018) Chokepoint sovereignty. *LIMN* 10. Available at: https://limn.it/articles/chokepoint-sovereignty/.

Easterling K (2016) *Extrastatecraft: The Power of Infrastructure Space.* London and New York, NY: Verso.

Engels F (1968 [1845]) *The Condition of the Working Class in England*. Stanford, CA: Stanford University Press.
Fawcett JA (2007) Port governance and privatization in the United States: Public ownership and private operation. *Research in Transportation Economics* 17(1): 207–235.
Fox-Hodess K (2020) Building labour internationalism 'from Below': Lessons from the International Dockworkers Council's European Working Group. *Work, Employment and Society* 34(1): 91–108.
Fox-Hodess K (2019) Worker power, trade union strategy, and international connections: Dockworker unionism in Colombia and Chile. *Latin American Politics and Society* 61(3): 29–54.
Fox-Hodess K (2017) (Re-)locating the local and national in the global: Multi-scalar political alignment in transnational European dockworker union campaigns. *British Journal of Industrial Relations* 55: 626–647.
Grappi G (2018) Asia's era of infrastructure and the politics of corridors: Decoding the language of logistical governance. In: Neilson B, Rossiter N and Samaddar R (eds) *Logistical Asia: The Labour of Making a World Region*. Singapore: Palgrave Macmillan, 175–198.
Great Britain House of Commons (1962) *Report of the Committee of Inquiry into the Major Ports of Great Britain (1961–1962)*. Great Britain House of Commons, Parliamentary Papers.
Hammami M, Ruhashyankiko JF and Yehoue EB (2006) Determinants of public-private partnerships in infrastructure. *IMF Working Paper* 99: 1–37.
Harvey D (2007) *A Brief History of Neoliberalism*. Oxford: Oxford University Press.
Harvey D (2005) *Spaces of Neoliberalization: Towards a Theory of Uneven Geographical Development*. Stuttgart: Franz Steiner Verlag.
Hatzopoulos P and Kambouri N (2018) Pireus Port as a machinic assemblage: Labour, precarity, and struggles. In: Neilson B, Rossiter N and Samaddar R (eds) *Logistical Asia: The Labour of Making a World Region*. Singapore: Palgrave Macmillan, 155–174.
Hayuth Y (1987) *Intermodality, Concept and Practice: Structural Changes in the Ocean Freight Transport Industry*. Colchester: Lloyd's of London Press Limited
Hesse M (2004) Land for logistics: Locational dynamics, real estate markets and political regulation of regional distribution complexes. *Tijdschrift voor economische en Sociale Geografie* 95 (2): 162–73.
Hilling D and Hoyle BS (1984) Spacial approaches to port development. In: Hilling D and Hoyle BS (eds), *Seaport Systems and Spacial Change*. Chichester: John Wiley, 20–37.
Hoyle B and Pinder D (1992) Cities and the sea: Change and development in contemporary Europe. In: Hoyle B and Pinder D (eds) *European Port Cities in Transition*. London: Belhaven, 1–19.
Ince OU (2014) Primitive accumulation, new enclosures, and global land grabs: A theoretical intervention. *Rural Sociology* 79(1): 104–131.
Khalili L (2020) *Sinews of War and Trade: Shipping and Capitalism in the Arabian Peninsula*. London and New York, NY: Verso Books.
Linebaugh P and Rediker M (2000) *The Many-Headed Hydra: Sailors, Slaves, Commoners, and the Hidden History of the Revolutionary Atlantic*. Boston, MA: Beacon Press.
Martin C (2012) Containing (dis)order: A cultural geography of distributive space. Unpublished doctoral dissertation, Royal Holloway, University of London.
Mezzadra S and Neilson B (2019) *The Politics of Operations: Excavating Contemporary Capitalism*. Durham, NC: Duke University Press.
Moody K (2017) *On New Terrain: How Capital is Reshaping the Battleground of Class War*. London: Haymarket Books.
Mu R, De Jong M and Koppenjan J (2011) The rise and fall of Public-Private Partnerships in China: A path-dependent approach. *Journal of Transport Geography* 19(4): 794–806.
Nam HS and Song DW (2011) Defining maritime logistics hub and its implications for container port. *Maritime Policy Management* 38(3): 269–292.
Ng A, Ducruet C, Jacobs W, Monios J, Notteboom T, Rodrigue JP, Slack B, Tam KC and Wilmsmeier G (2014) Port geography at the crossroads with human geography: Between flows and spaces. *Journal of Transport Geography* 41(2014): 84–96
Ng A and Liu JJ (2014) *Port-Focal Logistics and Global Supply Chains*. New York, NY: Palgrave Macmillan.
Notteboom TE (2006) Strategic challenges to container ports in a changing market environment. *Research in Transportation Economics* 17: 29–52.
Notteboom TE (2009) Complementarity and Substitutability among Adjacent Gateway Ports. *Environment and Planning A* 41(3): 743–762.
Olivier D and Slack B (2006) Rethinking the port. *Environment and Planning A* 38(2006): 1409–1427.

Parola F, Satta G, Penco L and Profumo G (2013) Emerging Port Authority communication strategies: Assessing the determinants of disclosure in the annual report. *Research in Transportation Business & Management* 8: 134–147.

Polanyi K (1944 [2001]) *The Great Transformation: The Political and Economic Origins of Our Time*. Boston, MA: Beacon Press.

Robinson R (2002) Ports as elements in value-driven chain systems: The new paradigm. *Maritime Policy and Management* 29: 241–255.

Sekula A (1996) *Fish Story*. Düsseldorf: Richter Verlag.

Sheppard E (2016) *Limits to Globalization: The Disruptive Geographies of Capitalist Development*. Oxford: Oxford University Press

Si K (2021). Ningbo-Zhoushan retains world's busiest cargo handling port crown in 2020. *Seatrade Maritime News*. Available at: www.seatrade-maritime.com/ports-logistics/ningbo-zhoushan-retains-worlds-busiest-cargo-handling-port-crown-2020.

Slack B (1993) Pawns in the game: Ports in a global transportation system. *Growth and Change* 24: 579–588

Smith A (1989) Gentrification and the spatial constitution of the state: The restructuring of London's docklands. *Antipode* 21: 232–260.

Smith N (1984 [2008]) *Uneven Development: Nature, Capital, and the Production of Space*. Athens, GA: University of Georgia Press.

Stenmanns J (2019) Logistics from the margins. *Environment and Planning D: Society and Space* 37(5): 850–867.

Subrahmanyam S (2006) Imperial and colonial encounters: Some comparative reflections. In: Calhoun C, Cooper F and Moore K (eds) *Lessons of Empire: Imperial Histories and American Power*. New York, NY: New Press, 217–228.

Taussig M (2002) The beach (a fantasy). In: Mitchell WJT (ed) *Landscape and Power*. Chicago, IL: University of Chicago Press, 317–346.

Tracy JD (1990) *The Rise of Merchant Empires: Long-Distance Trade in the Early Modern World, 1350–1750*. Cambridge: Cambridge University Press.

Wang J and Olivier D (2006) Port-FEZ bundles as spaces of global articulation: The case of Tianjin, China. *Environment and Planning A* 38(8): 1487–1503.

Wang C and Ducruet C (2013) Regional resilience and spatial cycles: Long-term evolution of the Chinese port system (221 BC–2010 AD). *Journal of Economic and Social Geography* 104(5): 521–538.

World Bank (2007) *Port Reform Toolkit, Module 3: Alternative Port Management Structures and Ownership Models*. Available at: http://rru.worldbank.org/Documents/Toolkits/ports_fulltoolkit.pdf.

Zeiderman A (2016) Submergence: Precarious politics in Colombia's future port-city. *Antipode* 48(3): 809–831.

# 11

# CONTAINERS

## The shipping container as spatial standard

*Matthew Heins*

### Introduction

The shipping container is a seemingly mundane object, one whose appearance is banal and generic, yet its impact on global freight transportation has been revolutionary. Essentially a giant steel box with doors at one end, it does not possess any inherent technological capabilities or other remarkable qualities whatsoever. What makes the container so successful is not what it can do on its own, but how it is used. Containers move intermodally, which means they can travel by multiple forms of transportation on one journey; they are carried by shipping, trucking and railroad infrastructures, and are transferred easily and quickly between these otherwise very different modes of transportation. Since the container itself can be shifted from one transportation system to another through the use of cranes and other mechanised devices, there is no need to load and unload cargo in this transition between modes, thereby making the worldwide movement of freight dramatically faster and cheaper. Crucial to this is the standardisation of the container's size and other physical characteristics, in particular its length, width and height, at the global level, so that it can fit into and work with infrastructures everywhere. As I have described previously (Heins, 2015, 2016), containerisation has a particular spatial logic, based on the size and dimensions of the container itself, which is intertwined with the spatial qualities of the transportation modes that hold containers as they move on their globe-spanning itineraries. The container, then, is a fundamentally spatial device.

Though there are several variations, the most common shipping container is 40 feet long, 8 feet wide, and 8 feet 6 inches high (in metric units, these dimensions are 12.19 metres long, 2.44 metres wide, and 2.59 metres high). Another popular size is a 20-foot-long container with the same width and height, and this is convenient since two 20-foot containers can fit in the same space as one 40-foot container. Containers 45 feet long, with the same width and height, have also been introduced, but they are rare. A 'high-cube' container is one that is 9 feet 6 inches high (a foot higher than the normal container), and these have become very widespread. High-cube containers come in any length, though the 40-foot version is most often seen. There are containers of other sizes that only move in particular countries and regions, rather than globally; in the US and Canada, for example, 'domestic containers' that are 53 feet long, 8 feet 6 inches wide, and 9 feet 6 inches high are frequently used. The fittings at each corner of the container, known as corner castings – for cranes and other lifting devices to grip,


DOI: 10.4324/9781315111643-14

and for attachment to other containers – are standardised. A container normally has two doors at one end that open outward.

## The historical development and standardisation of the container

The shipping container originated in the early twentieth century in North America and Europe, with precursors going back to the nineteenth century and arguably earlier. Containers were in sporadic, though not widespread, use from the 1920s to the 1950s, their sizes gradually expanding over the years. Then, as now, their primary purpose was to make freight movement more efficient by eliminating the need to laboriously transfer cargo between different forms of transportation. The dimensions and other characteristics of a container were often specific to its maker or to a particular transportation provider, as universal standards for containerisation were not well recognised, and did not exist at all on the global level.

The first truly global use of containerisation appears to have been a remarkable network created by the United States military in the late 1940s and 1950s, which carried the CONEX container, as it was known, on routes around the world.[1] Compared to today's containers, the CONEX was quite small, as most containers were until the mid-1950s, being 8 feet 6 inches long, 6 feet 3 inches wide, and 6 feet 10.5 inches high (Department of the Army, 1957). Although less globally oriented, the private sector was not far behind military developments. During the 1950s a few transportation providers in the Pacific Northwest region spanning the US and Canada began to move containers systematically in large quantities, interchanging them between ships, trucks and trains (Donovan and Bonney, 2006: 42; Norris, 1992). Some of the major American railroad companies were also innovating with containerisation, as they had done previously in the 1920s and 1930s, but now generally with containers of a larger size. These containers were intended primarily for interchange between trains and trucks within North America, rather than to be carried by ship. One of the first examples was the 20-foot-long container introduced by the Illinois Central Railroad in 1948, and such operations became common in the late 1950s when several railroads embarked on containerisation. The largest network was that of the New York Central Railroad, whose Flexi-Van containers, in lengths of 36 and 40 feet, entered service in 1958 (Norris, 1994: 118–126; Solomon, 2007: 61–70).

The larger containers increasingly common in the late 1950s were generally between 20 and 40 feet long, about 8 feet wide, and about 8 feet high, these dimensions deriving from the largest possible object that could be carried on a flatbed trailer hauled by a truck. This logic was not entirely new, as containers in earlier years were sometimes designed to maximise what a truck (or even a wagon, in the days before the motor vehicle) could carry, but it became more prevalent in the 1950s. The goal was to make the container as large as possible, in order to maximise its capacity, and it was the trucking infrastructure that imposed the upper limit on size. In short, it was the spatial affordances of trucking that generated the container's dimensions – a key point to understand. The shipping container is widely associated with ships and ports, where its presence is most evident to the ordinary observer, and its revolutionary impact on shipping and port operations is undeniable. Yet despite this powerful connection to maritime transportation, "it was, somewhat ironically, conditions ashore that determined the maximum width, height and length of the box [container]", as one scholar of containerisation explains (Broeze, 2002: 13). The centrality of trucking to containerisation is further demonstrated by the reality that most container journeys begin and end on a truck. While this portion of the container's overall movement is often comparatively short, since rail and shipping are more efficient, the truck possesses the essential ability to traverse the 'last mile' – to reach any location accessible by road.

The container's dimensions stemmed from trucking because the truck was, compared with the train and ship, the mode of transportation that imposed the tightest *spatial constraints* in all three dimensions. But it also reflected trucking's great importance by this time in the land-based movement of freight. The motor vehicle had become largely dominant in personal transportation by the end of the 1950s in the wealthy industrialised nations, thanks to cars and buses, and in freight transportation the truck was also overtaking the train. "As admirable a vehicle as the box car is, there is no denying that much of American commerce has found itself more comfortably secured inside a truck trailer", a railroad industry trade journal admitted in 1960 (Morgan, 1960: 36). While major industrial facilities were still served by railroad spurs, it was increasingly the case that ordinary factories and warehouses could only be reached by road. In fact, a great deal of industry was migrating from cities to more suburban or exurban locations, a pattern of decentralisation supported by the construction of new highways.

Yet though the container owed its dimensions to the truck, it was brought into use mainly by railroad and shipping companies. The American trucking industry in the 1950s was highly fragmented with innumerable small players, while the rail and shipping sectors were more concentrated and tended to be dominated by large corporations, which could afford to strategise for the long-term future and spend money developing new ideas. Railroads were motivated to pursue containerisation by the growing realisation that they needed to involve truckers in order to reach most freight customers, while shipping lines wished to quicken the loading and unloading of cargo and reduce the labour associated with this work. It was the shipping companies who would ultimately make the container globally ubiquitous because their operations spanned the world rather than being limited to particular countries or regions. Before the mid-1950s, containers were transported over the water on occasion – larger containers were typically placed on ships' decks, while smaller containers could be held in the hold – but rarely on a systematic basis or in large quantities. This changed forever in 1956, when the Sea-Land service was launched by Malcom McLean, a former trucking executive, for the purpose of carrying containers by coastal shipping in the eastern US. Although a few other shipping lines had already experimented with containers both in North America and Europe, McLean's historic initiative kicked off the wider use of containerisation by demonstrating its efficiency and profitability on a large scale. A revolution was underway.

The voyages made by Sea-Land ships constituted most of the distance of the cargo's journey, but the containers were hauled by truck, using flatbed trailers modified for this purpose, between the ports and the cargo's initial origin or final destination. Hence McLean, like most of the innovators in containerisation, had to take into account the spatial limitations trucking imposed. But another spatial factor was also critical. Initially Sea-Land containers were carried on the ships' decks, and so McLean decided to use a 33-foot-long container because that length would fill all of the available area on the decks as efficiently as possible with no wasted space left over (Donovan and Bonney, 2006: 59; Levinson, 2006: 49). However, the container's width of 8 feet and height of 8 feet 6 inches were chosen due to trucking standards: 8 feet was the maximum width that a motor vehicle or trailer could be, and 8 feet 6 inches was the maximum height possible to allow the container, resting on the modified flatbed trailer Sea-Land provided to truckers, to fit under road clearances.

Sea-Land began operating ships that were *designed specifically* to hold containers in 1957. At that time McLean switched to a container that was 35 feet long, 8 feet wide, and 8 feet 6 inches high. Now the container's length, like its width and height, was determined by the spatiality of trucking, for 35 feet was the maximum allowable length of a truck's trailer on certain roads in the eastern part of the US, the region where Sea-Land containers were moving by truck (Donovan and Bonney, 2006: 68). Actually, many of the eastern states of the US had raised the permissible length to 40 feet by this time, but in Pennsylvania, where the Pennsylvania Railroad held powerful sway and sought to impede trucking, the limit remained 35 feet for a while longer. This was important to Sea-Land, since many of its containers moved through that state (Gibson, 1998: 38).

In 1958 another shipping line, Matson, introduced a container system for the transport of cargo between the US West Coast and Hawaii, using containers 24 feet long, 8 feet wide, and 8 feet 6 inches high. Many states in the western portion of the US, where Matson's containers would move by truck, allowed truckers to haul two trailers one behind the other, each trailer in this setup being a maximum of 24 feet long. Matson hoped to exploit this practice when it decided on a 24-foot container length (Harlander, 1997: 16–17). Another shipping company, Grace Line, was also cognisant of trucking's spatial limitations when it introduced 17-foot-long containers for its service between the US and Venezuela, as Grace judged this to be the maximum length feasible for a truck trailer travelling on the less developed Venezuelan roads (Levinson, 2006: 130).

As containerisation grew more common in the late 1950s and 1960s, it became evident that it made no sense for each transportation provider to use a different container with its own particular size and other qualities. The need for universal standards, a globally consistent spatiality, was clear. European container standards had existed since the 1930s, but these were primarily used by railroads in Europe, and now some believed standardisation should be implemented on a worldwide basis. This process began with efforts to standardise the containers used by American shipping lines, and in 1961 the American Standards Association (now the American National Standards Institute) came up with a set of standards for this purpose. It was decided that the standard container lengths would be 10 feet, 20 feet, 30 feet and 40 feet, the standard width would be 8 feet, and the standard height would be 8 feet (Levinson, 2006: 130–137).

While establishing these standardised dimensions was no easy task, it seems to have been obvious to all concerned that the spatial constraints inherent in trucking had to influence – to determine, really – the container's size. It was widely agreed that 8 feet was a reasonable standard for the width, as this was how wide an American motor vehicle could be, but the length and height were harder to agree on. While some pioneers of containerisation like Sea-Land and Matson had containers 8 feet 6 inches high, it was necessary to use a gooseneck type of flatbed trailer to hold these containers low enough to fit under the typical vertical road clearance in the US. In comparison, the use of an ordinary flatbed trailer necessitated the container be only 8 feet high to squeeze under these clearances. Therefore an 8-foot height was established as the standard, though only after bitter debate (Hinden, 1962: 2; Levinson, 2006: 133). The greatest length that the new standards established, 40 feet, seems to have derived from the longest allowable trailer on American highways, for road regulations had evolved since the mid-1950s, when Sea-Land selected its 35-foot container length, and 40-foot trailers were now permitted and in use across the country (Levinson, 2006: 135). These new standards were promulgated over the vociferous protests of a few shipping lines, most notably Sea-Land and Matson, which objected to the decision with regard to both length and height. Sea-Land and Matson were not prevented from continuing to use their own containers, but the standards put them at a disadvantage.

Soon after the American standards were set, the International Organization for Standardization (commonly referred to as the ISO) embarked on worldwide container standardisation, an endeavour that lasted for most of the 1960s. Once again, the spatialities imposed by trucks and roads were central to the ISO's decision-making process, but now trucking had to be considered in a worldwide context. The 8-foot maximum truck width was fairly consistent globally, though in some nations it was a bit larger, and in a few countries such as Switzerland and

Paraguay it was slightly smaller (Egyedi, 1996: 12; Rath, 1973: 266). Accordingly, it was fairly simple for the ISO to agree on this as the standard container width. The decisions on height and length involved heated debates similar to those of the American standardisation process, and with some of the same participants, but ultimately the ISO decided to follow the American standards of an 8-foot height and lengths of 10, 20, 30 and 40 feet (Levinson, 2006: 137–138). (Container size was not the only issue addressed, as several other characteristics of the container were also standardised by the ISO, such as the strength of containers, their corner connection points, and the means by which they were held and lifted [Levinson, 2006: 138–144].) But these dimensions were not mandatory, and while most shipping lines quickly adopted container lengths of 20 feet and 40 feet (the 10-foot and 30-foot lengths never became popular), and a width of 8 feet, they generally chose to use a height of 8 feet 6 inches rather than 8 feet. Faced with this rebellion, the ISO backpedalled in 1969 and made 8 feet 6 inches an additional standard height, and soon the 8-foot height was all but forgotten and nearly all containers became 8 feet 6 inches high (Levinson, 2006: 144–146). In later years an alternative height of 9 feet 6 inches for 'high-cube' containers was created by the ISO, and an alternative length of 45 feet also came into existence.

## The intertwined spatiality of transportation infrastructures and the container

As this account shows, the standard shipping container, now widely used around the world and a key device in the global economy, owes its size and dimensions to a particular source: the spatial characteristics of the American trucking and road infrastructure in the early 1960s. This reflects the extent of American dominance in the post-war years, and demonstrates how spatial standards are enmeshed with other, earlier spatial standards in a historically contingent fashion. Yet it is also the case that this American-based standard became universal in large part because it did not conflict excessively with the spatial affordances of trucking in other developed nations.

In its creation and early evolution, as previously described, the shipping container was designed to fit into the *existing* transportation modes of shipping, trucking and railroads. These infrastructures were well established, of course, and to make any fundamental alterations to them – such as changing a spatial dimension pervasive throughout a system – would have been expensive, difficult and time-consuming. There was little incentive to implement such transformations merely to suit the container, at this time an unproven device not yet widely adopted. As such, it had to work within existing practices. Accordingly, containers were often carried on the decks of conventional cargo ships, on ordinary flatbed trailers hauled by trucks, and on flatcars (i.e., flatbed railcars) in trains. But as the container took hold and came to be globally standardised, transportation systems were themselves increasingly modified to accommodate it. Now it was the infrastructures of transportation that were being redesigned around containerisation, since freight operators wished to carry containers as efficiently as possible. Devices and practices were introduced to hold the container securely in place on ships, trucks and trains, while also allowing the container to be swiftly placed onto or detached from these transportation modes. Specialised cranes and other machines were developed for moving containers on and off transportation vehicles, and for carrying them around port terminals, rail yards, and other places where they were stored.

In histories of containerisation, the alterations made to ocean shipping to accommodate the container are given the most emphasis, and understandably so. Containers initially were typically carried on the decks of ships whose primary purpose was transporting breakbulk freight (breakbulk refers to cargo loaded and hauled in the traditional manner, as separate individual items), with the container an added cargo. Even when the pioneers of containerisation began to load ships mostly or entirely with containers, they were still sometimes placed only on the decks because they could not be fitted into the holds. This was the case with Sea-Land's original container service. But as containerisation grew common, container ships specially designed to carry containers entered use: a container ship's hold consists of cells in which containers are stacked, with additional containers usually stacked on the deck – strictly speaking, on the hatch covers – above. These vessels, which first appeared in the 1950s, were a complete break in design and operation from conventional cargo ships.

There is no consensus as to the identity of the first container ship: candidates include the *Clifford J. Rogers* and *Susitna*, both operating in the Pacific Northwest, and ships run by a Danish line in European coastal shipping (Levinson, 2006: 31, 292n.25; Norris, 1992: 24). Container ships began to travel on worldwide routes in the 1960s. Since then, of course, they have become pervasive in ocean shipping, and over the years have been continuously upgraded and improved, growing dramatically larger in size. Today the vast majority of the world's marine freight, apart from bulk cargoes like oil, coal, grain and so forth, is transported by container ship. The container is inevitably the spatial module around which container ships are designed. The objective being to carry as many containers as possible, the container's dimensions, weight and other physical qualities are the controlling factors, and thus the length, width and height of the vessel derive largely from multiples of the dimensions of the container. If one seeks to hold nine containers across the width of the ship, for example, then it makes sense for the ship to be just wide enough to accomplish this, and no wider. Aside from such spatial concerns, container ships are also designed to serve containerisation in other respects, such as by allowing the loading and unloading of the containers to be as rapid as possible. Compared to earlier generations of freighters, designed to hold a variety of cargoes in their holds and on their decks, the container ship has a certain simplicity in its design – one might even call it boring.

The container's impact on trucking has been much more subtle, yet nevertheless significant. In the early days of containerisation it was usually a flatbed trailer, sometimes slightly modified for this purpose, that was used to carry the container when it moved by truck. Over time these trailers were increasingly customised in order to secure the container firmly in place, to allow it to be attached and removed quickly, and to hold it as low as possible to fit beneath vertical clearances. The trailers were also progressively made lighter, to save on fuel costs. Such a trailer has come to be known as a 'trailer chassis', and it has a skeletal appearance quite different from a flatbed trailer, with the container supported by an open web of steel components rather than a continuous flat surface.

By the early 1960s, trailer chassis were already being built in this skeletal form, at least in the US and presumably other places as well (Ogden, 1961: 2). Initially there were two types of skeletal designs, referred to as 'parallel frame' and 'perimeter frame', but by the 1970s the parallel frame was pervasive (American Bureau of Shipping, 1987: 2; Tabak, 1970: 109–117). In addition, American trailer chassis came to be designed with a 'gooseneck' at the front, where the trailer is attached to the truck (properly speaking, the 'tractor', at least in American terminology) that pulls it, allowing the container to be held several inches lower and thus to fit below vertical clearances. Today, whether trailer chassis are of the gooseneck design varies from country to country. The trailer chassis is designed around the container's particular qualities, most notably its dimensions but also its means of connection at each corner. Yet while the trailer chassis is customised for the container, it is also designed to work within the norms of trucking in whatever country it is used; in particular, the way it attaches to the truck (the tractor) is the same as for any ordinary trailer. As I have argued previously, the trailer chassis can thus be understood as an interface between the global system of container movement and the national infrastructure of trucking (Heins, 2016: 102).

Containerisation has wrought comparable changes to the rail infrastructure. Until the 1960s, American railroad companies used containers sporadically but rarely on a large scale, and so they did little to create railcars specifically for containers. Instead, ordinary flatcars were generally used to carry containers, which were tied down or otherwise attached in a conventional fashion. One exception was the aforementioned Flexi-Van container system of the New York Central Railroad, which had railcars designed specifically for it. Flexi-Van containers were shifted between this specialised railcar and a truck by a unique method of sliding and rotation, which though clever ultimately proved impractical (DeBoer, 1992: 63–65; Norris, 1994: 122–126; Solomon, 2007: 61–70). In the early 1960s the Matson shipping line experimented with a new hydraulic cushion frame, mounted on a flatcar, intended to provide a smoother ride for containers (Gutridge, 1962: 5–6). Flatcars specially designed to carry either containers or truck trailers were introduced by the Trailer Train Company and other railcar makers in the late 1960s, and began to be widely used in the rail industry (DeBoer, 1992: 106; Norris, 1994: 201).

As with the trailer chassis developed for trucking, these customised flatcars were an instance of how the American rail infrastructure adjusted to accommodate the container rather than simply pushing it through the existing system. Such a flatcar functions as an interface between containerisation's worldwide infrastructure and the national railroad network. In the decades that followed, the design of these flatcars continued to evolve, as the difference grew between them and traditional flatcars, and new models were introduced for containers only rather than for both trailers and containers. The biggest breakthrough, however, came in the early 1980s with the arrival of radically new railcars designed for the 'double-stacking' of containers, with each railcar able to hold two 40-foot containers, one placed atop the other. This would lead to a significant spatial reconfiguration of the US rail infrastructure.

The vertical clearances on American rail corridors vary significantly, tending to be higher in the western regions of the nation. At the dawn of the 1980s, there were only a few corridors with the clearance necessary to accommodate railcars with containers double-stacked (Heins, 2016: 111). Such a railcar of two stacked containers measures 18 feet 3 inches in height, while the standard vertical clearance of American rail has traditionally been sized for a maximum railcar height of 15 feet 1 inch[2] (Heins, 2016: 110–111). However, the profits that would accrue from double-stacking, which effectively allows a train to carry almost twice as many containers as before, were substantial. Consequently, in the early 1980s several companies began moving containers by stacktrain (a stacktrain is a train consisting of double-stacked railcars), on certain rail corridors that possessed exceptionally high clearances, from West Coast ports into the American heartland (Solomon, 2007: 85–89).

The success of double-stacking, at a time when the American railroad business was otherwise doing poorly and battered by competition from trucking, led the rail companies to embark on construction projects to raise vertical clearances on many of their major corridors. To adjust such clearances is expensive and laborious, as it involves raising bridges and other obstacles, lowering track beds, and (hardest of all) enlarging tunnels which may be several miles long, yet the railroads deemed these efforts worthwhile. It was a watershed moment. *The shipping container, originally designed to fit spatially within existing, entrenched infrastructural systems, was now altering and reconfiguring the spatial dimensions and qualities of a land-based infrastructure. The spatiality of the global was, in a sense, reshaping that of the nation-state.* Initially, most of the clearance raising projects were carried out in the western US, allowing containers stuffed with imports from Asia to be hauled by train from West Coast ports to the Midwest or other central locations, with Chicago the most common destination. But soon clearances were likewise raised in the eastern parts of the country, enabling stacktrains to traverse the entire US and also improving the connectivity of East Coast ports.

A railcar designed for double-stacking differs greatly from the flatcars previously used (and still seen occasionally) to hold containers. Indeed, the double-stacking railcar was a novel type of railcar, and several variations were proposed in the 1960s and 1970s before the first example was brought into service in 1981 by the Southern Pacific Railroad (Cudahy, 2006: 163–164; DeBoer, 1992: 108, 139; Traffic World, 1971: 77). Such a railcar holds the bottom container as low as possible, close to the rails, and consequently the distance from the front wheels to the rear wheels must be longer, so the container can be placed between the wheels rather than above them. Over time, the design of these railcars was tweaked and improved. For instance, the railcar first introduced for double-stacking had bulkheads to keep the upper container in place, but subsequent designs relied on connecting the containers to each other at the corner points, thereby dispensing with the bulkheads and lightening the railcar significantly (Malone, 1985: 62–64; Solomon, 2007: 116–118). As with the trailer chassis that trucks use to haul containers, the double-stack railcar supports the container with a web of steel rather than a flat surface; flatcars by contrast typically offer an entire surface able to support various types of freight.

In its early years especially, the container's dimensions and size sometimes bumped up against existing spatial constraints. When containerisation was being introduced by Sea-Land at the port of Bremerhaven in Germany, for example, a low bridge over a crucial railroad corridor to the port blocked containers from being carried on flatcars. The city and Sea-Land shared the expense of raising the bridge (Boylston, 1998: 15–16). More commonly, trains were routed around such problems, though important lines generally had their clearances raised eventually. This was the case in upstate New York, where a line known as the Water Level Route is the major east–west rail corridor spanning the region but was originally built with clearances too low for double-stacking. Hence the Erie Line, an alternate corridor with a higher clearance, was used by stacktrains for several years in the 1980s until clearances had been raised throughout the Water Level Route (Cudahy, 2006: 166–167; Solomon, 2007: 87–88, 173).

## Conclusion

Each transportation mode that carries the shipping container has particular spatial qualities, often standardised at the national or regional level. Possessing its own dimensions and other physical characteristics, the container works spatially with all these systems – little more than a giant steel box, there is nothing remarkable about it and the key to its success lies in its simple spatiality. The concept of the 'spatial regime' illuminates this crucial element of containerisation. A spatial regime is a set of spatial measures that plays some sort of controlling or governing role in a system and has been standardised through regulation or tradition (Heins, 2016: 34–35). Such a spatial regime can be the required size of a corridor as mandated by fire codes, the minimum turning radius for a highway exit ramp as established by engineering practice, the standardised sizes of nuts and bolts that fit together, the typical width of an airplane seat, or many other examples. The tangible materiality of a spatial regime is important, especially since its physicality makes the spatial system difficult to alter once it has been established, and this is particularly true of infrastructures, which often possess such a substantial physical presence and inertia in the built world (Heins, 2016: 35–36). Spatial regimes may be seen as ideal technical solutions resulting from unbiased decision-making by their proponents and users, but political and social factors, or mere historical contingency, are often embedded within them.

The container, then, has its own spatial regime. The same is true for the shipping, trucking and railroad infrastructures that transport the container, and the cranes and other mechanised devices that move it about. Those spatial regimes and that of the container are necessarily interlocking, since the container as a tangible object must be physically carried and held by the transportation systems upon which it depends. (The container's spatial regime is also linked to the spatiality of the boxes, packages, crates, pallets and other things that are loaded inside it, as shippers strive to maximise the use of its interior space by adjusting the dimensions of these items.) Absolutely crucial to this is the *imposition* and *acceptance* of worldwide standards for the container, governing its length, width and height in particular; this standardisation turns the container's inherent spatiality into a spatial regime, and makes it successful.

Shipping containers circulate on global networks, crossing and thereby blurring traditional boundaries between nations, between land and water, and between different transportation modes. In this sense, the container encompasses a vast geographic space through its movement. Meanwhile, the tangible presence of containers is most obvious when they are stacked in massive quantities on container ships, and in even larger assemblages at port terminals. These are the sort of spatialities people commonly associate with containerisation. But the container's most powerful spatiality lies in its own size, and in how its dimensions are inextricably tied to the transportation infrastructures that carry it.

## Notes

1 The account of containerisation presented in this chapter focuses largely on the United States, because that nation played such a prominent role in the container's evolution (see also Heins, 2016).

2 The vertical clearance of the bridge or tunnel must be six inches higher than the railcar, to allow an adequate safety margin; a double-stacked railcar with a height of 18 feet 3 inches therefore requires a clearance of 18 feet 9 inches.

## References


American Bureau of Shipping (1987) *Marine Container Chassis Test Report* (pamphlet).

Boylston J (1998) Interview by A Donovan. Containerization Oral History Collection 1995–1998. Archives Center, National Museum of American History, Washington, DC. December 7.

Broeze F (2002) *The Globalisation of the Oceans: Containerisation From the 1950s to the Present* [Research in Maritime History No. 23]. St. John's, Newfoundland: International Maritime Economic History Association.

Cudahy B (2006) *Box Boats: How Container Ships Changed the World*. New York, NY: Fordham University Press.

DeBoer D (1992) *Piggyback and Containers: A History of Rail Intermodal on America's Steel Highway*. San Marino, CA: Golden West Books.

Department of the Army, Office of the Chief of Transportation. (1957) CONEX: a milestone in unitization. *TC in the Current National Emergency: The Post-Korean Experience*. March 27.

Donovan A and Bonney J (2006) *The Box That Changed the World: Fifty Years of Container Shipping – An Illustrated History*. East Windsor, NJ: Commonwealth Business Media.

Egyedi T (1996) The standardized container: Gateway technologies in the system of cargo transportation. Helsinki Workshop on Standardisation and Transportation, Helsinki. August 23–24.

Gibson A (1998) Interview by A Donovan. Containerization Oral History Collection 1995–1998. Archives Center, National Museum of American History, Washington, DC. April 28.

Gutridge JE (1962) Technical advances in the railroad industry leading to integrated transportation. In: *Ship Design Implications of Recent Marine Cargo Handling Research*. Conference Proceedings, University of California, Berkeley. May 2–4.

Harlander L (1997) Interview by A Donovan and A Gibson. Containerization Oral History Collection 1995–1998. Archives Center, National Museum of American History, Washington, DC. June 19.

Heins M (2015) Globalizing the nation-state: The shipping container and American infrastructure. *Mobilities* 10(3): 345–362.

Heins M (2016) *The Globalization of American Infrastructure: The Shipping Container and Freight Transportation.* New York, NY: Routledge.

Hinden E (1962) Factors affecting container design. In: *Ship Design Implications of Recent Marine Cargo Handling Research.* Conference Proceedings, University of California, Berkeley. May 2–4.

Levinson M (2006) *The Box: How the Shipping Container Made the World Smaller and the World Economy Bigger.* Princeton, NJ: Princeton University Press.

Malone F (1985) APL Linertrains stack up profits. *Railway Age* 186(4): 62–64.

Morgan D (1960) What price piggyback? *Trains* 20(7): 30–42.

Norris F (1992) Cargoes north: Containerization and Alaska's postwar shipping crisis. *Alaska History* 7(1): 17–30.

Norris F (1994) *Spatial Diffusion of Intermodal Rail Technologies.* Unpublished doctoral Dissertation, University of Washington.

Ogden EB (1961) What the operator wants in containers. Society of Automotive Engineers (1961 SAE International Congress and Exposition of Automotive Engineering, Detroit, January 9–13, 1961), New York.

Rath E (1973) *Container Systems.* New York, NY: John Wiley and Sons.

Solomon B (2007) *Intermodal Railroading.* St. Paul, MN: Voyageur.

Tabak H (1970) *Cargo Containers: Their Stowage, Handling and Movement.* Centreville, MD: Cornell Maritime Press.

Traffic World (1971) CN car "double-decks" containers [services and products]. 146(11): 77.

# 12

# SEAFARERS

## The force that moves the global economy

*Maria Borovnik*

### Introduction

If shipping is described as "the life blood of the global economy", responsible for circulating 90 per cent of world trade, then the currently 1.65 million merchant seafarers working on ships, at sea, are a force that keeps this global economy in flow (ICS, 2021). Indeed, within a geography of globalisation, seafarers are a vital part, entrenched in global networks and processes. At the start of 2020, The United Nations Conference on Trade and Development reported there to be a fleet of 98,140 commercial ships transporting 11.08 billion tons of cargo across the globe (UNCTAD, 2020). Seafarers enable these operations on container ships, bulk carriers, oil tankers, and are also working on cruise ships. Considering these crucial numbers behind processes that link places of production and our everyday consumption, it is not surprising that the work of seafarers has been in steady global demand.

And yet, seafarers are also manoeuvred by the neoliberal processes of the current global economy. With economic progress and global competition as driving factors, the shipping industry is an assemblage of networks working within and across ocean spaces. Such shipping assemblages include material and social connections, performing "the fluidities of liquid modernity … and of capital" (Hannam et al., 2006: 3). Seafarers, embedded in this speed-driven assemblage, facilitate the ever-more fluid processes that are characteristic of such time–space compressions on board ships and in ports. Yet they are also living their lives in ever-more fragmented ways through the interactions that come hand in hand with these fluidities (Borovnik, 2011a). The fragmentation of seafarers' everyday lives occurs in both place and time: ships traverse across nation states and time-zones; various ship types operating in different speeds and rhythms are similarly dependent on economic highs or lows; temporal contracts differ depending on seafarers' origins; terms of contracts are always exposed to cost saving measures of shipping management; and seafarers circulate between contracts and off-time at home – or between periods of income and no income. In other words, seafarers exemplify what Büscher would describe as 'liquid' labour that are parts of "larger technological, socio-economic and political contexts" (2014: 224). They are "trapped in mobility" (Büscher, 2014: 224). Quite in contrast to the assumed notion and principles of the "freedom of the sea" (UNCLOS, 1982, Article 87), neoliberal globalisation articulates ongoing uncertainties, temporalities, and contradictory forms of 'freedoms'. Bastos et al. (2021: 158) contend that "the


DOI: 10.4324/9781315111643-15

'freedom' to move about and sell one's labour is produced by the lack of freedom to withhold one's labour", a notion that was also recognised by Rediker's (1987) historical analysis as a consequence of seafarers' embeddedness in capitalist free market global systems. With accelerating economic globalisation, not only are employment conditions looser, ship safety is also jeopardised (Bloor et al., 2000, Borovnik, 2011b). The coronavirus pandemic years of 2020 and 2021, however, have been presenting a slowing down of economic trade, with an exacerbating effect on a lack of freedom for seafarers. Some have been stranded on ships, others remained stuck in a foreign country unable to return home, and yet others are still waiting at home for a chance of re-employment. What is identified as 'crew-change crisis' (De Beukelaer, 2020a), has led to disruptions, stuckness, and supply shortages globally.

This chapter is concerned with the relations between spatial conditions and seafarer lives. It constructs seafarers along two lines. Firstly, they are explored in their function as labour force that drives the global economy. In this view seafarers are objectified as enablers of merchant shipping, maritime transport, or cruise ship tourism.[1] Secondly, seafarers are also comprehended as subjective workers, as affective persons with everyday needs. They mingle and operate within the multicultural, social ship environment, in a constrained space, while using their physical power and bodies to work and engage in the economic activities a ship requires. All this occurs on watery 'grounds' while ships move across a number of politically exclusive economic zones and beyond, or while cargo is loaded and unloaded in ports. This chapter aims to review these two ways of understanding seafarer worlds to open up the complex ways they relate to and engage with ocean space.

## Driving the global economy – or being driven?

Embedded in shipping, which has historically always been pioneering ever-increasing measures of effectiveness, speed, and innovation, seafarers are driving global flows. Yet, they are also living within these flows and are affected by global change. Moreover, globalisation also drives shipping owners, who have a number of different strategies in order to stay competitive in the continuing acceleration of global economic processes. To keep crewing costs down, owners and managers draw on a multi-national supply of seafaring labour worldwide. The option to 'flag out' ships, or in other words, to use second registries within a 'Flags of Convenience' system, is a major driver of the maritime side of the global economy. It allows companies to circumvent restrictions and to keep manning costs low (Chapman, 1992). The same system has also generated the establishment of numerous recruiting agencies in countries, such as the Philippines, China, Indonesia, the Russian Federation, and Ukraine, which are, including ratings and officers, the five largest supply pools of seafaring labour (ICS, 2021). Despite a continuing demand, especially for qualified seafarers, global competition for labour remains high, a situation that continues the options for shipping companies to choose low-cost labour and offer minimal employment conditions.

Like other global industries, branches of shipping companies and their recruiting agencies, may be located outside of their places of origin, for example the Isle of Man, or Cyprus, or in the Global South. Shipping companies may also use second ship registers, with whole fleets 'flagging out' to countries such as Panama, Liberia, Malta, and the Bahamas, currently the largest so-called 'Flags of Convenience' states. These cost-saving strategies make use of lower tax requirements and proximity to cheaper labour. Throughout the 2010s, at the rear of the economic crisis in 2008–9, a reversal of some of these strategies had occurred. Some nations aimed to revive the industry, offering tax-free packages for fleets registered under the original states' 'first' registers in exchange for training packages for locals. One example is Britain where, as Gekara (2021) notes, the incentive to introduce a minimum training obligation in return for strongly tax reduced regimes did not turn out to be as successful as anticipated. A training official in Gekara's study explained the reason for people not taking up this qualified training opportunity was a "strong competition from other countries internationally" (2021: 44). From the perspective of a shipping manager, "the number of British people wanting to go back to sea seems to be drying up" and seafaring "is not an interesting career anymore" (Gekara, 2021: 45). And yet, the British labour union noted that this training scheme needed to offer a "secure stable employment within the UK fleet" (Gekara, 2021: 45). However, an employment obligation (or duty to employ) was not included in the scheme, and as a result, shipping owners continued to bypass requirements, drawing instead on the available large pool of international labour. Not linking certified training with an employment obligation is a problem world-wide, leaving many trainees on waiting lists at home and without income.

Before the introduction of second registries, seafarers were employed as "citizens of the nations represented in their ship's flags and ports of registry" (Alderton et al., 2004: 1). Even though Flags of Convenience have a long history – for example, during war times they were used to "circumvent neutrality laws" (Chapman, 1992: xxiv) – foreign flagged ships had a fairly small component in the global fleet before the 1960s (Lillie, 2004: 51). 'Flagging out' accelerated between the 1970s and 1990s, in accordance with an enormous growth in tonnage transported during these decades. Today, the industry draws on a large pool of international labour, with mixed-crew assemblages that can account for up to 20 different nationalities on cruise ships and more than four nationalities on container ships, depending on the crewing policy of shipping companies. The complexities of these structural changes that were brought into the maritime labour market are recognised by the Maritime Labour Convention (MLC) in 2006 (reinforced in 2013). The MLC requires that seafarers shall have a fair employment agreement, be paid for their services, and demands that shipowners must enable seafarers to remit their earnings to families at home (ILO, 2021). However, these binding agreements are phrased in fairly general language, hence allowing companies – at least to some degree – to interpret these conditions according to their needs and to suit the mobile and flexible work environment on ships.

Many seafarers love their jobs despite the confinements and unpredictability it entails. But it is earnings and remittances that are the most important reasons for seafarers to take on what is one of the most stressful and dangerous occupations. In contrast to these needs, however, keeping wages low and contracts long are the cost-saving measures that were brought into shipping with these internationalised second registries. And there is a global disparity in these measures. Seafarers from the Global North are usually paid more and have shorter contracts than those from the Global South, and this gap applies to both officers and ratings. The International Transport Workers' Federation (ITF) had launched a Flag of Convenience campaign already well before the 1970s (in 1948).[2] This campaign aims at protecting seafarers with minimum acceptable standards and supports them in bargaining for wages and working conditions (ITF, 2021). This need for support must be seen in the context that the flag state presents the jurisdiction over a ship, which affects everyone working on board. Finding a voice to support regulations was particularly important as Flags of Convenience allow for "lenient regulatory requirements" (Lillie, 2004: 49). However, the ITF negotiates on behalf of a number of labour supplying countries both in the Global North and South. Their global inter-union consensus agreement, aimed at accommodating different local settings, is based on a total crew cost concept that permits different wage scales (ITF, 2020). A strong difference within these agreements is the type of contract. As Markkula (2021: 172) explains, European seafarers have usually permanent contracts, while other nationalities, such as Filipinos or other South-East

Asian, Asian and Pacific seafarers work on flexible and temporary contracts. The somewhat staggered wage system has placed seafarers of different nationalities in a situation where the same rank, working side by side with each other, may be paid quite differently depending on the nationality and outcome of negotiations. This represents just one of the unequal geographies wrought through seafarer lives.

Another relates to reducing the length of contracts, which is one more cost-saving option for shipping companies. The MLC guidelines, which were addressed by the ILO (2021) and the UN Global Compact et al. Report (2021) on safe ship crew changes, stress eleven months (or less than twelve) as the maximum length on board and there are possibilities to extend these with back-to-back contracts.[3] It should be said, depending on the ship type, shipowners can decide to have shorter maximum times (e.g. on tankers). That said, as observed by Turgo (2020), there are ways of working around maximum time agreements. Extended contracts maintain cost-saving measures for staffing up to a point where seafarers are worn out and not productive anymore. Shipping companies have argued that they are willing to support employment from the Global South as long as they can find strategies to compensate for visa and transport costs. Consequently, these costs are included in fees that some recruiting agencies demand from seafarers (Markkula, 2021; Turgo, 2020; Zhao, 2021). European seafarers usually work between ten weeks and six months (Borovnik, 2011a; Devereux, 2021; Oldenburg et al., 2009); and Turgo (2020: 7) observed that Filipino junior officers can choose to stay between four and nine months on vessels, while ratings work usually for nine months. However, those from Kiribati or Tuvalu have nine to eleven months contracts and often stay longer. My own observations (Borovnik, 2011a) are in line with Devereux (2021) who explained: "[i]t is not uncommon to find two seafarers on board the same ship, working at the same rank but having completely different tour lengths". Both in Devereux's (2021) and my own research it could be observed that contractual differences, and this includes tour lengths as well as discrepancies in payment, can potentially lead to resentment and "a lack of understanding of the amount of time individuals spend on board [and I would add the amount being paid] in comparison to those from more economically developed countries" (Devereux, 2021: 80).

Then again, with the start of the COVID-19 crisis in 2020, all these crew regulations are now in jeopardy. Countries started going into lockdown and closed their borders, and "hundreds of thousands" of seafarers were (and at the time of writing still are) "trapped on ships as routine crew changes cannot be carried out", or they are "stranded on land, prevented from re-joining ships" or from being permitted to transit, and even from access to health-care on shore (UN Global Compact et al., 2021: 3; see also, De Beukelaer, 2020a, 2020b). Overall, these lockdowns have also caused a slowdown of the global economy, with restrictions on goods being transported to certain areas. These developments impinge on job losses in the maritime industry, affecting both the cruise ship tourism industry and merchant shipping. Being stuck on board their vessels has prolonged usual contract times extensively. De Beukelaer (2020a) voiced concern that this situation, where seafarers are working on ships well beyond 17 months, is classified as forced labour by the ILO. This situation has caused anxiety, depression, insomnia, and other forms of distress (Obeleke and Aponjolosun, 2021).

Many seafarers who have continued operating throughout the pandemic have been dealing with the fear of being infected by the coronavirus and not knowing when or how they will be able to return home. These disruptions have also led to hopelessness and mistrust (BBC News, 2021), as some governments are struggling to decide on financing a return of their citizens (Garbe, 2021). For example, a number of seafarers from Kiribati are "stranded in northern Germany" and "haven't been able to see their families on the other side of the world for nearly two years" (BBC News, 2021: no page). On the other side of this situation are those who want to return to their ship-based jobs and are unable to as the Kiribati government "lacks the ability to conduct Polymerase Chain Reaction Covid-19 testing" (RNZ News, 2021: no page). For countries, such as Kiribati and Tuvalu, these issues are pressuring on their economic income. Considering the enormous world-wide competition, it would be easy for companies to replace anyone, and consequentially countries with high dependency on seafarer remittances potentially experience economic hardship. The immediate effects are felt by seafarers themselves, families and the wider communities (Borovnik, 2007).

In their special issue on Pandemic (Im)mobilities, Adey et al. (2021) highlighted discriminations and stigmatism caused by border restrictions in light of the COVID-19 crisis, which can be confirmed by the situation that seafarers are facing. While throughout the first few months of 2021 a large number of countries have finally allowed seafarers to cross international borders, there are still challenges to overcome with off-and-on COVID-19 restrictions, quarantine requirements in different regions, and high costs of travel (Bailey et al., 2021). Therefore, the international community, led by the International Maritime Organisation (IMO), the ILO, the United Nations High Commissioner for Human Rights (OHCHR), also UNCTAD, and the United Nations Global Compact initiative, have declared the COVID-19 related governments' restrictions that keep seafarers stuck as a humanitarian and safety crisis. The IMO has accredited the year 2021 as the "year of action for seafarers", calling for recognition of seafarers as essential "key workers" (IMO, 2021a).

## Living a sailor's life

Having examined seafarers as a labour force that drives the global economy – and the global inequities and (geo)politics of this role – this second part of the chapter turns to seafarers' subjective lifeworlds. Recognising seafarers as key workers seems overdue, considering the key role the shipping industry plays in moving and shifting the bulk of global trade. And yet, even outside a global health crisis, seafarers are exposed to a number of stresses. The Seafarers International Research Centre (SIRC) has documented the many different aspects of seafarers' lives, and a significant list of publications can be accessed through the centre's webpage.[4] One observation is notable:

> Seafarers constantly live and work under threat of severe sanctions whether these relate to instant dismissal (a potential consequence of the displeasure of a captain or chief engineer) or to the imposition of personal fines and imprisonment by port authorities charged with enforcing international regulations.
>
> *Sampson, 2021a: 4*

This stress is well recorded (Alderton et al., 2004; Borovnik, 2011a, 2017; Chapman, 1992; Langewiesche, 2004; Rediker, 1987). Marcus Rediker (1987), drawing on historical accounts, explained how seafarers have been "workers of the world" for centuries, increasingly so since the seventeenth century. The accelerating manufacturing speed and needs for resources had also mobilised recruitment for seafaring labour. Within these processes, seafarers have become a commodity, or the 'hands on deck' of a capitalist, free-sailing labour market, turning the ship assembly into a 'mobile community'. Rediker, also remarked that within these ship communities, bonds and collaboration were one way of approaching their occupational risks:

> Bonds among seamen arose from the very conditions and relations of cooperative work, not least from sailing a frail and isolated vessel that was surrounded by the perils

of the deep. Other perils originated from within the ship, usually the captain's cabin. Hazards natural and unnatural, whether accidents, disease, or abusive mistreatment, made sailoring a dangerous calling.

*Rediker, 1987: 290–291*

Therefore, when comparing Helen Sampson's recent observation with Rediker's historical analysis, a continuing culture of hazards on ships can be observed, some of which have to do with social relations among crew, others with external risks. Although ships are not so frail anymore and technology has become quite sophisticated – sailing is safer now after the introduction of large containerships – "continuous mobility can … be very isolating" as argued by Bastos et al. (2021: 158). This isolation has persisted and is exacerbated with improved technology and security standards in specialised ports. Turnarounds are fast and with ports located well away from shores, such shore facilities that could be delighted in, are often at far distance. In addition, controlling patrols to assure ship security, port security and international regulations are complied with, are now conducted by a number of agencies, including port states and the ITF. The process is long and complex. In Turgo's study, one of the captains explained at arrival into a British port that "his vessel had to complete 26 documents prior to arrival, 24 upon arrival and 22 prior to departure in a part that they last visited" (2020: 9). Through my own longstanding research into seafarer lifeworlds, I found it was not only captains but also crew members who were affected by these security requirements. They were drenched in frequent patrols' requirements and security check-ups that interrupted their sleep. Considering that many ships are equipped with minimum crew numbers, and that watch-keeping turns come in short circulation, fatigue is a chronic circumstance among seafarers.

Depending on the type of ship and cargo-contracts, port and cargo operations can also be frequent; for example roll-on-roll-off vessels travel short distances and have less time in ports; ships with cargo-contracts that have regular, self-repeating loop systems also do not normally have more than a few days between ports.[5] Under these circumstances, having to wait for pilots to guide a vessel into port, and being frequented by controls, adds to the overall fatigue the job involves and further isolates seafarers. When looking at possible shore leave, they must make a choice between badly needed rest or pleasure. Consequently, many seafarers do not leave their ship for prolonged periods. Isolation, lack of sleep, lack of enjoyment, hierarchical power pressures, and distance from loved ones at home make the seafaring job tough, physically, mentally, and emotionally (Borovnik, 2011b; Sampson, 2021b). Other risks are thefts or pirate attacks, in these days executed with use of sophisticated technology. One example of a theft was narrated to me by this seafarer:

*We were in anchor and I heard a metal click so I knew it was this metal hook. And I had this big knife with me all the time, and I just look over the side and cut the rope, you know, that they were climbing up. … People come to steal things from the ship when everybody is sleeping. We were lucky, because on the other ship they sprayed inside, you know, the door. There is a space, a very small space, they spray something, and it makes you unconscious and they pick the lock on the door open.*

While this unexpected incident was a memory of this experienced seafarer, similar situations were confirmed by others. Ships remain exposed to interception and infiltration as well as subject to natural conditions – ever-changing sea currents, wind, and at times dangerous weather, that are to be conquered and moved with, which requires flexibility and stamina (Borovnik, 2017, 2019, 2021). The old idea about 'a sailor's life' as adventurous, cosmopolitan, with a 'work hard play harder' type of existence, has faded and has left behind a rather monotonous, job-focused essence.

Getting to know the world in these days of seafaring does not occur on shore anymore. Rather, seafarers explore the world in a more limited way and directly aboard vessels, where a majority are male, are working with a multicultural peer-group and are dealing with different customs and communication styles. The mutual experience of distance from home, homesickness at times, concerns about what is going on with families and whether kids, parents and partners are okay, the constant movements of ocean currents and ship kinaesthesia, and the shared limitations of a vessel, often leads to bonding between multi-national peers (Borovnik, 2007; Sampson, 2003). It is possible to compare the connections being made among ship-crews with other mobile networks because, as Urry argues, the continuing (re)formation of networks also reassures power relations: "[m]ovement makes connections and connections make [or confirm pre-existing] inequalities" (2012: 24). On ships, inequalities are intersectional with post-colonial structures continuing and indeed repeated on ships, especially where officers are from the Global North and ratings from anywhere else.

Tensions may occur when higher ranks use bullying and physical abuse against lower ranks. These practices might be exacerbated by post-colonial cultural differences among crews. Bullying on ships is described by Helen Sampson (2021b) as fairly common. She argues that abusive treatment can be taken for granted as part of a ship-job culture, where it is expected that seafarers 'toughen up' and not report abuse, and where abuse from leadership might be covered up by companies. These and similar scenarios to a ship culture in Sampson's view are "entwined with European notions of masculinity" that reflect on "the historical roots of the merchant navy and the current European and OECD dominance of ownership of the global merchant fleet", a 'can do' culture that encourages "stoicism, humour, and emotional toughness" (2021a: 96). In other words, all seafarers are expected to literally 'take one on the chin'.

And so, despite the sharing of mutual experiences, seafarers clash. Sometimes this is because of misunderstandings related to cultural differences, sometimes it is because of built-up resentment against overbearing ship-leaders, and also because of resistance against the constraining ship environment and the lack of shore time. Sometimes all of this can lead to an intensifying of physical energy. During one of my research projects on seafaring geographies a research participant said: "What do you expect? Where men come together in a close environment over a longer time, there will be fights". He was referring to occasional, yet common, violent outbursts that he had related to the male-culture on ships. Other seafarers would refer to ships as so constrained that it felt like being in prison. During my time conducting seafaring ethnography, I quietly joined one of the men on the containership I travelled with, whilst he was watching the TV show *Prison Break*. It felt unsettling not being able to see a positive ending as the DVDs in the entertainment cabin only had some of the episodes available. Seafarers agree that it depends on the ship's 'master' – the captain, the chief officer or 'first mate', and the chief engineer – whether a ship community works well together. These three leaders determine how everyone else, such as 'deckhands', who are ordinary and able-bodied seafarers, or motormen in engine rooms, are treated. Indeed, this masculine, hierarchical language for roles and positions has persisted, signposting the paternalistic and gendered ship environment.

Within such a structured milieu, shipping owners and managers aim at ideal crew constellations. Crews are put together with consideration to choosing nationalities that work best together. Handbooks explaining the diverse communication styles among various cultures are available for officers. Markkula (2021) notes that these handbooks may be intended to achieve intercultural understanding on ships, but at closer look may imprint racial differences, such as describing Filipinos as 'easy to instruct and they accept the "white man" as their superior'

(Markkula 2021: 172, drawing on The Swedish Club, 1998: 10). These perceptions promoted and reinforced in handbooks then underline the officers' authorities more than leading to greater cultural understanding. In arranging crews that work well together, mature, and calm tempered bosuns (boatsmen) are a preference among employers, as their role is to direct deck workers and negotiate between officers and deck crew. Employers explained to me that their aim is at avoiding conflict among crew members and hostility against officers. Consequently, Sampson (2021b) observed that mixed nationality crews are often a preference. In my own research, I heard from some managers that this intermingling might include ratings from one nationality and usually is contrasted with officers from other nationalities. Nevertheless, Sampson (2021b) provided a detailed analysis of the complexities of different crew constellations, where belonging to the same nationality and/or different crew positions complicate relationships. Maintaining harmony aboard vessels and a smoothly cooperating crew promises best outcome for safe and efficient voyage and cargo transport. Here again, facilitating transport remains priority over workers' well-being and a sense of equality and justice.

From the perspective of many seafarers, the shipping industry is gendered and racialised (Fajardo, 2011; Kitada, 2021; Markkula 2021; Stanley 2016; Turgo, 2020). Decisions on crewing constellations and crewing numbers are no exception, as these decisions are often based on cultural perceptions and stereotypes, where some nationalities are preferred by employers as they are seen as more feminine than others and therefore as more compliant and service oriented (Fajardo, 2011). Decisions also depend on the physical strength of seafarers in the overall masculine ship environment. Of all seafarers there are currently estimated just over 1 per cent women employed globally (Kitada, 2021: 66; see also IMO, 2021b). The majority of these are in the passenger sector, including cruise ships and ferries (IMO, 2021b) and approximately 0.12 per cent operate on cargo ships (Kitada, 2021: 66, drawing on BIMCO and ICS 2016). Kitada (2021) notes that the number of women seafarers has decreased in the last few years. Most are from Northern European countries. Both Kitada (2021) and Stanley (2016) found that the gender segregation on ships and the masculine occupational culture, regardless of the advanced ship technology, still foregrounds the physical aspects of ship work. Despite women's recognised teamwork and leadership skills, women are also disregarded in terms of their mental ability based on gendered constructions: "[t]he idea that women tend to be emotional and cry promotes a negative image about women's capacity to work at sea" (Kitada, 2021: 70). Being stereotyped and condescended upon are reasons for women to be reluctant to stay in this occupation. The IMO (2021b), however, explains the importance of enhancing women in the maritime industry as necessary towards achieving gender equality.

Within the last two decades the negative stereotype of women on ships has not changed, considering that Belcher et al. (2003) observed how women tended to have less opportunities to access ship-based employment and had to continuingly negotiate their positionalities among the dominating male culture. All these observations are startling, considering the rich history of women in the maritime industry (MITAGS, 2020). Early accounts show women were even involved as pirates in the eighteenth century, and during the nineteenth century women worked as merchant seafarers disguised as men or were assisting their captain husbands (Stanley, 2016). Some of the earliest women on ships were stewardesses who accompanied rich female travellers. There are accounts of women from the UK and Ireland sailing as deck workers during the 1950s and '60s, and from China during the 1970s (Stanley, 2016: 69). The first US female captain was certified in 1972 (MITAGS, 2020), and one of the first British captains also sailed already during the 1970s.

During interviews with some companies during my various research on seafarers, I found that women's employment was regarded differently among management. One old-school shipping owner, whom I interviewed in 2012, voiced his strong opposition to women working on board his ships, as they did not 'belong' on ships in his view and would only bring trouble to the industry. Quite differently, another shipping owner, from a similar long line of shipping ancestry, had been willing to give women employment opportunities, including women from the Global South, as he felt that they could bring a positive aspect to the job. Considering, however, the hierarchical and hard-working conditions among male crews, Chin (2008) explained there was a cultural class system, where women were placed on the bottom. Guo and Liang found that a friendly working environment and lower "degree of hegemonic masculinities" would make it more possible for women to work on ships (2012: 200). Yet, there is still a long way to go for women to be accepted as committed and suited for the seafaring occupation.

There may be different reasons why seafarers go back to their jobs repeatedly, rather than finding a job ashore. Aside from the salaries, which are usually higher than those in other jobs, the most cited reasons amongst mostly younger seafarers from Kiribati and Tuvalu was 'seeing the world for free'. A similar sentiment could be seen in Jo Stanley's reflection on women seafarers' reasons to take up jobs that are in so many ways restrictive and uncomfortable (Stanley, 2016). Stanley also points out the "joy of doing it, together..." (Stanley, 2016: 235), where the ship community itself can be a strong focal point for seafarers to return. Circulating between home and ship communities, then, has constructed a dual life for many seafarers, where each side must be negotiated. It is easier to enjoy the seafaring adventure as a single, free young person, who may take this opportunity to trade the restrictions at home with those on board a ship. At the beginning, a sailor's life could be perceived as exciting and new. Yet, the reality may wear off. Accompanying couples also may find the seafaring life interesting and at least bearable. Being a seafarer with partners and children back home is more burdensome as responsibilities must be met, and the circulatory lifestyle negotiated. One of the more experienced seafarers I interviewed told me enthusiastically that he loved sailing and was happy to be in his job. But there was one year, he said, where he only spent one week at home. His wife rung up the shipping company to ask "what is the matter? Why is he going again without any time at home?" But they needed his specialisation, and so he was in demand. And so, he left after an only one-week holiday. Another seafarer explained the strangeness of circulation. I noted the following in my ethnographic journal:

> *He finds that he has nothing to talk about when he is at home. Like, he says "what can you talk about? When you travel, all you see is the ocean. And again, the ocean". ... It is also going to be difficult for him to find a job at home. ... He knows of some men who tried to stop being seafarers and did not cope with the life on land and so they went back to sea. I ask: "Why is that?" He explains: It's because there is a different mentality on sea or on land. People have nothing to talk about to each other because when you are on sea you don't really experience the daily life of people, so you have nothing to contribute to the conversation. He mimics to me how he often sits there while others are talking with him having no idea what they are talking about. "It's boring", he says.*

In Asia and the Pacific, where there is a lack of employment opportunities at home and the attraction of higher salaries on ships, both men and women take on these jobs despite the price of being apart. Regardless of the toughness of their jobs, many seafarers enjoy going to sea. The beauty of the ever-changing ocean space, and even the exposure to currents, wind and weather have an attraction that many are drawn back to after some time on shore. Those who have been in their jobs for many years found that they are in a conundrum. Going back to the job again and again is paradoxical, it is "situated between the known and the unknown" (Breyer et al., 2011, in Amit and Knowles, 2017: 168). On one hand, seafarers have a bizarre sense of 'being crazy' – putting up with what one person in my research explained, the most dangerous and unpredictable of jobs. And on the other hand, seafarers enjoy the same fluid, dangerous unpredictability about their jobs. Seafarers have compared themselves with astronauts or to soldiers (Borovnik, 2019: 138–9). Both comparisons involve notions of strongly structured – if not overpowering – ship hierarchies, but they also give insight into the vulnerability of ship-life, where one has to deal with a criss-crossing, watery way of life. While working on their structured daily tasks, the everyday lives of seafarers are constantly adjusting to the crossing time-zones and climate zones, and any unpredictable occurrences. Occupational hazards identified can be related to specific cargo – there could be poisoning fluids leaking out of containers, or containers could be damaged, or loosened. There are regular firefighting exercises on all ships to prevent the most dangerous hazards. On deck, one of the most dangerous jobs is handling robes, but it is also always possible to fall from a ladder, or to slip on the surfaces that are frequently sprayed with ocean water. Those working in engine rooms deal with heat and noise, and slipping hazards on oily surfaces, smells from strong substances, and injuries related to handling machineries. One engine worker told me: "[t]here is hardly anything that is not difficult and uncomfortable. This is part of the job". Although engineers do not have a direct sea view during most of their everyday, they do have the important task of steering the ship from below when leaving ports. Cooks and stewards must do their jobs while ships frequently tip over in any direction at any time, with hot pots, fragile dishes and sharp knives in need of fastening and vigilant looking after. 'Cookies' as they are endearingly labelled often have to put up with grumpy, unsatisfied remarks at time, especially when the food they have cooked is of a different style and flavour to what most of the crew is used to. On the other hand, cooks are often those that everyone comes to for moments of comfort and the homeliness of galley space. These moments and also any more unusual events are somewhat appreciated. They may break up the boredom of the same old journey, where ports have the same design, ship operations repeat themselves, and time sometimes seems to crawl.

## Conclusion

This chapter has provided an overview of seafarer lives in respect of the many and multiple places, spaces and interpersonal and professional relationships that they negotiate. It has examined the relations between seafarers and the global socio-economic geographies of the shipping industry and in turn how those geographies shape intimate seafarer worlds in often deeply uneven and unequal ways. Indeed, seafarers criss-crossing the global seascape, while working on the mountainous tankers, cargo and cruise ships, are not instantly visible to outsiders – that is, most of us. Yet, they are essential for our everyday lives, as they move both resources for manufacturing and products for consumption in this world's economy. As ocean dwellers seafarers gaze at the world from a sea-to-land perspective. Their work is different from everyone else's. Seafarers operate on the wobbly maritime surface and the continuous jittery ship-kinaesthetic, in the darkened engine space, or in a messroom without a view, or on the high-up bridge with an amazing overview, or out in the open on deck side by side with the unpredictable wildness of the sea, the scorching sun or icy, splattering rain. Whether it is working on a container ship or a cruise liner, life aboard ships often entails climbing ninety or so steep footsteps several times every day: from the bunkroom to the workspace, onto the deck, down to the engine room, up on the bridge, back to the messroom, or cruise liner restaurant or guest swimming pool, back on deck, down to the laundry room, and so on. Life on-board vessels journeying maritime spaces entails the circular movements between home and the ship, and it involves transversal ship operations.

These operations are rarely adventurous and more often repetitive and monotonous, much dependent on the decision making by an assemblage of companies and institutions. Living and working aboard ships requires compliance to hierarchical structures and other rules and regulations. Working in a multinational environment requires willingness to be openminded. In a sailor's life physical and emotional resilience, intuition and creativity, and flexibility matter. This is because, for seafarers, the unexpected is bound to happen.

## Notes

1 Not included in this chapter are those involved in military operations or those who work on fishing vessels. There is hardly literature on the *lived* experiences of naval seafarers, see for example Brown (2012) and Stanley (2018). There are two kinds of literature about seafarers on fishing vessels, Gao et al. (2021) look at the intersection between local fishing and environmental concerns and Yuliantiningsih and Barkhuizen (2021) have written widely on the extreme constraints and often abuse that working on international fishing vessels can involve.

2 It must be noted that one of the strongest advocates for seafarers, the International Labour Organisation (ILO), was established in the aftermath of the Titanic disaster in 1912, which revealed how seafarer were not included in the ship's safety standards. As a consequence, the first International Convention for the Safety of Life at Sea was established in 1914 and confirmed in 1929, and the in 1919 established ILO convention to protect a minimum age at sea, unemployment indemnity regarding shipwreck, and recruitment and placement of seafarers. The International Maritime Organisation (IMO) created its first International Convention on Standards of Training, Certification and Watchkeeping for Seafarers (STCW) in 1978, which requires minimum qualification standards for work on ships. The most recent update was in 2010 (see EduMaritime webpage: www.edumaritime.net/stcw-code).

3 The ILO (2015: C2.1j-k), described possibilities to extend contracts as dependent on the agreement of the nation state, under which the ship is flagged, which can mean maximum extensions of different lengths.

4 See www.sirc.cf.ac.uk/

5 A few days of voyage without port contact are welcomed by seafarers, but any longer periods without a chance to leave the ship are difficult. This also applies to longer journeys across continents when there is no change for weeks.

## References


Adey P, Hannam K, Sheller M and Tyfield D (2021) Special Issue: Pandemic (im)mobilities. *Mobilities* 16(1): 1–19.

Alderton T, Bloor M, Kahveci E, Lane T, Sampson H, Thomas M, Winchester N, Wu B and Zhao M (2004). *The Global Seafarer: Living and Working Conditions in a Globalized Industry*. Geneva: International Labour Office.

Amid V and Kowles C (2017) Improvising and navigating mobilities: Tacking in everyday life. *Theory Culture and Society* 34(7–8): 165–179.

BBC News (2021) Covid-19: The Kiribati sailors stranded 8,000 miles from home. *BBC News*, 8 March 2021. Available at: www.bbc.com/news/av/world-56308350.

Bailey R, Borovnik M and Bedford C (2021) Seafarers stranded at sea: An unfolding humanitarian crisis. *Development Policy Centre*, 3 June 2021. Available at: https://devpolicy.org/seafarers-in-a-covid-world-20210603/

Bastos C, Novoa A and Salazar N (2021) Mobile labour: An introduction. *Mobilities* 16(2): 155–163.

Belcher P, Sampson H, Thomas M, Veiga J and Zhao M (2003) *Women Seafarers: Global Employment Policies and Practices*. Geneva: International Labour Office.

Bloor M, Thomas M and Lane T (2000) Health risks in the global shipping industry: An overview. *Health, Risk & Society* 2(3): 329–340.

Borovnik M (2007) Labor circulation and changes among seafarers' families and communities in Kiribati. *Asian and Pacific Migration Journal*, 16(2): 225–249.

Borovnik M (2011a) The mobilities, immobilities and moorings of work-life on cargo ships. *SITES* 9(1): 59–82.

Borovnik M (2011b) Occupational health and safety of merchant seafarers from Kiribati and Tuvalu. *Asia Pacific Viewpoint*, 52(3): 333–346.

Borovnik M (2017) Night-time navigating: Moving a container ship through darkness. *Transfers: Interdisciplinary Journal of Mobility Studies* 7(3): 38–55.

Borovnik M (2019) Endless, sleepless, floating journeys: The sea as workplace. In: Brown M and Peters K (eds) *Living with the Sea: Knowledge, Awareness and Action*. Abingdon: Routledge, 131–146.

Borovnik M (2021) Seafarers and weather. In: Barry K, Borovnik M and Edensor T (eds) *Weather: Spaces, Mobilities and Affects*. Abingdon: Routledge, 95–110.

Brown M (2012) *Enlisting Masculinity. The Construction of Gender in U.S. Military Recruiting Advertising During the All-Volunteer Force*. Oxford University Press: Oxford, New York.

Büscher M (2014) Nomadic work: Romance and reality. *Computer Supported Cooperative Work* 24(2): 223–238.

Chapman PK (1992) *Trouble on Board. The Plight of International Seafarers*. Ithaka, NY: ILR Press.

Chin CBN (2008) Labour flexibilization at sea. 'Mini U[nited] N[ations]' crew on cruise ships. *International Feminist Journal of Politics* 10(1): 1–18.

De Beukelaer C (2020a) Crew-change crisis risks supply chains – and lives. *The Interpreter*, 26 October 2020. Available at: www.lowyinstitute.org/the-interpreter/crew-shift-crisis-risks-supply-chains-and-lives

De Beukelaer C (2020b) Stranded at sea: The humanitarian crisis that's left 400,000 seafarers stuck on cargo ships. *The Conversation*. Available at: https://theconversation.com/stranded-at-sea-the-humanitarian-crisis-thats-left-400-000-seafarers-stuck-on-cargo-ships-150446

Devereux H (2021) Transitions and adjustments made by seafarers whilst at sea. In: Gekara VO and Sampson H (eds) *The World of the Seafarer*. London: Springer, 79–86.

Fajardo KB (2011) *Filipino Crosscurrents. Oceanographies of Seafaring, Masculinities, and Globalization*. Minneapolis, MN and London: University of Minnesota Press.

Gao Q, Xu H and Yan B (2021) Environmental change and fishermen's income: Is there a poverty trap. *Environmental Science and Pollution Research* https://doi.org/10.1007/s11356-021-14254-1

Garbe E (2021) COVID-19 pandemic triggered seafarers' odyssey back to the Pacific Islands. *Toda Peace Institute Global Outlook*. Available at: https://toda.org/global-outlook/covid-19-pandemic-triggered-seafarers-odyssey-back-to-the-pacific-islands.html

Guo JL and Liang GS (2012) Sailing into rough seas: Taiwan's women seafarers' career development struggle. *Women's Studies International Forum* 35: 194–202.

Gekara VO (2021) Can the UK Tonnage Tax minimum training obligation address declining cadet recruitment and training in the UK? In: Gekara VO and Sampson H (eds) *The World of the Seafarer*. London: Springer, 37-49.

Hannam K, Sheller M and Urry J (2006) Mobilities, immobilities and moorings. *Mobilities* 1(1): 1–22.

International Chamber of Shipping (ICS) (2021) *Shipping and World Trade: Global Supply and Demand for Seafarers*. Available at: www.ics-shipping.org/shipping-fact/shipping-and-world-trade-global-supply-and-demand-for-seafarers/

International Labour Organisation (ILO) (2015) *Maritime Labour Convention 2006. Frequently Asked Questions*. Available at: www.ilo.org/wcmsp5/groups/public/---ed_norm/---normes/documents/publication/wcms_238010.pdf

International Labour Organisation (ILO) (2021) *Maritime Labour Convention, 2006*. Available at: www.ilo.org/dyn/normlex/en/f?p=NORMLEXPUB:91:0::NO::P91_INSTRUMENT_ID:312331

International Maritime Organisation (IMO) (2021a) *The World Maritime theme for 2021 is dedicated to seafarers, highlighting their central role in the future of shipping*. International Maritime Organisation News, 16 February 2021. Available at: www.imo.org/en/MediaCentre/PressBriefings/pages/IMO-launches-a-year-of-action-for-seafarers.aspx

International Maritime Organisation (IMO) (2021b) *Women in Maritime IMO's Gender Programme*. Available at: www.imo.org/en/OurWork/TechnicalCooperation/Pages/WomenInMaritime.aspx

International Transport Federation (ITF) (2021) *About the FOC Campaign*. Available at: www.itfseafarers.org/en/focs/about-the-foc-campaign

International Transport Workers' Federation (ITF) (2020) *ITF Uniform Total Crew Cost ("TCC") Collective Bargaining Agreement ("CBA") for Crews on Flag of Convenience Ships*. 1 January 2019–2020. Available at: www.itfseafarers.org/sites/default/files/node/resources/files/UNIFORM%20TCC%20FINAL%202019-2020.pdf

Kitada M (2021) Women seafarers: An analysis of barriers to their employment. In: Gekara VO and Sampson H (eds) *The World of the Seafarer*. London: Springer, 65–78.

Langewiesche W (2004) *The Outlaw Sea. A World of Freedom, Chaos, and Crime*. New York, NY: North Point Press.

Lillie N (2004) Global collective bargaining on Flag of Convenience shipping. *British Journal of Industrial Relations* 42(1): 47–67.

Markkula J (2021) 'We move the world': The mobile labor of Filipino seafarers. *Mobilities* 16(2): 164–177.

The Maritime Institute of Technology and Graduate Studies (MITAGS) (2020) *Women in the Maritime Industry*. Available at: www.mitags.org/women-at-sea/

Okeleke UJ and Aponjolosun MO (2021) A study on the effects of COVID-19 pandemic on Nigerian seafarers. *Journal of Sustainable Development of Transport and Logistics* 5(2): 135–142.

Oldenburg M, Jensen HJ, Latza U and Baur X (2009) Seafaring stressors aboard merchant and passenger ships. *International Journal of Public Health* 54(2): 96–105.

Rediker M (1987 [2007]) *Between the Devil and the Deep Blue Sea*. Cambridge: Cambridge University Press.

RNZ News (2021) Kiribati seafarers losing jobs due to lack of PCR Covid-19 testing. *RNZ [Online]*. Available at: www.rnz.co.nz/international/pacific-news/442734/in-brief-news-from-around-the-pacific

Sampson H (2003) Transnational drifters and hyperspace dwellers: An exploration of the lives of Filipino seafarers aboard and ashore. *Ethnic and Racial Studies* 26(2): 253–277.

Sampson H (2021a) Introduction. In: Gekara VO and Sampson H (eds) *The World of the Seafarer*. London: Springer, 1–8.

Sampson H (2021b) The rhythms of shipboard life: Work, hierarchy, occupational culture and multi-national crews. In: Gekara VO and Sampson H (eds) *The World of the Seafarer*. London: Springer, 87–98.

Stanley J (2016) *From Cabin 'Boys' to Captains: 250 Years of Women at Sea*. Stroud: The History Press.

Stanley J (2018) *A History of the Royal Navy: Women and the Royal Navy*. London and New York, NY: LB Tauris.

Turgo N (2020) Temporalities at sea: Fast time and slow time onboard ocean-going merchant vessels. *Ethnography* http://doi:10.1177/1466138120923371

United Nations (UN) (2020) Secretary-General highlights seafarers' essential role in supply chain, calls for enhanced sustainability, on World Maritime Day. *United Nations Statement and Messages SG/SM/20274, 23 September 2020*. Available at: www.un.org/press/en/2020/sgsm20274.doc.htm

United Nations (UN) Global Compact, United Nations Office of the High Commissioner for Human Rights (OHCHR), the International Labour Organisation (ILO) and The International Maritime Organisation (IMO). (2021) *Maritime Human Rights Risks And The COVID-19 Crew Change Crisis. A Tool to Support Human Rights Due Diligence*. Available at: www.itfglobal.org/en/reports-publications/maritime-human-rights-risks-and-covid-19-crew-change-crisis

United Nations Conference on Trade and Development (UNCTAD) (2020) *Review of Maritime Transport 2020*. New York, NY: United Nations Publications. Available at: https://unctad.org/system/files/official-document/rmt2020_en.pdf

UNCLOS (1982) United Nations Convention on the Law of the Sea. Available at: www.un.org/depts/los/convention_agreements/texts/unclos/unclos_e.pdf

Urry J (2012) Social networks, mobile lives and social inequalities. *Journal of Transport Geography* 21: 24–30.

Yuliantiningsih A and Barkhuizen J (2021) Modern slavery in fishing industry: The need to strengthen law enforcement and international cooperation. *Yustisia*, 10(1): 1–15.

Zhao Z (2021) Recruiting and managing labour for the global shipping industry in China. In: Gekara VO and Sampson H (eds) *The World of the Seafarer*. London: Springer, 23–36.

# 13
# (DE)GROWTH
## The right to the sea

*Maria Hadjimichael*

### Introduction

This chapter explores the blue economy and the idea of economic growth (and degrowth), building on the theories developed in Henri Lefebvre's *Production of Space* and David Harvey's *The Right to the City*. It uses these geographical, spatial, ideas to analyse the ways in which the ocean has been (re)produced through the (re)evaluation of its spatial economic realities. This chapter focuses on the European Union (EU) Blue Growth strategy to explore how such an economic plan and approach for tapping oceanic potential is part of a process of increased commodification and privatisation of the sea. The sea and the coasts are described in the strategy as 'drivers of the economy' and the strategy itself is promoted as "an initiative to harness the untapped potential of Europe's oceans, seas and coasts for jobs and growth" (European Commission, 2012: no page). In turn, the EU strategy is one that does political work by producing the space of the ocean as a market space. Accordingly, constructions built through ocean policies and plans – such as the EU Strategy – change how we understand and use the sea. Such strategies are part of what is now known as the developing 'Blue Economy'.

In this chapter, I argue that the EU's Blue Growth strategy is part of an appropriation of the maritime commons that is becoming institutionalised at an EU level. To develop this argument the chapter unfolds by first examining how ocean space is constructed as an economic space that is an object of economic growth and often subject to institutional intervention and privatisation. The chapter then uses the frameworks and theories provided by Lefebvre and Harvey to discuss how the ocean is *produced* as an economic space of potential, use and exploitation and then, how this raises questions of ownership, property and *rights* to the sea. The chapter ends by considering the march towards institutionalised modes of privatisation and alternative arguments that stress the need for blue *de*growth.

### Constructing an ocean of economic growth

It is impossible to define ocean space in a single way, given it is a space of vastness and heterogeneity as well as a space of multiple imaginings. Indeed, ocean space in time has taken on different forms; it has been a space of physical, political and social struggle (see Featherstone, this volume; Griffin, this volume), as well as manifesting as an imaginative space of poetic or

DOI: 10.4324/9781315111643-16

artistic inspiration (see Crawley et al., this volume; Jones, this volume). The ocean is also a migratory route and home for both humans as well as non-human species. Moreover, theories about the ocean, such as 'wet ontology' (Steinberg and Peters, 2015) and 'more-than-wet ontology' (Peters and Steinberg, 2019) highlight the different states of the ocean, from liquid, to solid (ice) and air (mist) which one must take on board when thinking with the sea/about the sea. Such ideas provide "means by which the sea's material and phenomenological distinctiveness can facilitate the reimagining and re-enlivening of a world ever on the move" (Steinberg and Peters, 2015: 248). Such work opens up new frames of thinking about spatial experiences that derive from an ontology that builds on the character of the sea as a more-than-wet space (Peters and Steinberg, 2019: 20) allowing us to not be constrained by our 'landed lives' and territorial frames of thinking. Thus, a new 'Thalassology' (Anderson and Peters, 2014: 17) of water–human interactions can be conceptualised that helps us think of ocean spaces, and our relations with them, anew.

This chapter will not attempt to provide a deeper evaluation of differing oceanic definitions and, linked to this, the struggles over and about ocean space that arise with competing ideas, uses and visions of the seas. Rather, it will 'ground' its analysis on discussions over the meanings of ocean space being produced around its economic definition as a space of growth, and how that definition is shaping its future in laying the foundations of struggles for producing counter meanings moving forwards. The chapter does so by thinking of ocean space through an ontology alerted to the very difference that water makes to its emergent economic use. Indeed, ocean space as economic space has been viewed, imagined and constructed in multiple ways, whether this is through the articulation of alternative cosmopolitics (Childs, 2020), as a valuable 'real estate' product the battle for which resembles the 'gold rush' (as the Pacific Ocean is described by Conway et al., 2010), or a commodity frontier (Campling, 2012). Currently, marine space has become the latest element from which a vast amount of economic wealth can be extracted to ensure that economic growth is ensued; something which has been picked up across geographic scales, whether under the name of Blue Growth in the European Union or the Blue Economy per se (Ertör and Hadjimichael, 2020).

In his 2001 analysis, Steinberg is explicitly constructivist in his history of ocean space, giving attention to the social structures, individual behaviours, institutional arrangements, and natural features that have intersected to create, or construct, specific spaces, both on land and sea (Steinberg, 2001: 21). Central to Steinberg's analysis is a discussion of the ways in which the ocean has been constructed in various ways for economic growth, from mercantile to postmodern capitalism. Here, however, I focus on Lefebvre's notion of the 'right to the city' (1991) in the context of the ocean in view of how space is *produced* under capitalism. Political geography and political ecology have already taken up discussion, to some degree, on the economic-spatialities of the ocean. Campling and Colás (2018) suggest, through the term 'terraqueous territoriality', for example, that capital accumulation uses experiences from land in order to territorialise the sea, yet though capitalism the sea provides distinctive spatial and juridical forms aimed at reconciling the production, appropriation and distribution of value at sea.

Crucial to economic growth at sea, under capitalism, and to the increased privatisation of the ocean, has been the definition of the sea as an empty space (see also Germond, this volume). The term *space* per se, is commonly specified as "an empty area that is available to be used" (Cambridge Dictionary, no date). Although such an understanding of space is misleading at best, the definition is itself a manifestation of dominant western ideas of (ocean) space that have been projected to society (see also Davies, this volume; Fawcett et al., this volume). This links to what Lefebvre describes as 'abstract space'. Abstract space is a space which becomes instrumentalised in the sense that it is manipulated by all kinds of authorities whilst it has the power to maintain specific social relations, to dissolve others, and stand opposed to others through the use of technology, applied sciences and knowledge bound to power.

According to Wilson,

> [T]he concept of abstract space should be understood as an attempt to grasp the ways in which the space of capital embodies, facilitates and conceals the complex intertwining of structural, symbolic and direct forms of violence that Lefebvre refers to as the 'violence of abstraction', and it is in this sense that the concept offers a unique contribution to our understanding of the capitalist production of space.
>
> *Wilson, 2013: 517*

Just as space in the form that it is socially produced can be a means of production, it can also be a mean of control and domination, and as (social) space can be 'abstract', it can also be real just as in the way concrete abstractions such as a commodity can be real (Lefebvre, 1991: 26). The 'abstraction' of ocean space has an apparent political symbolism, which in the more recent years, this chapter argues, has been filled with the ideology of growth – projecting the ocean as the new commodity frontier, ripe for development, often through modes of privatisation and institutional orderings that promote its use for economic gain (Campling, 2012; Ertör and Ortega-Cerdà, 2019). In understanding this abstraction of the ocean, it is important to link the ecological transformation undermining marine ecosystems with the commodification of marine resources and consequently of ocean space (Longo et al., 2015). For example, Ranganathan (2019) suggests that the law has turned the seabed over to the extractive interests of states and corporations, creating a new legal imaginary of the seabed which constructs the seabed as primarily an economic site and a private site of mining.

This construction of ocean space is creating injustices over its ownership, use and exploitation (Bennett et al., 2020). Thus, beyond an analysis of ocean space and the way it becomes abstracted, emptied and reconstructed in political, economic and social terms in the light of new strategies and economic realities, we must also examine who is affected and how: who becomes dispossessed over their rights to this space (or resource) and who 'wins'. This discussion has taken place already in terms of fisheries, where the term 'ocean grabbing' has been used by the Transnational Institute to show the way in which certain economic processes and dynamics of private capital are impacting people and communities whose way of life, cultural identity and livelihoods depend on activities, for example such as small-scale fishing (Franco et al., 2014). The term 'grab', referring to the grab of (often) private companies (but also states) for economic spoils, has been used to illustrate similar processes taking place at the coastal realm (Bavinck et al., 2017) as well as deeper and further from one's 'view', the ocean floor/seabed (Mallin, 2018; Ranganathan, 2019). The chapter will now build from this discussion of the economic growth of ocean space to explore the EU Blue Growth Strategy as a window for critically interrogating the commodification and privatisation of the sea and challenges to this.

## The production and re-production of the seascape

Lefebvre in his discussion about the 'Production of Space' suggests that capitalism now has laid "its imprint upon the total occupation of all pre-existing space and upon the production of new space" (1991: 326). He puts forward that capital can actively shape space not just to achieve ideological results or to sustain appropriate conditions for production in other spheres, but because of "the market in spaces themselves" (1991: 86). As he notes,

> When it comes to space, can we legitimately speak of scarcity? The answer is no – because available or vacant spaces are still to be found in unlimited numbers, and even though a relative lack of space may have left its mark on some societies (particularly in Asia), there are others where just the opposite is true – where, as in North America, society bears the clear traces of the vastness of the space open to its demographic and technological expansion. Indeed, the space of nature remains open on every side, and thanks to technology we can 'construct' whatever and wherever we wish, at the bottom of the ocean, in deserts or on mountaintops – even, if need be, in interplanetary space.
>
> *Lefebvre, 1991: 330*

Lefebvre's 'trialectics of space' theory is at the core of arguments for reclaiming the seascape and the coastal environment under capitalist regimes of, for example, privatisation. This 'conceptual triad' is comprised of three 'spatial moments' that affect each other simultaneously: (i) perceived space, refers to space in its real, physical form, as it is perceived and generated; (ii) conceived space, refers to space in its imagined, mental form, as it is conceived and imagined; (iii) lived space, refers to space as it is lived and modified over time through its use (Dhaliwal, 2012). 'Perceived space' is where the routine of everyday practices and perceptions are played out, whilst 'conceived space' is space as it is conceived once imagined and often links as the space of cartographers, urban planners and property speculators. 'Lived space', finally, represents the spatial imaginary of the time which has the power to transcend and refigure conceived space.

Currently, the push of the neoliberal dogma for the creation of markets in the sea and the seashore highlights the battle between "those who produce a space for domination against those who produce space as an appropriation to serve human need" (Molotch, 1993: 889). By applying the 'trialectics of space' to the seascape and the seashore, we can gain insights into how the Blue Growth strategy is not solely an attempt to exploit markets within new spaces, but is also a strategy that *re-defines our relations with the sea*, through the effects of such a strategy on our spatial imaginary. For example, the Exclusive Economic Zone (EEZ) can be understood as a strategic political technology, enrolled to normalise ocean space and secure it as a space for national endeavours (Sammler, 2016) – or what has also been termed as a "classic case of enclosure" (Buck, 1998: 84). Through a lens of political economy of ocean space we can examine how ocean space has been re-evaluated and reproduced under different terms of reference, through a transfiguration of the trialectics of ocean space. Though "global capitalism is a seaborne phenomenon" (Campling and Colás, 2021: 16), ocean space has been re-evaluated and reproduced, particularly since the institutionalisation of the EEZ, with licensing procedures for activities and developments in/on ocean space becoming simplified. Blue Growth as an economic strategy builds on the societal growth imaginary, and has allowed for the creation of a new 'lived space', thus normalising and thus allowing for ocean space to be reconceived.

Indeed, the Blue Growth strategy can be analysed through the understanding of the 'treadmill approach', with which sustainable development becomes synonymous to sustainable growth, while 'development' tends to be measured in terms of the expansion of gross national product (Baker et al., 1997). Schnaiberg (1997) suggests that the 'treadmill' has different components, each of which has different goals. Its economic component has the goal of expanding industrial production and economic development, and concomitantly the increase of consumption. Its political one has a public confluence of interests among private capital, labour and government in promoting this expansion. Its social component advances public welfare primarily through economic growth. In the current climate of the global financial crisis, developed countries are increasingly focused on economic growth and development, and the crisis is continuously being used to entrench a neoliberal agenda allowing for further deregulation of the economy and privatisation of public assets. The expansion of the privatisation of space for corporate interests has moved from primarily on land to marine space with a range of policies and strategies. Here is where the danger lies with the EU's Blue Growth strategy, in which the sea and the coasts are described as "drivers of the economy" and the strategy as "an initiative to harness the untapped potential of Europe's oceans, seas and coasts for jobs and growth" (European Commission, 2012).

## From the *Right to the City* to the *Right to the Sea*

When David Harvey discusses the concept of accumulation by dispossession, he builds on Marx's concept of primitive accumulation and frames it around the example of "capital accumulation through urbanization" (Harvey, 2012: 42). In this discussion, urbanisation is put forward as having a particular function in the dynamics of capital accumulation, as well as a geographical specificity, such as the production of space and of spatial monopolies, which become integral to this accumulation. Thus, whilst in the work of Lefebvre, abstract space allows for different sites to be strategically arranged in a way that capital accumulation and everyday life can be unfolded, further work has attempted to interpret this in a way through which tactics of primitive accumulation are articulated (i.e. through ideological manoeuvres, scientific discourse, and through legal tools [Mels, 2014]). In the same way, terms and strategies like 'blue growth' and the 'blue economy' have a role in the accumulation of capital through the production of new economic realities of the ocean space.

The idea for reclaiming the right to the *sea* arose from the idea put forward by Harvey (2008) in his essay *The Right to the City* (following 'the right to the city' proposed by Lefebvre [1996]) in his book *Le droit à la ville*), which suggested that the 'right to the city' is a human right. Lefebvre discusses the right to the city as the "demand… [for] a transformed and renewed access to urban life" (1996: 158). For David Harvey, the idea of the right to the city can be taken literally, and give a theoretical framework to the fight of urban commoners in reclaiming the squares, the streets, the right to housing and rights to exist in space.

But what can 'the right to the sea' imply? It is important to take on the notion of rights carefully, as the usefulness of rights comes to an end when they lose their aim of resisting injustice. As Costas Douzinas states in his essay, *The End of Human Rights*:

> they are an expression of the human urge to resist public and private domination and oppression. Their force unites Chinese dissidents, the defenders of refugees, immigrants and detainees of the war on terror as well as schoolkids in Greece. In the hands of western governments however they have become the latest version of the civilising mission.
>
> *Douzinas, 2008: no page*

Attoh (2011) has critically examined the concept of the right to the city and uses discussions and theories on *rights* to explore what the different theorists who use the concept mean by it. Disputes on rights, for example, discuss the rights of individuals and groups, but also what it means when these rights are violated. Hohfeld (as discussed in Attoh, 2011), for example, argues that all legal entitlements can be understood as either one or a combination of four basic rights: claim rights, liberty rights, powers and immunities. A claim right is a right which entails responsibilities, duties or obligations on other parties regarding the right-holder. In contrast, a liberty right is a right which does not entail obligations on other parties, but rather only freedom or permission for the right-holder. To envision the right to the city, or the sea as a claim right would mean something very different than to envision it as an immunity or liberty right.

Attoh then discusses Waldron's (1993) generational rights, where first generation rights refer to traditional liberties and privileges of citizenship (free speech, religion etc.), second generation rights refer to socioeconomic rights such as housing, fair wage and their links to the welfare state and the third generation rights are rights attached to communities, peoples and groups such as minority language rights, environmental values, integrity of culture etc. Finally, Attoh addresses Dworkin's essay (1977), *Taking Rights Seriously* which focuses on the relationship between moral rights and constitutional law, whilst understanding rights in a democracy as, "the majority's promise to minorities that their dignity and equality will be respected" (Attoh, 2011: 673). He suggests that when laws infringe upon our dignity or our equality, the language of rights allows us to challenge but also to break such laws.

In sum, Attoh states that "rights [...] are sites of struggle, [...] (and) how we define them surely matters" (2011: 670). If we are to define the right to the sea, we would not be able to do so universally. It is apparent that the right to the sea can be defined differently for different groups and communities. For small-scale fishers and fishing communities, for example, it can be understood also as a claim right, where the right holder has a responsibility to protect the natural resource. For broader society, the right to the sea can be understood as a liberty right, one of freedom to access and enjoy the sea. How I understand it primarily, however, is a provocation. Just like for the right to the city, the openness of the notion is something to be welcomed as it can "serve to unify the struggles of various marginalized groups around a common rallying cry" (Attoh, 2011: 674, drawing on Mitchell and Heynen, 2009).

## Blue growth and the EU's strategy: Production and rights

The two frameworks outlined – the production of space and the right to the sea – are helpful in making sense of EU Blue Growth Strategy and in turn, the ocean as a commodity frontier (see also Fawcett et al., this volume). Indeed, they inform the ways that ocean space has been conceived by politicians, multinational organisations and technocrats, and have given rise to new possibilities, the name of which has become known as 'blue growth' or the 'blue economy'. The latest economic fantasy has been given a blue colour and has materialised in ocean space, across continents and oceans (for a more detailed outline of the blue economy strategies across continents and oceans see Winder and Le Heron, 2017; Ertör and Hadjimichael, 2020).

The EU's Blue Growth strategy envisioned ocean space as an element for re-starting the Community's economy and suggested that the 'blue' economy represents 5.4 million jobs and a gross added value of just under €500 billion a year, focusing on five key sectors which have been identified via an analysis of the job-creation potential, as well as the potential for research and development to deliver technology improvements and innovation. These sectors, marine aquaculture, coastal tourism, marine biotechnology, ocean energy and seabed mining will be supported by fostering investment in research and innovation, promoting skills through education and training and by removing the administrative barriers that hamper growth.

In previous work, I have used the Environmental Impact Assessments (EIAs) conducted by the European Commission for these sectors, as well as existing literature, to specifically look at the five sectors and the potential socio-ecological impacts of their expansion (see Hadjumichael, 2018). Although it is recognised that the actual social, environmental and economic impacts of this strategy are difficult to analyse at such an early stage, an initial appraisal becomes possible through past experiences. Such an understanding of the impact of these sectors, the way they *change spatial realities* as well as social and economic realities, allows for an understating of precisely the way the production of the ocean works, according to the blue growth paradigm. At the same time, an understanding of the struggles over the space – the rights to sea – allows us to explore the possibilities of a collective right to the sea, to become a right over the exploitation of the sea for the few.

Marine aquaculture, for example, is a sector whose economic importance has led to its elevation in Europe at the EU and at national level. The facilitation of its development has occurred through deregulation of the sector and financial assistance (i.e. through subsidies) (Grigorakis and Rigos, 2011). The inclusion of marine aquaculture in the Blue Growth strategy suggests the intensification of these policies. Although there is environmental damage to coastal ecosystems from marine aquaculture (through wastes offloads, introduction of alien species, genetic interactions, disease transfer, release of chemicals, use of wild recourses, alterations of coastal habitats and disturbance of wildlife), which in turn lead to socioeconomic and environmental problems (Bohnes et al., 2018), the reproduction of marine space for aquaculture also involves socio-environmental conflicts with existing users (Aducci, 2009; Ertör and Ortega-Cerdà, 2015). A political ecology perspective on the expansion of marine aquaculture can suggest that within the current institutional and decision-making structures, there is the danger of continuously *undermining the rights* of coastal communities and other users of the sea through deregulation, privatisation and capital development – for further growth (Hadjimichael et al., 2014).

Moreover, marine biotechnology developments also present examples of a specific oceanic production – a spatial imaginary of an abstracted ocean open for privatisation and growth. Marine biotechnology refers to the development of new pharmaceuticals or industrial enzymes that can withstand extreme conditions, through the use of knowledge acquired from the exploration and exploitation of marine biodiversity. The sector is presented as a niche sector with the potential to produce mass-market products in the long run. No significant environmental impacts of this activity are yet identified, and there appears only a few tensions between the sector and other marine activities (Ecorys, 2012), something that is supported by a study into the literature. However, one of the points in the European Commission's (EC) Communication (2012) is that intensified efforts to manufacture biofuels and other chemicals from algae arose from concerns about growing crops for biofuels on land. Thus, it is yet to be seen whether the land grabbing which has been taking place in countries in Africa, South and Central America and Asia for the production of biofuels will be intensified in the form of *ocean grabbing* as such activities shift from land to the 'abundant' space of the sea (Borras et al., 2011, 2012).

The production of energy from the ocean is identified as a sector whose expansion at sea can serve the increasing energy needs of European societies. Besides the widely known offshore windfarms, ocean energy technologies are also being developed to exploit the energy potential of tides and waves as well as differences in temperature and salinity. The ocean is thus being reproduced as a vast energy hub, without issues such as ocean grabbing being sufficiently addressed. Additionally, timely and politically important questions that have to do with ways in which this renewable energy transition ensuring more democratic and just renewable energy futures (as discussed in Burke and Stephens, 2018). In this framework, questions around right(s) to the sea, must address the objectives of this spatial reproduction, and analyse them in terms beyond economic growth.

Furthermore, tourism is a sector which has already received attention with regards to its social and ecological impacts. The Blue Growth strategy promotes maritime, coastal and cruise tourism through the focus on its growth and jobs potential in its relevant strategy (European Commission, 2014). Literature suggests however that it is local communities and environments that will have to face the burden of such growth (Boissevain and Selwyn, 2004; Gössling, 2002). Tourism is an industrial activity which consumes (sometimes over-consumes) resources, creates waste and has specific infrastructural needs, whilst it is a private sector whose investment decisions are predominantly based on profit maximisation. Investigations of tourism-related interactions often point to conflicts between different interest groups contesting ownership and control of the coast with examples along the European coastline varying from issues such as the privatisation of the Mediterranean coastline due to tourist-related developments, conflicts between fisheries, aquaculture, marine protected areas and tourism as well as socio-cultural and economic impacts (Boissevain and Selwyn, 2004). Accordingly, in a recent Special Issue on 'Tourism and Degrowth', Fletcher et al. (2019) bring together discussions on degrowth and the tourism industry, in an attempt to take the opportunity to re-politicise the discussion of tourism development. The Special Issue puts forward that though tourism is commonly portrayed as a relatively clean sector, its growth requires vast amounts of materials and energy to be transformed into capital and its questioning is imperative to challenge the growth imperative. Thus, tourism is in many ways a sector that offers much for reflecting on rights to the sea and the production of the sea as a private space of capital accumulation. Indeed, a different approach to tourism and economics has recently been proposed, following alternative forms of development, and aiming towards a just and materially responsible society (Hall, 2009). The idea of tourism degrowth (Fletcher et al., 2019) calls for scholars to be attentive to the demands and struggles of social movements so that another tourism becomes possible within planetary boundaries. Policy-makers (from different institutions at different levels) must do the same.

Finally, the most controversial sector from those promoted within the EU Blue Growth strategy which opens up spatial imaginaries around ocean economic growth and rights, is seabed mining. The sector is promoted with the premise that advances in underwater technology means that mining companies can exploit the seafloor and contribute towards the growing global demand for non-energy raw materials. It has been calculated that the global annual turnover of marine mineral mining can be expected to grow from virtually nothing to €5 billion in the next ten years and up to €10 billion by 2030 (European Commission, 2012). The environmental impacts disclosed in the relevant Blue Growth study on marine mineral resources acknowledge that seabed mining has "considerable environmental concerns and uncertainties" and, the "effects on the ecosystems are difficult or even near impossible to predict" and that "processing demands high energy input and the use of chemicals" (Ecorys, 2012). The EC suggests that to overcome this hurdle funding should also flow towards assessing and minimising environmental impacts of seabed mining. Van Dover et al. (2017), however, suggest that despite potential technological advances, mining-induced impacts cannot be remedied given the very slow natural rates of recovery in affected ecosystems and thus focus must be on avoiding and minimising harm. Seabed mining by European Companies will not have its only focus in EU waters but rather within the high seas and third country waters. Specifically, in the relevant Blue Growth study (Ecorys, 2012: 28), a specific project called the Nautilus project in Papua New Guinea (PNG), which won the first lease to mine the ocean floor for gold and copper, is presented as a successful pilot. Further research, however, has shown a grassroots counter-movement opposing the project with vocal opposition by custom land owners, non-government groups, faith-based groups and scientists (Deep Sea Mining Campaign, 2013) .

## Conclusion: Towards blue *de*growth?

We live in an era when ideals of human rights have moved centre stage both politically and ethically. A great deal of energy is expended in promoting their significance for

> the construction of a better world. But for the most part the concepts circulating do not fundamentally challenge hegemonic liberal and neoliberal market logics, or the dominant modes of legality and state action. We live, after all, in a world in which the rights of private property and the profit rate trump all other notions of rights.
>
> *Harvey, 2008: 1*

Reflecting on Harvey's perhaps rather grim quote above, Lefebvre offers perhaps more hope. He argues that the political utility of a concept isn't that it should tally with reality, but that it enables us to experiment with reality, that it helps us glimpse another reality, a virtual reality that's there, somewhere, waiting to be born inside us. The right to the city, "implies nothing less than a new revolutionary conception of citizenship" (Lefebvre, 2014: 205). Thus, as Andy Merrifield asks, if we are to argue of a right to the city, it is important to ask ourselves the right to *what* city. Consequently, the right to what *sea*?

So how can we move forward? This chapter has explored the idea of a Blue Growth, focusing on the EU Blue Growth Strategy which emerges from an understanding of, or production of space, that imagines the sea somewhere empty, abstract, and through such abstraction, a space of potential capitalist accumulation through activities from aquaculture to seabed mining. But such growth raises questions of rights and such a question of rights acknowledges inequalities and inequities (growth for what and for who) and the need to question rights and realities – and perhaps imagine new ones, such as *degrowing* and re-imagining the sea as commons. The idea of degrowth, defined as an equitable downscaling of production and consumption that increases human well-being and enhances ecological conditions at the local and global level, in the short and long term, is an important counter-concept.

The first step to reclaim the sea is the deconstruction of the reasons behind its appropriation. By highlighting our increasing (constructed) energy needs (ocean energy), needs for fuels (marine biotechnology), fish (marine aquaculture), minerals (seabed mining) and a certain type of entertainment (tourism), it reinforces a treadmill of production, consumption and accumulation. It demonstrates an increased claiming of the sea through privatisation and reveals its pitfalls. The focus on economic growth based on the belief that it is the core for public welfare is highly problematic. Under Blue Growth, workers will have to prove they can work hard, consumers will need to keep consuming to support this new economy, and companies will need to accumulate more profit to prove the success of the strategy. The seascape and the coastal environment become another abstract market space and their importance in both environmental and social terms becomes unimportant. As the sea becomes one of the new spaces, which as Lefebvre argued, has the functional necessity able to "save capitalism from extinction" (1991: 335), its prior importance to the people and the communities will transform, and so will people's relations with it.

The Blue Growth strategy is a tool which whilst it is enclosing the marine commons through acts of privatisation, incentivised by institution, it is, at the same time, changing people's understanding and relations to the sea to one of domination and exploitation. Institutions such as the EU that underscore this shift, alongside the companies that benefit from financial incentives and subsidies, are not inherently bad; it is the amalgamation of the values and ideologies which drive them, and the power dynamics and social hierarchies they create, which makes them so (see also Heller et al., this volume). Considering the erosion of democratic rights at the expense of restoring a status quo focusing on economic growth, the role of degrowth strategies can be interpreted as a reaction against traditional centres of public authority (such as the EU, national governments, and the market), which have been driving the economic growth ideology, inculcating it as the ultimate goal for achieving social well-being (D'Alisa et al., 2013).

The idea of the *commons* suggests radically democratic solutions that don't create conflicts between environmental concerns and social justice, because economic growth is not an element they require in order to thrive. As Harvey notes, "[t]he political recognition that the commons can be produced, protected, and used for social benefit becomes a framework for resisting capitalist power and rethinking the politics of an anticapitalist transition" (Harvey, 2012: 87). And this is what reclaiming the seascape suggests; the refusal to shift what Castoriadis (1987) defined as our "social imaginary" of the seascape, to not therefore allow it to become another space which capitalism can exploit in order for our economies to grow.

## References


Adduci M (2009) Neoliberal wave rocks Chilika Lake, India: Conflict over intensive aquaculture from a class perspective. *Journal of Agrarian Change* 9(4): 484–511.

Anderson J and Peters K (2014) 'A perfect and absolute blank': Human geographies of water worlds. In: Anderson J and Peters K (eds) *Water Worlds: Human Geographies of the Ocean*. Abingdon: Routledge, 21–38.

Attoh KA (2011) What kind of right is the right to the city? *Progress in Human Geography* 35(5): 669–685.

Baker S, Kousis M, Richardson D and Young S (1997) Introduction: The theory and practice of sustainable development in EU perspective. In: Baker S, Kousis M, Richardson D and Young S (eds) *The Politics of Sustainable Development: Theory, Policy and Practice Within the European Union*. London and New York, NY, 1–40.

Bavinck M, Berkes F, Charles A, Esteves Dias AC, Doubleday N, Nayak P and Sowman N (2017) The impact of coastal grabbing on community conservation–a global reconnaissance. *Maritime Studies* 16(8) https://doi.org/10.1186/s40152-017-0062-8.

Bennett MM, Stephenson SR, Yang K, Bravo MT and De Jonghe B (2020) The opening of the Transpolar Sea Route: Logistical, geopolitical, environmental, and socioeconomic impacts. *Marine Policy* 121 https://doi.org/10.1016/j.marpol.2020.104178

Boissevain J and Selwyn T (2004) Contesting the foreshore. *Tourism, Society, and Politics on the Coast*. MARE Publication Series No. 2. Amsterdam: Amsterdam University Press.

Bohnes FA, Hauschild MZ, Schlundt J and Laurent A (2018) Review of environmental sustainability assessments of aquaculture systems: main findings and outlook. In: *2018 International Conference on Sustainable Global Aquaculture*.

Borras Jr SM, Hall R, Scoones I, White B and Wolford W (2011) Towards a better understanding of global land grabbing: an editorial introduction. *The Journal of Peasant Studies* 38(2): 209–216.

Borras Jr SM, Franco JC, Gómez S, Kay C and Spoor M (2012) Land grabbing in Latin America and the Caribbean. *The Journal of Peasant Studies* 39(3–4): 845–872.

Buck SJ (1998) *The Global Commons: An Introduction*. New York, NY and Abingdon: Routledge.

Burke MJ and Stephens JC (2018) Political power and renewable energy futures: A critical review. *Energy Research and Social Science* 35: 78–93.

Campling L (2012) The tuna 'commodity frontier': Business strategies and environment in the industrial tuna fisheries of the Western Indian Ocean. *Journal of Agrarian Change* 12(2–3): 252–278.

Campling L and Colás A (2018) Capitalism and the sea: Sovereignty, territory and appropriation in the global ocean. *Environment and Planning D: Society and Space* 36(4): 776–794.

Campling L and Colás A (2021) *Capitalism and the Sea: The Maritime Factor in the Making of the Modern World*. London: Verso Books.

Castoriadis C (1987 [1975]) *The Imaginary Institution of Society*. Cambridge, MA: Polity Press.

Childs J (2020) Performing 'blue degrowth': Critiquing seabed mining in Papua New Guinea through creative practice. *Sustainability Science* 15(1): 1–13.

Conway F, Stevenson J, Hunter D, Stefanovich M, Campbell H, Covell Z and Yin Y (2010) Ocean space, ocean place: The human dimensions of wave energy in Oregon. *Oceanography* 23(2): 82–91.

D'Alisa G, Demaria F and Cattaneo C (2013) Civil and uncivil actors for a degrowth society. *Journal of Civil Society* 9(2): 212–224.

Deep Sea Mining Campaign (2013) Available at: www.deepseaminingoutofourdepth.org

Dhaliwal P (2012) Public squares and resistance: The politics of space in the Indignados movement. *Interface: A Journal for and about Social Movements* 4(1): 251–273.

Doyzinas C (2008) The 'end' of human rights. *The Guardian [Online]*. Available at: www.theguardian.com/commentisfree/2008/dec/10/humanrights-unitednations

Ecorys (2012) Blue Growth: Scenarios and drivers for sustainable growth from the oceans, seas and coasts. Marine Sub-Function Profile Report. Marine Mineral Resources (3.6). Third Interim Report. Final. Call for Tenders No. MARE/2010/01

Ertör I and Ortega-Cerdà M (2019) The expansion of intensive marine aquaculture in Turkey: The next-to-last commodity frontier? *Journal of Agrarian Change* 19(2): 337–360.

Ertör I and Ortega-Cerdà M (2015) Political lessons from early warnings: Marine finfish aquaculture conflicts in Europe. *Marine Policy* 51: 202–210.

Ertör I and Hadjimichael M (2020) Blue degrowth and the politics of the sea: Rethinking the blue economy. *Sustainability Science* 15(1):1–10.

European Commission (2012) Communication from the Commission to the European Parliament, the Council, the European Economic and Social Committee and the Committee of the Regions. Blue Growth – opportunities for marine and maritime sustainable growth, COM(2012) 494, final.

European Commission (2014) Communication from the Commission to the European Parliament, the Council, the European Economic and Social Committee and the Committee of the Regions. A European Strategy for more Growth and Jobs in Coastal and Maritime Tourism, COM(2014) 86, final.

Fletcher R, Murray Mas I, Blanco-Romero A and Blázquez-Salom M (2019) Tourism and degrowth: An emerging agenda for research and praxis. *Journal of Sustainable Tourism* 27(12): 1745–1763.

Franco J, Buxton N, Vervest P, Feodoroff T, Pedersen C, Reuter R and Barbesfaard M (2014) The global ocean grab: A primer. Available at: www.tni.org/en/publication/the-global-ocean-grab-a-primer

Grigorakis K and Rigos G (2011) Aquaculture effects on environmental and public welfare: The case of Mediterranean mariculture. *Chemosphere* 85(6): 899–919.

Gössling S (2002) Global environmental consequences of tourism. *Global Environmental Change* 12(4): 283–302.

Hadjimichael M, Bruggeman A and Lange MA (2014) Tragedy of the few? A political ecology perspective of the right to the sea: The Cyprus marine aquaculture sector. *Marine Policy* 49: 12–19.

Hadjimichael M (2018) A call for a blue degrowth: Unravelling the European Union's fisheries and maritime policies. *Marine Policy* 94: 158–164.

Hall CM (2009) Degrowing tourism: Décroissance, sustainable consumption and steadystate tourism. *Anatolia* 20(1): 46–61.

Harvey D (2012) *Rebel Cities: From the Right to the City to the Urban Revolution*. London: Verso.

Harvey D (2008) The right to the city. *New Left Review*, 53. Available at: https://newleftreview.org/issues/ii53/articles/david-harvey-the-right-to-the-city

Lefebvre H (1991) *The Production of Space*. Oxford: Blackwell.

Lefebvre H (2014) Dissolving city, planetary metamorphosis. *Environment and Planning D: Society and Space* 32(2): 203–205.

Lefebvre H (1968[1996]) *Writings on Cities* (ed. and trans. E Kofman and E Lebas). Oxford: Blackwell.

Longo SB, Clausen R and Clark B (2015) *The Tragedy of the Commodity: Oceans, Fisheries, and Aquaculture*. New Brunswick, NJ: Rutgers University Press.

Mallin MAF (2018) From sea-level rise to seabed grabbing: The political economy of climate change in Kiribati. *Marine Policy* 97: 244–252.

Mels T (2014) Primitive accumulation and the production of abstract space: Nineteenth-century mire reclamation on Gotland. *Antipode* 46(4): 1113–1133.

Molotch H (1993) The space of Lefebvre. *Theory and Society* 22(6): 887–895.

Peters K and Steinberg PE (2019) The ocean in excess: Towards a more-than-wet ontology. *Dialogues in Human Geography* 9(3): 293–307.

Ranganathan S (2019) Ocean floor grab: International law and the making of an extractive imaginary. *European Journal of International Law* 30(2): 573–600.

Sammler KG (2016) The deep pacific: Island governance and seabed mineral development. In: Stratford E (ed) *Island Geographies: Essays and conversations*. Abingdon: Routledge, 24–45.

Schnaiberg A (1997) Sustainable development and the treadmill of production. In: Baker S, Kousis M, Richardson D and Young S (eds) *The Politics of Sustainable Development: Theory, Policy, and Practice within the European Union*. New York, NY: Routledge, 72–88.

Steinberg PE (2001) *The Social Construction of the Ocean*. Cambridge: Cambridge University Press.

Steinberg PE and Peters K (2015) Wet ontologies, fluid spaces: Giving depth to volume through oceanic thinking. *Environment and Planning D: Society and Space* 33(2): 247–264.

Van Dover CL, Ardron JA, Escobar E, Gianni M, Gjerde KM, Jaeckel A, Jones DOB, Levin LA, Niner HJ, Pendelton L, Smith CR, Thiele T, Turner PJ, Watling L and Weaver PPE (2017) Biodiversity loss from deep-sea mining. *National Geoscience* 10(7): 464.

Wilson J (2013) "The devastating conquest of the lived by the conceived": The concept of abstract space in the work of Henri Lefebvre. *Space and Culture* 16(3): 364–380.

Winder GM and Le Heron R (2017) Assembling a Blue Economy moment? Geographic engagement with globalizing biological-economic relations in multi-use marine environments. *Dialogues in Human Geography* 7(1): 3–26.

# 14
# RESOURCES
## Feminist geopolitics of ocean imaginaries and resource securitisation

*Amanda Thomas, Sophie Bond and Gradon Diprose*

### Introduction

Given the increasingly common understanding of oceans as the 'last frontier' for resources – of rare minerals, including oil and gas (see Fawcett et al., this volume; Steinberg, 2018; Zalik, 2018) – it is increasingly important to recognise the highly contested and political nature of resource geographies, and to investigate not only economic dimensions but how oceans are imagined and in turn how they are 'secured'. Security is a key element to how the 'rules, norms and geographic divisions' in ocean spaces are re-articulated through 'competing social forces' (Steinberg, 2011: 14). Dominant framings are concerned with securing access to resources in the name of economic progress. More so, an imaginary of oceans as 'stable spaces' of 'functional zones' that are benignly defined by 'international law' frames security as an apolitical necessity (Steinberg, 2011: 14). This version of security is distributed in the name of a common good, for peace and conservation. Yet key questions around what is being 'secured', for whom and by whom are raised in a framing that is attentive to flows and fluxes in security. Important are alternate ocean imaginaries that contest the colonial and capitalist roots and implications of such frontier constructions, and highlight the complex geopolitics of securing ocean resources.

In this chapter, we draw on issues of resource extraction through the frames of sovereignty, and Indigenous and climate justice imaginaries to highlight these points. We tease out the geopolitics of ocean security, focusing first on how we understand resource geographies from a feminist geopolitical perspective. Feminist geopolitics takes a multi-scalar approach and interrogates how security of resources is defined, by who, and for who. We then turn to how oceanic resource frontiers are secured by drawing on a number of examples from the territorial waters and Exclusive Economic Zone (EEZ) of Aotearoa New Zealand. We focus on the contested nature of securing ocean spaces by exploring a case study of contestation over deep sea oil exploration and extraction. This case study demonstrates the different meanings of both resources and security, and the importance of different kinds of sovereignty in mobilising these meanings.

DOI: 10.4324/9781315111643-17

## Framing resources and security

Resource geographies focus on the material and social relationships that construct and shape the very idea of, and value attributed to, a thing or 'resource' (Bridge, 2014). Such valuing involves a mix of the material world, constructions of need, and social relationships that shape these constructions. Different things become resources at different times according to perceived demand, and the capacity of that thing to deliver a service (Huber, 2018). The value of a resource can be thought of in economic terms but also according to the cultural and ethical values attributed to it, and the way these social dynamics shape that economic value. As Huber (2018: 154, emphasis in original) states, "while value is abstract and fetishizes exchange value it *cannot escape* the cultural particularities of use value".

Critical resource geographies have examined how turning objects into resources requires the enclosure of common goods: "resource making, then, is a form of taking or theft in which the material and cultural attachments of existing resource users are alienated" (Bridge, 2010: 824). Suchet (2002) connects these processes of enclosure to colonisation and colonialism. She describes the way Eurocentric views often parcel the world into objects that are constructed as "resources to be developed or conserved to fuel scientific and capitalistic processes" (Suchet, 2002: 147). Resource making is also, therefore, an act of asserting power over space. Moreover, resource making is often also state-making – states are typically at the centre of these assertions of power. The state is, however, a moving beast, made up of many different relationships and specific ways of claiming and regulating territories (Bridge, 2014).

In contrast to a world of discrete objects that are made into resources through enclosure, a wealth of relational thinking understands a world of 'things'. The value of these things exists within relationships with other things and people. Indigenous communities and theorists have detailed the complex, nuanced ways different communities understand, know and value things around them (see Coombes et al., 2012; Waiti and Wheaton, this volume). So, while a river may be a resource in that it provides basic necessities for life, it is also a resource that fulfils a range of cultural, societal, spiritual, ecosystem needs. What it means to count as a 'resource' is specific to the different *worlds* that exist. For example, Blaser (2009) describes a supposed agreement between an Indigenous Yshiro community and the Paraguayan government related to hunting and species management that ultimately failed. Blaser argues that it failed not because of different cultural understandings about a singular natural world that exists, but rather because entirely different worlds, with different truths and realities, exist at the same time. Sustainable management of hunting failed because the animals being hunted, and the worlds they were part of, were different for the government and Indigenous people. Each party ignored the regulations and norms of the other because they seemed so totally incoherent in relation to their own.

Similarly, and in relation to oceans, recent contestations about ocean space in Aotearoa New Zealand can be understood as part of a long trajectory of an expanding colonial and (post) colonial state. Since Polynesian people navigated the Pacific to arrive in Aotearoa, iwi and hapū (Māori tribal groupings) have affirmed mana moana – authority over waters that extends outwards from land-based territories, which may overlap or be seasonal. There is no segmentation of land and sea, "rather, the ecosystem is considered as a seamless whole" (Cram et al., 2008: 152; see also Erueti and Pietras, 2013; Waiti and Wheaton, this volume). Examples like this demonstrate how issues that may appear to be 'mere resource conflicts' are connected to sovereignty and self-determination, different ways of organising the economy, and, entirely different worlds (Coombes et al., 2012).

Like resources, definitions of security are not static, as indicated in the introduction to this chapter. Security can be thought of as being made up of practices and processes that are intended to make (people, nonhumans, environments, places) safe, to address fears about the present or future, and to provide certainty for those seeking security. Fears and what counts as safety and certainty are specific to time and place (Dalby, 1993), and the subjects and agents of security change across time and space. While conventional geopolitics has typically examined security as something that concerns sovereign states and is enforced by governments, militaries and police, critical and particularly feminist, geopolitics have taken more nuanced approaches to exploring security (Massaro and Williams, 2013). Rather than primarily being about relations between states, critical and feminist theorists have argued that numerous different actors are part of creating, undermining and experiencing security. Geographers have examined the many scales security moves within and through. By interrogating the scales of security, feminist geographers have pointed to the uneven effects of securitisation and questioned how personal or intimate practices shape and are shaped by national and international relations (Botterill et al., 2019; Pain and Staeheli, 2014). Focusing on how security is felt and experienced at a bodily scale connects people's everyday lives with state practices and global relations, and demonstrates the way securing resources also disciplines particular bodies (Botterill et al., 2019; Schoenberger and Beban, 2018).

Feminist geopolitics have also examined the way different territories are constructed as in need of security. This is done through the way that threats are defined (Hyndman, 2007); via practices of surveillance, monitoring and intervention; and the introduction of legislation and regulations (Massé and Lunstrum, 2016). These are not inherently negative practices, nor are they linear. For example, in the context of climate change, securitisation practices might be mobilised to address social injustices, or they might be motivated by the threat that climate change presents to capitalism (McDonald, 2013, 2015). Similarly, Kristoffersen and Young (2010) describe the way one set of resources, specifically oil, may simultaneously be at the heart of national security, energy security and climate security discourses. Each of these understandings of security implies very different concerns, courses of action, different experiences of security and involves the construction of different spaces in need of security. A feminist geopolitics approach to security therefore raises important questions about how security is defined, by who, and for who.

In the context of oceans, the different answers to these questions are evident in the tensions in international maritime law. The United Nations Convention on the Law of the Sea (UNCLOS) determines how different parts of ocean space are amenable to territorialisation and thereby secured by different claims by states or groups. As Zalik (2018: 344, emphasis in original) notes, the very tensions within the text of the UNCLOS and its implementation demonstrate "competing approaches to maritime law: the deep seabed is divergently represented as *null* versus *common* property". Framing ocean spaces as *terra nullius* invokes the kinds of frontier discourses used to justify "sites of colonial expansion" (Steinberg, 2018: 237).

In the next section we turn to the tactics and practices that are used by states to 'secure' such ocean territories. Adopting a feminist geopolitical lens we focus in on examples of resource politics from Aotearoa New Zealand. These examples highlight how threats to security are defined and how territory is secured through the introduction of legislation and regulations to facilitate access (for some) to resources. In doing so, we examine different constructions of both resources and security to examine relations of power and how they play out in and through ocean space.

## Securing ocean territories, resources

As a signatory to the UNCLOS, Aotearoa New Zealand has a clearly defined territorial sea that extends to 12 nautical miles from the coast, over which it has sovereignty, and a well-developed set of domestic laws and regulations. Nevertheless, sovereignty in this zone remains contested. One of Aotearoa New Zealand's founding documents, Te Tiriti o Waitangi,[1] signed in 1840, promises continued Māori sovereignty over "their lands, their villages and all their treasured possessions" (Mutu, 2010: 24; see also Erueti and Pietras, 2013). Given the lack of a territorial boundary at the coastline that distinguishes between land and sea in te ao Māori (the Māori world), for many Māori, sovereignty over lands includes mana moana, or authority over coastal waters. Yet the Crown spent the decades following Te Tiriti o Waitangi disregarding mana moana, and expanding its possession of land, sea and resources.

The often violent rolling out of private property rights and the enclosure of resources in (post)colonial states and liberal democracies has been at the expense of communal access, use (Huber, 2019) and Indigenous sovereignty. For example, petroleum was nationalised in Aotearoa New Zealand in 1937. This alienated Māori from the resource and its future development potential (Ruckstuhl et al., 2013). A group of hapū, iwi and Māori interests who brought a claim to the Waitangi Tribunal[2] for this breach stated: "in terms of customary law, Māori, as part of the natural world, have proprietary rights in the resources of their universe, including the petroleum within their lands" (quoted in Ruckstuhl et al., 2013: 30). While the Tribunal found in favour of the claimants in 2003 and recommended redress, the government at the time did not accept the recommendation (Erueti and Pietras, 2013). Rather, the government argued that the ownership of these precious minerals and the energy and economic security they provide are administered by the state for the benefit of the nation as a whole.

In a similar example in the late 1990s and early 2000s, Māori authority and ocean space was at the forefront of controversies in Crown–Māori relationships. In 1997, iwi from a region at the top of the South Island applied to the courts for a determination of customary title and rights to the foreshore and seabed (the inter-tidal area, and the seabed of the territorial sea). Gradually, through questions of jurisdiction and appeals, the case passed through to the Court of Appeal, and was determined in favour of iwi in 2003. The government quickly responded and generated huge controversy by passing legislation in 2004 that vested ownership of the foreshore and seabed with the Crown to "quell public concern about tribes obtaining ownership of New Zealand beaches" (Erueti and Pietras, 2013: 51). Unlike the land, the foreshore and seabed had not been extensively confiscated from Māori by the Crown nor privatised and 'ownership' had not been clearly defined – it was "therefore open to contest" (Smith, 2010: 213). Through the Act, the territory was confiscated and replaced with full Crown title (Jackson, 2004). In response, 20,000 people marched on parliament in protest at such large-scale dispossession. Te Ururoa Flavell, Māori Party MP, later described upset "that the Crown is allocating space that is not theirs, making decisions about marine spaces that tangata whenua [Indigenous people of a place] should rightly be involved in" (Flavell, 2008).[3]

These examples demonstrate the ongoing nature of colonialisms, through the extension of legislation to provide for and give certainty to the rights of some groups over others. In addition, they render certain imaginaries of ocean spaces more visible, alienating others, and prioritising particular worlds. We turn now to a case study of oil and gas in the deep sea surrounding Aotearoa New Zealand that further highlights the rationalities and tactics that secure boundaries, jurisdiction, sovereignty and control in respect of resources (broadly defined).

## Extending the ocean frontier – the deep sea

In the 2000s, the central government began to reorganise oil and gas governance in Aotearoa New Zealand. Although oil and gas extraction in the country's EEZ has been underway since the 1960s, production is confined to one area. The centre-right government that was elected in

2008 took a more active role in enabling mineral exploration (see Bond et al., 2015). This was justified as "catching up" to Australia (Diprose et al., 2016) and offering considerable economic development opportunities given that the marine environment is "vast, diverse and mostly unexplored" (New Zealand Government, 2012: 27). Through various initiatives, it is clear the New Zealand Government at the time was buying into various 'frontier' discourses in the "scramble for maritime territory" (Nicol, 2017: 78; see also Ruckstuhl et al., 2013).

An early investor in this 'scramble' was Petrobras, a Brazilian company who were awarded a five-year permit in 2010 to explore for reserves off the East Coast of the North Island. This permit was granted without free, prior and informed consent from Te Whānau-ā-Apanui, the iwi with mana moana.[4] The iwi voiced strong objections:

> … because deep sea oil exploration; from surveying through to drilling, compromises the environmental integrity of the tribal territory. The environmental risk poses a risk to the survival of the indigenous people who depend on the lands, seas and natural resources to sustain themselves, and future generations. Te Whānau-ā-Apanui oppose the deep sea oil drilling because it threatens their survival and their way of life. The NZ Government refused to enter dialogue to correct the situation, and Te Whānau-ā-Apanui have been forced to engage in physical defence of their territory.
>
> *Te Whānau-ā-Apanui, 2012*

Te Whānau-ā-Apanui and Greenpeace New Zealand joined together in 2011 to resist Petrobras' activities and assert Te Whānau-ā-Apanui's sovereignty. In April of that year, after a range of land and sea actions, an iwi fishing boat sailed into the path of a seismic testing boat and refused to move. By this point, the police, navy and airforce had all been deployed in the area. The police, supported by the navy, boarded the vessel and arrested the skipper on charges under a piece of domestic legislation – the Maritime Safety Act. However, a lack of jurisdictional clarity led to a drawn-out court case.

The arrest of the fishing boat skipper occurred in the EEZ which begins at 12 nautical miles offshore and extends out to 200 nautical miles. It was unclear whether domestic legislation applied as the EEZ extends beyond the sovereign state's boundaries marked by the 12-mile limit of the territorial sea. Aotearoa New Zealand's EEZ is 20 times the size of its land area and one of the largest EEZs in the world. As noted above, it is this area that the then Government referred to in frontier terms as 'vast' and ripe for exploration. Yet, in this zone the state has only limited authority according to the UNCLOS. Specifically, states have exclusive rights over resources in the EEZ but not full sovereignty (Steinberg, 2011).

In this context and while the court case was being heard, discussions between government and the oil and gas industry ensued. The oil and gas industry sought increased investment security against active communities (Bond et al., 2019). As a result, the government sought to tighten the regulation of the EEZ. This was done through two pieces of legislation. The first, passed in 2012, was the EEZ and Continental Shelf (Environmental Effects) Act.[5] The regulations under this Act, which sets out the permitting processes for oil and gas exploration activities (and other forms of sea bed mining), gives few opportunities for public engagement and participation in decisions about exploration. There is also no requirement for companies to appropriately consult with iwi and hapū in relation to their planned activities (Erueti and Pietras, 2013). Moreover, there is a significant shift away from the way the Act addresses Treaty of Waitangi rights, compared to what had become common practice in Aotearoa New Zealand legislation since the 1980s. A wide reference to Treaty principles has been replaced with a narrow prescriptive formula stating how the Act meets Treaty obligations. Arguably, this narrows the grounds by which iwi and hapū or other actors can make a claim. It is a fundamental shift in approach that secures ongoing colonialisms, and in the context of the oceans, narrows the imaginary of ocean spaces as one of a colonial political economy of extractive resources.

The second piece of legislation was an amendment to the Crown Minerals Act 1991, the main piece of legislation governing mineral extraction in Aotearoa New Zealand. The 2013 amendment included a clause, dubbed the "Anadarko Amendment" by environmental groups because of that company's involvement in the sector at the time. This clause created a new offence of interfering with ships, structures and activities relating to offshore mining in the EEZ and territorial waters and made it a criminal offence to enter a designated 500-metre exclusion zone around oil and gas vessels. The police and defence force were also granted new powers to arrest civilians who enter the exclusion zone, a notable extension of military powers to detain civilians.

This amendment was justified through appeals to safety and security. The (then) Minister of Energy and Resources stated that excluding people was in the "interests of New Zealanders" and: "what it says is that it will be stopping people out there [in the] deep sea, in rough waters, dangerous conditions, doing dangerous acts, damaging and interfering with legitimate business interests with ships, for example, seismic ships" (TVNZ, 2013: no page). The Minister argued that most New Zealanders "would agree with me, I think, that it [protest at sea] should be treated as criminal behaviour" (TVNZ, 2013: no page). While he states that people can still protest at sea and in the EEZ, he qualifies this by saying just not on "rough, choppy seas" and near moving vessels (TVNZ, 2013: no page). This framing depicts a particular version of ocean space and the types of activities that are considered appropriate in it. Ocean space is framed as inherently dangerous, and 'legitimate' activities are related to business interests and science work associated with mapping the resources of the space. This version of ocean space ignores the wealth of international law that protects the right to protest at sea (Currie, 2013; Devathasan, 2013) and Indigenous connections to their ocean territories.

In addition, the exclusion zone may be an over-reach of the limited sovereignty that UNCLOS provides coastal states with in the EEZ. The Convention provides for coastal states to establish up to a 500m exclusion zone around installations and structures for the purpose of securing their access to resources. Phillips (2018) argues that creating and enforcing these exclusion zones around infrastructure extends state sovereignty into parts of the EEZ because the zone becomes an area of enhanced state control. However, UNCLOS does not provide for exclusion zones around vessels (Currie, 2013). This is because ships in an EEZ or on the high seas are subject to the laws of the state they are flagged to, whereas installations and structures are subject to the sovereign laws of the coastal state. Without visible markers of these boundaries (Peters, 2014) the 'Anadarko Amendment' creates clashing zones of jurisdiction between enhanced New Zealand state control and the rules of a vessel's flag state (Currie, 2013). Through the Amendment, vessels involved in the exploration for oil and gas resources create moving zones of enhanced New Zealand state power. As they shift, so does the territory designed to exclude citizens engaged in protest (Devathasan, 2013) and a territory where the military enjoys increased powers of arrest and detention of the public. Aotearoa New Zealand's floating, moving exclusion zones are a unique construction.

However, the exclusion zone shouldn't only be understood as a construction of the state. Rather, it demonstrates the complex relationships between state and global capital in enclosing and securing resources. The government had several meetings with the oil and gas industry in advance of these reforms. The exclusion zone and limitations on public participation were designed in the interests of economic growth, and to entice the continued investment of global capital in Aotearoa New Zealand's offshore resources (Ministry for the Environment, no date).

As it stands, the legislative framework around deep sea oil and gas resources extends state sovereignty in ways that run counter to international law, to Māori self-determination and mana moana (Erueti and Pietras, 2013), and to the spirit of Te Tiriti o Waitangi. Particular ways of understanding the ocean space, who has access to it, who has authority over it, and the resources contained within it, draw on particular constructions of security. Here security of investment, energy independence and the security of an Aotearoa New Zealand economy based on oil came to the fore (Diprose et al., 2016).

## Contestation – securing alternatives?

The enclosure of parts of ocean space through exclusion zones and heightened security by the state was, however, contested. From the time of Te Whānau-ā-Apanui-led protests in 2011, community groups against offshore oil exploration and for climate justice became more visible and active. These groups were networked together nationally and motivated by joint concerns to prevent oil spills and confront climate change. For climate justice advocates there is inherent value in a healthy, functioning planet and the economic, cultural and social relations that underpin the human life that this enables (Routledge et al., 2018). Fossil fuels threaten the fundamental 'resource' of a healthy planet. A healthy planet, accessed equitably, is not valued by capitalism. It is more profitable for emissions to be externalised – to be borne by the public in the form of a degraded climate-commons. Climate justice activists construct a different kind of 'resource security'; one that, by valuing environmental and social health, is typically more aligned with the kinds of Indigenous theories and worlds described above.

One of the groups that emerged in Aotearoa New Zealand was the Oil Free Seas Flotilla. The Flotilla was organised in 2013 to contest offshore exploration for fossil fuels and the legislative changes limiting protest at sea. They summoned Pacific and national histories of protest at sea, and in particular, they drew on the legacy of anti-nuclear and peace activism from the second half of the twentieth century. The mobilisation against nuclear testing in the Pacific, and later, against US Navy ships entering New Zealand ports in the 1980s, are seen as formative events for a more independent national identity. In particular, this time reflected an increasing sense of Pākehā New Zealanders' belonging to the Pacific region, rather than Europe. The Oil Free Seas Flotilla held a very different idea of 'most New Zealanders' to that referred to by the Minister of Energy and Resources quoted in the section above. They stated:

> We have a long tradition of non-violent peaceful protest at sea. This includes the peace flotilla that eventually stopped nuclear testing at Mururoa, the nuclear free flotillas that stopped plutonium shipments through the Tasman, and, most recently, the Stop Deep Sea Oil flotilla off East Cape, that, together with the Iwi and Greenpeace, managed to chase away Brazilian oil giant Petrobras from our shores.
>
> *Oil Free Seas Flotilla, 2013a*

In one of the Flotilla's first actions, in November 2013, six vessels sailed off the west coast of the central North Island to challenge Anadarko's exploration activities. After a period of waiting, the oil and gas exploration vessel and its support boats arrived. One ketch, the *Vega*, purposely sailed into the 500-metre exclusion zone around the Noble Bob Douglas survey vessel and stayed there, tacking back and forth and openly challenging the state and the industry's construction of that ocean space. They posted to their website:

> We intend to protest peacefully and safely. We are fully aware of the law, but we cannot sit on shore in silence and watch, we have a history and it is part of our culture, to take to the waters and let our voices be heard.
>
> *Oil Free Seas Flotilla, 2013a: no page*

At one point, a boat supplying the *Noble Bob Douglas* squeezed the *Vega* up against the large exploration vessel and ignored radio calls from the flotilla. This behaviour, which put the *Vega* and her crew at serious risk, violated a code of ethics that shapes behavioural norms in the ocean space. Even when vessels are in adversarial situations, for example during confrontations between peace flotillas and the French military conducting nuclear testing at Mururoa in the mid-nineties, there is an etiquette expected of fellow sailors (Taylor, 2006).

While the *Vega* was within the exclusion zone, the other five boats stayed nearby and blogged:

> *100 miles offshore from Raglan, in the Tasman sea onboard R/V Tiama*
>
> It is midnight, a dark evening, some stars, no moon. A light sailing breeze makes it easier for the 5 oil free seas flotilla boats to stay close, circling around the Anadarko drilling ship. The drilling ship sits on the very same spot that we occupied with the flotilla boats for a few days before they arrived. It was a nice peaceful place then, with lots of marine and bird life around us, now it is an industrial factory site, complete with stinking exhaust gases drifting our way, loudspeakers bellowing and the ocean lit up like downtown Auckland.
>
> *Dunford, 2013: no page*

By reporting in this way and through daily updates during the ten days at the exploration site, they took "the fight back to land" (Oil Free Seas Flotilla, 2013b: no page). In this way they drew the ocean space closer to people (Peters, 2018), demonstrated by the thousands of people who gathered for events at 45 beaches across the country to support the flotilla on their penultimate day at sea.

Despite the then Prime Minister stating that the navy could be deployed in a police operation to enforce the exclusion zone (Gardner, 2013), the crew of the *Vega* were not arrested or charged. Later, in 2017 three Greenpeace activists swam in front of the *Amazon Warrior*, another exploration ship. They were charged after breaching the exclusion zone but were either given diversion, or discharged without conviction. Given the strengthened laws enabling military-supported arrest and prosecution, these two examples lie in stark contrast to Te Whānau-a-Apanui's protest on the East Cape. The militarised state response targeting Te Whānau-ā-Apanui compared to the Oil Free Seas Flotilla and Greenpeace, demonstrates the way different (raced and classed in this case) bodies are constructed as threatening, and how they experience security (Pain and Smith, 2008). The selective militarisation of the ocean space shows the importance of understanding colonial governance, and the persistence of certain forms of (violent) rule (Bargh, 2012), in relation to how resources are constructed, secured and defended.

## Contestation over whose security and what a resource is

The case study of offshore oil and gas in Aotearoa New Zealand illuminates the contested meanings of both resources and security, and the role of different kinds of sovereignty in mobilising these meanings. It also demonstrates the way ocean space – its materialities, and the cultural, political and economic ways it is valued – shapes resource securities.

Relationships between resource making and state making are complex (Bridge, 2014). In the oil and gas example outlined above, state sovereignty was expanded to secure access to fossil fuels under the seabed through frontier narratives, legislation, state–industry relationships, and the workings of global capital. Specifically, sovereignty was extended to give *certainty* to investors interested in exploring Aotearoa New Zealand's vast EEZ. But this vastness makes reregulation, and enforcing new regulations, challenging. So, the areas with exceptional state sovereignty are limited – just 500-metre zones around a small number of mobile exploration and extraction vessels at any one time. Enhanced state sovereignty moves in relation to the mobile presence of global capital in the ocean. Salmond (2015: 321), writing about the reregulation of Aotearoa New Zealand's ocean space, states that the logics of ever-expanding state sovereignty into waters undermines the complex interconnections of non-human and human life; it "fragments the sciences, detaches people from 'the environment' and makes the well-being of other life-forms contingent".

However, contestation against the enclosure of resources such as oil and gas, or the extinguishing of customary title in the case of the foreshore and seabed, question the underpinning truths of international and domestic laws about ocean sovereignty. Recognising, questioning and contesting the dominant logics that are utilised in frontier discourses makes visible the power-laden ways that ocean spaces are secured, which in turn carves openings for alternative forms of sovereignty, notions of security and resource imaginaries. The Indigenous assertions of rights, authority and self-determination disrupt the idea of a stable and universally agreed-upon definition of sovereignty (Nicol, 2017), and arbitrary boundaries where territorial waters and exclusive economic zones begin and end. Instead different worlds exist where sovereignty and resources mean quite different things.

Climate justice and environmental activists were not claiming sovereignty or self-determination in quite the same way as Te Whānau-a-Āpanui (and other iwi and hapū) have, and so were not necessarily policed and disciplined in the same way. However, in contesting what counts as a resource, the spectrum of activists – from some iwi and hapū, as well as the Oil Free Seas Flotilla and other community environmental groups – also contested the dominant construction of the sovereign state. The resource these groups want to secure is a healthy, functioning planet, freed from enclosures and commodification, and all the life forms this enables. These claims – focused on local territories and global scales rather than states – work past the state-based ways ocean spaces are currently delineated and bounded (Dalby, 1993). These groups, therefore, construct very different terrains to be securitised, and security itself then, looks very different. This kind of security is about addressing fears for the current and future climate, working for justice for those who bear the brunt of the greatest changes to the climate, and also radically changing dominant ways of organising the economy and society.

The case study explored in this chapter on oceanic resources draws us back to the fundamental questions posed by feminist geopolitics: who is defining security, who is security defined for, and who bears the costs? It also speaks to the forward-looking, transformative aims of this approach (Massaro and Williams, 2013). How might relations shaping security in the oceans, security of resources, be intervened in and reconfigured to enhance social and ecological justice?

## Notes

1 A Declaration of Independence was signed in 1835 by a number of rangatira (chiefs). Te Tiriti o Waitangi refers to the version of the Treaty written in te reo, or the Māori language which most rangatira signed and did not cede soveriegnty. The Treaty of Waitangi refers to the version in English (see Mutu, 2010).

2 The Waitangi Tribunal was established in 1975 and is a body that investigates claims by Māori on breaches to Te Tiriti o Waitangi and makes recommendations to the Government of redress.
3 The Foreshore and Seabed Act was repealed in 2011 and replaced. The new 'no title' regime provide a means for Māori to apply for customary rights but the threshold for 'securing' such rights is so high that for many Māori it has extinguished title (see Erueti and Pietras, 2013; Joseph, 2012).
4 Free prior, informed consent (FPIC) is a principle articulated in the United Nations Declaration on the Rights of Indigenous Peoples (UNDRIP), to which Aotearoa New Zealand is a signatory. Similarly, consultation is required as a principle of the Treaty of Waitangi. See Erueti and Pietras (2013) for a detailed discussion.
5 The purpose of the Act is the sustainable management of the EEZ and continental shelf of Aotearoa New Zealand. While the Act addressed a gap in existing legislation at the time, we argue that there are weaknesses in the regime that demonstrated the Government's more facilitative orientation to oil and gas.

## References


Bargh M (2012) Community organizing: Māori movement-building. In: Choudry A, Hanley, J and Shragge E (eds) *Organize! Building from the Local for Global Justice*. Oakland, CA: PM Press, 123–131.

Blaser M (2009) The threat of the Yrmo: The political ontology of a sustainable hunting program. *American Anthropologist* 111(1): 10–20.

Bond S, Diprose G and McGregor A (2015) 2Precious2Mine: Post-politics, colonial imaginary, or hopeful political moment? *Antipode* 47(5): 1161–1183.

Bond S, Diprose G and Thomas AC (2019) Contesting deep sea oil: Politicisation-depoliticisation-repoliticisation. *Environment and Society C: Politics and Space* 37(3): 519–538.

Botterill K, Hopkins P and Sanghera GS (2019) Young people's everyday securities: Pre-emptive and pro-active strategies towards ontological security in Scotland. *Social and Cultural Geography* 20(4): 465–484.

Bridge G (2010) Resource geographies I: Making carbon economies, old and new. *Progress in Human Geography* 35(6): 820–834.

Bridge G (2014) Resource geographies II: The resource-state nexus. *Progress in Human Geography* 38(1): 118–130.

Coombes B, Johnson JT and Howitt R (2012) Indigenous geographies I: Mere resource conflicts? The complexities in Indigenous land and environmental claims. *Progress in Human Geography* 36(6): 810–821.

Cram F, Prendergast TA, Taupo K, Phillips H and Parsons M (2010) Traditional knowledge and decision-making: Māori involvement in aquaculture and biotechnology. In: *Proceedings of the Traditional Knowledge Conference* (2008) Te Tatau Pounamu: The Greenstone Door Auckland: Te Pae o te Maramatanga. Auckland: Knowledge Exchange Programme of Ngā Pae o te Māramatanga (New Zealand's Māori Centre of Research Excellence), 147–157).

Currie D (2013) *Proposed Amendments to Crown Minerals (Permitting and Crown Land) Bill Under International Law*. Available at: www.greenpeace.org/new-zealand/Global/new-zealand/P3/publications/other/Legal_opinion_proposed_crown_minerals_act%20_amdts.pdf

Dalby S (1993) The 'Kiwi disease': Geopolitical discourse in Aotearoa/New Zealand and the South Pacific. *Political Geography* 12(5): 437–456.

Devathasan A (2013) The Crown Minerals Amendment Act 2013 and marine protest. *Auckland University Law Review* 19: 258–263.

Diprose G, Thomas AC and Bond S (2016) 'It's who we are': Eco-nationalism and place in contesting deep-sea oil in Aotearoa New Zealand. *Kōtuitui: New Zealand Journal of Social Sciences Online* 11(2): 159–173.

Dunford J (2013) We are making it very clear to them that they are not welcome here. *Oil Free Seas Flotilla*. Available at: https://oilfreeseasflotilla.org.nz/400/

Erueti A and Pietras J (2013) Extractive industry, human rights and Indigenous rights in New Zealand's exclusive economic zone. *New Zealand Yearbook of International Law* 11: 37–72.

Flavell T (2008) Speech on aquaculture legislation. *Scoop [Online]*. Available at: www.scoop.co.nz/stories/PA0808/S00068/te-ururoa-flavell-aquaculture-legislation-speech.htm

Gardner C (2013) Oil ship begins drilling. *Stuff [Online]*. Available at: www.stuff.co.nz/business/industries/9441095/Oil-ship-begins-drilling

Huber M (2018) Resource geographies I: Valuing nature (or not). *Progress in Human Geography* 42(1): 148–159.
Huber M (2019) Resource geographies II: What makes resources political?. *Progress in Human Geography* 43(3): 553–564.
Hyndman J (2007) The securitization of fear in post-tsunami Sri Lanka. *Annals of the Association of American Geographers* 97(2): 361–372.
Jackson M (2004) An analysis of the Foreshore and Seabed Bill. *Peace Movement Aotearoa website*. Available at: www.converge.org.nz/pma/fsbprimer.doc
Joseph R (2012) Unsettling Treaty settlements: Contemporary Māori identity and representation challenges. In: Wheen N and Hayward J (eds) *Treaty of Waitangi Settlements*. Wellington: Bridget Williams Books, 151–165.
Kristoffersen B and Young S (2010) Geographies of security and statehood in Norway's 'Battle of the North'. *Geoforum* 41: 577–584.
McDonald M (2013) Discourses of climate security. *Political Geography* 33: 42–51.
McDonald M (2015) Climate security and economic security: The limits to climate change action in Australia? *International Politics* 52(4): 484–501.
Massaro VA and Williams J (2013) Feminist geopolitics. *Geography Compass* 7(8): 567–577.
Massé F and Lunstrum E (2016) Accumulation by securitization: Commercial poaching, neoliberal conservation, and the creation of new wildlife frontiers. *Geoforum* 69: 227–237.
Ministry for the Environment (no date) *Regulatory Impact Statement: Activity Classification of Exploration Drilling Under The Exclusive Economic Zone and Continental Shelf (Environmental Effects) Act 2012*. Available at: www.mfe.govt.nz/sites/default/files/ris-for-activity-classification-of-exploration-drilling-under-the-eez-act.pdf
Mutu M (2010) Constitutional intensions: The Treaty of Waitangi texts. In: Mulholland M and Tawhai V (eds) *Weeping Waters. The Treaty of Waitangi and Constitutional Change*. Wellington: Huia Publishers, 13–40.
New Zealand Government (2012) *Building Natural Resources. The Business Growth Agenda Progress Report*. Wellington: New Zealand Government.
Nicol HN (2017) From territory to rights: New foundations for conceptualising Indigenous sovereignty. *Geopolitics* 22(4): 794–814.
Oil Free Seas Flotilla (2013a) About us: The flotilla. *Oil Free Seas Flotilla*. Available at: https://oilfreeseasflotilla.org.nz/about-us/
Oil Free Seas Flotilla (2013b) Blog. *Oil Free Seas Flotilla*. Available at: https://oilfreeseasflotilla.org.nz/blog/
Pain R and Smith S (eds) (2008) *Fear: Critical Geopolitics and Everyday Life*. Burlington, VT: Ashgate.
Pain R and Staeheli L (2014) Introduction: Intimacy-geopolitics and violence. *Area* 46(4): 344–360.
Peters K (2014) Tracking (im)mobilities at sea: Ships, boats and surveillance strategies. *Mobilities* 9(3): 414–431.
Peters K (2018) *Sound, Space and Society: Rebel Radio*. London: Palgrave Macmillan.
Phillips J (2018) Water – order and the offshore: The territories of deep-water oil production. In: Peters K, Steinberg PE and Stratford E (eds) *Territory Beyond Terra*. London: Rowman and Littlefield, 51–67.
Routledge P, Cumbers A and Driscoll Derickson K (2018) States of just transition: Realising climate justice through and against the state. *Geoforum* 88: 78–86.
Ruckstuhl K, Carter L, Easterbrook L, Gorman AR, Rae H, Ruru J, Ruwhiu D, Stephenson J, Suszko A, Thompson-Fawcett M and Turner R (2013) *Māori and Mining*. Dunedin: Te Poutama Māori, University of Otago, The Māori and Mining Research Team.
Salmond A (2015) The fountain of fish: Ontological collisions at sea. In: Bollier D and Helfich S (eds) *Patterns of Commoning*. Amherst, MA, Jena and Chiang Mai: The Common Strategies Group, 309–329.
Schoenberger L and Beban A (2018) 'They turn us into criminals': Embodiments of fear in Cambodian land grabbing. *Annals of the American Association of Geographers* 108(5): 1338–1353.
Smith K (2010) Māori political parties. In: Mulhooland M and Tawhai V (eds) *Weeping Waters The Treaty of Waitangi and Constitutional Reform*. Wellington: Huia Publishers, 207–217.
Steinberg PE (2011) The Deepwater Horizon the *Mavi Marmara*, and the dynamic zonation of ocean space. *The Geographical Journal* 177(10): 12–16.
Steinberg PE (2018) Commentary: The ocean as frontier. *International Social Science Journal* 68(229–230): 237–240.

Suchet S (2002) 'Totally wild'? Colonising discourses, indigenous knowledges and managing wildlife. *Australian Geographer* 33(2): 141–157.

Taylor M (2006) Moruroa protest. *New Zealand Geographic [Online]*. Available at: www.nzgeo.com/stories/moruroa-protest/

Te Whānau-ā-Apanui (2012) *Statement of Te Whānau-ā-Apanui*. Internatioanl Maori Affairs website. Available at: https://iwimaori.weebly.com/uploads/8/1/0/2/8102509/te_whanau_a_apanui_statement_16_may_2012_1.pdf

TVNZ (2013) Q+A: Transcript of Simon Bridges. *Scoop [Online]*. Available at: www.scoop.co.nz/stories/PO1303/S00360/qa-jessica-mutch-interviews-simon-bridges.htm

Zalik A (2018) Mining the seabed, enclosing the Area: Ocean grabbing, proprietary knowledge and the geopolitics of the extractive frontier beyond national jurisdiction. *International Social Science Journal* 68(229–230): 343–359.

# SECTION IV

# Ocean histories, ocean politics

# 15
# SECURITY
## Pragmatic spaces and the maritime security agenda

*Christian Bueger*

### Introduction

The rise of the maritime security agenda in the light of global security issues, such as piracy, extremist violence, smuggling or illegal fishing has led to profoundly new thinking about the oceans. In this chapter I ask in what ways the new maritime security agenda is productive of ocean spaces and novel spatial thinking. Identifying a range of examples of new spaces, the chapter shows how these spatialities enable different forms of governance and international collaboration.

Traditionally the seas have been understood as governed through a dual approach as laid out in the UN Convention on the Law of the Sea (UNCLOS), in which the oceans are either subjected to a zone or governed by the idea of the free seas (Steinberg, 2011; Tanaka, 2016). Yet, there is increasing evidence that there is a proliferation of third types of spaces (Bremner, 2013; Ryan, 2013). These are neither territorial (belonging to a distinct nation state), nor global and 'free'. They are constructed through largely technical practices of surveillance, policing and protection. These zones are here discussed as 'pragmatic spaces', reflecting spatial ideas that have been discussed through the concepts of 'assemblages' (Allen and Cochrane, 2007; Müller, 2015), 'technological zones' (Barry, 2006), or 'zones of exception' (Ong, 2006).

I start out with some general considerations concerning the contours and character of the maritime security agenda and a speculation how security is linked to the production of space. I then review a number of empirical examples. Firstly, I discuss the case of piracy off the coast of Somalia, and how counter-piracy operations produced a new kind of maritime space, the so-called High-Risk Area, and associated with it, a new type of map. I then turn to the production of maritime regions as the outcomes of maritime security politics drawing on the case of two regional codes of conduct. Next, I review a type of space that is constructed through the consideration of a smuggling route, the so-called Southern Route for Afghan Heroin, and investigate the form of international cooperation (the Southern Route Partnership) it spurs. Finally, I turn to a more technological zone: the so-called Areas of Interest and Common Operating Pictures as they are established in recent maritime domain awareness structures. I show how maritime surveillance projects lead to a new form of representing ocean space (see also Chapter 4). In summary, the chapter points to several new empirical examples of spaces

DOI: 10.4324/9781315111643-19

which are the effect of the maritime security agenda and opens up an empirical agenda for the study of pragmatic spaces.

## The oceans and the new maritime security agenda

A conventional reading of the governance of the oceans is that of a dual approach established through the conclusion of UNCLOS. Following the convention, the sea is governed through two major types of spatial construction: spaces which are under governance of nation states (the territorial sea and the 200 nm Exclusive Economic Zone (EEZ)) and a global space of the high seas (Steinberg, 2011; Tanaka, 2016). This dual approach has increasingly been challenged by researchers who document how several additional legal regimes also provide governance spaces. This includes the zones established by the search and rescue regime (Aalberts and Gammeltoft-Hansen, 2014; Bremner, 2015), by regional fishery management organisations (Sydnes, 2002) or safety zones to protect offshore installations (Pesch, 2015). Increasingly, we are gaining an understanding of the oceans as a space consisting of various multiple overlapping zones of governance.

Adding to this discussion, the starting point of the following observations is that legal regimes are not the only or primary forms of constituting such spaces. The cases of spatiality investigated below are constituted by security practices rather than legal ones. With the term 'security practices', I refer to patterns of doings and sayings organised by a distinct problem-atisation of issues as 'security problems' often involving instruments of the military or police (Bueger, 2016).

Since at least the 1940s, security practices have been primarily concerned about national security and the territorial integrity of the nation state. This has implied to think of oceans as territory whose integrity needs to be protected and controlled through varieties of military instruments, in particular navies (see Depledge, this volume). This 'seapower' thinking focuses on how to control maritime territory, how to deny its use by an adversary, and how to project power (Germond, this volume; Till, 2004). National security practices led to the construction of ocean space as partial sovereign territory. They also centre on focal points of particular strategic significance for national economy and trade, as expressed in conceptions such as 'sea lines of communication' or 'chokepoints'.

As part of the general revolution in security thinking which implies a wider and broader focus on other objects and actors than the nation state, security at sea is increasing understood through the concept of *maritime security*. While in many ways fuzzy as a concept (Bueger, 2015), maritime security stands for significant attention given to transnational issues such as maritime terrorism, piracy, smuggling or various forms of other "blue crimes" (Bueger and Edmunds, 2020). A good indicator for the salience of these issues is the agenda of the UN Security Council. As Wilson (2018) notes, between 2008 and 2017 the Security Council adopted 50 resolutions related to maritime security, implying no less than one new resolution every 2.5 months. The majority of global security actors have devoted, since the mid-2000s, substantial resources for maritime security for patrolling, interceptions, or capacity building. As argued by Bueger and Edmunds (2017), the rise of maritime security and the new emphasis on it by states as well as regional organisations indicate the emergence of new thinking about security at sea and that maritime space is increasingly problematised from a security perspective. What kind of spaces is maritime security productive of?

The spaces discussed in the following are productions of maritime security practices. They are here, moreover, considered as 'pragmatic spaces'. With this concept I refer to spaces *created* to address a particular securitised problem and to develop special regulatory regimes, forms of measurement and other technical responses.[1] The concept of pragmatic spaces can be usefully contrasted with a range of other closely related concepts: assemblage, technological zones, and zones of exception.

Similar to recent notions of 'assemblage', the idea of pragmatic spaces aims at a relational, process-oriented understanding of space as an effect of symbolic and material activities (Allen and Cochrane, 2007; Bueger, 2018; Müller, 2015). Humans and non-humans are given equal weight. Understanding how space is produced and performed is the primary objective. Assemblage is a general concept and structural metaphor. The concept grasps wholes of heterogenous parts and as such operates on a very generic level. To speak of pragmatic spaces, by contrast, is to refer to a distinct kind of, or sub-set of, assemblages that arises in the context of responses to a particular problem or fixing a certain concern.

Pragmatic spaces are particular kinds of assemblages made to respond to a problem and address a particular issue. This brings the concept close to what Andrew Barry (2006) calls 'technological zones'. For Barry (2006) these are spaces constituted by distinct regimes of regulation and measurement. As he argues, such zones are often characterised by the lack of territorial reference or representation. In contrast, many of the spaces discussed in the following are represented on maps and in other artefacts as distinct territories.

Another concept of space that offers similarities are 'zones of exception'. As discussed in anthropology, such zones are temporary fixations of extraordinary rules in order to enable neoliberal practices (Ong, 2006) or the global circulation of goods (Cowen, 2014). Such spatialities share with the notion of pragmatic spaces the limited temporality and problem orientation. Yet, pragmatic spaces do not necessarily imply the exception from rules, but often are just a re-interpretation or complementation of existing rules.

In the following I use the concept of pragmatic spaces – as differing from assemblage or 'zones of exception' – as an open sensitising concept to discuss the emergence, performance and stabilisation of spaces in response to maritime security concerns. I discuss four kinds of such pragmatic spaces, each of which reveals different features and trajectories.

## Piracy and high-risk areas

When, from 2008, piracy attacks off the coast of Somalia escalated to levels that required international security actions, new maritime security spaces were created to organise and coordinate the response. Two spatial configurations became the most important means: a transit corridor and a high-risk area.

As one of the first operational measures, the international naval coalitions that started to respond to piracy in the area installed, the so-called International Recommended Transit Corridor (IRTC). The corridor aimed at offering better protection for merchant and recreational vessels against piracy attacks in the Gulf of Aden, close to Somali shores. The establishment of the corridor was endorsed by the International Maritime Organization (IMO). Transiting vessels were asked to register in advance with the EU's Maritime Security Center Horn of Africa (MSC-HoA) and to transit at agreed times. As Deborah Cowen (2014: 153) argues, "the creation of this corridor is literally the production of a new political space" since it establishes new forms of authority and legal regulations.

The IRTC was also calculated space. It was based on operational analysis – "including spatial analysis of piracy attacks; forecasting of piracy risk based on historical rates of attack, density of traffic and weather conditions; and definition of patrol areas" (MacLeod and Wadrop, 2015: 3). Feeding this kind of data to algorithms allowed the naval coalitions to maximise the amount of surveyed traffic, while minimising the overall mission costs (Fabbri et al., 2015: 5). It also significantly reduced the response time of navies to any incident, as described by two operational analysts working in one of the counter-piracy missions:

> A simple model was developed to calculate the recommended patrol area size. The method was based on the need for coverage of the patrol area to be dense enough that a military asset would be able to intervene within a critical time period from the start of an attack. The process would involve the warship receiving a distress call from a merchant vessel, then directing a helicopter to the vessel's position. On arrival warning shots were expected to be sufficient to deter the attack. The dimensions of the patrol boxes allowed a typical helicopter to reach the targeted vessel within 30 minutes of a distress call. The warship often could subsequently intercept the pirate vessel.
>
> *MacLeod and Wadrop, 2015: 3*

The corridor proved effective. Yet, pirates simply moved their operations out further into the Indian Ocean. This necessitated further measures and led to the construction of an additional space complementing the IRTC. In a historically unique constellation of actors, the international shipping associations started a discussion with the IMO, Interpol, naval operations and maritime crime experts in order to identify how shipping could be better protected (Bueger, 2018; Hansen, 2012). This led to a series of guidance documents for the shipping industry, known as Best Management Practices (BMP). The first version was published in 2009, with a series of revised editions published over the years. Starting from version three, the spatial construct of a High Risk Area (HRA) was introduced. As the document describes it,

> the High Risk Area for piracy attacks defines itself by where the piracy attacks have taken place. For the purpose of the BMP, this is an area bounded by Suez to the North, 10 degree South and 78 degree East.
>
> *BMP3, 2010: 3*

This area, in essence, comprised all of the Western Indian Ocean. It was the space in which the shipping industry should apply the guidelines. The BMP prescribe situational measures including, pre- and post-boarding measures and vessel hardening measures (e.g. barb wire, or additional lookouts). At the heart of the BMP is, however, the close coordination between the shipping industry and naval actors. According to the document, a transiting vessel is to report to the MSC-HoA which could assess the risk of a particular vessel, track it while in transit through the area, and pass on this information to the naval headquarters coordinating the counter-piracy missions.

The BMP, and with it the HRA, while not legally binding were endorsed by several international bodies. This included the UN Security Council and a series of states through a declaration and the informal global governance body addressing piracy: the Contact Group on Piracy off the Coast of Somalia. In this sense the HRA became the core spatial definition for the area in which the fight against piracy would take place. A unique set of relations between industry, navies, states and international organisations stabilised it as such (Bueger, 2018). The status of the HRA was re-enforced through a series of material inscriptions and representations. Print copies of the BMP were produced in a pocket-size format, thousands of copies distributed for free and a movie produced to be used in training of seafarers. Moreover, a new type of map was produced for the promulgation of the BMP.

The United Kingdom's Hydrographic Office (UKHO), an executive agency of the UK's Department of Defence in providing navigational aids, published a chart that marked the borders of the HRA in red colours. The chart also listed the core content of the BMP including the contact details for where shippers should register. The map was initially called the 'Anti-Piracy Planning Chart' and later renamed to the 'Maritime Security Chart Q6099 – Red Sea, Gulf of Aden and Arabian Sea'. The map is noteworthy in that it was the first map produced by the office that, as it explicitly states on the chart, should not be used for navigation. It also created an entirely new genre of charts, maritime security charts, or the so-called Q Series that contain 'Security Related Information to Mariners' (UKHO, 2019). A series of similar maps were produced for the Mediterranean Sea, the Persian Gulf and Arabian Sea, Karachi to Hong Kong, Singapore to Papua New Guinea, and West Africa including Gulf of Guinea. Each of these marks a high-risk area, lists guidelines for shippers as well as contact details for reporting centres.

HRAs, although inscribed in maps, are fragile spaces in the sense that they are frequently reviewed. Indeed, the original HRA has been, in recent years, frequently revised and with it the map. They are also contested spaces. The category of risk is dependent on epistemic work, but also the ownership and authority to define risk is contested. The size of the HRA in the Western Indian Ocean has been a frequent source of controversy (Bueger, 2018). In particular, countries whose territorial sea is part of the HRA have questioned the authority of the maps. They argued that representation of their waters as risky has consequences for trade volumes and also insurance premiums, since insurers, such as the Lloyds War Committee, use HRAs as a reference point in defining war risk zones.

Hence, the problem of how to protect shipping from piracy incidents and improve naval operational coordination in the Western Indian Ocean, established new spatialities – transit corridors and High Risk Areas. These, in turn, became used across different regions and shipping lanes, and became manifested and represented in a new genre of maps. The next section looks at a different example: the case of regions.

## Insecurity, capacity building and new maritime regions

Regions are not only the outcome of social practices and institutionalisation processes (Paasi, 2002, 2004, 2009), but also of distinct political strategies that empower certain actors and allow them to participate in governance processes differently (Gruby and Campbell, 2013). Gruby and Campbell (2013) for instance, describe the case of the Pacific Region. As they argue, it is a region that has been deliberately 'performed' to enable the small islands of the Pacific to strengthen their position within environmental governance.

In interesting ways, maritime security practice is productive of spaces that can also be understood as a means by which regions empower particular actors. The international response to piracy reveals several such instances. Starting from 2008, the IMO facilitated an agreement through which countries in the vicinity of Somalia would be better positioned to share information about piracy and organise joint capacity-building activities (Menzel, 2018; Warbrick et al., 2008). The Code of Conduct concerning the Repression of Piracy and Armed Robbery against Ships in the Western Indian Ocean and the Gulf of Aden, known as the Djibouti Code of Conduct, was signed in 2009. It brought together a unique combination of countries with little prior official relations or cooperation experience: Southern and Eastern African states and the states of the Arabian Peninsula. In sum, it created a new region.

The non-legally binding code contained a commitment to cooperate in addressing piracy and installed a regional architecture of information-sharing centres and a training centre. In practice the Code provided primarily a framework for technical cooperation between the states of the region and the IMO's Maritime Safety Division. In particular, training and workshops on maritime surveillance and data analysis were organised. Although the new regional construct did not develop many genuine forms of interactions outside the capacity-building work of the IMO and other international actors, it was further institutionalised. In 2016 the participatory states signed an amendment that broadened the focus of the Code to include other maritime crimes than piracy. It also included a provision to consider turning the code into a legally binding instrument. Hence, the ongoing capacity building work of the IMO led to the stabilisation of this new regional construct. States were incentivised to use the regional structure given the financial and resource benefits they would receive from participating in it.

When a piracy-related crisis situation started to evolve in West Africa a similar spatial construct was developed. The IMO facilitated a regional agreement, directly copying provisions from the Djibouti Code (Ralby, 2017). The Yaoundé Code of Conduct Concerning the Repression of Piracy, Armed Robbery against Ships, and Illicit Maritime Activity in West and Central Africa was signed in June 2013 and came to be known as the Yaoundé Code of Conduct. Similar to the case of the Djibouti Code, a unique range of states was assembled to form a region. In contrast to the Djibouti Code the region was formed as a supra-entity providing an umbrella for work that was already carried out within existing regional organisations (Ralby, 2017). The signatory states of the Code are the members of the Economic Community of West African States (ECOWAS), the Economic Community of Central African States (ECCAS) and the Gulf of Guinea Commission (GGC). Going beyond the focus on piracy and aiming to address other maritime crimes as well, the primary goal of the region was to increase regional cooperation as well as information sharing. For that purpose, an Interregional Coordination Center was created; the region was split into several technical subzones, named alphabetically (zones A–G, but omitting B and C), with each having a new maritime operations centre. A complex region was created including a range of technical zones. Again, the primary problem that the region addressed was to build the capacity of countries so that they would be able to respond to and prevent piracy incidents to occur.

Both of the spatial constructs are new regions produced in maritime security practice. The regions were created through inter-state agreement and brought to life through information sharing centres and regular capacity-building activities organised by international actors. As regions, they placed – in particular – the IMO into the centre of attention, and situated this civil international organisation as a core maritime security actor.

## Smuggling, routes and partnerships

In 2014 the UN Office on Drugs and Crime's (UNODC) Global Maritime Crime Programme (GMCP) initiated a forum for law enforcement officials from the Indian Ocean region. The basis was a joint proposal by Australia, Seychelles, Tanzania, Sri Lanka and the US-led Combined Maritime Forces.[2] The core objective of the regional forum was to facilitate information sharing between officials, in particular prosecutors, but also to organise joint capacity building and training activities. The so-called Indian Ocean Forum on Maritime Crime meets on a regular basis in different formats. It is organised in working groups related to three issues (narcotics, fishery crime, and regional sanction violations), as well as a cross-cutting prosecutors' network. As one of the most successful offspring of the forum, in 2016 an agreement was signed which institutionalised the working group on narcotics as the so-called Southern Route Partnership.

The spatial reference is here the concept of 'routes'. The partnership is structured through the *route* that smugglers are using to transport narcotics. The Southern Route is a colloquial term that drug enforcement practitioners and analysts have started to employ to refer to the smuggling of Afghan opiates through the Indian Ocean. In particular, the analytical work of UNODC and the collation of seizure data has made this route visible. The UNODC World Drug Report for 2015 lists the southern route as one of the main three routes for Afghan opiates, defining it as "southwards through Iran or Pakistan" (UNODC, 2015: 43). As one of the UNODC reports, prepared for the first major meeting of the partnership, states:

> The route to the eastern coast of Africa has been visible since 2010, with a considerable number of seizures carried out in both international and territorial waters and onshore. Seizures in the central section of the Indian Ocean have confirmed there are multiple maritime heroin trafficking routes. Interceptions confirm a range of landing points from those on the Swahili Coast that runs along the seaboard of much of Eastern Africa, to the central section of the Indian Ocean in the Maldives and Sri Lanka.
>
> *UNODC, 2016:4*

As the quote documents, the route is made visible through a number of reference points, which are mainly the location of 'seizures' at sea, as well as at 'landing points'. In addition, the concept of 'exit points' (from Afghanistan), as well as regular vessel 'transit routes' and 'metronomic data' is used throughout the report (UNODC, 2016). Constructed in such a way, the route becomes a reference for states along this space that are affected by the influx of opiates. The 2016 meeting, which led to an inter-governmental declaration for collaboration (UN, 2017), lists 18 countries from Eastern and Southern Africa, the Arab Peninsula, Asia and Australia as members of the partnership.[3]

Similar to the cases of Codes discussed above, a new form of inter-governmental space is constructed through this agreement. Identifying the quality of law enforcement at sea as the main problem to respond to (McLaughlin, 2016), the main activities within the Southern Route Partnership are capacity-building projects, geared at improving prosecutions, information sharing, and skills such as boarding, inspection or evidence collection.

## Maritime Domain Awareness and areas of interest

The concept of Maritime Domain Awareness (MDA) refers to a set of practices through which security actors have started to monitor and surveil the sea. Data is collected and fused from different sources to develop what is called a 'common operational picture' of marine activities. Part of the practices is also to assign threat levels to maritime behaviour through patterns of life analysis and anomaly detection algorithms. As a form of knowledge production about security at sea, MDA has become one of the core tools in maritime security responses (Boraz, 2009; Bueger, 2020; Doory, 2016). The wish to know more about what happens at sea, and compile statistics and trend analysis is in many ways a core component of the maritime security agenda, and its success in presenting the oceans as a transnational security space. A global network of national and regional centres conducting MDA has emerged in the past decade, with centres in the Mediterranean and in Southeast Asia the most widely known.

The MDA agenda is driven by the availability of new sensors (Nyman, 2019). Through the global space-based Automated Identification System (AIS) large vessels can be tracked in real time. Vessel monitoring systems are increasingly used to monitor smaller vessels, in particular fishing fleets. Such data is enhanced through availability of other data sources relevant for the maritime, for example, from customs and border agencies. MDA is also informed by ideas of intelligence-led policing at sea. The analysis of incident data is used to identify patterns where and when an offence is likely to occur and what vessels are potential offenders (Mcgarrell and Freilich, 2007). The associated hope is to move beyond reactive responses and develop strategies that allow to employ naval assets more efficiently in patrol and through targeted interceptions. Indeed, the operational analysis informing the IRTC discussed above is one example for such a form of intelligence-led operations. MDA has been widely promulgated through international capacity-building activities, including the UNODC and IMO, but also security actors such as the US and the EU, both of which have developed their own technical systems for MDA: the SeaVision platform and the Indian Ocean Regional Information Sharing (IORIS) system.

The core spatial references for MDA is that of the Area of Interest (AoI) and the Common Operating Picture (COP). The AoI defines what data an MDA centre collects and analyses. The majority of national MDA centres define their AoI as going well beyond the borders of their territorial waters and their EEZ. Australia's AoI, for instance, stretches far into the Indian Ocean (Brewster, 2018). Likewise, regional centres establish a quite large area. For instance, the Information Fusion Center based in Singapore, that is the core MDA center for Southeast Asia, has an AoI that stretches from the Maldives in the West, to Australia in the East. To some degree regional MDA centres have carved up ocean space through their AoIs. For instance, the MDA center for the Western Indian Ocean – the Regional Maritime Information Fusion Center – has designed its area so it directly borders the IFC to the East and the Mediterranean Center to the North (Jeulain, 2019).

The AoI is used as the template for constructing the COP. The COP is an onscreen reality in which all incidents and historical and real-time data on movements at sea are presented on an interactive digital map. As a technical officer from the US Coast Guard describes it,

> at its core, the COP is a geographic display that contains position and amplifying information about contacts (called tracks). Tracks in the common operational picture are discovered by various sensor sources. The COP provides the network infrastructure to exchange, share, and manipulate the track data.
>
> *Hannah, 2006: 66*

As Hannah describes it, the COP is the visualisation of all data available in the AoI. This onscreen reality also allows for users of the picture to interact and exchange data, to add data, but also to communicate through the platform:

> Technically the COP is a display of relevant information shared by more than one command. It provides a shared display of friendly, enemy/suspect, and neutral tracks on a chart, with geographically referenced overlays and data enhancement. [It] contains a decision-maker toolset, fed by track and object databases. Each user can filter and contribute to these databases according to his or her area of responsibility or command role. [It] includes distributed data processing, data exchange, collaboration tools, and communication capabilities.
>
> *Hannah, 2006: 65*

Through MDA the oceans are not only carved up in AoIs, but become virtual zones of interaction of law enforcement professionals. The oceans are rendered into a plane on which objects are tracked, colour-coded and are allocated risk levels. Similar to the on-screen realities of financial markets (Knorr Cetina and Bruegger, 2002), the COP allows law professionals to interact and share maritime space in a collective experience and to agree on what is a danger requiring response and what is not.

The technologically enhanced maritime space produced in MDA arguably dehumanises maritime space, so that it is no longer humans and people, but objects, which populate the space. Yet it becomes re-humanised as it provides for new interactions between professionals across agencies and borders.

## Conclusion

The starting point for this chapter was the question of 'if' and 'how' the rise of the maritime security agenda has led to new forms of spatialities. To address this question, I adopted the concept of pragmatic spaces. The concept integrates insights from other recent spatial metaphors, such as Assemblages or technological zones. Pragmatic spaces are firstly deeply relational. They depend on relations between people, objects and technologies established in practices. They are secondly made in and through practices. Practices, I have identified, include calculating optimal response times for naval vessels, developing guidelines for the self-protection of shipping vessels, information sharing and capacity building, operational coordination between navies, or attempts to know activities at sea by turning vessels into objects to be tracked. Pragmatic spaces are, thirdly, designed to respond to particular problems. The spaces I discussed are all responses to maritime security issues and attempts to repress and prevent incidents that threaten goods and populations. This included piracy attacks, but also the smuggling of narcotics and other forms of maritime crime. Pragmatic spaces are, fourthly, fragile in that they are weakly institutionalised. They tend not to rely on legally binding rules and norms, but are driven by informal guidelines, information-sharing networks, partnerships or technical apparatuses. They are not only open to revision, such as the HRA and the Q map series, but also need to be enacted, as the examples of the two regional codes, the Southern Route Partnership or MDA centres highlight. Without doubt, many more spatial constructs can be identified in tracking responses to maritime insecurity drawing on this conceptual framework.

Maritime security presents a profound shift in terms of how the oceans are problematised and governed. Maritime security is also, notably, *productive of new spaces*. These add to the complexity of how oceans today are ordered and governed through zones and other forms of spatialities. Only some illustrative cases could be investigated in this chapter. It is likely that studying the response to other maritime insecurities (such as illicit fishing) in other parts of the world than those focused upon here, will reveal further formations of new spatialities of governance. As the maritime security agenda gains in salience and is increasingly related to other spaces at sea – such as those established by the conservationist agenda (for example, Marine Protected Areas and maritime peace parks), as well as extended to cover new issues, such as critical maritime infrastructures (e.g. the global submarine data cable and electricity network) – this complexity is only likely to increase.

## Acknowledgements

Research for this chapter has benefitted from a grant by the Economic and Social Research Council [ES/S008810/1] and the Danish Ministry of Foreign Affairs administered by DANIDA. For comments and suggestions on earlier drafts I am grateful to Kimberley Peters, Tim Edmunds, Scott Edwards, the participants of the 2019 SafeSeas Ideaslab on Maritime Spaces and 2019 Sydney Maritime Security Forum.

## Notes

1 Contrary to Glück (2015: 644), I do not want to limit the concept of security space to "the production of secure spaces for the circulation of certain 'desirable' elements (in this case cargo vessels, commodities, and capital) and the suppression of other 'undesirable' elements (that is, piracy and the interruption of commodity and capital flows)". The concept of pragmatic spaces leaves it undecided what is secured, desirable and undesirable, and rather starts out from a description of the form of spatiality, relations and interactions security practices produce.

2 The Combined Maritime Forces are a US-led naval partnership comprised of task forces working on counter-terrorism, counter-piracy and counter-narcotics missions. For an overview and discussion see Percy (2016).

3 Bangladesh, Comoros, India, Maldives, Mozambique, Sri Lanka, Tanzania, Australia, Iran, Mauritius, Qatar, Pakistan, Indonesia, Kenya, Madagascar, Seychelles, South Africa and Thailand.

## References


Aalberts TE and Gammeltoft-Hansen T (2014) Sovereignty at sea: The law and politics of saving lives in Mare Liberum. *Journal of International Relations and Development* 17(4): 439–468.

Allen J and Cochrane A (2007) Beyond the territorial fix: Regional assemblages, politics and power. *Regional Studies* 41(9): 1161–1175.

Barry A (2006) Technological zones. *European Journal of Social Theory* 9(2): 239–253.

BMP3 (2010) *Best Management Practices 3. Piracy off the Coast of Somalia and the Arabian Sea Area* (Version 3 June 2010) Edinburgh: Witherby Seamanship.

Boraz SC (2009) Maritime Domain Awareness. Myths and realities. *Naval War College Review* 62(3): 137–146.

Bremner L (2013) Folded ocean: The spatial transformation of the Indian Ocean world. *Journal of the Indian Ocean Region* 10(1): 18–45.

Bremner L (2015) Fluid ontologies in the search for MH370. *Journal of the Indian Ocean Region* 11(1): 8–29.

Brewster D (2018) Give light, and the darkness will disappear: Australia's quest for Maritime Domain Awareness in the Indian Ocean. *Journal of the Indian Ocean Region* 14(3): 296–314.

Bueger C (2015) What is maritime security? *Marine Policy* 53: 159–164.

Bueger C (2016) Security as practice. In: Balzacq T and Dunn Cavelty M (eds) *Handbook of Security Studies*. Abingdon: Routledge, 126–135.

Bueger C (2018) Territory, authority, expertise: Global governance and the counter-piracy assemblage. *European Journal of International Relations* 24(3): 614–637.

Bueger C (2020) A glue that withstands heat? The promise and perils of maritime domain awareness. In: Lucas ER, Rivera-Paez S, Crosbie T and Jensen FF (eds) *Maritime Security: Counter-Terrorism Lessons from Maritime Piracy and Narcotics Interdiction*. Amsterdam: IOS Press, 235–245,

Bueger C and Edmunds T (2017) Beyond seablindness: A new agenda for maritime security studies. *International Affairs* 93(6): 1293–1311.

Bueger C and Edmunds T (2020) Blue crime: Conceptualising transnational organised crime at sea. *Marine Policy* 119 https://doi.org/10.1016/j.marpol.2020.104067

Cowen D (2014) *The Deadly Life of Logistics. Mapping Violence in the Global Trade*. Minneapolis, MN: University of Minnesota Press.

Doorey TJ (2016) Maritime Domain Awareness. In: Shemella P (ed) *Global Responses to Maritime Violence. Cooperation and Collective Action*. Stanford, CA: Stanford University Press, 124–141.

Fabbri T, Vicen-Bueno R, Grasso R, Pallotta G, Millefiori LM and Cazzanti L (2015) Optimization of surveillance vessel network planning in maritime command and control systems by fusing metoc & AIS vessel traffic information. *MTS/IEEE OCEANS 2015 - Genova: Discovering Sustainable Ocean Energy for a New World* 2015: 1–7.

Gruby RL and Campbell LM (2013) Scalar politics and the region: Strategies for transcending pacific island smallness on a global environmental governance stage. *Environment and Planning A* 45(9): 2046–2063.

Hannah R (2006) The common operational picture: The coast guard's window on the world. *The Coastguard Journal of Safety & Security at Sea* 63(3): 65–68.

Hansen SJ (2012) The evolution of best management practices in the civil maritime sector. *Studies in Conflict & Terrorism* 35(7–8): 562–569.

Jeulain A (2019) Learning from the IFC? The Information Fusion Centre in Madagascar. In: Bueger C and Chan J (eds) *Paving the Way for Regional Maritime Domain Awareness*. Singapore: S. Rajaratnam School of International Studies, Nanyang Technical University Singapore, 28–31.

Knorr CK and Bruegger U (2002) Global microstructures: The virtual societies of financial markets. *American Journal of Sociology* 107(4): 905–950.

MacLeod MR and Wadrop WM (2015) Operational analysis at combined maritime forces. *32nd International Symposium of Military Operational Research*, July 2015. Available at: www.lessonsfrompiracy.net/files/2015/12/32ismor_macleod_wardrop_paper.pdf

Mcgarrell EF and Freilich JD (2007) Intelligence-led policing as a strategy for responding to terrorism. *Journal of Contemporary Criminal Justice* 23(2): 142–158.

McLaughlin R (2016) Towards a more effective counter-drugs regime in the Indian Ocean. *Journal of the Indian Ocean Region* 12(1): 24–38.

Menzel A (2018) Institutional adoption and maritime crime governance: The Djibouti Code of Conduct. *Journal of the Indian Ocean Region* 14(2): 152–169.

Müller M (2015) Assemblages and actor-networks: Rethinking socio-material power, politics and space. *Geography Compass* 9: 27–41.

Nyman E (2019) Techno-optimism and ocean governance: New trends in maritime monitoring. *Marine Policy* 99: 30–33.

Ong A (2006) *Neoliberalism as Exception: Mutations in Citizenship and Sovereignty*. Durham, NC: Duke University Press.

Paasi A (2002) Place and region: Regional worlds and views. *Progress in Human Geography* 26: 802–811.

Paasi A (2004) Place and region: Looking through the prism of scale. *Progress in Human Geography* 28: 536–546.

Paasi A (2009) The resurgence of the 'region' and 'regional identity': Theoretical perspectives and empirical observations on regional dynamics in Europe. *Review of International Studies* 35: 121–146.

Percy S (2016) Counter-piracy in the Indian Ocean: A new form of military cooperation. *Journal of Global Security Studies* 1(4): 270–284.

Pesch ST (2015) Coastal state jurisdiction around installations: Safety zones in the law of the sea. *The International Journal of Marine and Coastal Law* 30: 512–532.

Ralby I (2017) Approaches to piracy, armed robbery at sea, and other maritime crime in West and Central Africa. In: Reitano T, Jesperson S and Bird Ruiz-Benitez de Lugo L (eds) *Militarised Responses to Transnational Organised Crime: The War on Crime*. Basingstoke: Palgrave Macmillan, 127–149.

Ryan BJ (2013) Zones and routes: Securing a western Indian Ocean. *Journal of the Indian Ocean Region* 9(2): 173–188.

Steinberg PE (2011) Free sea. In: Legg S (ed) *Spatiality, Sovereignty and Carl Schmitt: Geographies of the Nomos*. Abingdon: Routledge, 268–275.

Sydnes AK (2002) Regional fishery organisations in developing regions: Adapting to changes in international fisheries law. *Marine Policy* 26(5): 373–381.

Tanaka Y (2016) A dual approach to ocean governance: The cases of zonal and integrated management in international law of the sea. *The International Journal of Marine and Coastal Law* 19(4): 483–514.

Till G (2004) *Seapower. A Guide for the Twenty-First Century*. Abingdon: Routledge.

United Kingdom Hydrographic Office (UKHO). (2019) *Security Related Information to Mariners*. Available at: www.admiralty.co.uk/maritime-safety-information/security-related-information-to-mariners

UN Office on Drugs and Crime (UNODC) (2015) *World Drug Report 2015*. Vienna: UN Office on Drugs and Crime.

UN Office on Drugs and Crime (UNODC) (2016) Drug trafficking on the southern route and impact on coastal states. Conference paper published for the High Level Meeting of Interior Ministers of the Indian Ocean region, Colombo, Sir Lanka, 29 October 2016, Available at: www.southernroute.org/download/Drug%20Trafficking%20on%20the%20Southern%20Route.pdf

United Nations. (2017) Commission on Narcotic Drugs, 60th session, Vienna 13–17 March 2017. Columbo Declaration, submitted by Sri Lanka, UN Doc E/CN.7/2017/CRP.3.

Warbrick C, McGoldrick D and Guilfoyle D (2008) Piracy off Somalia: UN Security Council Resolution 1816 and IMO regional counter-piracy efforts. *International and Comparative Law Quarterly* 57(3): 690–699.

Wilson B (2018) The Turtle Bay pivot: How the United Nations Security Council is reshaping naval pursuit of nuclear proliferators, rogue states, and pirates. *Emory International Law Review* 33(1): 1–90.

# 16

# NAVIES

## Military security and the oceans

*Duncan Depledge*

### Introduction

There are currently more than 150 navies in operation around the world. The vast majority protect maritime resource rights, police internal and territorial waters, and defend coastlines. Being able to operate in the high seas or off hostile shores requires bigger ships, advanced shipbuilding capabilities, sophisticated doctrines, lengthy logistical tails and experienced crews. Only a handful of countries therefore are able to project 'sea power' globally. Since the Cold War ended, the United States Navy has stood out from its rivals as the dominant military force in the oceans (Lehman, 2018).

This chapter is about how naval thinkers have conceptualised and spatialised the sea, as well as the relationship between what happens at sea and what happens on land.[1] Readers will find a strong bias towards Anglo-American naval strategy and scholarship. That is not because these sources have divined universal truths, but because the Anglo-American experience of sending their navies to all of the world's oceans is illustrative of what naval thinking about the sea can look like in the most expansive and ambitious terms, notwithstanding the serious thinking that went on in the Soviet Union during the Cold War, and the growing interests of China and India in the sea (Pant, 2012; Rowlands, 2017; Xie, 2014).[2] Anglo-American naval thinking and teaching about the sea has also been closely aligned. That is in part because, since the nineteenth century, American thinkers have looked to the history of the Royal Navy (and its rise to ocean dominance) to help guide US naval doctrine. However, it is also a result of the close relationship that formed between the Royal Navy and the US Navy during and after the Second World War (Wells, 2017). As the international maritime law professor James Kraska (2009: 117) has observed, between them, the United Kingdom (UK) and the US have had an "outsized influence on the shape of maritime law and its effect on war prevention, naval warfare, and grand strategy".

The chapter's core argument is that the key objectives of navies – in the broadest sense – have remained largely unchanged for centuries. However, the environment in which navies operate, and how naval objectives are pursued, has been transformed by changing economic and legal geographies, technological advances, shifts in global geopolitics and growing scientific knowledge about the (changing) materiality of the sea.

 DOI: 10.4324/9781315111643-20

The chapter begins by locating the origins of Anglo-American naval thinking in Elizabethan England (1558–1603), when several royal advisers argued for English naval ambitions to expand from 'safeguarding the seas' around the British Isles, to 'command of the ocean'. Second, the chapter introduces the thought of two 'founding fathers' of naval strategy: Alfred Thayer Mahan (1840–1914) and Sir Julian Stafford Corbett (1854–1922). They were among the first to produce comprehensive works on the nature and purpose of navies. Third, the chapter examines how naval thinking turned volumetric with the expansion of submarine operations and naval aviation. Fourth, the chapter explores how significant commercial, legal and geopolitical developments in the twentieth century impacted military thinking about the sea. Finally, the chapter outlines future directions for the study of military thinking about the sea, and how they should be joined up with recent social science thinking about maritime geographies. A key provocation throughout is whether a new concept of 'sea-power', as a more forceful form of 'geo-power' distinct from the traditional military definition of 'sea power' as naval power projected at or from the sea,[3] warrants further analytical attention, potentially as the basis for a distinctive form of 'mar-politics' that would complement the social sciences' traditional pre-occupation with geopolitics. That would of course have implications for militaries in terms of how they make sense of the maritime environments in which navies operate.

## Origins

The roots of western thinking about the sea are traceable to Ancient Greece (Heuser, 2017).[4] Ancient Greeks like Thucydides (c. 472–400 BC) wrote of 'θαλασσοκρατία' ('thalassocracy') to convey the way in which Athenian maritime strength supported, in combination with land power, the possibility of indefinite empire (Hornblower, 2016). Proponents argued that the sea offered several strategic advantages over land for the forming and holding of empires, the organisation of military campaigns, the forestalling of hostile conditions, the accumulation of wealth through trade and the disruption of enemy trade (Momigliano, 1944). These themes have reverberated through Anglo-American naval thinking about the sea ever since (Speller, 2019; Till, 2018).

Thucydides' work may well have shaped how some key advisors to Queen Elizabeth I in England thought about the sea (Heuser, 2017). They wanted the Crown to use England's growing maritime prowess to exploit the sea for military advantage during the Anglo–Spanish War (1585–1604). They were among the first to articulate how naval dominance in the seas surrounding the British Isles would allow the English to besiege Spanish ports and trade routes, while at the same time protecting England and ending any threat to English trade. Among their most ambitious proposals was that Elizabeth should claim the title '*Regina Maris*' and establish 'command of the seas', not just in the waters around Britain, but across the Atlantic Ocean. This defied earlier Spanish–Portuguese attempts – through the Treaties of Tordesillas (1494) and Saragossa (1529) – to claim sovereignty over the world's oceans. As it happened, Elizabeth rejected the title, insisting instead that the use of the sea should be common to all and implicitly not made subject to military authority (Heuser, 2017).

Debates in Europe over the legal principles of *mare liberum* and *mare clausum* intensified after the publication of Hugo Grotius' (1583–1645) book *Mare Liberum* in 1609 (Steinberg, 1999; Theutenberg, 1984; Vieira, 2003). In a struggle that has persisted to the present day, whenever naval power has been localised in near waters *mare clausum* has tended to dominate naval thinking. That was, for example, the case when the English needed to protect its fisheries from Dutch fleets following the Anglo–Spanish War. However, since the breakdown of

Spanish–Portuguese imperial dominance,[5] when states have sought to project their power into and across the oceans, they have tended to favour *mare liberum*, not least because the ability to do so implies that dominance has already been achieved in waters closer to home. For example, it was the ability of Britain to uses its naval forces to dominate its rivals in European waters that allowed it to send out smaller naval squadrons across the world's oceans to open the way for the expansion of foreign trade and the colonisation of distant shores (Rodger, 1998, 2004).[6] Even so, early European inquiries into maritime legal geographies hardly amounted to anything like the comprehensive works of naval strategy that would appear in the latter part of the nineteenth century (Heuser, 2017).

## Ruling the waves: Sea power and maritime strategy

Naval strategy, as it is recognised today, was a 'modern' project, born in the latter part of the nineteenth century, both to inform and respond to a period of naval resurgence among the major maritime powers of the time: notably Britain and Germany in Europe, and Japan and the United States beyond (Speller, 2019). Among the leading proponents of naval strategy were Alfred Thayer Mahan and Sir Julian Stafford Corbett. Both had specific agendas. As an American, Mahan wanted the US to build up a capable naval force to counter those of Europe. Corbett, an Englishman, sought reforms that would enable the Royal Navy to benefit from the technological advances of the 'Second Industrial Revolution' (such as more sophisticated steam engines, new weapons systems and steel hulls). Nevertheless, their key works continue to provide the bedrock for contemporary Anglo-American naval thinking and teaching about the sea.

Of Mahan's work, *The Influence of Sea Power Upon History* (1890) is the most well-known, especially for its early elaboration of the main elements of 'sea power'. Mahan argued that the need for sea power was derived from vulnerabilities created by the sea. The difficulty of securing an entire coastline meant that potential seaborne enemies could always threaten an invasion from the sea (Westcott, 1999). However, Mahan recognised that the sea also created opportunities, as it promised cheaper, faster and safer trade routes, as well as a way to launch surprise attacks of one's own. Maritime nations therefore needed navies to establish control of the sea lines of communication. Diminishing the threat of invasion, whilst enabling commerce to flourish, would in turn boost national prosperity and support further maritime activity, whether commercial or naval, which could subsequently be used to weaken potential enemies. In the most extreme scenario, a navy could be used to drive the enemy from the sea.

Mahan came to be associated with the idea that the best way to wage war at sea was to concentrate one's own naval force, seek out the enemy's fleet, and destroy it in a decisive battle. To be able to fight a decisive battle anywhere on the world's oceans, Mahan argued that maritime states needed to build up networks of points of refuge, refuelling and resupply overseas. These often took the form of friendly ports or footholds in other lands which were often later transformed into colonies and permanent overseas bases (Westcott, 1999). In other words, military power at sea remained inextricably connected to the land. The contests that have been fought by the Americans, British, Spanish, French and other states with ocean-going navies over the islands that dot the world's oceans, as well as major ports such as Gibraltar and Singapore, are testimony of their importance as territorial nodes that underpin the exercise of sea power. So too are contemporary disputes between the UK and Mauritius over ownership of the Chagos Islands/British Indian Ocean Territory (where the US has a major naval base) and reported Chinese interest in investing in Atlantic port infrastructure in places such as Greenland, Iceland and West Africa.

Like Mahan, Julian Stafford Corbett also published extensively on how the military could exploit the sea, most notably in his 1911 work *Some Principles of Maritime Strategy*. Corbett's most recognisable contribution was to argue that military activity at sea should always be intimately connected to a state's objectives on land, because naval action alone would not win wars (Corbett, 1988). Corbett's interest in 'maritime strategy' went well beyond Mahan's argument that the purpose of navies was to seek out and destroy the enemy fleet in a decisive battle *at sea*. Rather, Corbett stressed that naval forces had a critical role to play in projecting power *from the* sea (for example, in the form of naval bombardments and amphibious landings) in support of broader military aims. Modern definitions of the 'maritime environment' as consisting of the seas *and* the 'littoral'[7] owe much to this Corbettian line of thought which extended the geographical limits of naval operations into land areas as far as ship-based weaponry, intelligence gathering, naval aircraft and amphibious forces could reach.

Corbett also made a clear distinction between 'command of the sea' and 'control of the sea'. Like Mahan, Corbett recognised that the need for naval power was principally linked to the need for maritime states to control sea lines of communication (navies could only effect events ashore once such control was assured). Contrary to Mahan's call for the pursuit of total *command* of the sea, Corbett argued *control* of the sea could only ever be limited (in space) and temporary (in time). This was an important distinction because in Corbett's mind, the sea was materially different to the land. For starters, armies can survive much longer without resupply by subsisting on the land they occupy; it is much harder for a navy to live off the sea. It is also much harder to exclude an enemy from the sea because there are so many potential entry points and because ships are so much harder to locate and track in vast maritime spaces. Consequently, whereas wars on land usually involve occupying *territory*, wars at sea tend to be about controlling *movement*. The latter meant navies needed to focus on securing strategic points from which authority can be exerted over who has access to the world's oceans. If that was not possible, then control of those strategic points had, at the very least, to be denied to others to prevent them gaining unfettered access to the world's oceans and the prosperity and security that would flow from it. Contrary to Mahan, Corbett argued that whether the enemy had a fleet, or where it was in the world's oceans, mattered far less than whether it was actually able to threaten whatever control of the sea was necessary to achieve one's own objectives (be that military or commercial).

What Mahan and Corbett held in common though was that they both viewed the sea as a single, geographically contiguous space (although neither seems to have made mention of whether the frozen Arctic Ocean was included in that), and that this in turn created opportunities and challenges for maritime nations.[8] However, they both appear to have also taken a 'flat' view of the sea. Despite the developing relationship between navies and scientists as they voyaged together across the world's oceans in the nineteenth century, Corbett and Mahan seem hardly to have touched upon questions relating to the materiality of the sea, beyond references to its vastness, winds, chokepoints and potentially dangerous weather. For example, none of Mahan's six principle material conditions affecting sea power refer to the sea itself (Westcott, 1999). For both men, it seems the maritime theatre was a surface affair.

## Volumetric sea power

The first military human-powered submarines were developed in the 1700s. Steam-powered submarines were introduced in the 1800s. Towards the end of the nineteenth century, 'modern' diesel-electric powered submarines began to be deployed. However, it was not until 1914 that the implications for naval strategy became clear. As the historian of naval science Gary Weir (2001: xii) has observed, during the First World War "the effectiveness of the German U-Boat changed everything".

Suddenly, naval warfare had a potent third dimension. To modify Stuart Elden (2013: 1), the sea had become "much more complicated and multi-faceted". Indeed, before the two world wars, both the US Navy and the Royal Navy preferred their officers to demonstrate "experience and common sense to academic education", gleaned from experiential knowledge gained from going to sea, or 'sea-sense' (Speller, 2019: 4). However, to operate under the sea, new basic and applied oceanographic knowledge related to the physical nature of the ocean (and the sea-bed) was urgently needed (Weir, 2001; Robinson, 2018). The scientific advances that followed produced a proliferation of sub-sea naval activity during the Cold War. Before long both the United States and the Soviet Union had even learned to navigate under, hide beneath and surface through the pack ice in the otherwise inaccessible Arctic Ocean (Leary, 1999).[9] The improbability of finding a nuclear-armed submarine, especially in the Arctic, created the foundations of continuous-at-sea-deterrence.

Defence planners responded with anti-submarine warfare operations involving undersea sensors, surface ships and supporting aircraft in the increasingly volumetric naval battlespace. For example, from WW1 onwards, the US and the UK were forced to invest heavily in developing and building various kinds of underwater listening systems to detect and guard against submarine, which were increasingly used to target sea lines of communication. Above the sea, naval aviation created another way to search for and attack enemy submarine forces. The development of naval aviation – pioneered by the UK and Germany during WW1 – over the course of the twentieth century, expanded the reach of navies above both sea and land (Haslop, 2018). In the Corbettian tradition, the ability to attack a land-based target from the sea was dramatically enhanced by the construction of aircraft carriers. As warships and submarines started to be armed with ballistic missiles and cruise missiles, and fitted with advanced intelligence, surveillance, reconnaissance and communications systems, the aerial dimension of naval power reached higher still, all the way into space.

## Changing maritime geographies

Major changes in the world's maritime geographies after the end of the Second World War inspired other changes in thinking about sea power and maritime strategy that dove-tailed with a more volumetric perspective of naval power. Growing interest in offshore economic resources, for example, was evident in the claims made by several states to the subsoil and seabed of continental shelves contiguous to their coastlines, and to exclusive economic jurisdictions for the harvesting of living and non-living marine resources. These claims have become ever-more urgent as technological advancements opened up the possibility of commercial mining of the seabed and subsoil, while the growing range of distant water fishing fleets have raised concerns about the depletion of 'local' fish stocks by foreign vessels.

For newly decolonised states, in particular, the pursuit of exclusive maritime economic rights became a national imperative, even if this rubbed up against the emerging system of globalised seaborne trade and the free use of the sea by military forces (Harrison, 2007). Since WW2 there has been a proliferation of small navies as newly independent states have sought to assert their coastal rights, defend their economic rights and enforce new maritime enclosures. The spread of ever-more advanced anti-ship and sea-denial technologies (mines, torpedo boats, anti-ship missiles, diesel submarines) among littoral states around the world increased the threat to those navies that previously took freedom of navigation for granted (Osgood, 1976). This kind of 'mini-navalism', as Ken Booth termed it, which has its antecedent in the strategies of 'small naval war' pursued historically by weaker navies, is far more likely to render the sea as a source of threat, and stands in stark contrast to the opportunism of the 'blue water tradition' of mainly Anglo-American naval strategy (Booth, 1985; Speller, 2019).

As the Cold War wore on, the international community recognised that a new settlement for the seas was needed to at least partly redress the balance between the minority of powerful maritime states with extensive histories of seafaring and ocean-going naval forces, and the much larger (and still growing) group of states with far more limited naval capabilities and maritime ambitions, and which (especially those that had experience of being colonised), were more like to see the sea as a source of threat (Glassner, 1990). The negotiations that followed eventually produced the 1982 United Nations Convention on the Law of the Sea (UNCLOS), which divided maritime space into several legal zones. Essentially, UNCLOS rebalanced the principles of *mare liberum* and *mare clausum* in such a way as to allow littoral states to extend their authority further out and deeper into the sea.

What UNCLOS did not do (and still has not done) was change the law around military mobility, with the exception of subjecting warships to a regime of 'innocent passage' in extended territorial waters. However, the strengthening of the exclusive principles behind *mare clausum* has over time offered a pretext for coastal states to test the international community's commitment to upholding the freedom of the seas by invoking security concerns in the broadest sense. That has led some commentators to take a deeply pessimistic view of the long-term prospects for freedom of navigation (Klein, 2011; Kraska, 2011; Osgood, 1976; Pirtle, 2000; Young, 1974). Such was the level of concern in the US about the potential for *de facto* closure of maritime spaces to its navy that, in 1979, it started a Freedom of Navigation Program to "aggressively exercise" its rights to navigation and overflight wherever they might be under threat from other nations (Pirtle, 2000: 32).

Interest in the use of warships for 'strategic communication' was another feature of Cold War thinking about navies (Booth, 1977; Speller, 2019). The term 'Gunboat Diplomacy' has traditionally been associated with using naval forces to threaten another state, but as Cable (1989) has argued, this is the extreme end of a spectrum of peacetime (in other words, any action short of war) naval activity that ranges from 'showing an unfriendly flag' to compel or deter an opponent's behaviour, to 'showing a friendly flag' in order to provide reassurance and support to allies. Ranging along the entire spectrum is the idea, as those interested in 'affect' in the emotional sense will recognise, that the physical presence of a warship (or warships) can create atmospheres that inspire allies or drive fear into their enemies (Anderson, 2014). At the same time, the relative ease with which a warship can be deployed or withdrawn, and made visible or invisible (under the sea or over the horizon), means that navies can be used to pose what Booth described as a 'vague menace' with (de)escalatory potential that arguably carries far less risk than moving troops around on land (which can look inherently more threatening and take longer to withdraw) or circling aircraft overhead (which necessarily involves violating sovereign airspace) (Booth, 1977, 1985; Cable, 1989; Grove, 1998). The ability to use naval forces in this way speaks volumes for the flexibility of navies (especially when compared to land-based and air-based forces), which is rooted in the mobilities that can be exploited at sea.

## From 'sea power' to 'maritime security'

After the Cold War, military thinking about the sea changed again. In particular, the emerging concept of 'maritime security' drew attention from multiple disciplines (Bueger, 2015, this volume; Germond, 2015; Klein, 2011). Several terrorist attacks against the United States (including the suicide bombing of the USS *Cole* in 2000 and the 9/11 attacks on the US mainland in 2001) prompted the US to rethink its approach to maritime law and policy, a shift that later filtered through to its allies (Bueger, 2015; Klein, 2011). The growing potential for seaborne terrorism, weapons smuggling, and the rise of piracy off the coast of Somalia between 2008 and 2011 added to these concerns, and over the course of the next decade or so, NATO (2011), the European Union (2014), the African Union (2014), France (2015) and India (2015), among others, all developed their own 'maritime security' strategies to sit alongside more traditional maritime doctrine. In each case, the maritime security strategies emphasise a comprehensive approach to national security that goes beyond traditional military interest in the sea and emphasises the need for closer cooperation internationally to address a wider range of challenges.

Yet the emphasis on a wider concept of security was not entirely new. With regards to policing, the reduction of international tensions and the rise of claims to extend territorial waters and exclusive economic zones (that would later be given legal weight by UNCLOS) led several analysts to note an expansion of constabulary functions during the 1970s détente between the United States and the Soviet Union (Booth, 1977; Osgood, 1976; Young, 1974). New tasks included the protection of offshore facilities, fisheries, resource claims and the environment. 'Constabulary' naval forces therefore required different ships and capabilities to traditional navies, complicating maritime strategies and force development (including whether to have a separate non-military coastguard), and orientating them towards more local maritime geographies (Speller, 2019).

For the purposes of this chapter though, maritime security also offers a different way of bordering maritime spaces, with implications for military mobilities and other maritime security actors. The transnational nature of many maritime security challenges, such as the trafficking of weapons, people and narcotics, illegal migration, transboundary environmental pollution, illegal, unreported and unregulated (IUU) fishing, has helped to soften maritime borders and promote greater international cooperation between maritime security forces, including navies. A striking example of this occurred when the recognised government of Somalia invited a coalition of international naval forces to contribute to counter-piracy operations in Somalian waters. Following a UN Resolution to remove Somali territorial waters to allows warships to operate inside the 12 nautical mile limit, a combined maritime force – or "maritime security community" (Bueger, 2015: 163) – that included warships from NATO, the EU, Japan, Russia, China, Iran, Pakistan and India – engaged in a level of international cooperation across maritime borders that would normally be unheard of among these nations (Stavridis, 2018).

However, maritime security apparatuses can also lead to the hardening of maritime borders and the extension of sovereign authority further out and deeper into the sea as states seek to secure economic resources, maritime environments and deter illicit activities. Naval forces from potentially hostile states may be treated as being as much of a threat to these maritime security interests as those of non-state actors. More generally, the endurance of exclusive sovereign maritime interests continues to produce resistance to using international law to improve maritime security (Klein, 2011). Fundamentally, despite the transnational nature of many challenges and threats, maritime security is still about the forming, controlling and crossing of *bounded* maritime spaces (whether hard or soft) and, as such, represents a departure from the more traditional relationship between sea power and control/command of *unbounded* maritime spaces.

While naval forces remain key to maritime security, their role and function has been expanded to include new tasks and new geographies (Till, 2018). At the same time, maritime security has brought an array of other actors (private security contractors, crime and law enforcement agencies, non-government organisations, among others) out to sea and navies are no longer the sole arbiters of security in the maritime domain beyond coastal jurisdictions (the establishments of coast guards predated modern concepts of maritime security which are relevant to a far more expansive ocean geography).[10] This is perhaps best exemplified by US Navy Admiral Mike Mullen's 2005 proposal for a '1,000 ship navy' made up of multiple forces, agencies and private operators around the world. By extension, the geographies of twenty-first-century naval battlespaces look increasingly hybrid and, as discussed earlier, volumetric in nature, reflecting competing national and international maritime interests and associated networks of sites, actors and practices.

## Future directions

While military thinking about the sea has never been entirely isolated from other battlespaces, or been the exclusive preserve of naval planners and analysts, the next generation of debates about maritime security and sea power will have to account for a much larger – and more distributed[11] – assemblage of actors, sites, materials, practices and technologies that may be situated a long way from the sea, precisely at a time when 'sea sense', 'fluid ontologies' and 'embodied maritime experiences' are being demanded by those trying to foreground the importance of the sea to human society at large.

In many respects, military thinking about the sea has remained unchanged for centuries. Protecting coastlines and resources in local waters, intimidating rival navies, securing sea lines of communication, denying those same lines of communication to 'hostile' states, and projecting power ashore in support of land, and later air, operations are as central to naval operations today as they were to the Ancient Greeks. Even Geoffrey Till's global internationalist 'post-modern navy' echoes less well-known aspects of Mahan's writing, suggesting there has only been evolution rather than revolution in the fundaments of military thinking about the sea (Sumida, 1999; Till, 2018). That evolution was perhaps most evident during the Cold War as naval interest grew in the sub-sea environment, aviation, land-attack capabilities, and the strategic use of naval forces in peacetime, but is also apparent in the recent turn towards 'maritime security'.

Yet while the fundaments of military thinking about the sea have remained the same, naval geographies have been utterly transformed. As this chapter has shown, changing patterns of maritime commerce, new technologies, developments in international law, and shifts in global geopolitics have considerably altered – and continue to alter[12] – the maritime environment in which navies operate. As we turn towards the future, it is increasingly clear that climate change with its associated impacts on oceans, coastlines, maritime infrastructure, weather conditions, and living resources, must be added to this list of influences shaping naval geographies (Germond and Mazaris, 2019). The US Navy's 2010 Climate Change Road Map and subsequent papers indicate that concerns about the implications of climate change for maritime operations are in fact already beginning to be recognised, although further analysis is needed about the extent to which naval doctrine is changing in response. Key textbooks on maritime power and strategy, such as Ian Speller (2019) and Geoffrey Till (2018), continue to make scant reference to climate change.

The upsurge in interest in 'maritime security' stems from a rediscovery of maritime space by what Germond (2015: 141) has described as a "very eclectic group comprising political scientists, geographers, lawyers, economists, criminologists, anthropologists, etc, resulting in different ontologies, epistemologies and methodologies". However, as Bueger (2015) notes, 'maritime security' is still being defined, although arguably it will always be 'becoming'. As a discourse it continues to evolve and there is a need to look more closely at the kinds of knowledges, sites, actors and practices that are being enrolled as the concept is taken up by planners in both national and international policy settings. Bueger and Edmunds (2017) set out a useful agenda for the future of maritime security studies and Germond (2015) has added a call for explicit attention to the geopolitical dimension of 'maritime security' that is attentive to the influence of geographical features and geopolitical discourses (as shapers of ideas, interests, intentions and constraints), and the linkages between them, on the pursuit of maritime security (also see other chapters in this collection).

There is scope for the emerging maritime security (and geopolitics) agenda to be more ambitious still. The maritime 'turn' in security studies has coincided with efforts by maritime geographers to position the sea more centrally within the discipline and develop a greater sensitivity to more fluid (or 'wet') ontologies (Anderson and Peters, 2014; Steinberg and Peters, 2015).[13] Key here is the idea that the sea changes more readily than the land, foregrounding a "world of flows, connections, liquidities and becomings" that challenges more conventional geographical and geopolitical notions of a static, enduring, and easy-to-measure landscape (Steinberg and Peters, 2015: 248). However, as the effects of climate change become more apparent, the maritime will matter not only because it will expand (as ice turns to water, and land is submerged by the sea), but also because 'maritime' academics and practitioners might be able to inform thinking about more dynamic geopolitical *land*scapes in both military and non-military contexts.

At the same time, for humans at least, the sea is still fundamentally uninhabitable without access to solid land, ice or some form of artificial surface. What humans can do at sea depends on what they can do on land, and that applies as much to material activities as it does to (geo) political ones: "[the] state ontology… is profoundly terrestrial", which helps to explain why the sea has traditionally been excluded from the territory of the state (Steinberg and Peters, 2015: 254). As Mahan also recognised, it is only by fixing onto specific physical features such as seabeds, coastlines, minerals, marine life and ships that states have been able to extend their authority out to sea.[14] Absent solid features, the sea remains difficult to control. Maritime security and sea power – in this case, as expressed through the deployment of navies for constabulary, diplomatic and military purposes – can be similarly reconceptualised as being directed at the control of solid features of the marine-scape rather than the control/command of the sea (as a volume of water) itself. If scholars and planners want to anticipate how sea power and maritime security might be directed in the future (or reconsider its past), they would likely gain much from thinking about what other solid features of the marine-scape sovereignty might be *attached* to (such as artificial islands, robots, underwater infrastructure, and newly accessible/ commercialisable bio-/mineral resources), and subsequently enforced through (bounded) maritime security. As the impacts of climate change re-shape coastlines, subsume islands, deepen waters, melt ice, cripple coastal and under-sea infrastructure, and decimate sea life, it will also be increasingly necessary to consider how sovereignty might be *detached* from marine spaces, potentially necessitating greater demand for (unbounded) sea power.

In such a setting, as noted from the outset of this chapter, a different way of defining 'sea' power – as an even more forceful form of 'geo' power – warrants analytical attention and could become the basis for elaborating a 'mar-politics' that is distinguishable from 'geo-politics', or which at the very least could foreground the sorts of material dynamism – in an elemental rather than, or preferably in addition to, a technological sense – that will become more relevant on land, particularly as the effects of climate change become more pronounced and humanity is forced to contend with more fluid geographies (Depledge, 2015, 2019; see also Yusoff et al., 2012). Such an endeavour may not be entirely alien to leading naval thinkers such as Till, who has noted recently that the word 'seapower' is also "a reminder of the fact that it is a form of power that derives from the attributes of the sea itself" (Till, 2018: 27; see also Speller, 2019).

As the instruments through which maritime security and sea power are pursued, navies themselves also warrant greater scholarly attention in the future, particularly as they take on a more visibly assemblage-like form. Thinking with assemblages is another recent trend in geography and international relations that can add depth to the new maritime security studies (Acuto and Curtis, 2013; Anderson and McFarlane, 2011; Bear, 2013; Davies, 2013). As legend has it, during the Battle of Trafalgar, Lord Horatio Nelson did not issue a single order to his fleet. He relied instead on his commanders to use their own initiative during the tumult of battle. That spirit of 'independence in command' has remained central to the culture of many modern navies (Stavridis, 2018). However, twenty-first-century navies are beginning to look very different as new technologies and doctrines are taken up that introduce and make use of more advanced anti-ship weapons systems, cyberspace and so-called 'hybrid warfare' tactics (Stavridis, 2018; Speller, 2019). As Speller (2019: 1, emphasis removed) contends, the emerging security environment – including the maritime – is expected to be characterised by "complexity, instability, uncertainty and pervasive information". Instantaneous communications, global networking, mass data gathering and processing, and artificial intelligence all emphasise greater connectivity (and thus potential for disruption) between warships, aircraft, submarines and satellites, often via land-based command and control hubs, and increasingly supported by fleets of unmanned aerial, surface and underwater 'drones'.

This volumetric, partly digitalised and, in some cases, remotely operated, naval assemblage – which extends the Navy into every other 'battlespace' (air, land, space, cyber) and allows operations to be organised and executed from ever greater distances – would be unrecognisable to the 'founding fathers' of naval strategy, let alone the Ancient Greeks.

## Notes

1 As is the case with other bodies of scholarship in the humanities and social sciences, far less has been written about the military and the sea than the military and the land (Speller, 2019).

2 For a useful summary of other sources of non-Anglo-American thinking about navies and the sea, see Till (2018) and Speller (2019).

3 Although, for a more nuanced definition of 'sea power' in the military sense, see Till (2018).

4 Based on Heuser's rough translation.

5 No other state has since sought to claim ownership of the oceans.

6 The only major battle fought by the main British fleet outside European waters before the Second World was the Battle of the Saintes in 1782.

7 The UK Ministry of Defence defines the 'littoral' as "those land areas (and their adjacent areas and associated air space) that are susceptible to engagement and influence from the sea [and] conversely… those areas of the sea susceptible to engagement from the land, from both land and air forces" (Ministry of Defence, 2017: 5).

8 Belief that the 'Sea is One' remains a key part of Anglo-American naval thinking today (Stavridis, 2018).

9 Stealing a march on their Soviet rivals, the USS *Nautilus* (1958) and the USS *Skate* (1959) were the first submarines to demonstrate this in practice. The Soviet *Leninski Komsomolets* reached the North Pole in 1962, and the Royal Navy's HMS *Dreadnaught* joined the 'under-ice' club in 1971.

10 In fact, as Bueger (2015) observes, as is the case for the military more broadly when it comes to non-traditional security issues, navies are not necessarily the most effective instrument for dealing with maritime security challenges such as illegal migration.

11 Note the US Navy's recent development ideas around 'Distributed Lethality' and 'Distributed Maritime Operations'.

12 For example, the recent decision by the United Nations General Assembly to develop a treaty on marine biodiversity of areas beyond national jurisdiction (BBNJ) is illustrative of the ongoing trend for high seas activity to be brought under some form of regulation.

13 For earlier treatments of the sea by maritime geographers, see Glassner (1990) and Steinberg (2001).

14 That applies as much to commercial and other uses of the sea as it does military ones. See, for example, Anim-Addo's (2016) work on the Royal Mail Steam Packet Company.

## References


Acuto M and Curtis S (2013) *Reassembling International Theory: Assemblage Thinking and International Relations*. Basingstoke: Palgrave Macmillan.

Anderson B (2014) *Encountering Affect: Capacities, Apparatuses and Conditions*. Farnham: Ashgate.

Anderson B and McFarlane C (2011) Assemblage and geography. *Area* 43(2): 124–127.

Anderson J and Peters K (eds) (2014) *Water Worlds: Human Geographies of the Ocean*. Abingdon: Routledge.

Anim-Addo A (2016) 'With perfect regularity throughout': More-than-human geographies of the Royal Mail Steam Packet Company. In: Anderson J and Peters K (eds) *Water Worlds: Human Geographies of the Ocean*. Abingdon: Routledge, 163–176.

Bear C (2013) Assembling the sea: Materiality, movement and regulatory practices in the Cardigan Bay scallop fishery. *cultural geographies* 20 (1): 21–41.

Booth K (1977) *Navies and Foreign Policy*. London: Croom Helm.

Booth K (1985) *Law, Force and Diplomacy at Sea*. London: George Allen and Unwin.

Bueger C (2015) What is maritime security? *Marine Policy* 53: 159–164.

Bueger C and Edmunds T (2017) Beyond seablindness: A new agenda for maritime security studies. *International Affairs* 93(6): 1293–1311.

Cable J (1989) *Navies in Violent Peace*. Basingstoke: Macmillan.

Corbett J (1988) *Some Principles of Maritime Strategy*. London: Brassey's Defence Publishers.

Davies A (2013) Identity and the assemblages of protest: The spatial politics of the Royal Indian Navy Mutiny, 1946. *Geoforum* 48: 24–32.

Depledge D (2015) Geopolitical material: Assemblages of geopower and the constitution of the geopolitical stage. *Political Geography* 45: 91–92.

Depledge D (2019) Geopower and sea ice: Encounters with the geopolitical stage. In: Zubrow E, Meidinger E and Connolly KD (eds) *The Big Thaw: Policy, Governance and Climate Change in the Circumpolar North*. Albany, NY: SUNY Press, 181–200.

Elden S (2013) Secure the volume: Vertical geopolitics and the depth of power. *Political Geography* 34: 35–51.

Germond B (2015) The geopolitical dimension of maritime security. *Marine Policy* 54: 137–142.

Germond B and Mazaris A (2019) Climate change and maritime security. *Marine Policy* 99: 262–266.

Glassner M (1990) *Neptune's Domain: A Political Geography of the Sea*. London: Unwin Hyman.

Grove E (1998) Principles of maritime strategy. In: Grove E and Hore P (eds) *Dimensions of Sea Power: Strategic Choice in the Modern World*. Hull: University of Hull Press, 26–31.

Harrison J (2007) Evolution of the law of the sea: Developments in law-making in the wake of the 1982 Law of the Sea Convention. Unpublished doctoral dissertation, University of Edinburgh.

Haslop D (2018) *Early Naval Air Power: British and German Approaches*. Abingdon: Routledge.

Heuser B (2017) *Regina Maris* and the command of the sea: The sixteenth century origins of modern maritime strategy. *Journal of Strategic Studies* 40(1–2): 225–262.

Hornblower S (2016) Sea power, Greek and Roman. *Oxford Classical Dictionary*. https://doi.org/10.1093/acrefore/9780199381135.013.5775

Klein N (2011) *Maritime Security and the Law of the Sea*. Oxford: Oxford University Press.

Kraska J (2009) Grasping the influence of law on sea power. *Naval War College Review* 62(3): 113–135.

Kraska J (2011) *Maritime Power and the Law of the Sea: Expeditionary Operations in World Politics*. Oxford: Oxford University Press.

Leary WM (1999) *Under Ice: Waldo Lyon and the Development of the Arctic Submarine*. College Station, TX: Texas A&M University Press.

Lehman J (2018) *Oceans Ventured: Winning the Cold War at Sea*. New York, NY: WW Norton and Company.

Ministry of Defence. (2017) *Joint Doctrine Publication 0–10: UK Maritime Power*. Shrivenham: DCDC.

Momigliano A (1944) Sea-power in greek thought. *The Classical Review* 58(1): 1–7.

Osgood RE (1976) Military implications of the new ocean politics. *Adelphi Papers* 16(122): 10–16.

Pant H (ed) (2012) *The Rise of the Indian Navy: Internal Vulnerabilities, External Challenges* Abingdon: Routledge.

Pirtle C (2000) Military uses of ocean space and the law of the sea in the new millennium. *Ocean Development & International Law* 31(1–2): 7–45.

Robinson S (2018) *Ocean Science and the British Cold War State*. Basingstoke: Palgrave.

Rodger NAM (1998) Sea-power and empire, 1688–1793. In: Marshall PJ and Low A (eds) *The Oxford History of the British Empire: Volume II: The Eighteenth Century*. Oxford: Oxford University Press, 169–183.

Rodger NAM (2004) Queen Elizabeth and the myth of sea-power in English history. *Transactions of the Royal Historical Society* 14: 153–174.

Rowlands K (ed) (2017) *21st Century Gorshkov: The Challenge of Seapower in the Modern Era*. Annapolis, MD: Naval Institute Press.

Speller I (2019) *Understanding Naval Warfare*. Abingdon: Routledge.

Stavridis J (2018) *Sea Power: The History and Geopolitics of the World's Oceans*. New York, NY: Penguin.

Steinberg PE (1999) Navigating to multiple horizons: Toward a geography of ocean-space. *The Professional Geographer* 51(3): 366–375.

Steinberg PE (2001) *The Social Construction of the Ocean*. Cambridge: Cambridge University Press.

Steinberg PE and Peters K (2015) Wet ontologies, fluid spaces: Giving depth to volume through oceanic thinking. *Environment and Planning D: Society and Space* 33: 247–264.

Sumida J (1999) Alfred Thayer Mahan, geopolitician. *The Journal of Strategic Studies* 22(2–3): 39–62.

Theutenberg BJ (1984) Mare clausum et mare liberum. *Arctic* 37(4): 481–492.

Till G (2018) *Seapower: A Guide for the Twenty-First Century*. Abingdon: Routledge.

Vieira MB (2003) Mare liberum vs. mare clausum: Grotius, Freitas, and Selden's debate on dominion over the seas. *Journal of the History of Ideas* 64(3): 361–377.

Weir G (2001) *An Ocean in Common: American Naval Officers, Scientists, and the Ocean Environment*. College Station, TX: Texas A&M University Press.

Wells A (2017) *A Tale of Two Navies: Geopolitics, Technology, and Strategy in the United States Navy and the Royal Navy, 1960–2015*. Annapolis, MD: Naval Institute Press.

Westcott A (ed) (1999) *Mahan on Naval Warfare: Selections from the Writings of Rear Admiral Alfred T. Mahan*. Mineola, NY: Dover.

Xie Z (2014) China's rising maritime strategy: Implications for its territorial disputes. *Journal of Contemporary East Asia Studies* 3(2): 111–124.

Young E (1974) New laws for old Navies: Military implications of the Law of the Sea. *Survival* 16(6): 262–267.

Yusoff K, Grosz E, Clark N, Saldanha A, and Nash C (2012) *Geo*power: A panel on Elizabeth Grosz's *Chaos, Territory, Art: Deleuze and the Framing of the Earth*. *Environment and Planning D: Society and Space* 30: 971–988.

# 17

# DISCIPLINE

## Beyond the ship as total institution

*Isaac Land*

### Introduction

"There is no justice or injustice on board ship, my lad. There are only two things: Duty and Mutiny—mind that. All that you are ordered to do is duty. All that you refuse to do is mutiny" (Eastwick, 1891: 25). This pithy observation carried potentially vast implications, as a list of prohibitions from 1636 indicates. At the start of the voyage, James Slade, the master of the *Mary* – an East India Company vessel – ordered the public reading of a lengthy catalogue of strictures. This included practical considerations such as keeping the gun-deck washed, but the list extended to "attendance at morning and evening prayers before the mainmast, with a fine of 12d. for any absence, fines for drunkenness, cursing, selling 'anie Strange drinke of what sort soeuer'..." Other prohibited behaviour aboard the *Mary* included smoking tobacco, engaging in quarrels, and gambling (Massarella, 2017: 425, see also Creighton, 1995: 104). Considering that the voyage to Surat would take six months, Slade anticipated not only holding the crew to these expectations, but keeping their impulses bottled up for quite some time.

We may question (as Louis Sicking charmingly put it, in Bogucka et al., 2002: 361) whether crews really imbibed the soup of discipline as hot as it was served. Historians enamoured of quantification have attempted to count how many strokes of the lash were administered per sailor per voyage (Byrn, 1989). A more sophisticated and comprehensive approach to discipline is the total institution concept. The term originated with the sociologist Erving Goffman to describe the special situation in prisons, insane asylums, monasteries, and boarding schools. Another sociologist, Vilhelm Aubert (1982), applied it to ships. Aubert noted that ships, like Goffman's examples, were "physically isolated from the family and the national and local community... frequently for years at a time" (Aubert, 1982: 260). While Aubert acknowledged some similarities with other work environments, such as the factory and the seasonal labour camp, he drew attention to the ship's 24-hour, tightly scheduled environment as well as the structured hierarchy of roles on ships, to the extent that a person might spend the whole voyage referred to only by their job title, such as "Cook" (Aubert, 1982: 270). While it has some obvious limitations (not least that the preponderance of sea voyages in human history were much shorter than the long ones undertaken by the Norwegian oil tankers where Aubert did his fieldwork), the total institution concept nonetheless has been helpful in stimulating debate and suggesting new avenues of inquiry.

 DOI: 10.4324/9781315111643-21

In this chapter, I examine how far the search for control and uniformity went. The first section of the chapter considers the total institution thesis in relation to punishment at sea, considering disciplinary practices, but also cultures of restraint. The middle section examines the overall environment of the ship as a disciplinary space, considering deck plans, hygienic regimes, ergonomic considerations, as well as different forms of social segregation. Finally, I tackle the question "how total was the total institution" from the opposite direction, inquiring into privacy, concealment, and disobedience even in the confined and invigilated spaces of a sea-going vessel.

What I offer here is, necessarily, an abbreviated treatment that includes many generalisations. Vessels varied greatly in size and purpose; some were under civilian, others under naval command; legal regimes differed; the names of ranks and roles aboard ship do not always translate well across languages, or mean the same thing in the same language in different historical eras; discipline itself took varying forms, as did the indiscipline it sought to stifle.[1] With this in mind, my approach is thematic, rather than chronological; I offer examples meant to be suggestive, rather than strictly representative, or in any sense exhaustive. The major focus is on problems of method, and influential lines of interpretation. I hope that my short discussion will prompt other comparisons or contrasts to situations that I did not think to include here.

## Punishment and its contexts

During a round of conscription in 1589, a petty official in Crete recorded the effusions of grief: The "wives, mothers and children wail… and claw at their faces as their menfolk say they would rather be beheaded" than serve on a Venetian galley (Panopoulou, 2017: 397, see also Earle, 1998: 147 for a balanced discussion of interpretations). It is unlikely that this reaction arose because whippings and beatings were alien to their experience. Rather, populations acquainted with corporal punishment developed the sort of acute discrimination that we associate with connoisseurs. They gossiped about forms of punishment, made jokes about it, and took the time to formulate careful comparisons. In the seventeenth and eighteenth centuries, many sailors preferred merchant ships because of their "disgust at the harsh discipline onboard naval vessels", although they would have encountered the lash in both environments (Witt, 2001: 372, see also Hope, 1990: 243).

At the same time, there is ample evidence that sailors understood and respected the need for strong leadership. A prominent naval historian has emphasised how "any seaman knew without thinking that at sea orders had to be obeyed for the safety of all" (Rodger, 1986: 207). Margaret Creighton, in her study of the nineteenth-century whaling industry, notes that "the deep, cold ocean and the jaws and flukes of whales exacted fast, fatal punishment" from those who failed to absorb the basic lesson that at sea, all "were in the same boat" (Creighton, 1995: 81–82). Until relatively recent times, the ocean-going vessel was remote from the usual apparatus of law enforcement and judicial review. Many ships sailed with close to the minimum number of sailors required for safe operation. Locking up a malefactor was rarely a viable or attractive option. An ideal punishment would be proportionate to the offence, but also quick and decisive. Therefore, summary punishments were one of the practical necessities of seafaring, whether naval, mercantile, or otherwise. Jurists respected the testimony of seasoned mariners as to what constituted a normal and appropriate "custom of the sea" in this area (Blakemore, 2015). However, Yrjö Kaukiainen inserts a cautionary note here (in Bogucka et al., 2002). Norms undoubtedly existed for each type of voyage or industry, but Kaukiainen urges us to consider whether these "mainly unwritten rules" were truly "homogenous", and whether we are confident "that both masters and ordinary sailors understood [them] in much the same way" (in Bogucka et al., 2002: 350). Circular reasoning predicated upon norms cannot account for contestation and historical change, yet most cultures and time periods involve both.

Part of the appeal of Marcus Rediker's approach to discipline and shipboard culture was that it put both contestation and historical change front and centre. It was Rediker's *Between the Devil and the Deep Blue Sea* (1987) which introduced the total institution concept to a wider readership, although his emphasis was somewhat different from Aubert's. Rediker argued that the eighteenth-century Atlantic merchant ship, "prefiguring the factory", involved workers "confined within an enclosed setting to perform, with sophisticated machinery and under intense supervision, a unified and collective set of tasks" (1987: 83). Sailors resembled proletarians in other ways; referred to merely as "hands", they laboured under captains with "near-dictatorial powers" in "a system of authority best described as violent, personal, and arbitrary" (Rediker, 1987: 212–226). If, in many traditional societies, a voyage had been a "a relatively communal and egalitarian undertaking" (Rediker, 1987: 209), the transition to larger vessels with bigger crews going on longer voyages broke the old pattern of captains sailing with men they knew previously, tipping the balance toward a more austere style of command (Bogucka et al., 2002: 348).

In the aftermath of Rediker's influential book, scholars sought to clarify the ways in which the rule of law and the strictures of custom empowered captains, but also constrained them. As Rediker himself acknowledged, his findings align with a particular historical period and not with all seafaring situations. European merchant ships did not sail with 'captains' until the early eighteenth century, when the term emerged as "an analogy to commanders of warships" (Witt, 2001: 165). The earlier term was 'master'. Social and legal norms anticipated consultation between master and crew whenever possible. Medieval law codes were various, but one stipulated that crews were not obligated to sail if the master changed the voyage itinerary from the one originally agreed upon, and another even guaranteed the crew's right to set sail without the master on board, if he was drunk on shore and did not return on time (see particularly Coureas, 2017: 380 and also Jahnke, 2017: 574, 577). In a similar spirit, the *Consulado del Mar* – a famous compilation of maritime law from the western Mediterranean – instructed sailors to stand and take both scolding and beatings, but also stipulated that a sailor could flee and stand "beside the anchor chain", where the master was not allowed to follow (Perez Mallaína, 1998).[2]

For the early modern period itself, Richard Blakemore (2015: 108) has drawn attention to the paradox that our best evidence of harsh discipline comes from lawsuits brought by sailors, suggesting there was some recognition that certain behaviour was, indeed, insupportable and that sailors had some say in demarcating the outer boundaries of acceptable punishments. Some historians (e.g. Bogucka et al., 2002; Witt, 2001) have questioned Rediker's premise of class conflict aboard ship, noting that everyone in the merchant service started their career before the mast (though promotions went disproportionately to those who were themselves captain's sons, or related to ship owners). In the Royal Navy, however, sailors observed that some of the toughest, most vindictive punishments came from those promoted from the ranks (Land, 2009: 128; McKee, 2002: 61). It should not come as a surprise that upwardly mobile individuals often serve as enforcers in grossly unequal societies, and do so with enthusiasm.

"Can't a man ask a question here without being flogged?" The sadistic captain from Richard Henry Dana Jr's memoir *Two Years before the Mast* (1844) answered this challenge with blows and a monologue worthy of a Quentin Tarantino film, adding with relish, "[d]on't call on Jesus Christ. He can't help you" (Dana, 1844: 127). Reflecting on Dana's anecdote, it is interesting to note that for most of the nineteenth century, American merchant captains sailed with a copy of *The Shipmaster's Assistant and Commercial Digest*, packed with detail about their legal rights and duties (de Oliveira Torres, 2014). The image of officers thumbing through reference works complicates any picture of unchecked megalomaniacs. So also, do the testimonials from later in the century about the impact of telegraph wires and undersea cables on officers' authority. A captain from that era lamented that he was little more than "an underpaid first-class clerk" (Bogucka et al., 2002: 357, or see Gerstenberger, 2001). Was the captain the first among equals, a boastful demigod, or a surly middle manager? It depends, in part, on the century.

In some ways, it is misleading to focus on captains, since most punishments were informal, and administered by others. It is impossible to quantify, since the action left no written record, but it appears that stinging blows from the "rope's end", administered by a boatswain or boatswain's mate, made up the bulk of corporal punishments at sea for many centuries (Perez-Mallaína, 1998: 204–205). In English, striking a sailor with this short length of rope was known as 'starting', as if to suggest that the aim was to get a lazy man moving.[3] Some men ran from this flurry of blows in such haste that they hurt themselves. It would be, once again, impossible to quantify what percentage of falls and other injuries at sea resulted from sailors trying to *escape* the reach of the rope's end. Starting was officially prohibited in the Royal Navy in 1809, though autobiographies by sailors averred that the practice continued for many years afterward. A recent study using court martial evidence corroborates the sailors' claim that the forbidden practice persisted, and was treated as unexceptional (Malcolmson, 2016. For a similar, ambiguous period following the formal abolition of certain practices by the US Congress, see Creighton, 1995: 95). In his extensive study of lower deck sentiments in the early twentieth century, Christopher McKee noted that the most common word used to refer to a petty officer was "bastard" (2002: 128), reminiscent of the attitudes toward boatswain's mates in an earlier period.

A more subtle, but potentially quite effective, form of discipline was the manipulation of small incentives that could be amped up or withheld as needed. The grog ration is a classic example; shore leave is another (c.f. McKee, 2002: 37). Sailors who worked on slave ships considered themselves entitled to sexual access to women slaves, and officers may have winked at this behaviour for their own reasons, considering it "a sop involving no financial outlay to keep these infamously rebellious men placid" (Christopher, 2006: 190–191). Some sailors cherished the privilege to ship a few small items themselves for "private trading" (Rediker, 1987; see also Stöckly, 2017: 162–163). In many seaborne endeavours, the main incentive was the shares concept, in which the whole crew could expect a proportional division of the profits at voyage's end. The idea of shares appeared in the medieval Mediterranean, where crews were not 'hands' but companions (*compañeros*). It turned up again aboard privateers, and in navies that divvied out prize money for captured vessels. It was standard on nineteenth-century Pacific whaling voyages (Creighton, 1995: 28-30; Pérez-Mallaina, 1998: 194). Some individuals were entitled to only a tiny percentage through the shares system, but it offered a plausible path for optimistic daydreaming, and made the success of the voyage a matter of interest to all.

Naval officers and ship's masters, well aware that their reputation might precede them, had good reasons to cultivate an image that might aid in attracting the best sailors, rather than repelling them. Remarkably, similar considerations applied earlier even with oarsmen on ancient Greek triremes, who were often skilled mercenaries who might desert if not treated well (Gabrielsen, 2017: 437–438, also Unger, 2017: 95). Lord Sandwich, vouching for the value of permitting naval officers to recruit their own volunteer crews, expressed the relationship in this way:

> The [officer] is in some measure bound to act humanely to the man who gives him a preference of serving under him; and the [seaman] will find his interest and duty

> united, in behaving well under a person from whom he is taught to expect every present reasonable indulgence, and future favour.
>
> *quoted in Rodger, 2004: 398*

Conversely, the restraints of shame might give officers pause. One stern disciplinarian, returned from the sea, encountered an old woman in a marketplace:

> "Be you Captain ----- ?"
> "Yes, my good woman", he said. "What can I do for you?"
> "Take *that*, for flogging my son", she cried, whipping out a hake-fish and 'letting him have it' across the face.
>
> *quoted in Winton, 1977: 82*

This public humiliation was, surely, an extreme example, yet many officers retained local ties and expected to return to their communities of origin. The whaling captain Richard Weeden, taunted by his crew that they would report his floggings when they got home, felt enough concern on this point that he sought to prevent his crew from mailing letters (Creighton, 1995: 100–101, 106).

Knowing that officers potentially faced both legal and social sanctions for disciplinary excesses, why would any crew stand for mistreatment? As one historian has noted, "[c]rew members needed above all to make a living and probably had little time, occasion or money to initiate a lawsuit against their employer; besides, such action would probably reduce their chance of getting employment again" (Bogucka et al., 2002: 361). While desertion offered a cheaper, and quicker, method of expressing dissatisfaction, a known deserter also had reasons to worry about his future, which could be true even in civilian contexts (Creighton, 1995: 94). In communities where the widows of sailors (or the wives of sailors long absent) could expect to rely on charity boards composed of local ship owners and captains, petitioners invariably emphasised the loyal and deferential service of the sailor to named individuals. It is not difficult to imagine what sort of reception a troublemaker's widow would receive (Land, 2014: 103–105). Thus, although social pressures might stay the hand of an abusive authority figure, they might also leave sailors feeling unable, or unwilling, to put up much resistance to abuse.

Not all crews enjoyed the mantle of customary social or legal protections, or, indeed, any expectation that paths might cross again. It was precisely in these situations that behaviour strayed into truly extreme territory. Seemingly anything would do on East India Company ships, from "kicks or punches" to blows with "objects within reach of their officers", including umbrellas (Jaffer, 2015: 34, see also Creighton, 1995: 109). The use of the dried penis of a bull as "a whip or flogging instrument" was widespread enough that it is enshrined in the *Oxford English Dictionary* under the word "pizzle", with recorded instances aboard ships going back to the sixteenth century.[4] The symbolism was not exactly subtle, although to unpack all the different registers of humiliation here might require the skills of a psychoanalyst. The sailor who complained of being beaten "upon the head with an Elephant's dry'd Pizle" had a sense of additional violation, perhaps, from being consciously at the mercy of a superior's improvisatory whims and embellishments (Rediker, 1987: 217).

Certain punishments made little sense except as attempts to annihilate the identity of the individual, assaulting the visible signs of their manhood, or of their faith community. A Spanish court case from the 1550s involved a ship's master who punished a sailor by holding him down and using his bare hands to tear out most of his beard (Pérez-Mallaína, 1998: 208). Once again, East India voyages offer some of the most disconcerting examples. Officers seeking to bully and humiliate Muslim lascar crews zeroed in on their dietary prohibitions as a weak spot. While the practice of addressing a Muslim sailor as "Hog" in an effort to insult him may have come across as a *non sequitur* at first, the intent was clear, and sailors pushed back against this behaviour. It could, however, get worse from there; some officers escalated by "ramming pig tails into their mouths and festooning them with entrails from the same animal" (Jaffer, 2015: 36). On a Dutch East India Company voyage, one sailor was forced to march around the deck singing in Latin, with a dead pig hung around his neck (Jaffer, 2015: 35). Three mutineers, convicted by a court in Penang in 1851, all cried out the name of the Prophet Mohammed from the gallows; it is possible that they wished to position themselves as martyrs to the faith, and the existence of faith-based punishments provide some context for such sentiments (Jaffer, 2015: 51).

If shock and awe constituted one venerable tradition of command, paternalism also has a pedigree stretching back centuries. While the fight against scurvy is remembered today as an episode in the history of vitamins and nutrition, eighteenth-century discussions of the disease considered whether the ship was clean and well managed. In keeping with the view that the despondent as well as the "indolent and sluggish" were especially susceptible to scurvy, the argument ran that since scurvy's causes were "general in their generation", therefore the remedy must follow the same, holistic, lines (Lawrence, 1996: 84; for paternalism in a different context, see Creighton, 1995: 92). Captain James Cook's method of coaxing his crew to eat sauerkraut (thought to be an antiscorbutic) by letting them see the officers eat it at every meal was a canny deployment of what later generations would call emotional intelligence (Lawrence, 1996: 88). Captain Rory O'Conor's handbook *Running a Big Ship* (first published in 1937), offered dozens of insights in this vein:

> If a man is caught leaning on the paintwork… then put him as a sentry on the upper deck after hours, and leave him there until he catches someone else offending against the ship… Setting a thief to catch a thief is the policy which works well, and has a sporting element about it.
>
> *O'Conor, 2017: 20–21*

Simply reasoning with the crew could also work. In 1946, at a time when anxieties ran high about nationalist or even Communist agitators spreading their ideas in the Royal Indian Navy, one officer remarked: "a rating will not write *Jai Hind* on a bulkhead if it is explained to him that somebody else will have to rub it off" (Davies, 2014: 396). Of course, the tone in these situations mattered at least as much as the substance of the communication. *Running a Big Ship* cautioned against nagging, since "there is nothing to surpass it for making an intelligent man feel insubordinate". While "some people have nagging voices", it went on, "[t]ell them about it, and insist on a change of voice" (O'Conor, 2017: 71).

The success of such methods formed a standing reproach to harsher regimes, but the transition was slow and sometimes uneven. Critics of corporal punishment met with resistance from traditionalists. To consider only the case of the Russian Empire, the debate over phasing out the knout and the practice of running the gauntlet went on both inside the Navy and outside it, in an overlapping and reciprocal relationship with parallel debates over modernisation, and the status of serfs (Violette, 1978). Further, Russian sailors abroad compared notes with each other about other social norms and legal arrangements they had encountered, resulting (according to the naval reformer Pavel Glebov) in a "sense of shame" (quoted in Violette, 1978: 588). Glebov added: "[t]he vigour and strength of military forces everywhere are undermined when self-respect begins to suffer" (quoted in Violette, 1978: 588; see also Land [2009] for parallel British debates). As this quotation suggests, naval reform involved a prolonged process of introspection calling into question both the methods of punishment, and punishment's ultimate aims. The long-term, worldwide trend away from corporal punishment at sea deserves additional scholarly attention and, as the Russian example suggests, it would benefit from a comparative, transnational analysis.

## Deck plans and the built environment

Vilhelm Aubert remarked that the ship was unique among total institutions in that it had no design or mission "to change, model, and reshape individuals" (1982: 248). Recent scholars disagree with Aubert here; indeed, it has become one of the most productive areas of study. To begin with some of the starkest examples, Marcus Rediker notes that the slave ship "not only delivered millions of people to slavery, it had prepared them for it" through a host of degrading rituals and enactments (Rediker, 2007: 350). The confined spaces of the lower decks on the Middle Passage are well known, but other parts of the vessel also served to disconcert and intimidate. Robert Barker, a ship's carpenter, related how he built "a barricade seven feet high, just behind the main-mast, with spike-nails at the top, pointing up, and two port-holes for swivel guns, to hold men slaves in awe" (Land, 2014: 109). Hamish Maxwell-Stewart (2013) proposes that convict vessels, similarly, took in one sort of person, but disgorged another. On ships bound for Australia, "[c]onvicts were subjected to a system of regimentation from the moment they were delivered on board. They were divided into messes, grouped in turn into divisions, each under the eye of a 'captain' handpicked by the surgeon" (Maxwell-Stewart, 2013: 185). The hours, days, and weeks on convict ships ran according to a set rotation of tasks and activities. The physical space of the ship reinforced this message, from "strengthened bulwarks, supplies of leg irons and handcuffs, and hatches that could be guarded" to "the solitary confinement box that was secured to the deck" (Maxwell-Stewart, 2013: 186, 190).

Another area where the ship took on an ambitious social engineering mission was in the area of sanitation, as public health doctrines and biomedicine grew more influential in the mid-nineteenth century. The Royal Navy's Medical Department, for example, devoted a great deal of effort to improve laundry facilities on board, phased in special hospital ships equipped with hot water systems, and introduced zinc chloride as a disinfecting agent. Creating and maintaining an antiseptic environment on board, of course, required a regimentation of minds as well as bodies; crews "were increasingly instructed in personal hygiene", and officers learned that the ship's laundry schedule was not beneath their attention (McLean, 2010: 203).

The early ironclad *USS Monitor*, launched in 1862, placed demands on the crew that foreshadowed a host of later developments. Herman Melville described the *Monitor* as a "welded tomb" (Mindell, 2000: 125). Although its 11-inch-thick hull performed admirably at deflecting enemy attacks, the crew spent most of its time in a sealed compartment below the waterline where temperatures could soar as high as 156° Fahrenheit in the summer, and the air supply "depended on the integrity of ventilators and blower belts" (Mindell, 2000: 65, 147). Notwithstanding the scolding of officers – who made light of the arduous environment – the confined space, limited lighting, excessive heat, and uneven air quality really did take a toll on the crew's physical and mental health, and nearly made this path-breaking vessel useless in combat.

After World War Two, the US Navy commissioned *A Survey Report on Human Factors in Undersea Warfare* with chapters such as 'Human factors in panel design', 'The sleep-wakefulness cycle of submarine personnel', and 'Psychophysiology of stress'. While the term "human factors" implied that the machine needed to accommodate crews, such research also reflected the ongoing project to mould the "operators" into the configuration required by the equipment's demands – even in cramped and poorly lit conditions – and to stay alert and focused for prolonged periods (Stellar, 1949: 159). New cultures of discipline and old ones collided in the 1960s, during the final debate over the rum ration in the Royal Navy. Advocates of abolition invited the testimony of professors and efficiency experts on hours wasted, and "the effect on tasks requiring skilled manipulation and quick, accurate decisions, or involving electronic displays" (Moore, 2017: 72). Without suggesting that the new equipment dictated what people did, we can observe that it produced a "swerve", in the sense outlined by Bruno Latour (Ross, 2014). This suggests some intriguing pathways for future research.

While a submarine lurking in the dark waters beneath a polar ice cap may offer one of the more plausible examples of a total institution, it is worth considering examples of the opposite phenomenon, in which the sea-going vessel was reshaped to better approximate the norms of terrestrial life. Passengers on Atlantic Ocean liners operated by Cunard and other companies came from a world that included parlours, drawing rooms, and smoking rooms. Therefore, each company plying the Atlantic route had to devise regulations with an eye to urgent questions such as where men and women could socialise, where gender segregation should prevail, and where smoking was permitted. The ocean liner's deck plan evolved in dialogue with templates of respectable behaviour derived from river steamboats, railway carriages, and grand hotels (Hart, 2010).

Meanwhile, the Royal Navy – anxious over the influx of parvenu middle class men into the officer corps in the final decades of the nineteenth century – furnished officers' living spaces to evoke (variously) the neo-Georgian wood panelling of a country house, the public school common room, the lodging of an undergraduate university student, and the gentlemen's club. In a stimulating article, Quintin Colville (2009: 506) proposes that this reshaped the warship into "a gymnasium of authorized corporate masculinity". Oddly enough, this development involved asserting the forms of one total institution (the boarding school) as the supreme template in the setting of another (the ship).

Finally, racial segregation offers a particularly telling example of a terrestrial norm that eventually imposed itself on shipboard life. After experiencing widespread acceptance as sailors in the 1700s, African-Americans found themselves restricted to the roles of stewards and ship's cooks by the end of the nineteenth century (Harrold, 1979). Ocean liners operating in the South Pacific, with racially mixed crews and occasional Pacific Islander passengers camped out on the deck, faced complaints about odours, the placement of toilet facilities, and unwelcome encounters (Steel, 2011). To whatever extent the ship was, indeed, a total institution, the exigencies of white supremacy proved more total still.

## Concealment, indiscipline, and resistance

One feature of total institutions like prisons and asylums is the aspiration to a very nearly complete system of surveillance. This might appear easy to accomplish at sea, but despite the obvious difficulties, it was possible to achieve a surprising degree of privacy aboard ships. A poignant example from the 1850s underscores this. British authorities in charge of the conveyance of indentured migrants from South Asia imposed an unprecedented, intrusive regime of record-keeping, medical regulation, and "bureaucratic state involvement" (Brown and Mahase, 2009: 196). In practice, western male doctors had difficulty winning the trust of their charges. An investigation into the *Salsette*, which lost a staggering 42 per cent of its indentured passengers during its voyage from Calcutta to Trinidad, revealed how "female indentured immigrants actively sought to evade medical observation", reporting symptoms only when on the verge of death (Brown and Mahase, 2009: 201–203). The captain's wife suggested that female nurses should become a standard feature on such ships in the future, but her idea did not fall on receptive ears.

Similarly, sailors who were "loath to snitch on their comrades" could form an impenetrable screen (Massarella, 2017: 427). The Royal Navy banned political discussions, but we know that some went on anyway (McKee, 2002: 98–101). If the aim of this rule was in part to prevent conversations that could end in unseemly squabbling, it is worth reflecting on the British sailor from the World War One era who recalled, "I have seen the mess almost in blows over the question 'Is marmalade jam?'" (McKee, 2002: 84). Likewise, if any object interrupted the officer's line of sight, concealment was possible. Sailors in the early-twentieth-century Royal Navy astutely noted the difference in disciplinary styles between smaller vessels such as submarines and destroyers versus the larger ones, where the distance between officers and men might be stern and more rigidly enforced, but also where there were more opportunities to escape the unceasing invigilation of a superior (McKee, 2002: 102). For example, any room with a door that locked could potentially offer a safe haven for gambling (McKee, 2002: 147). Aboard the British battleship *Iron Duke*, the carpenter's shop developed a reputation as a place with sufficient privacy that men who sought the sexual company of other men could go and do as they wished (McKee, 2002: 192).

Surely, to some extent, transgressions took place with the active connivance of superiors who understood the need to let off steam. In the case of illegal gambling, some petty officers took advantage of the situation by breaking up the game and lining their own pockets with the gambling proceeds, confident that the guilty parties had no recourse (McKee, 2002: 147–148). Some customs could only survive if officers were willing to overlook a temporary lapse in discipline, as in the case of the "birthday ration" of rum, in which a sailor was allowed to drink all the grog he wanted that one day of the year, and while he slept off the spree, his absence would go officially unnoticed (McKee, 2002: 155).

If we can discern a firmly established shipboard culture of looking the other way, it is interesting to consider what sorts of situations would elicit the opposite behaviour. The court martial for sodomy aboard the *HMS Africaine* in 1815 resulted from the unusually public nature of the sex acts. One witness commented with indignation that the men were "copulating in plain view like dogs", standing up against, or leaning over, the ship's cannons (Burg, 2007: 142–143). Comparing the *Africaine* case with other court martial records from the same era does demonstrate that couples had a number of more discreet options for their sexual encounters, ranging from doing it in the ship's galley, behind a canvas partition, or simply down in the space in between the cannons (Burg, 2007).

If prosecutors sometimes found it difficult "to distinguish between consent and coercion", the advances of an older man towards a teenage boy struck many shipmates as egregious and probably non-consensual, and these made up the preponderance of cases that actually were reported to officers (Conley, 2019, 85). At times, the cross-examination in what appeared to be an isolated case uncovered a much more extensive pattern of activity. A witness overheard William Sutton remarking that he had caught gonorrhoea from having "a lump of ass of a boy" (Conley, 2019: 84). This testimony prompted a line of questioning about whether such language was unusual, to which the response came: "it was a sort of a remark made in the head night after night". The 16-year-old at the heart of this prosecution further alarmed the officers by stating that he had accepted money in exchange for sexual acts. This incident provoked moral panic in the Admiralty offices, as the Navy's leadership reflected on implications of the discovery that everyday conduct aboard their own ships was largely opaque to them.

The shock of discovering hidden networks of sexual transgression paled in comparison to witnessing a mutiny that officers had not even suspected was in the offing. Mutiny – and more so resistance – relate to the subject of other chapters in this volume, but it is worth mentioning that the ability to conceive, plan, and execute a mutiny under the officers' noses suggests that paternalist claims, like some other fantasies of control, were overblown.

## Conclusion

Not surprisingly, since many concepts from the early and mid-twentieth century (including 'totalitarian') now seem dated, most scholars today would draw attention to the shortcomings of the total institution concept. Any approach that divorces the vessel from its contexts, for example, is probably dead on arrival. It would be difficult to explain the lawsuit against the captain who tore out a sailor's beard without reference to concepts of masculinity that originated in a non-maritime setting. The bizarre punishments imposed on Muslim lascars are unintelligible without reference points such as empire, capitalism, or globalisation.

Despite its limitations, the total institution framework is hard to dismiss entirely. As with other settings, such as dictatorships, it is useful to differentiate between the ambition for total control, and the somewhat messier, uneven implementation of the control project. If slave ships in the Middle Passage sought to 'make' slaves, we know that not everyone submitted, or reacted in the same way. Those in charge of Britain's transports for indentured servants promised that no passenger would escape medical inspection and monitoring, but the statistics and anecdotal evidence establish that many did. If, however, "authority rested ultimately on the threat of violence, the sort of casual violence that does not get into logbooks", then our scepticism may be overdone (Earle, 1998: 159). Historians may catch the total institution napping from time to time, but that does not mean it did not pose a real threat when it was alert and on the prowl.

A different approach might argue that the ship was never a total institution, and that omission was by design. Inconsistency offered a kind of safety valve, even in the most austere and hierarchical environments. We see hints of this as early as the medieval law code that admonished sailors to take their beatings, but also offered them a place to retreat where the master could not follow. Reactions to mutiny provide similar evidence of ambiguity and equivocation. Even when confronted with the ultimate act of disobedience, officers often took care to escalate with caution. During the mutiny on the battleship *Potemkin*, an officer called for a squad of marines, and a tarpaulin. The spread-out tarpaulin would "protect the deck from bloodstains" during an execution, so it was a credible method of signalling that the next step – if it came – would involve gunfire (Guttridge, 1992: 189). In September 1931, Chilean military aircraft suppressed the mutiny aboard the battleship *Almirante Latorre* using a bomb dropped close enough "to shower her people with spray and shrapnel" but not targeting the deck (Guttridge, 1992: 189). Thus, retribution could fall short of what the law actually authorised.

This behaviour extended to the period after the mutiny, when it was time to prosecute. Many scholars have raised an eyebrow at the frequent circumlocutions adopted (by naval officers and merchant captains alike) to avoid admitting that any mutiny had occurred (Guttridge, 1992; Jaffer, 2015; Rose, 1982). Legal definitions that hinged on ambiguous terms such as 'mutinous assembly' offered a great deal of room for manoeuvre. One result of this is that many mutineers escaped the severe punishment corresponding to the crime. To some extent, clearly, this was a self-serving decision, as officers who had lost control of their own ships suffered in the estimation of their peers. However, relenting in the pursuit of justice may reflect a deeper wisdom. Charitably reinterpreting unpleasant events represents a survival – well into the modern era – of the old practice in which masters would exercise their singular and arbitrary power, but in the interests of forgiveness (see, for example, Pérez-Mallaína, 1998: 193). The potential for clemency formed part of the constellation of discipline. More broadly, the practice of overlooking certain forms of indiscipline could itself serve as an oblique method of control, permitting a tempering of the necessities of command with accommodation and realism.

## Notes

1 For example, consider the numerous caveats and objections raised in Witt (2001).

2 For the original text of this law, see Capítulo 164: "Cómo debe el marinero soportar a su patron" which appears in *Libro del Consulado del Mar: Edición y texto original catalán y traducción al castellano de Antonio Capmany* (Barcelona: Cámara Oficial de Comercio y Navegación, 1965), page 144.

3 See the *Oxford English Dictionary* s.v. "start" (verb), senses 7, 8b, and 9a (b) www.oed.com/view/Entry/189183?rskey=PvpnlK&result=5

4 *Oxford English Dictionary*, s.v. "pizzle". Accessed 30/6/2019. www.oed.com/view/Entry/144847?redirectedFrom=pizzle

## References


Aubert V (1982) *The Hidden Society*. London: Transaction Books.

Blakemore RJ (2015) The legal world of English sailors, c. 1575–1729. In: Fusaro M, Allaire B, Blakemore RJ and Vanneste T (eds) *Law, Labour and Empire: Comparative Perspectives on Seafarers, c. 1500–1800*. Basingstoke: Palgrave Macmillan, 100–122.

Bogucka M, Bruijn JR, Davids K, Gerstenberger H, Kaukiainen Y, Krieger M, Müller L, Sicking L and Heerma van Voss L (2002) Reviews of Jann Markus Witt, Master Next to God? Der nordeuropäische Handelsschiffskapitän vom 17. bis zum 19. Jahrhundert with a Response *International Journal of Maritime History* 14(2): 331–366.

Brown L and Mahase R (2009) Medical encounters on the *Kala Pani*: Regulation and resistance in the passages of indentured Indian migrants, 1834–1900. In: Haycock DB and Archer S (eds) *Health and Medicine at Sea, 1700–1900*. Woddbridge: Boydell Press, 195–212.

Burg BR (2007) *Boys at Sea: Sodomy, Indecency, and Courts Martial in Nelson's Navy*. London: Palgrave Macmillan.

Byrn JD (1989) *Crime and Punishment in the Royal Navy: Discipline on the Leeward Islands Station, 1784–1812*. Aldershot: Scolar Press.

Christopher E (2006) *Slave Ship Sailors and Their Captive Cargoes, 1730–1807*. Cambridge: Cambridge University Press.

Colville Q (2009) Corporate domesticity and idealised masculinity: Royal naval officers and their shipboard homes, 1918–39. *Gender and History* 21(3): 499–519.

Conley M (2019) The admiralty's gaze: Disciplining indecency and sodomy in the Edwardian fleet. In: Colville Q and Davey J (eds) *A New Naval History*. Manchester: Manchester University Press, 70–88.

Coureas N (2017) The Lusignan Kingdom of Cyprus and the sea. In: Balard, M (ed) *The Sea in History*, vol. 2 *The Medieval World*. Woodbridge: Boydell Press, 369–381.

Creighton M (1995) *Rites and Passages: The Experience of American Whaling, 1830–1870*. New York, NY: Cambridge University Press.

Dana RH Jr. (1844) *Two Years before the Mast: A Personal Narrative of Life at Sea*. New York, NY: Harper and Brothers.

Davies AD (2014) Learning large ideas overseas: Discipline, (im)mobility and political lives in the Royal Indian Navy Mutiny. *Mobilities* 9(3): 384–400.

de Oliveira Torres R (2014) Handling the ship: Rights and duties of masters, mates, seamen and owners of ships in the nineteenth-century merchant marine. *International Journal of Maritime History* 26(3): 587–599.

Earle P (1998) *Sailors: English Merchant Seamen, 1650–1775*. London: Methuen.

Eastwick RW (1891) *A Master Mariner* (ed. H Compton). London: Fisher Unwin.

Gabrielsen V (2017) Financial, human, material and economic resources required to build and operate navies in the classical Greek world. In: de Souza P and Arnaud P (eds) *The Sea in History*: vol. 1 *The Ancient World*. Woodbridge: Boydell Press, 426–442.

Gerstenberger H (2001) The disciplining of German seamen. *International Journal of Maritime History* 13(2): 37–50.

Guttridge LF (1992) *Mutiny: A History of Naval Insurrection*. Annapolis, MD: Naval Institute Press.

Hart, D (2010) Sociability and 'separate spheres' on the North Atlantic: The interior architecture of British Atlantic liners, 1840–1930. *Journal of Social History* 44(1): 189–212.

Harrold FS (1979) Jim Crow in the Navy, 1798–1941. *United States Naval Institute Proceedings* 105: 46–53.

Hope R (1990) *A New History of British Shipping*. London: John Murray.

Jaffer A (2015) *Lascars and Indian Ocean Seafaring, 1780–1860*. Woodbridge: Boydell Press.

Jahnke C (2017) The maritime law of the Baltic Sea. In: Balard, M (ed) *The Sea in History*, vol. 2 *The Medieval World*. Woodbridge: Boydell Press, 572–584.

Land I (2009) *War, Nationalism, and the British Sailor, 1750–1850*. New York, NY: Palgrave Macmillan.

Land I (2014) Patriotic complaints: Sailors performing petition in early nineteenth-century Britain. In: Paisley F and Reid K (eds) *Critical Perspectives on Colonialism: Writing the Empire from Below*. New York, NY: Routledge, 102–122.

Lawrence C (1996) Disciplining disease: Scurvy, the Navy, and imperial expansion, 1750–1825. In: Miller DP and Reill PH (eds) *Visions of Empire: Voyages, Botany, and Representations of Nature*. Cambridge: Cambridge University Press, 80–106.

McKee C (2002) *Sober Men and True: Sailor Lives in the Royal Navy, 1900–1945*. Cambridge, MA: Harvard University Press.

McLean D (2010) *Surgeons of the Fleet: The Royal Navy and its Medics from Trafalgar to Jutland*. London: IB Tauris.

Massarella D (2017) '& thus ended the buisinisse': A buggery trial on the East India Company ship *Mary* in 1636. *Mariner's Mirror* 103(4): 417–430.

Malcolmson T (2016) *Order and Disorder in the British Navy, 1793–1815: Control, Resistance, Flogging and Hanging*. Woodbridge: Boydell Press.

Maxwell-Stewart H (2013) 'Those lads contrived a plan': Attempts at mutiny on Australia-bound convict vessels. *International Review of Social History* 58: 177–196.

Mindell DA (2000) *War, Technology, and Experience aboard the USS Monitor*. Baltimore, MD: Johns Hopkins University Press.

Moore R (2017) 'We are a Modern Navy': Abolishing the Royal Navy's rum ration. *Mariner's Mirror* 103(1): 67–79.

O'Conor R (2017) *Running a Big Ship: The Classic Guide to Managing a Second World War Battleship*. Oxford: Casemate Publishers.

Panopoulou A (2017) At the centre of the sea routes: Maritime life in Crete. In: Balard M (ed) *The Sea in History*, vol. 2 *The Medieval World*. Woodbridge: Boydell Press, 382–400.

Pérez-Mallaina PE (1998) *Spain's Men of the Sea: Daily Life on the Indies Fleets in the Sixteenth Century* (trans. C Rahn Phillips). Baltimore, MD: Johns Hopkins University Press.

Rediker M (1987) *Between the Devil and the Deep Blue Sea: Merchant Seamen, Pirates, and the Anglo-American Maritime World, 1700–1750*. Cambridge: Cambridge University Press.

Rediker M (2007) *Slave Ship: A Human History*. New York, NY: Viking.

Rodger NAM (1986) *The Wooden World: An Anatomy of the Georgian Navy*. New York, NY: WW Norton.

Rodger NAM (2004) *The Command of the Ocean: A Naval History of Britain, 1649–1815*. New York, NY: WW Norton.

Rose E (1982) The anatomy of a mutiny. *Armed Forces and Society* 8(4): 561–574.

Ross S (2014) 'History, mystery, leisure, pleasure': Evelyn Waugh, Bruno Latour, and the ocean liner. In: Tally RT Jr. (ed) *Literary Cartographies: Spatiality, Representation, and Narrative*. London: Palgrave Macmillan, 111–125.

Steel F (2011) *Oceania under Steam: Sea Transport and the Cultures of Colonialism, c. 1870–1914*. Manchester: Manchester University Press.

Stellar E (1949) Human factors in panel design. In: *A Survey Report on Human Factors in Undersea Warfare*. Washington DC: Committee on Undersea Warfare, 153–176.

Stöckly D (2017) 'Quod vita et salus nostra est quod galee nostre navigent': Les gens de mer à Venise du XIIIe au XVe Siècle. In: Balard M (ed) *The Sea in History*, vol. 2 *The Medieval World*. Woodbridge: Boydell Press, 158–169.

Unger R (2017) The maritime war in the Mediterranean, 13th–15th Centuries. In: Balard M (ed) *The Sea in History*, vol. 2 *The Medieval World*. Woodbridge: Boydell Press, 90–100.

Violette AJ (1978) Judicial reforms in the Russian navy during the 'Era of Great Reforms': The Reform Act of 1867 and the abolition of corporal punishment. *Slavonic and East European Review* 56(4): 586–603.

Winton J (1977) *Hurrah for the Life of a Sailor! Life on the Lower-deck of the Victorian Navy*. London: Michael Joseph.

Witt JM (2001) 'During the voyage every captain is monarch of the ship': The merchant captain from the seventeenth to the nineteenth Century. *International Journal of Maritime History* 13(2): 165–194.

# 18
# PROTEST
## Contested hierarchies and grievances of the sea

*Paul Griffin*

### Introduction

As Linebaugh and Rediker (2000) note, the sea holds great theoretical and empirical potential as a scholarly interest for those seeking 'histories from below', or contemporary accounts of protest politics. Their much-celebrated work *The Many Headed Hydra* illuminates, amongst other hidden histories of struggle, the revolutionary narratives of protest, disobedience and organising forged within and across the Atlantic Ocean to foreground previously downplayed acts of sailors, workers and pirates. Such works indicate the illuminating nature of sea-based protest and the specific need to situate this resistance in relation to the dynamism of the sea. Their work points to the ever-transforming notion of maritime geography, and the need to position resistance in relation to the changing processes of domination that grievances are made in response to. Such efforts link with wet ontologies, what Steinberg (2013: 165) describes as a perspective whereby "the ocean becomes the object of our focus not because it is a space that facilitates movement – the space across which things move – but because it is a space that is constituted by and constitutive of movement". This spatial understanding of the sea can be deepened through a greater sensitivity to power relations and an acknowledgement of the processes of domination and resistance made and found within such spaces (Sharp et al., 2000). Here, this chapter considers examples of sea-based resistances to extend understandings of maritime spaces.

The chapter surveys existing academic works on maritime protest and in a latter section briefly draws upon my own archival research on sailor organising and protests associated with early-twentieth-century radicalism in Glasgow, UK. Bringing together these contributions allows the chapter to consider those acts of protest found at sea, protests constructed through movement across the sea and landed protests articulating grievances of the sea (see also Featherstone, this volume). This spatial approach to maritime protest can be understood through the lens of radical geography but also complicates and extends some understandings of protest conceptually. Thus, the chapter will begin with a brief theoretical reflection on spatial and temporal conceptualisations of protest to shape the engagements that follow. It will then consider three scales of maritime protest with reference to historical and contemporary examples. The chapter concludes with some wider comment on the enduring and contested nature of ocean protests.

DOI: 10.4324/9781315111643-22

## Theorising maritime protest

Routledge (2018) has understood protest through 'emergence', identifying that resisting acts relate to, emerge from and shape particular places and spaces. His conceptualisation asserts the merits of a radical geography that engages with how space influences or shapes the dynamics of contentious politics. Such comments follow a longer tradition of radical geographers who have engaged with radical, alternative, protest geographies. A search through the associated and influential radical geography journal *Antipode*, though, would suggest that the vast majority of protest geographies have concerned themselves with landed matters (for exemptions see Dunnavant, 2021; Menon et al., 2016; Stierl, 2018). Here, it is argued that the ocean provides plentiful opportunity for extending these debates, particularly around territorial and emergent understandings of protest, and more broadly developing conversations around the spatial politics of protest (see also Halvorsen, 2015).

As such, the relational power dynamics of the sea must be interrogated. Linebaugh and Rediker's (2000) work famously notes the potentiality of the pirate ship to position the sea to be viewed a site whereby power relations were inverted and transformed. They draw upon the term 'hydrarchy', borrowed from the upper-class member Richard Braithwaite who used the term in the seventeenth century in his description of the mariner, to consider the dynamics between the maritime state in its pursuit of control 'from above' and "the self-organization of sailors from below" (Linebaugh and Rediker, 2000: 144). This positioning of the ship and the ocean more broadly, as holding possibilities for protest even in the most extreme circumstances of structural control, discipline and punishment, allows for recognition of protest through numerous means including 'small acts', collective political organising and demands articulating aggregating grievances (Linebaugh and Rediker, 2000; see also Land, this volume, on 'discipline'). Whilst Linebaugh and Rediker's work is integral to Featherstone's chapter, as he has noted elsewhere, there is a tendency in their account to "treat space as a fixed backdrop to political activity" whereby "[i]deas, tactics and radical experiences flow and move across space, but these circuits remain unchanged through these processes" (Featherstone, 2005, 392–393) Such comments inform the approach taken below, whereby three 'cuts' of maritime protest are engaged with to indicate the significance of spatial relations when unpacking resistance at sea.

Although the spatial elements of agency provide a structuring device for the chapter, it is also essential to engage with the temporal elements of such acts. As such, Chakrabarty's (2000: 66) notion of "history 2s" is helpful here, foregrounding pasts that "may be under the institutional domination of the logic of capital and exist in proximate relationship to it, but they also do not belong to the 'life process' of capital". By engaging with protest through 'other ways of being', the chapter illuminates the possibilities for protest to be constructed as inclusionary and progressive (such as the alternative political visions and humanitarian acts considered below) but also potentially exclusionary (such as the violence and racialised hostilities also considered below). To consider this diversity of positions, the chapter draws upon and blends together historical examples with contemporary issues to indicate the enduring, variable and contested nature of protest within ocean spaces. This allows the chapter to suggest a temporality of protest that complements the spatial approach, and allows for recognition of intense moments of action alongside the longer trajectories of movements, as part of wider movements articulating aggregative grievances with potential for unintended outcomes (see Hughes, 2020).

Such temporalities suggest a need to think of maritime protest beyond associations with spontaneity. Guha has stressed this in his work on *Elementary Aspects of Peasant Insurgency in Colonial India* (1983) identifying how spontaneity is often wrongly ascribed to particularly disruptive forms of protest from a top-down perspective, creating a potential 'moral outrage' which fails to consider the political elements that shape and mould such acts (Nolan and Featherstone, 2015). Although focusing on insurgency located on-shore, Guha's comments, which he links to the work of Gramsci, remain important for a framing of maritime protest:

> there is no room for pure spontaneity in history. This is precisely where they err who fail to recognize the trace of consciousness in the apparently unstructured movement of the masses. The error derives more often than not from two nearly interchangeable notions of organization and politics.
>
> *Guha, 1983: 5*

This questioning of spontaneity is important for the wider aims of the chapter, particularly in the following section, and raises questions over how subaltern agency is considered and represented. Guha was constructing a direct reply to Hobsbawm's (1959: 5) notion of the "pre-political" and "social banditry" of protesting rural peasants whom he considered to have "no organization or ideology" and to be "totally inadaptable to modern social movements". This portrayal carries a pre-conceived concept of politics and organisation, which undermines the possibilities within Guha's work for illuminating agency from below. In what follows, resistance is considered through a continuum of maritime protest acts, including those that might appear seemingly marginal or unorganised, to consider multiple examples of ocean protest and to indicate the sustained resistances to dominant powers.

There is thus a need for "imaginative connections" to be made between events which may be portrayed as "minor topics" and those considered "important issues" (Searby et al., 1993: 20) as asserted by EP Thompson and others who have utilised the 'history from below' tradition to uncover a variety of protest acts (see Featherstone and Griffin, 2016). To pursue such constructions of maritime protest, this chapter considers these theoretical influences in relation to three scales of maritime protest. Firstly, the chapter considers protests acts found at sea; secondly, the chapter considers protest acts as connected and shaped by experiences of the sea; and thirdly, the chapter considers landed protests articulating sea-based grievances. This multi-scalar approach to maritime protest is developed to engage with a wide repertoire of protest acts whilst the multiple examples drawn upon move across historical examples and contemporary disputes to indicate the enduring presence of ocean-based protest. These examples are deployed to illustrate the particularity of disputes and protest strategies but should be read in conversation to acknowledge the connections between the scales of protest considered below.

## Maritime protest at sea

Hasty notes how the materialities of the ship reflect the complexities of power relations found at sea. His work on seventeenth-century pirates shows how the ship itself was manipulated and moulded to create a ship space that:

> existed as a real, lived and dynamic space, one crafted by pirates in their own image with their own ends in mind. The ship functioned as a technology of mobility and speed, as a locale for piratical politics and as a space of multiple contestations, and revealing their spatial practices in modifying this space sheds much needed light on their intriguing way of life.
>
> *Hasty, 2014: 364*

Such contestation held radical possibilities as Hasty considers through the remodelling of the ship deck following a pirate takeover to foster more horizontal hierarchies. Hasty notes how these radical possibilities are similarly stressed by Rediker who suggested that the pirate ship provides a specific space whereby it was possible to view "the world turned upside down" (Rediker, 2004: 61), acknowledging how pirate captains and workers would coalesce and work in a co-operative manner. Whilst similarly acknowledging these radical possibilities, Hasty is keen to identify the uneven nature of such radical spaces, noting differences between pirate ships, whereby some ships would maintain more vertical hierarchical structures (e.g. clear distinguishing of captain space) and the prevalence of radicalisms and exclusions, of prisoners for example, on board certain ships. This uneven distribution of power within particular ships begins to indicate the contested nature of maritime spaces and more specifically the plurality of potential protest practices.

A challenge within these structures is to identify the multiplicity of resisting alternative acts from subaltern groups and individuals. Davies (2013) has utilised assemblage theory to consider such diversity through his study of colonial navy sailors during the 1946 Royal Indian Navy mutiny. He identifies the spreading of mutiny and violence following strike action on His Majesty's Indian Ship (HMIS) *Talwar*, a shore installation of the Royal Indian Navy (RIN) in Bombay. He notes how the grievances associated with this action "included issues ranging from the banal (the poor food served in the RIN), through to the overtly political (the continued British rule of the Indian subcontinent)" (Davies, 2013: 24). The initial mutiny was supported by subsequent strike action with Davies describing how a further 20,000 sailors 'mutinied' in military stations across South Asia and as far as the Andaman Islands and Aden (2013).

Davies' engagement with such a wide-ranging set of seafaring grievances and protests associated with the strike allow the seafarers' political agency in their own right, rather than the dilution of grievances through overarching narratives or attributions of spontaneity. Elsewhere, Davies has considered how such maritime anticolonial activisms reveal "how 'nationalist' ideas were inherently stretched beyond the territorial limits of the landed 'ocean' space'" (2019: 69–70). Through engagements with the shipping routes of the Swadeshi Steam Navigation Company he illustrates how "the dreams of industrial nationalist development that spurred much *swadeshi* organising were also often *inter*nationalist in nature" (Davies, 2019: 70, emphasis in original). Such moments are often difficult to measure in terms of their success, as they may be suppressed in the short term and restricted in enforcing immediate change. What they are indicative of, though, is the resistant nature of colonial labour and the plurality of protests emergent from maritime spaces.

This diversity of political positions within protests links to the diversity of strategies and places where such maritime acts might be found. Featherstone (2009) has also utilised Guha's conceptualisation of subaltern historiography to consider how subaltern protest might be found in unexpected places. He revisits the Royal Navy Court Martial from July 1797 whereby six mutineers from the *Grampus* ship, as part of a wider 26-ship Nore mutiny, faced trial for their protest actions, which included pay-related demands but also wider democratic political views. Featherstone (2009: 766) notes that the associated records document how the mutineers "circumvented the logics of the trial to assert the justice of their actions" through declarations reflecting their democratic cultures. Featherstone concludes that:

> Rather, this has located the court-martial as part of the ongoing struggles aboard the ship and as an element in the routes and connections that shaped sailors' mutinous

> cultures. These struggles were to continue to shape mutinous dispositions aboard the ship after the Nore mutiny.
>
> *Featherstone, 2009: 785*

Featherstone's detailing of the court martial narratives begins to reveal longer trajectories of protest acts, beyond the immediacy of an event, to indicate the makings of "assertive subaltern political identities" (Featherstone, 2009: 766). More broadly, these works indicate the possibility for disobedience to be framed as protest. Such constructions of disobedience as protest is particularly crucial to consider in scenarios where organised resistance may seem unlikely. Rediker (2007), for example, has considered the 'small acts' of resistance found on the seventeenth-century slave ship. In *The Slave Ship: A Human History* he notes how slaves resisted the inhuman conditions, violence and terror of their passage, through acts such as hunger strikes, jumping overboard and insurrections. He considers how these acts reflect practices of mutual aid and survival, positioned within "the beginnings of a culture of resistance, the subversive practices of negotiation and insurrection" (Rediker, 2007: 350). Accounting for this specificity in the nature, spatiality and articulation of protest acts is crucial for a critical exploration of maritime protest (see also Dunnavant, 2021). In doing so, the acts raised here connect with previous comments regarding the emergent spatial-temporal nature of protest.

In contemporary times, the sea remains a space of disobedience, activism and direct action. Couper et al. (2015: 163) note the prevalence of mutinies and exit practices in modern fisheries where fishers respond to "intolerable conditions". They identify the Indonesian island of Tual as being a space where such exits are particularly frequent, noting how it has held between "700–1000 Burmese (at any given time) who have fled from fishing boats" (Couper et al., 2015: 164). Couper et al. also document the continued prevalence of mutinies and violence within fishing industries. Exiting can be considered a protest strategy, particularly when framed within a punitive work regime. Such protest narratives can be read alongside strike action and viewed as contributing towards successes of fishers in courts, such as those in New Zealand (NZ) where the government developed increased regulation of working conditions and pay, through a Code of Practice in 2006 and a decision in 2016 that only NZ flag vessels would be allowed to fish in particular zones. These acts were made in response to the activism of trade unions and NGOs regarding the intolerable working experiences of fishers within these zones. Environmental groups and anti-whale hunting protest have also utilised disruptive acts in their efforts to prevent environmentally detrimental fishing and hunting practices. Sea Shepherd (no date) for example are a conservation group noted for using direct action, including ramming and disabling pirate whalers, in their efforts to protect marine life (see McKie, 2017). To extend and deepen accounts of the sea's role in capital accumulation and globalisation, these contemporary disruptions indicate the sea to be a continued site of direct action, resistance and contestation.

Humanitarian acts in more recent times have been similarly positioned as holding potential as political dissent. Stierl (2018: 704), for example, points to a "humanitarian spectrum" of NGO activities that seek to turn the sea into a less deadly space in the context of large-scale Mediterranean migration. The scale of such rescues is undoubtedly commendable for the thousands of lives saved from precarious and life threatening border crossings, but such acts might not immediately appear as a protest act. Stierl notes, though, the importance of political imaginaries in shaping the positionality of humanitarianism, as NGOs may be co-opted by the state (in this case the EU) or alternatively, might be considered as a vehicle to critically challenge border practices (see also Heller et al., this volume). The article considers those organisations that actively oppose existing regimes, and here the chapter considers these activities within the maritime protest repertoire as these actions are explicitly situated against EU refugee schemes. Sea-Watch, a search and rescue organisation operating in the Central Mediterranean, for example, have always envisioned their work as an outspokenly political intervention:

> [w]e have decided to fight for the humanisation of politics. Hospitality should once again be the norm. A civil sea rescue service must be created. The EU is not willing to do so. Therefore, we are taking the initiative.
>
> *Sea-Watch, 2015, cited in Stierl, 2018*

As Peters (2013) has noted, ocean spaces have remained a site of resistance to authority. The challenging of the EU state policy noted by Stierl and the potentiality for breaching maritime borders highlights the contested nature of ocean regulation that can also lead to conflict arising from border crossing and sea-based grievances. Menon et al. (2016), for example, illuminate the contested nature of boundary crossings of Indian and Sri Lankan fishers and the emerging conflicts that arise from such acts. Their article considers the associated artisanal activism in the Palk Bay Fisheries to consider the contested dynamics of capital accumulation associated with the fishing industry. Their work indicates how, at different times, artisanal fisher activism (following conflicts with trawl fishers) has been successful in establishing no-go zones for trawler fishers, yet conflict still remains with fishers refusing to follow regulations. Sinha (2012) has noted similarly successful organising practices from the Indian Fishworkers' Movement during the late twentieth century who were able to establish a 3 km exclusive fishing zone for artisanal fishers in the North Kerala region. These acts are indicative of the sea as a contested and dynamic space, but such acts are often informed by a wider set of connections beyond their particularity, and these connections are considered further below.

## Maritime spaces and connected protest

Whilst protest acts are found on the ship and across the ocean, maritime spaces also play a crucial role in connecting radical alternative visions and activisms. The ocean is integral to the connectivity of transnationalism and solidarity (see Featherstone, this volume) but also the experience of travel itself and multiculturalism at sea has played a crucial role in informing and shaping the geographies of protest cultures of individuals and collectives. Here, the chapter briefly reflects on two related elements to this, firstly the protest geographies of individual lives shaped through the sea and secondly the multi-scalar possibilities of shared protest geographies developed through the spatiality of the sea.

Gilroy has shown the possibilities of this approach through *The Black Atlantic*, which engages with "the long neglected involvement of black slaves and their descendants in the radical history of our country in general and its working-class movements in particular" (1993: 12). He considers the Atlantic as "one single, complex unit of analysis", forming part of a "webbed network, between the local and the global" (Gilroy, 1993: 15, 29). This positioning of a fluid and relational Atlantic facilitates a far more transformative sense of the spatial politics of maritime protest, and specifically those resistances associated with race and racism, than those that impose fixed notions and assumed politics on the basis of place, ethnicity and nationhood. More broadly, Linebaugh and Rediker note the shared influences of maritime cultures that informed political cultures on previously uninhabited or remote islands in the Caribbean, highlighting how resistances of the sea informed politics on land. In particular, they note how seventeenth-century radical Atlantic traditions informed the buccaneers in America whereby seafarer culture developed a "Jamaica discipline" that "boasted a distinctive conception of justice and class hostility towards shipmasters" and "featured democratic controls on authority and provision for the injured" (Linebaugh and Rediker, 2000: 158). There is an indication here of the significance of travel and exposure to difference as being a critical factor in the making of radical and alternative cultures.

Chris Braithwaite was a Pan-African seafarer and activist during the early twentieth century, whose life was profoundly shaped by experiences offshore. In his article 'Mariner, renegade and castaway', Høgsbjerg (2011) describes the political life of Barbadian Braithwaite to reveal a similarly connected and transatlantic life of organising and protest. Høgsbjerg notes Braithwaite's involvement in resistance activities, such as demonstrations in London on behalf of the Scottsboro boys and wider campaigning on "militant anti-capitalist and anti-imperialist" matters through the Negro Welfare Association (Høgsbjerg, 2011: 44). Such political sentiment and world-view can be attributed to his wide ranging maritime and transnational experiences. During the previously mentioned Royal Indian Navy mutiny, Davies (2013) similarly notes the influence of travel upon Balrai Chandra Dutt, which informed his involvement with the RIN mutinies previously discussed, identifying an "embodied cosmopolitanism" felt through the "lived experience of travelling, seeing other's struggles against grievances, and recognising that these political struggles have parallels with one's own" (Davies, 2013: 29). Whilst political biographies of activists such as these indicate the influence of mobility on the individual, it is also clear that a maritime geography of protest was also profoundly shaped by the particularity of ocean experiences. Here, Braithwaite's life is again revealing, with his Pan-Africanism reflecting networks and connections shaped through his seafaring life, such as the 1930s campaigns to defend Ethiopia under military threats from Italy. Quest (2009: 122) has noted how Braithwaite mobilised seafaring networks to facilitate "direct action to undermine the economy of Italy and smuggle weapons to Ethiopia". Such connections begin to reveal the spatiality of maritime protests and potentiality for wider geographies of protest activity through disruptive transnational influences, actions and solidarities, countering the wider mobilities of the state and capital.

A further strategy of resistance shaped by maritime spaces is evident through the distinctive sharing of seafarer objections through petitions, for example, which would highlight shared concerns and issues. These developed through novel strategic practices such as the eighteenth-century round robins whereby sailors would write petitions and letters signed in a circular manner to reflect horizontal organising practices and to make the attributing of leadership impossible. Rediker (1987: 234–235) describes such practices as an "instrument of protest" reflecting "a cultural innovation from below, an effort at self-defense in the face of nearly unlimited and arbitrary authority". The reality of the punitive conditions within which such acts emerged makes the moments and practices of solidarity and protest amongst seafarers even more noteworthy. Silverman, for example, notes practices of labour internationalism at sea whereby exiled seafaring unions during the Second World War received refuge and support from British trade unions seafarers, reflecting a "brotherhood of the oceans" (Silverman, 2000: 40). He also notes the limitations of such solidarities though, and how exclusions – particularly experienced by Chinese sailors – forged and maintained during these times, reflected a less open vision of protest geographies as is considered further below. Thus, whilst the sea contains numerous examples of protests that influenced a broad politics of transnationalism and solidarity, these spatial linkages often culminated in actions or held repercussions for events which took place onshore. The final section of this chapter considers such moments where maritime protests meet onshore communities and institutions.

## Onshore protests of maritime grievances

Sea-related protests are not only formed, connected and situated in the oceans. As hinted at above, they are also deeply connected to places on-shore, particularly the formation and evolution of port and fishing communities (Tabili, 2011). Early-twentieth-century grievances around ocean-related labour relations reveal the exclusionary potentiality for ocean disputes protested on land, whereby worker conflicts became racialised (Hyslop, 1999; Jenkinson, 2009). In this regard, maritime grievances can be viewed through their engagements with place-based institutions such as the state, trade unions, shipping companies and boarding houses. Thus, the associated ports and fishing communities have occasionally become sites of maritime-related resistance and conflict. This was particularly evident in Britain during the early twentieth century, where increasingly racialised articulations of seafarer grievances primarily around worker's rights, pay and working conditions, culminated in events such as the 1919 seaport riots that took place across British ports such as Cardiff, Liverpool, South Shields and Glasgow (Jenkinson, 2009). Tracing the longer trajectories of these grievances and protests provides an example of how maritime protest met landed places, such as the Albert Dock in Liverpool, Cardiff's Tiger Bay, Glasgow's Broomielaw and South Shields' Mill Dam.

My own archival research in Glasgow has shown how racialised grievances were prevalent in the early twentieth century. The *Forward* newspaper, for example, reported seafarer concerns in an explicitly racialised manner. J. O'Connor Kessack and R.F. Bell published a series of articles in 1911 that claimed to detail the grievances and demands of British seamen. Both of these figures were significant labour organisers within the National Sailors' and Firemen's Union (NSFU) and their articles illustrate the ideological nature of their concerns, with different sections of their articles entitled 'The Chow Invasion', 'The Cheap Asiatic' and 'The Asiatic Peril' detailing their views; one example being:

> The Asiatic in times of danger is a miserable cur. He may suit the convenience of an officer, whose boots he would lick, and perform the menial duties without demur, but he has to be kept in his place by fear and authority. When occasion arises he can be the most arrant villain, sneaking and bloodthirsty, with an utter disregard to all that is lawful and authoritative. His colour, religion, and all that belongs to him, is associated with all that is alien to us.
>
> *'Seamen's Demands', Forward, 13/5/1911: 8*

Bell was secretary of the Glasgow branch of the NSFU and his articles revealed some of the more overtly racist views within British seafaring unions at the time. These views were part of wider "war of words" that shaped events that followed. Routledge (2018: 137) has proposed a "war of words" as "the creative utilisation of activist media [that] can create protest cultures" which can become "a critical tool in generating sites of potential". Such actions and media cultures are integral to protest geographies, as vehicles for articulating demands, whilst also contributing towards direct actions themselves. Seafarer grievances such as these during the early 20th century were regularly found within trade union based publications, including those of their own printing press, such as *The British Seafarer.*

In this instance such hostile rhetoric, similarly found elsewhere and consistently present across the UK, informed exclusionary and hostile events and moments such as the 1919 seaport riots whereby violent clashes occurred between white and non-white sailors, including the deaths of black sailors in Liverpool and Cardiff. Such disruptions reflected tensions within ocean spaces and racialised protests in localities. They also had wider effects with a 'colour bar' established by shipping companies in collaboration with trade unions during this period (see Jenkinson, 2008). The violence noted here must also be viewed as being globally connected, such as those activisms considered in the previous sections, as Hyslop (1999) notes through his characterisation of an early twentieth-century 'white labourism' whereby imperial logic provided transnational solidarities, and consequently exclusionary policies towards non-white labour, amongst particular workers (see also Featherstone, this volume).

Within places defined by such exclusionary political practice, subordinated and marginalised workers had to develop their own strategies to resist the changing regulations of particular places that were integral to their employment. In Britain, these acts are hard to trace in the early twentieth century with non-white voices often missing (Bressey, 2006). However, fragments remain within the records of primarily white trade unions, such as those acts of Chinese sailors, being indicative of 'getting-by' strategies. Despite the extreme forms of disciplining, control and slavery, it is vital that non-white workers are not positioned simply as victims and without their own forms of agency, as it is clear that there are plentiful examples of resourcefulness from people of colour, where possible and strategies to circumvent authority. One example of this is their negotiation of language tests before boarding ships:

The language test – Merchant Shipping Act of 1906:

> British subjects were exempted from this language test: therefore all Chinamen hailing from Hong Kong and Singapore escaped examination under the Act. This led to a number of men from all of parts of China claiming to be born in Hong Kong and Singapore in order to escape the language test.[1]

In Cardiff, for example, 84 per cent of Chinese sailors claimed to be from Hong Kong or Singapore and it would be expected that similar responses would be made in other British ports. Manipulations of tests such as these posed a direct challenge to authority of both the ship-owners and the trade unions considered here (unions had previously applied pressure for a language test), whilst providing an example of the agency of the Chinese seafarers within these contentious maritime work spaces. Actions and co-ordinated decisions such as those to counter the language test illustrates the significance of 'unofficial' action amongst workers and also highlights the uneven contestation over employment. Tabili (1994) notes how sailors originating from colonial countries would also ask 'for British justice' (albeit often unsuccessfully) and campaign upon their grievances and issues from a British standpoint, given their previous service to the colonial government. Such campaigns, including those of the 'Delegates of Coloured Seamen in Glasgow', often took the form of petitions whereby sailors would protest their working conditions, living situations and experiences of welfare provision (see Griffin, 2015).

Despite such hostilities and conflicts, onshore protests of sea-based grievances were also, and continue to be, forged through more inclusive solidarities with like-minded groups and the sharing of concerns. During 1930, in South Shields, for example, large meetings and gatherings organised by the Minority Movement articulated a more progressive and inclusive vision of seafaring organising whereby white and Arab workers protested against rota systems that were used at the detriment of non-white workers. Satnam Virdee (2014: 93) notes how such actions provided a "working class solidarity against racism [that] was to resonate in the minds of working people throughout the South Shields area". Smaller acts were also evident in seaports, with boarding house owners and women for example often protesting on behalf of sailors and campaigning and demanding improvements to working and living conditions (see Lawless, 1995). In more recent times maritime protests have continued to engage key place-based institutions, such as the state. Co-operatives and trade unions in India for example, have played a crucial role in supporting fishworker movements to establish state implemented regulation protecting their rights to fish (Sinha, 2012; Menon et al., 2016).

Similarly, landed demonstrations have indicated the connectivity amongst sailors and international solidarity between workers. International connections were visible in Glasgow, for example, during labour demonstrations in the early twentieth century, reflecting solidarities and protests shaped through the connections of the sea. In May 1917, for example, over 200 Russian sailors, from a warship lying in the Clyde, participated in a march of over 25,000 Glasgow citizens at Glasgow Green. The demonstration, organised by the Glasgow Trades Council, the Glasgow Labour Party, and the Independent Labour Party, formed part of a broader movement against the impacts of the First World War (*Forward*, 2/6/1917: 3). During the demonstration the Russian sailors were presented with a red flag and they presented a memorial steamer to the Clydesiders in return. One of the Russian sailors also spoke at one of the platforms at Glasgow Green alongside Emanuel Shinwell (Chairman of the Trades Council and seafaring union leader). These periodic international connections within Clydeside combined with the specific activities of the organising labour bodies and introduce a previously downplayed diversity to the direct actions during this period. In more recent times, this potentiality for landed social justice campaigning related to the ocean has been illuminated through the work of Winchester and Bailey (2012) who highlight the role of the international conference in bringing together social justice campaigners from seafaring communities. Their work engages with the 'seafarer forum' as part of a wider international conference framework. In this instance, sailors had opportunity to articulate grievances over pay discrepancies between sailors of different nationalities. Such conferences provide opportunity for grievances to be articulated and heard by associated employers and governments. This onshore articulation is crucial for highlighting shared economic inequalities, human rights abuses and poor working conditions.

Moreover, the 'war of words' practised through sailor publishing practices continues, with regular attempts made to articulate grievances emerging from contemporary ocean spaces. Tang et al. (2016) note the multiple grassroots efforts to raise awareness of Chinese workers' rights through online activism. They highlight efforts to illuminate health and safety concerns for Chinese sailors, and the awareness raising practices of families who are connected to an injury and or death at sea. Such small acts of protests can be positioned alongside the wider and collective efforts of sea connected trade union movements. Sinha (2012), for example, identifies the role of trade union protest in Alleppey, India, during the 1970s whereby confrontational and militant actions, such as 'gheraos', and hunger strikes, were used to successfully fight for sea regulations to protect artisan fishers' ability to work within 3 km of the shore. Such wide-ranging and potentially intersecting maritime protest acts reflect important intersections between ocean-based grievances and landed protests.

## Conclusions

Maritime acts of protest are well placed to theoretically advance recent debates regarding resistance, and particularly those found within geography regarding spatial connections and political reach (see Harvey, 1996), linking with those that have stressed more relational constructions of protest and resistance (see Ahmed, 2012; Routledge, 2003). This geographical emphasis has challenged a conceptualisation of resistance as being potentially limited by its boundedness through the particularity of place or singular grievances and demands. Such thinking appears more appropriate for a protest geography of the sea as it caters for the spatially connected processes and influences evident within protest acts, such as those considered above.

The acts of resistance raised here are indicative of this spatial approach to maritime protest both in terms of contributing grievances and resultant ripple effects. As a result, they are difficult to separate neatly into categories, as might be suggested by the structure of this chapter. In contrast, it is perhaps more helpful to think of a repertoire of protest geographies that are distinctive yet often inherently connected. Thus, the three geographies of protest considered above should be viewed relationally. The seaport riots considered in the latter section were not isolated, discrete events but in contrast were shaped by experiences at sea and connected by the sharing of experiences on ships and in ports. Similarly, the contemporary humanitarian efforts of Sea Watch are found at sea but must be considered in relation to the place-based experiences of those involved and their grievances, articulated through various means, towards the landed governance of key actors such as the state.

These acts of protest, from the small acts to those that targeting political change, from progressive, inclusionary and co-ordinated moments to reactionary and exclusionary activisms, must be positioned within their wider context, through what Mitchell (2011: 567) describes as the "world as it really is". To simply consider the acts of resistance or organising practices, would be to ignore the dominating structural and conditioning influences, of colonialism, slavery and capitalism found within maritime geographies, and the associated exploitation of difference. Due to the focus of this chapter, protest acts have been foregrounded but such acts should not be considered in isolation. Thus, whilst acts of exiting, sabotage, violence, strikes, demonstrating, communicating, humanitarianism and charity might appear marginal within these contexts, it remains vital to uncover their presence to complicate the geographies of the sea and to acknowledge the presence of alternative and resistant visions that similarly illuminate wider controlling and structural influences.

## Note

1 Modern Records Centre, University of Warwick, National Union of Seamen Archives, MSS.175/3/14/1–2

## References

Ahmed W (2012) From militant particularism to anti-neoliberalism? The Anti-Enron movement in India. *Antipode* 44(4): 1059–1080.

Bressey C (2006) Invisible presence: The whitening of the black community in the historical imagination of British archives. *Archivaria* 61(1): 47–61.

Chakrabarty D (2000) *Provincializing Europe: Postcolonial Thought and Historical Difference*. Princeton, NJ: Princeton University Press.

Couper A, Smith H and Ciceri B (2015) *Fishers and Plunderers: Theft, Slavery and Violence at Sea*. London: Pluto Press.

Davies AD (2013) Identity and the assemblages of protest: The spatial politics of the Royal Indian Navy Mutiny, 1946. *Geoforum* 48: 24–32.

Davies AD (2019) *Geographies of Anticolonialism: Political Networks Across and Beyond South India, c. 1900–1930*. West Sussex: John Wiley and Sons.

Dunnavant J (2021) Have confidence in the sea: Maritime maroons and fugitive geographies. *Antipode* 53(3): 884–905.

Featherstone D (2005) Atlantic networks, antagonisms and the formation of subaltern political identities. *Social & Cultural Geography* 6(3): 387–404.

Featherstone D (2009) Counter-insurgency, subalternity and spatial relations: Interrogating court-martial narratives of the Nore mutiny of 1797. *South African Historical Journal* 61(4): 766–787.

Featherstone D and Griffin P (2016) Spatial relations, histories from below and the makings of agency: Reflections on the making of the English working class at 50. *Progress in Human Geography* 40(3): 385–393.

Gilroy P (1993) *The Black Atlantic – Modernity and Double Consciousness*. London: Verso.
Griffin P (2015) Labour struggles and the formation of demands: The spatial politics of Red Clydeside. *Geoforum* 62: 121–130.
Guha R (1983) *Elementary Aspects of Peasant Insurgency in Colonial India*. Oxford: Oxford University Press.
Halvorsen S (2015) Encountering Occupy London: Boundary making and the territoriality of urban activism. *Environment and Planning D: Society and Space* 33(2): 314–330.
Harvey D (1996) *Justice, Nature and the Geography of Difference*. Oxford: Blackwell.
Hasty W (2014) Metamorphosis afloat: Pirate ships, politics and process, c.1680–1730. *Mobilities* 9(3): 350–368.
Hobsbawm EJ (1959) *Primitive Rebels*. Manchester: University of Manchester Press.
Høgsbjerg C (2011) Mariner, renegade and castaway: Chris Braithwaite, seamen's organiser and Pan-Africanist. *Race and Class* 53(2): 36–57.
Hyslop J (1999) 'The imperial working class makes itself 'white'': White Labourism in Britain, Australia, and South Africa before the First World War. *Journal of Historical Sociology* 12(4): 398–421.
Hughes S (2020) On resistance within human geography. *Progress in Human Geography* 44(6): 1141–1160.
Jenkinson J (2009) *Black 1919: Riots, Racism and Resistance in Imperial Britain*. Liverpool: Liverpool University Press.
Lawless R (1995) *From Taizz to Tyneside: An Arab Community in the North East of England during the Early Twentieth Century*. Exeter: University of Exeter Press.
Linebaugh P and Rediker M (2000) *The Many Headed Hydra: Sailors, Slaves, Commoners and the Hidden History of the Revolutionary Atlantic*. Boston, MA: Beacon Press.
McKie R (2017) How Sea Shepherd lost battle against Japan's whale hunters in Antarctic. *The Guardian [Online]*. Available at: www.theguardian.com/environment/2017/dec/23/sea-shepherd-loses-antarctic-battle-japan-whale-hunters
Menon A, Bavinck M and Stephen J (2016) The political ecology of Palk Bay fisheries: Geographies of capital, fisher conflict, ethnicity and nation-state. *Antipode* 48(2): 393–411.
Mitchell D (2011) Labor's geography: Capital, violence, guest workers and the post-World War II landscape. *Antipode* 43(2): 563–595.
Nolan LJ and Featherstone D (2015) Contentious politics in austere times. *Geography Compass* 9(6): 351–361.
Peters K (2013) Regulating the radio pirates: Rethinking the control of offshore broadcasting stations through a maritime perspective. *Media History* 19(3): 337–353.
Quest M (2009) George Padmore's and C.L.R. James's International African Opinion. In: Baptiste F and Lewis R (eds) *George Padmore: Pan-African revolutionary*. Kingston: Ian Randle, 105–132.
Rediker M (1987) *Between the Devil and the Deep Blue Sea: Merchant Seamen, Privates, and the Anglo-American Maritime World, 1700–1750*. Cambridge: Cambridge University Press.
Rediker M (2004) *Villains of All Nations: Atlantic Pirates in the Golden Age*. London: Verso.
Rediker M (2007) *The Slave Ship: A Human History*. London: John Murray.
Routledge P (2003) Convergence space: Process geographies of grassroots globalization networks. *Transactions of the Institute of British Geographers* 28(3): 333–349.
Routledge P (2018) *Space Invaders: Radical Geographies of Protest*. London: Pluto Press.
Sea Shepherd (no date) *Mission Statement*. Available from: https://seashepherd.org/mission-statement/
Searby P, Rule M and Malcolmson R (1993) Edward Thompson as a teacher: Yorkshire and Warwick. In: Rule M and Malcolmson R (eds) *Protest and Survival, Essays for E.P Thompson*. New York, NY: New York Press, 1–23.
Sharp J, Routledge P, Philo C and Paddison R (eds) (2000) *Entanglements of Power: Geographies of Domination/Resistance*. Abingdon: Routledge.
Silverman V (2000) *Imagining Internationalism: In American and British Labour, 1939–49*. Urbana and Chicago, IL: University of Illinois Press.
Sinha S (2012) Transnationality and the Indian fishworkers' movement, 1960s–2000. *Journal of Agrarian Change* 12(2–3): 364–389.
Steinberg PE (2013) Of other seas: Metaphors and materialities in maritime regions. *Atlantic Studies* 10(2): 156–169
Stierl M (2018) A fleet of Mediterranean border humanitarians. *Antipode* 50(3): 704–724.
Tabili L (1994) *"We Ask for British Justice": Workers and Racial Difference in Late Imperial Britain*. Ithaca, NY: Cornell University Press.

Tabili L (2011) *Global Migrants, Local Culture: Natives and Newcomers in Provincial England, 1841–1939.* Basingstoke: Palgrave Macmillan.

Tang L, Shan D and Yang P (2016) Workers' rights defence on China's internet: An analysis of actors. *Information, Communication & Society* 19(8): 1171–1186.

Virdee S (2014) *Racism, Class and the Racialized Outsider.* Basingstoke: Palgrave Macmillan.

Winchseter N and Bailey N (2012) Making sense of 'global' social justice: Claims for justice in a global labour market. *Sociological Research Online* 17(4): 10. Available at: www.socresonline.org.uk/17/4/10.html

# 19

# SOLIDARITIES

## Oceanic spaces and internationalisms from below

*David Featherstone*

### Introduction

In 1975–6 following the coup in Chile which deposed Salvador Allende's democratically elected Popular Unity government, a handbill was circulated by militant seafarers in British ports. The Executive Council of the Union of Seamen (NUS) had instructed NUS members "[n]ot to go on ships sailing to Chile" because of "the atrocities committed against the Chilean Trade Unionists and Workers" by the regime of General Augusto Pinochet.[1] Noting that the response to the Executive Council's decision had been "tremendous" it observed that "over 700 Liverpool Seamen who have been unemployed for many weeks and months will not set foot on ships carrying Chilean cargoes". As a result, the Shipowners had "been forced to seek their Crew members from other ports" and the union appealed to members of the union elsewhere not to "take these ships thereby" and to support "the gallant stand made by the unemployed Liverpool lads" and to help "the Chilean People in their struggle against the military junta".[2]

The transnational solidarities in response to the coup in Chile and in opposition to Pinochet's murderous regime are well documented as are its links to the emergence of neoliberalism (e.g. Harvey, 2005; Jones, 2014). The profile of these solidarities has been raised by Felipe Bustos Sierra's excellent film *Nae Pasaran* and are of significant contemporary relevance given the significant repression of protests which were sparked in October 2019 by Metro ticket price hikes by the government of President Sebastian Piñera (Riethof, 2019). The film tells the story of how workers at the Rolls Royce plant in East Kilbride near Glasgow refused to work on Hawker Hunter jet engines sent from Chile for refurbishment at the factory in 1974. The actions of NUS seafarers in Liverpool and in other ports signal important aspects of the response of maritime workers to the coup. The solidarities of Liverpool seafarers with Chile demonstrate how workers' positions in maritime networks and connection could become strategically mobilised at key junctures and moments. That it was the Liverpool 'lads' whose actions are described as 'tremendous' also suggests some of the intersections between left organising and seafaring masculinities that have often been integral to maritime solidarities.

This chapter explores different articulations of maritime solidarities and the spatial relations they have both been shaped by and generated. It first outlines different approaches to theorising


DOI: 10.4324/9781315111643-23

the spaces of maritime solidarities, before discussing three different aspects of such solidarities; the relation between anti-colonial internationalisms and maritime solidarities, transnational maritime labour struggles and finally, anti-nuclear and anti-militarist protests which have taken place in different oceanic spaces. The chapter concludes with some thoughts about contemporary articulations of maritime solidarity constituted in opposition to the rise of far-right politics.

## Theorising the spaces of maritime solidarities

"The sea", Nikolas Kosmatopoulos writes, "has always been a special place for internationalism, solidarity and resistance, chiefly among those labouring on ships, loading the cargos or lying captured in their holds" (2019: 743). Kosmatopoulos uses the concept of 'terraqueous solidarity' to analyse some of the internationalist articulations of politics generated through maritime spaces. He uses this term to define "grassroots political movements that take to the sea to practice solidarity with those resisting or suffering from contemporary forms of colonial enclosures, state oppression and humanitarian neglect" (Kosmatopoulos, 2019: 741). In this respect Kosmatopoulos's approach builds on an important body of work in radical social history which has foregrounded radical maritime spaces and challenged long-standing associations of maritime histories with conservatism.

A key reference point here that he builds on is the work of Peter Linebaugh and Marcus Rediker who have positioned maritime spaces and actors as central to the formation of radical contestation of the emergence of capitalism in the early modern period (Linebaugh and Rediker, 2000; see also Campling, and Colás, 2017). Thus, their book *The Many Headed Hydra*, arguably their most influential contribution, uses an engagement with maritime spaces to challenge the limits of nation-centred histories from below (c.f. Linebaugh and Rediker, 1990; see also Hyslop, 2019). By positioning the ship, particularly its proletarian lower deck, as a "forcing house of internationalism", they position maritime spaces as key sites of multi-ethnic encounter, solidarity and exchange (Linebaugh and Rediker, 2000: 144; see also Scott, 2018). Foregrounding such maritime spaces can open up different ways of understanding histories from below and open up different perspectives on the formations of solidarities.

In this respect Marcus Rediker has used the term "terracentrism" to refer to the "unspoken proposition that the seas of the world are unreal spaces, voids between the real places, which are landed and national" (Rediker, 2014: 3–4). This is part of a broader challenge to nation-centred histories and geographies which has been central to scholarship that foregrounds oceanic spaces (see Gilroy, 1993; Shilliam, 2015; Steinberg and Peters, 2015). As Radhika Mongia has observed, the very "formulation of the trans*national* obliges if not shackles us to assumptions of space, state and subjectivity *already* conceived in *national* terms" (Mongia, 2019: 5, emphasis in original). The diverse routings and trajectories of maritime solidarities are a good example here, but also are often articulated across land–sea relations as much as being neatly contained by oceanic spaces (see also Anderson, 2012; Davies, 2019; Subramanian, 2014).

Kosmatopoulos's concept of 'terraqueous solidarities' implicitly references Rediker's critique of terracentrism, but the way in which it actively combines territory and ocean is a direct contrast to some of the binaries that have at times inflected Linebaugh and Rediker's work. In this way his work foregrounds diverse connections and trajectories that can contribute to useful cartographies of maritime solidarities. Attending to such connections and trajectories has been particularly important in drawing attention to some of the diverse translocal practices and dynamics shaped by subaltern/resistance movements in different contexts, especially in the Indian Ocean (Chari, 2019; Kothari, 2012). Thus, Ishan Ashutosh has recently argued that "[a]nti-colonial nationalism swelled across the ocean, its politics replenished by diasporic circulation" and that "its co-ordination and simultaneity with anti-colonial action in the homeland inherently projected outwards, towards engagement with other forms of oppression" (Ashutosh, 2019: 6). The next section explores some of the different forms of solidarity shaped at the intersection of maritime spaces and anti-colonial internationalisms in the early to mid-twentieth century.

## Anti-colonial internationalisms and maritime solidarities

In April 1914, Gurdit Singh a "prosperous Singapore-based labour-transport contractor" who was from the Amritsar District of what was then British Punjab, "chartered a ship to take a load of passengers to Canada from among the hundreds of his compatriots then awaiting passage in Hong Kong and other East Asian ports" (Ramnath, 2011: 47–48). The *Komagata Maru* had been built on the River Clyde by Charles Connell and Company, at Scotstoun in Glasgow in 1889, and before Singh acquired it had been known as the *Stubbenhuk* (Mawani, 2018: 88). Gurdit Singh conceived the ship's voyage to Vancouver as a "deliberate challenge to new immigration restrictions" in Canada and was to prove "a catalyst for radicalization on both sides of the Pacific after the passengers were refused entry to Canada" (Ramnath, 2011: 4). The *Komagata Maru's* voyage, as Maia Ramnath notes, "culminated in a violent standoff in the harbor before the ship turned back to sea and in a shoot-out on arrival in Calcutta in which more than twenty passengers were killed" Ramnath, 2011: 4.

In her compelling book on the *Komagata Maru*, Renisa Mawani draws on Gurdit Singh's own narrative of these events, *Voyage of the Komagatamaru or India's Slavery Abroad*, which offers "an acerbic critique of British colonial and imperial rule" (Mawani, 2018: 221). She notes that Singh narrates the ship's "unsuccessful journey, not as a single or exceptional incident, but as a tragic moment in a much longer trajectory of crimes committed by Britain against its colonial and racial subjects" (Mawani, 2018: 221). She suggests that through drawing "fleeting but compelling links between transatlantic slavery and systems of Indian indenture" Singh shaped particular imaginative geographies of solidarity. These imaginaries were informed by his view that "indenture was yet another system of bondage that grew directly from the abolition of slavery" (Mawani, 2018: 221–222; see also Mangru, 1993).

Mawani's discussion of the relations between indenture and slavery position critiques of the treatment of the *Komagata Maru* in relation to the broader unequal power-geometries and relations that shaped colonial maritime spaces (see also Anim-Addo, 2019). The "news of the *Komagata Maru's* fate lit a fuse" which had significant impacts particularly on the West Coast of the United States where the transnational anti-colonial radical networks of the Ghadar movement had a significant presence (Ramnath, 2011: 49). The ship's voyage was, however, only one of a number of attempts to challenge the colonial geographies that shaped maritime commerce producing solidarities between different anti-colonial struggles. Andrew Davies has recently drawn attention to the importance of the Swadeshi Steamship Navigation Company (SSNCo) which was established in October 1906 to challenge the monopoly of the British India Steam Navigation Company (BISNCo) on transit routes between South India and Ceylon (Davies, 2019: 69).

The SSNCo was established during the upsurge of *swadeshi* activism in the first decade of the twentieth century which "saw India as politically and economically under-developed, and argued that the development of a national economy would encourage a period of national 'self renewal'" (Davies, 2019: 73). Davies argues that the,

> SSNCo acted as an example of what could be possible for Indians as owners and developers of indigenous capital, and the alleged attempts by the BISNCo and other 'British' or 'foreign' interests to disrupt the *swadeshi* enterprise, whether true or not, played into the long-established repertoire of grievances that saw the British as exploiting and draining India's wealth.
>
> *Davies, 2019: 73, 84*

By using a discussion of the SSNCo to provide an 'archipelagic' reading of the Swadeshi movement, he excavates different geographies and routes through which the solidarities generated through the Swadeshi movement were constructed and articulated.

The politics of the SSNCo and the *Komagata Maru*, however, were articulated with relatively elite anti-colonial projects which were distanced from the dynamics that shaped labour relations aboard ships (see also Legg, 2020). In this respect the circuits of maritime colonialisms not only produced and intensified unequal global geographies they were also shaped by particular unequal racialised divisions of labour aboard ship (Ahuja, 2012; Balachandran, 2012; Tabili, 1994). These racialised divisions of labour were associated with shipping companies, but were also shaped and enforced by dominant maritime unions along the lines of what Jonathan Hyslop (1999) has termed 'white labourism', which generated racialised forms of solidarity, based on systematic exclusions particularly of Black and Asian seafarers from maritime labour markets. Such exclusionary organising practices had significant impacts on how maritime solidarities were articulated and envisioned.

Hyslop uses this term to refer to some of the mechanisms/spaces through which an "imperial working class" was formed in the later nineteenth/early twentieth century which "produced and disseminated a common ideology of White Labourism" (Hyslop, 1999: 414–415). He contends that 'white labourism' was produced, not by a top-down process, but through the formation of whiteness 'from below' (Hyslop, 1999: 414–415). Maritime unions in the Britain and the US such as the International Seamen's Union and the National Sailors and Firemen's Union, which became the National Union of Seamen in 1926 (NSFU/NUS) adopted such organising strategies and were involved in concerted campaigns against seafarers from racialised minorities which included uses of violence (Featherstone, 2019; Nelson, 1988). Such white labourism was not, however, uncontested or the only frame through which solidarities were constructed or envisioned in relation to maritime labour.

Peter Cole's work has demonstrated, for example, how organisations like the Industrial Workers of the World (IWW) were significant in shaping multi-ethnic organising cultures both in particular dockside spaces and aboard ship (Cole, 2007; see also White, 2017). His account of Local 8 of the IWW on the Philadelphia waterfront in the 1910s and 20s has demonstrated how it functioned as an integrated local which fought the existence of segregated work gangs, had strong black leadership and was part of broader circuits of transnational solidarity and syndicalist political cultures (Cole, 2007). Militant seafarers of colour, for example, challenged the racism that underpinned constructions of white labourism and shaped solidarities with anti-colonial struggles and imaginaries.

Thus, in the early 1930s seafarers from the Caribbean and West Africa who were in contact with the Trinidadian activist and agitator George Padmore were integral to organisations in London such as the Negro Welfare Association (NWA) (Adi, 2014; Høgsbjerg, 2011). Such maritime workers were involved in various activities including smuggling anticolonial literature and being involved in boycotts of Italian ships during the invasion of Ethiopia in 1935 (see Høgsbjerg, 2011; Weiss, 2020). While the sea afforded Gurdit Singh's "a wider perspective from which to chart the deep entanglements between different manifestations of imperial coercion" radical anti-colonial seafarers used these connections in more explicitly political terms (Mawani, 2018: 225). Thus, the Barbadian seafarer and pan-Africanist Chris Jones in his column 'Seamen's Notes' for the journal *International African Opinion* argued that the connections and movements of 'colonial seafarers' could be integral to left anti-colonial organising. He noted that "[i]t is up to us, therefore, as coloured seamen, to enlighten our fellow colonial workers during our travels that we underdogs have nothing to gain by fighting in the interests of the imperialist robbers" (Jones, 1938 cited by Featherstone, 2012: 89).

The mobilities of seafarers did not always, however, align neatly with leftist political projects. As Erik McDuffie makes clear in his account of the experiences of Audley Moore they could also reconfigure the terms on which maritime solidarities were envisioned. In 1946 Moore, a union-organiser for the National Maritime Union (NMU) which had strong links to the Communist Party of the USA, "crossed the Atlantic ten times aboard NMU affiliated merchant vessels while working as a steward for the US army's Civilian Army Department" (McDuffie, 2011: 152). Her commitment to Communism was, however, unsettled by travelling beyond the US to Europe as these experiences "compounded her frustration with her outside-inside status in the Communist Left" and she began "to see black people globally, both white workers, as the revolutionary vanguard" (McDuffie, 2011: 153). While the focus of much work on the unequal racialised articulations of maritime organising and solidarity has been the inter-war period, there were important struggles over the terms and practices of maritime organising and solidarity in the context of decolonisation which continued to challenge articulations of labour and coloniality.

A major strike of the Nigerian Union of Seamen in 1959, for example, challenged the continuing influence of the British NUS, which adopted a strongly anti-Communist position, and the racist attitudes and practices of European officers. Key figures such as Sir Thomas Yates, who was the General Secretary of the NUS between 1947 and 1960, and chairman of the Seafarers' section of the International Transport Workers' Federation worked "to curtail leftist leadership among seamen in Nigeria (and West Africa generally)" (Tijani, 2012: 94). The strike began, as Hakeem Ibikunle Tijani notes, in early June 1959 "on board the *MV Apapa* vessel on its northbound voyage from Lagos to Liverpool". By the time the crew of 77 seamen had arrived in Liverpool they "had compiled a list of grievances against their European officers – whom they accused of gross color discrimination, inequality and 'slavery'" (Tijani, 2012: 93). The action of the seafarers on the *MV Apapa* had significant effects "within two days their action had spread as hundreds of other seamen in Liverpool and abroad went on solidarity strikes; noteworthy among these were the crews of five cargo ships in Liverpool and eight ships in Lagos harbor" (Tijani, 2012: 93).

The Nigerian Union of Seamen's strike emphasises how articulations of coloniality and labour reproduced through forms of union organising and strongly informed by anti-Communism were brought into contestation (see Herod, 2000; Horne, 2005). The strike emphasises that such relations were challenged through various forms of maritime solidarity and these continued to be important, particularly in relation to the growing global opposition to apartheid. The next section uses a discussion of opposition to the white supremacist South African regime as a way into a discussion of the important role of dockworkers in transnational maritime solidarities.

## Maritime labour, dockers and oceanic constructions of solidarity

In his comparative study of dockers in Durban and the San Francisco Bay Area, *Dockworkers' Power*, Peter Cole has made a significant contribution to understandings of the forms of solidarity shaped by organising in these ports in the post-war period. He traces how through their activism they "translated their beliefs in the need for and possibilities of solidarity into tangible actions: boycotting ships to protest apartheid and other forms of authoritarianism" (Cole, 2018: 210). Key Locals of the International Longshore Workers Union (ILWU) in the San Francisco Bay Area, for example, "condemned apartheid and periodically undertook direct actions in solidarity with South Africa's black majority". In 1984 "shortly after Ronald Reagan's reelection, members of the ILWU Bay Area branches, Locals 10 and 34, refused to handle South African cargo for eleven days" (Cole, 2018: 187). This action was directed against the *Nedlloyd Kimberley*, which became the subject of broader mobilisation and William Allan noted in the *People's World* of 23 December 1984 that nearly "two hundred anti-apartheid protesters took to Oakland streets on December 22nd to speak out against apartheid and ships from there bringing in cargo made by slave labour" (Allan, 1991: 290). Cole locates understandings of such radical trade unionism in the dockers' transnational labour process. He demonstrates how dockworkers' role in shifting trade, their knowledge of the cargoes they handled and the connections they made between places through their work were integral to their political interventions and agency. In related terms Kosmatopoulos has drawn attention to the spatial practices of solidarity shaped by dockworkers in Greece through supporting the 'Ships to Gaza' organisation. This maritime solidarity activism supported the 'Gaza Freedom Flotilla' which was "violently attacked" in international waters by the Israeli navy leading to the "killing of nine and the arrests of 700 passengers" (Kosmatolpoulos, 2019: 741). He locates these solidarities in relation to the broader dynamics of what he terms "maritime settler colonialism" in Gaza. As Ronald Smith has noted, the waters of the coast of Gaza are integral to both Israel's occupation and blockade of the Palestinian territory with the "maritime borders of Gaza" being "enforced unilaterally by Israel" and Palestinian fishing vessels facing routine harassment from both Israeli and Egyptian military (Smith, 2016: 759).

Kosmatopoulos draws particular attention here to volunteers from the Dockworkers' Union in Piraeus in supporting the Ships to Gaza project, with approximately 60 of the union's 300 members being involved. The union, he notes, "arranged both the labour and the protection of the ship". The executive committee of the union "supervised the workflow, organized the shifts, support groups and security teams, and the cargo management" and used the union's "acquaintances in the port" to resolve issues with "customs procedures" which were particularly significant given the extremely contested character of Gaza's jurisdiction (Kosmatopoulos, 2019: 751). These solidarities also involved particular interventions in the space of the ship itself – "they arranged how to divide the cargo within the hold, make it fit, and secure it" (Kosmatopoulos, 2019: 751–752). This political activity had effects and he notes that the dockworkers in Piraeus view their engagement with "the ship as a political, pedagogical and highly emotional experience" and while "working on the ship they engaged more intensely with the geopolitics of the Middle East" (Kosmatopoulos, 2019: 752). As well as the support of the Dockers' Union the project was also "embraced by the public-owned Piraeus Port authority" and the "local administration at the port openly supported the Ships" (Kosmatopoulos, 2019: 752). The ongoing neoliberalisation of Greek ports which has seen the privatisation of both Piraeus and Thessaloniki clearly lessen the likelihood of such support from local officials (see Karaliotas, 2017).

Solidarities shaped by dockworkers, however, have often been the product of unofficial action against port authorities and shipping lines and crucially against or in open defiance of union officials and hierarchies. Such dynamics are a key theme of Jack Dash's *Good Morning Brothers!*, which recounts his time as a rank and file militant among the London dockers. In 1949 when Canadian seamen who were berthed in ports such as Bristol "quickly established picket lines around their ships" during the Canadian Seamen's Union strike, Dash recounts how the solidarities of dock workers in London, Bristol and Liverpool were "carried out in direct opposition to the officials of the Transport and General Workers' Union" (TGWU) (Davis, 2003: 186; Dash, 1987: 69). Dash recounts that stevedores refused to work on the Canadian ship *Argemont*, London's Surrey docks and that in "Avonmouth and Bristol, our West Country Brothers' stood firm on principle though every effort was made by the authorities to intimidate and coerce them" (Dash, 1987: 69). He recounts that even after a secret ballot called by TGWU officials "showed 646 in favour of strike action, 108 against", the officials "nevertheless rejected the outcome of their own democratic procedure and had the audacity to declare the strike unofficial" (Dash, 1987: 69).

In similar fashion during the mid-1990s the iconic Liverpool Dockers' Strike which lasted two years and, like the Canadian Seamen's Union dispute, garnered significant transnational solidarity, was never recognised as an official dispute by TGWU albeit under threat from severe anti-trade union legislation. Noel Castree has argued that this failure of the union to recognise the dispute with Merseyside Docks and Harbour Company had very strong impacts on the spatial politics of the strike (Castree, 2000). Thus, he considers the extensive international solidarities forged during the dispute as a response to the lack of local and national union support (Castree, 2000). It is, however, important to situate the impressive transnational support that the dockers received in relation to the broader internationalism that has shaped the political cultures and outlook of dockers' trade unionism and broader cultures of maritime solidarity (see also Kelliher, 2018). Billy Bragg's song about the dispute, 'Never Cross a Picket Line', captures this aspect of the strike and speaks more generally to some of the dynamics of maritime solidarities: "Look away, look away/Look away out west to San Francisco/ Look away, look away/ Look away down south to Sydney Harbour/ Where the dockers have organized/ The world's longest picket line" (Bragg, 1998).

## Maritime solidarities, anti-militarism and the spaces of the ocean

Since the late 1950s and early 1960s Holy Loch and Faslane Bay on Gare Loch on the West Coast of Scotland have been central spaces of anti-nuclear protest and contestation. The latter site, where the UK's own nuclear submarines are based, has been a consistent site of protest and Faslane Peace Camp 'has been in existence continually since 1982 and claims to be the "longest running permanent peace camp in the world"' (Eschle, 2017: 472). The location of US nuclear submarines at Holy Loch in the early 1960s catalysed a strong protest movement linked to the Committee of 100 which used assertive forms of direct action. As Brian P. Jamison notes "the physical presence of the US Polaris force" bolstered/catalysed the determination of organisations such as the Scottish Campaign for Nuclear Disarmament (SCND) "to have the sea-based deterrent removed from the Scottish lochs" (Jamison, 2003: 115).

The opposition to Polaris took the forms of demonstrations, which were often interlinked on both sea and land. In March 1961 when the *Proteus*, the first US naval vessel to arrive *at the base*, sought to sail "through the narrow entrance to Holy Loch" they were met by "the West of Scotland Canoe Club and like-minded souls" (McVicar, 2010: 99–100). As *The Scotsman* reported, "the sea-borne invasion of anti-Polaris demonstrators who tried to board the submarine depot ship *Proteus* in the Holy Loch" were repulsed by the US Navy with "a barrage of fire hoses" (*The Scotsman* 22/5/1961, cited by McVicar, 2010: 99–100). The *Proteus* had been met by a 'flotilla' which "included kayaks, dinghies, launches, and a motor house boat which bore a Red Cross symbol and the slogan Life Not Death" (*The Scotsman*, cited by McVicar, 2010: 99–100). In this way the ship *Proteus* emerged as a key, if mobile, site of grievance. This is indicated by the way it figures in songs such as Hamish Henderson's Anti-Polaris sequence written, or as he termed it – workshopped – through the Holy Loch protests: "We'll hae tae shift Polaris/ Proteus an aa" (Henderson, 2019: 288; see Gibson, 2015).

These solidarities were articulated through left internationalist imaginaries, most notably in Henderson's famous song the *Freedom Come All Ye* with its links between anti-nuclear protest and struggles against apartheid which has often been touted, albeit against Henderson's wishes, as an alternative Scottish national anthem. This internationalist sensibility also shapes some of his less well-known songs written for the marches: "You may come frae Odessa, mate/ Frae Baltimore or Perth/ But the threat o Polaris/ Maks ae country o the Earth" (Henderson, 2019: 287). These protests at Holy Loch and Faslane emphasise how oceanic spaces became importance sites of struggles and solidarities relating to ecological, peace, anti-Nuclear activism in the second half of the twentieth century. In this regard although the spatial and temporal contexts are different there are strong resonances between some of the forms of protest at Holy Loch in the early 1960s and more recent protests against militarism.

Thus, Sasha Davis has recently documented protests against the US and Japanese governments relating to the construction of a new US air base in Okinawa. He notes how the protester's tactics of using kayaks to disrupt construction work "have enabled activists in Okinawa to delay the landfilling of Oura Bay and the construction of the new base since it was first proposed in 1996" (Davis, 2017: 110). Drawing on assemblage approaches Davis considers some of the translocal spatialities that are integral to these protests/ solidarities. Thus, he notes that:

> … protesters in Okinawa have been able to block the construction of this new base because the kayakers paddling through the contested sea are not alone. The protesters in Okinawa are linked to anti-militarisation activists around the globe who have been responsible for closing military bases and blocking the construction of others.
>
> *Davis, 2017: 112*

Further, he uses this approach to argue that the activist occupation of Oura Bay is "not a local or isolated action". Davis suggests that it is

> misleading to think of spaces that are wrested form state control as being isolated, disconnected or 'local'. Instead of seeing occupied sites as merely autonomous islands and state-ruled areas as vast, unbroken seas, it is more accurate to think of both as interconnected archipelagos of sovereignty.
>
> *Davis, 2017: 118*

Such an approach can usefully help reframe ideas of maritime solidarities in ways that are alive to the intertwining of ecological and social relations in shaping oceanic spaces. This is of particular significance given the vast changes being wrought on oceanic spaces through climate change, but Davis's focus on particular spatial relations can help to attend to questions of specificity which can easily be lost in such global discussions. As Sunil S. Amrith notes, the "rising waters" of the Bay of Bengal are "due to global causes but it is at the level of the region that their effects will be felt" (Amrith, 2013: 275). He also argues, however, that the,

> region has the cultural resources to generate a new ethic of hospitality, and aid to strangers: a store of collective memories, intercultural understandings, and stories that allow the imagination of solidarities over long distances, though many of these have been forgotten or lie buried beneath the surface of official ideologies.
>
> *Amrith, 2013: 275*

For Amrith such solidarities may act as a *key* to whether the Bay of Bengal's coastal rim provokes an ecology of fear – or serves to "harness ecologies of hope" (Amrith, 2013: 276).

## Conclusions

In July 2019 thousands of people took to the streets of German cities and towns, including port cities such as Hamburg, to show their solidarity in support of Carola Rackete and the Sea-Watch rescue group. Rackete, the captain of the rescue ship *The Sea Watch 3* had been arrested for breaking an Italian naval blockade "that was trying to stop her from docking the vessel in Lampedusa", while it was carrying 40 people who the vessel had rescued from the Mediterranean (DW, 2019). Her actions defied a ban by the far-right politician Matteo Salvini, who at that juncture was Italy's interior minister (*Guardian*, 2019). As Rackete's arrest indicates maritime spaces have been rendered central to recent political struggles over different racialised populist imaginaries. Thus, Atul Bhardwaj has recently noted the ways in which rightist political projects such as Brexit have invoked articulations between empire and maritime spaces (Bhardwaj, 2019: 10–11).

The actions in support of Rackete and in defence of the rights of thousands of migrants at risk of death in the Mediterranean, however, emphasise that the demonisation of migrants by far-right movements and politicians such as Salvini is being vigorously challenged and contested. There are powerful ways in which these solidarities have been shaped by and draw on leftist maritime political cultures (see also Heller et al., this volume). In this spirit this chapter has sought to signal the importance of the diverse and multi-faceted uses and spaces of maritime solidarities. Drawing attention to the important histories and geographies of maritime solidarities can help to foreground aspects of the histories of left internationalisms and solidarities that have often been ignored or down-played by nation- and terra-centric approaches. As I have noted, these were tensioned and open to various forms of internal contestation and challenge. Recovering and honouring such struggles and histories of maritime solidarities can nonetheless help to counter the exclusionary ways of constructing oceanic spaces being mobilised by the far right.

## Notes

1 Liverpool Record Office, Merseyside Communist Party records 329COM/13/10.
2 Liverpool Record Office, Merseyside Communist Party records 329COM/13/10.

## References

Adi H (2014) *Pan-Africanism and Communism: The Communist International, Africa and the Diaspora, 1919–1939*. Trenton, NJ: Africa World Press.

Ahuja R (2012) Capital at sea, Shaitan below decks? A note on global narratives, narrow spaces, and the limits of experience. *Histories of the Present* 2(1): 78–85.

Allan W (1991) Anti-apartheid groups gird for all-out battle to halt new ship. In: Filling B and Stuart S (eds) *The End of a Regime? An Anthology Scottish-South African Writing Against Apartheid*. Aberdeen: Aberdeen University Press, 290–292.

Anderson C (2012) *Subaltern Lives*. Cambridge: Cambridge University Press.

Anim-Addo A (2019) Reading postemancipation in/security: Negotiations of everyday freedom. *Small Axe* 57: 105–114.

Amrith S (2013) *Crossing the Bay of Bengal: The Furies of Nature and the Fortunes of Migrants*. Cambridge, MA: Harvard University Press.

Ashutosh I (2019) The spaces of diaspora's revitalization: Transregions, infrastructure and urbanism. *Progress in Human Geography* 44(5): 898–918.

Balachandran G (2012) *Globalizing Labour? Indian Seafarers and World Shipping, c. 1870–1945.* Delhi: Oxford University Press.

Bhardwaj A (2019) Brexit and the continental fears of maritime Britain. *Economic and Political Weekly* 54(17) 27 April 2019: 10–11.

Bragg B (1998) Never cross a picket line. Originally on *Rock the Dock* benefit CD for the Liverpool Dockers. Available at: www.youtube.com/watch?v=ojPTz4VAOMA

Campling L and Colás A (2017) Capitalism and the sea: Sovereignty, territory and appropriation in the global ocean. *Environment and Planning D*, 36(4): 776–794.

Castree N (2000) Geographic scale and grass-roots internationalism: The Liverpool dock dispute, 1995–1998. *Economic Geography* 76(3): 272–292.

Chari S (2019) Subaltern sea? Indian Ocean errantry against subalternization. In: Jazeel T and Legg S (eds) *Subaltern Geographies*. Athens, GA: Georgia University Press, 191–209.

Cole P (2018) *Dockworker Power: Race and Activism in Durban and the San Francisco Bay Area.* Urbana, IL: University of Illinois Press.

Cole P (2007) *Wobblies on the Waterfront: Interracial Unionism in Progressive Era Philadelphia.* Urbana, IL: University of Illinois Press.

Dash J (1987) *Good Morning Brothers!* London Borough of Tower Hamlets: Wapping Standing Neighbourhood Committee.

Davies AD (2019) *Geographies of Anticolonialism: Political Networks Across and Beyond South India c. 1900–1930.* Chichester: Wiley Blackwell.

Davis CJ (2003) *Waterfront Revolts: New York and London Dockworkers, 1946–61.* Urbana, IL: University of Illinois Press.

Davis S (2017) Apparatuses of occupation: Translocal social movements, states and the archipelagic spatialities of power. *Transactions of the Institute of British Geographers* 42(1): 110–122.

DW (2019) Germans march in solidarity with sea watch. *DW [Online]*. Available at: www.dw.com/en/germans-march-in-solidarity-with-sea-watch/g-49499396.

Eschle C (2017). Beyond Greenham woman? Gender identities and anti-nuclear activism in peace camps. *International Feminist Journal of Politics* 19(4): 471–490.

Featherstone DJ (2012) *Solidarity: Hidden Histories and Geographies of Internationalism.* London: Zed Books.

Featherstone DJ (2019) Maritime labour, transnational political trajectories and decolonisation from below: the opposition to the 1935 British Shipping Assistance Act. *Global Networks* 19(4): 539–562.

Gibson C (2015) *'The Voice of the People': Hamish Henderson and Scottish Cultural Politics.* Edinburgh: Edinburgh University Press.

Gilroy P (1993) *The Black Atlantic: Modernity and Double Consciousness.* London: Verso.

The Guardian (2019) Captain defends her decision to force rescue boat into Italian port. *The Guardian [Online]*. Available at: www.theguardian.com/world/2019/jun/30/italy-refugee-rescue-boat-captain-carola-rackete-defends-decision

Harvey D (2005) *A Brief History of Neoliberalism.* Oxford: Oxford University Press.

Henderson H (2019) *Collected Poems* (ed. C Gibson). Edinburgh: Polygon.

Herod A (2000) *Labor Geographies* New York, NY: Guilford Press.

Høgsbjerg C (2011) Mariner, renegade and castaway: Chris Braithwaite, seamen's organizer and Pan-Africanist. *Race and Class* 53: 36–57.

Horne G (2005) *Red Seas: Ferdinand Smith and Radical Black Sailors in the United States and Jamaica.* New York, NY: New York University Press.

Hyslop J (2019) German seafarers, anti-fascism and the anti-Stalinist left: The 'Antwerp Group' and Edo Fimmen's International Transport Workers' Federation, 1933–40. *Global Networks* 19(4): 499–520.

Hyslop J (1999) The imperial working class makes itself 'white': White labourism in Britain, Australia, and South Africa before the First World War. *Journal of Historical Sociology* 12(4): 398–421.

Jamison BP (2003) Will they blow us a' tae hell? Strategies and Obstacles for the Disarmament Movement in Scotland. In: Jamison BP (ed) *Scotland and the Cold War.* Dunfermline: Cualann Press Limited, 113–144.

Jones A (2014) *No Truck with the Chilean Junta! Trade Union Internationalism, Australia and Britain.* Canberra: Australian National University Press.

Karaliotas L (2017) Performing neoliberalization through urban infrastructure: Twenty years of privatization policies around Thessaloniki's port. *Environment and Planning A* 49(7): 1556–1574.

Kelliher D (2018) Historicising geographies of solidarity. *Geography Compass* 12(9) https://doi.org/10.1111/gec3.12399

Kosmatopoulos N (2019) On the shores of politics: Sea, solidarity and the Ships to Gaza. *Environment and Planning D: Society and Space* 37(4): 740–757.

Kothari U (2012) Contesting colonial rule: Politics of exile in the Indian Ocean. *Geoforum* 43(4): 697–706.

Legg S (2020) Political lives at sea: Working and socializing to and from the India Round Table Conference in London, 1930–1932. *Journal of Historical Geography* 68: 21–32.

Linebaugh P and Rediker M (1990) The Many-Headed Hydra: Sailors, Slaves and the Atlantic Working Class in the Eighteenth Century. *Journal of Historical Sociology* 3(3): 225–252.

Linebaugh P and Rediker M (2000) *The Many Headed Hydra: Sailors, Slaves, Commoners and the Hidden History of the Revolutionary Atlantic*. Boston, MA: Beacon Press.

McDuffie E (2011) *Sojourning for Freedom: Black Women, American Communism, and the Making of Black Left Communism*. Durham, NC: Duke University Press.

McVicar E (2010) *The Eskimo Republic: Scots Political Song in Action, 1951–1999*. Linlithgow: Gallus Publishing.

Mangru R (1993) *Indenture and Abolition: Sacrifice and Survival on the Guyanese Sugar Plantations*. Toronto: TSAR Publications.

Mawani R (2018) *Across Oceans of Law: The Komagata Maru and Jurisdiction in the Time of Empire*. Durham, NC and London: Duke University Press.

Mongia R (2019) *Indian Migration and Empire: A Colonial Genealogy of the Modern State*. Raniket: Permanent Black.

Nelson B (1988) *Workers on the Waterfront: Seamen, Longshoremen and Unionism in the 1930s*. Urbana, IL: University of Illinois Press.

Ramnath M (2011) *Haj to Utopia: How the Ghadar Movement Charted Global Radicalism and Attempted to Overthrow the British Empire*. Berkeley, CA: University of California Press.

Riethof M (2019) Chile protests escalate as widespread dissatisfaction shakes foundations of country's economic success story. *The Conversation* Available at: https://theconversation.com/chile-protests-escalate-as-widespread-dissatisfaction-shakes-foundations-of-countrys-economic-success-story-125628

Rediker M (2014) *Outlaws of the Atlantic: Sailors, Pirates, and Motley Crews in the Age of Sail*. London: Verso.

Scott JS (2018) *The Common Wind: Afro-American Currents in the Age of the Haitian Revolution*. London: Verso.

Shilliam R (2015) *Black Pacific: Anti-Colonial Struggles and Oceanic Connections*. London: Bloomsbury Academic.

Smith R (2016) Isolation through humanitarianism: Subaltern geopolitics of the siege on Gaza. *Antipode* 48(3): 750–769.

Steinberg PE and Peters K (2015) Wet ontologies, fluid spaces: Giving depth to volume through oceanic spaces. *Environment and Planning D* 33(2): 247–264.

Subramanian L (2014) *The Sovereign and the Pirate: Ordering Maritime Subjects in India's Western Littoral*. New Delhi: Oxford University Press.

Tabili L (1994) The construction of racial difference in twentieth-century Britain: The special restriction (Coloured Alien Seamen) order, 1925. *Journal of British Studies* 33(1): 54–98.

Tijani HI (2012) *Union Education in Nigeria: Labor, Empire, and Decolonization since 1945*. New York, NY: Palgrave.

Weiss H (2020) 'Unite in international solidarity!' The call of the International of Seamen and Harbour Workers to 'colonial' and 'negro' seamen in the early 1930s. In: Bellucci S and Weiss H (eds) *The Internationalisation of the Labour Question*. London: Palgrave Macmillan, 145–162.

White M (2017) "The cause of the workers who are fighting in Spain is yours": The marine transport workers and the Spanish Civil War. In: Cole P, Struthers D and Zimmer K (eds) *Wobblies of the World: A Global History of the IWW*. London: Pluto Press, 212–227.

# 20

# MIGRATION

## Security and humanitarianism across the Mediterranean border

*Charles Heller, Lorenzo Pezzani and Maurice Stierl*

### Introduction

With hundreds of migrants on board, several overcrowded rubber dinghies departed from Libya in mid-January 2019. The precarious travellers left the shores of the northern African country the day after a shipwreck had occurred in the same region, with over one hundred people going missing and presumably dying. The rubber boats would need to master a significant distance to the Libyan coast in order to avoid being captured by its so-called coastguards who, financed, equipped and trained by their European Union (EU) allies, have put the figure of returned migrants at over 15,000 for the year of 2018 while only 23,000 made it to Italy (IOM, 2018). Though travelling through the night and making it relatively far, eventually the dinghies were detected, and their passengers returned to Libya, some by merchant vessels and others by Libyan authorities, a practice in serious violation of the principle of non-refoulement under international law (Mann, 2016; Moreno-Lax, 2018). From detention, some of the survivors reached out to voice their suffering and protest, speaking of sickness and injury, overcrowded conditions, and violent abuse (Stierl, 2019a).

Precarious migrations across the Mediterranean occur in a politically contested space. Though routinely folded into narratives within which the fate of migrant travellers seems to depend on their own struggle with biophysical forces at work – the winds, the currents, the waves, and the cold – the phenomenon of maritime migration needs to be viewed in light of shifting policies and practices of securitisation as well as the unabated desire and needs that underpin 'acts of escape' across the space of the sea (Mezzadra, 2004; Steinberg, 2001). Given the diverse background and places of origin of the people on the move, the drivers and conditions of flight via the Mediterranean differ considerably, as several scholars have noted (Crawley et al., 2017; Squire et al., 2017). What people on the move in the Mediterranean share, however, is the experience of increasingly securitised migratory routes, to a large part the consequence of restrictive and preventative approaches to certain forms of human mobility, by the member states and institutions of the EU.

In this chapter, we highlight scholarly and activist insights into the interplay between migratory movements and forms of border governance in the Mediterranean. We point to the migratory dynamics which have resulted in the crossing of about two million individuals between

DOI: 10.4324/9781315111643-24

2014 and the end of 2019 alone, as well as the political processes that have turned the sea into a space of severe human rights violations and a 'graveyard', with tens of thousands of people counted as having died during attempted crossings. Many of those who disappear at sea will not feature in official statistics of death at Europe's maritime borders but remain missing without a trace (Heller and Pécoud, 2018). With reference to our own research practice and activist engagement in this contested space, we emphasise the significance of deploying a *spatial lens* when interrogating the Mediterranean mobility conflict (Heller et al., 2017, 2019).

Our chapter is organised into three main parts. We begin by offering a brief historic overview of policies and practices of border governance in the Mediterranean region vis-à-vis ongoing struggles for movement, pointing to the ways in which restrictions on legal migration have been productive of sea migration in the first place, while also emphasising migratory dynamics and agency pivotal in the evasion and reconfiguration of containment policies and practices. We further discuss the particular political geography of the sea, and the role that its overlapping jurisdictions play in migrant deaths. Next, we explore the humanitarianisation of the border in the Mediterranean due to a complex entanglement of security and humanitarian rationales through which border enforcement operations are regularly portrayed as acts of saving precarious lives. Finally, we discuss different modalities of contestation of the humanitarian border: migrant struggles and nongovernmental action at sea, but also the policies of containment implemented by states over the last years which have resulted in a shift towards the *de-humanitarianisation* of the border.

## The Mediterranean mobility conflict

The Mediterranean has long been a space of friction (Tsing, 2005), across which illegalised migratory trajectories have evolved in light of increasingly militarised means of policing, deployed by governments of the 'Global North' and their allies in the attempt to bridle turbulent movements from the 'Global South' into orderly and governable mobilities. Though we focus on the contemporary period, one needs to situate such uneven forms of mobility control in a much longer history. The work of historians reveals that a selective and unequal mobility regime was an intrinsic part of European imperial expansions towards the Mediterranean's southern shores in the nineteenth century. While during the nineteenth century it was mostly European settlers who migrated towards colonised territories, as of the beginning of the twentieth century, the northbound movement of colonised populations towards metropolitan territories took prominence. This phase was marked by successive moments of partial opening and closing of borders, with restrictions always leading to forms of evasion by migrants and early cases of deaths at sea (Borutta and Gekas, 2012; Clancy-Smith, 2010).

It is, however, more recently that illegalised migration and deaths across the Mediterranean have become a structural and highly politicised phenomenon. At the end of the 1980s, in conjunction with the consolidation of freedom of movement within the EU through the Schengen Agreement, visas were increasingly denied to citizens of the 'Global South' (Bakewell and De Haas, 2007). As scholarly research on the history of migration towards Europe has repeatedly shown, restrictions on legal migration did not prompt an end of Europe-bound migration but were, to the contrary, rather productive of unauthorised attempts to cross into Europe, including attempts via the sea (Boswell and Geddes, 2010; Samers, 2010). These 'EUropean'[1] forms of migration governance have thus been constitutive factors in the production of Mediterranean migration, which, in particular during the 1990s, became an increasingly spectacularised and fatal phenomenon (Mountz and Hiemstra, 2012; Weber and Pickering, 2011), "reflected not only in images of human misery and suffering that dominated newspapers, TV screens and social media feeds but also in growing public fears about the perceived economic, security and cultural threats of increased migration to Europe" (Baldwin-Edwards et al., 2018: 2).

During the 2000s, EU member states and institutions established a range of measures to prevent irregularised forms of migration, most notably by establishing the EU border agency Frontex (Neal, 2009) and by fostering new agreements with North African and other countries through which these countries would gradually turn into what Ataç et al. (2015: 3) have referred as the "(post-colonial) wardens of the European border regime" (see also, Bialasiewicz, 2012; Klepp, 2010, 2011). These alliances between Europe and North Africa, however, were considerably ruptured from 2011 onward, when the Arab Uprisings prompted a 're-opening' of the North African migration corridor. The fall of the Ben Ali regime in Tunisia and the Qaddafi regime in Libya in early 2011 allowed migrants to seek out maritime routes to escape to the European continent while the war in Syria led to an exodus that would contribute decisively to the mass crossings via the Aegean Sea in 2015 and early 2016. During this period, migratory movements across the central Mediterranean route, between Tunisia, Egypt, or Libya and Italy or Malta, have increased considerably, particularly from Libya: from an estimated annual average of 23,000 between 1997 and 2010 (McMahon and Sigona, 2016), to approximately 64,000 in 2011, and further to an annual average around 156,000 between 2014 and 2017. This was followed by a drastic decline between 2018 and 2020, when the annual average of migrant arrivals dropped to approximately 25,000 people, not least due to increased containment operations carried out by Libyan authorities (UNHCR, 2021). This mobility conflict has come at an exorbitant human cost: it is not only crossings that have intensified over recent years, but also migrant deaths at sea. While more than 40,000 deaths have been documented since the early 1990s, over 21,000 deaths have been counted between 2014 and 2020 alone (UNITED, 2021), with the real figure of migrant fatalities estimated to be much higher as many people disappear without ever being accounted for.

The political geography of the sea is essential to understanding the way the mobility conflict plays out across it and the specific form of violence that is exercised against migrants. At sea, the moment of border crossing is expanded into a process that can last several days and extends across an uneven and heterogeneous territory that sits outside the exclusive reach of any single polity (Steinberg, 2011; Suárez de Vivero, 2010). The spatial imaginary of the border as a line without thickness dividing isomorphic territorial states is here stretched into a deep zone "in which the gaps and discrepancies between legal borders become uncertain and contested" (Neilson, 2010: 126). Maritime territory constitutes, then, a space of 'unbundled sovereignty' in Saskia Sassen's terms (2006), one in which sovereign rights and obligations are disaggregated from each other and extended across complex and variegated jurisdictional spaces.

As soon as a migrant boat starts navigating, it passes through the jurisdictional regimes that crisscross the Mediterranean: from the various areas defined in the UN Convention on the Laws of the Sea (UNCLOS) to Search and Rescue (SAR) regions, from ecological and archaeological protection zones to areas of maritime surveillance. At the same time, it is caught between legal regimes that depend on the juridical status applied to those onboard (refugees, 'economic' migrants, illegals, and so on), based on the rationale of the operations that involve them (rescue, interception, and such like) and on several other factors. These overlaps, conflicts of delimitation, and differing interpretations, are not malfunctions but rather a structural characteristic of the maritime 'frontier' that has allowed states to simultaneously extend their sovereign privileges through forms of mobile government and elude the responsibilities that come with it (Gammeltoft-Hansen and Alberts, 2010; Steinberg, 2001).

For instance, the strategic mobilisation of the notion of 'rescue' has allowed coastal states to justify police operations in the high seas (Andersson, 2012), but overlapping and conflicting

SAR zones have led to recurrent cases of non-assistance to migrants in distress, as was the case for the 'left-to-die boat' which occurred in March 2011 and has been reconstructed in detail elsewhere by Forensic Oceanography (Heller and Pezzani, 2012). Here 72 people were left to drift for 14 days, despite having sent out several distress calls to maritime rescue agencies and having interacted with at least one helicopter and one military ship deployed as part of NATO's intervention in Libya. As a result, only nine people survived (see Figure 20.1). This incident, which has been the basis for several legal challenges (which are still ongoing) against states taking part in NATO's operation, exemplifies the way states increase the radical precarity and uncertainty of migrant journeys across the maritime frontier. The sea's 'geopower' (Grosz, 2012) is made to ambivalently oscillate between offering a medium enabling migrant movement and constituting a threatening liquid mass that risks swallowing their lives at any moment. Water then is turned into a deadly liquid that inflicts violence in indirect ways, mediating between state policies and practices on the one hand, and the bodies and lives of migrants on the other.

Within the Mediterranean frontier's overlapping and conflicting jurisdictional zones and legal norms, illegalised migrants are thus constituted as highly ambivalent subjects, and framed both as "a life to be protected *and* a security threat to protect against" (Vaughan-Williams, 2015: 3, emphasis in original). In this sense, the mobility conflict is one in which the conflicting logics of security and humanitarianism are deeply enmeshed. To understand how they are assembled,

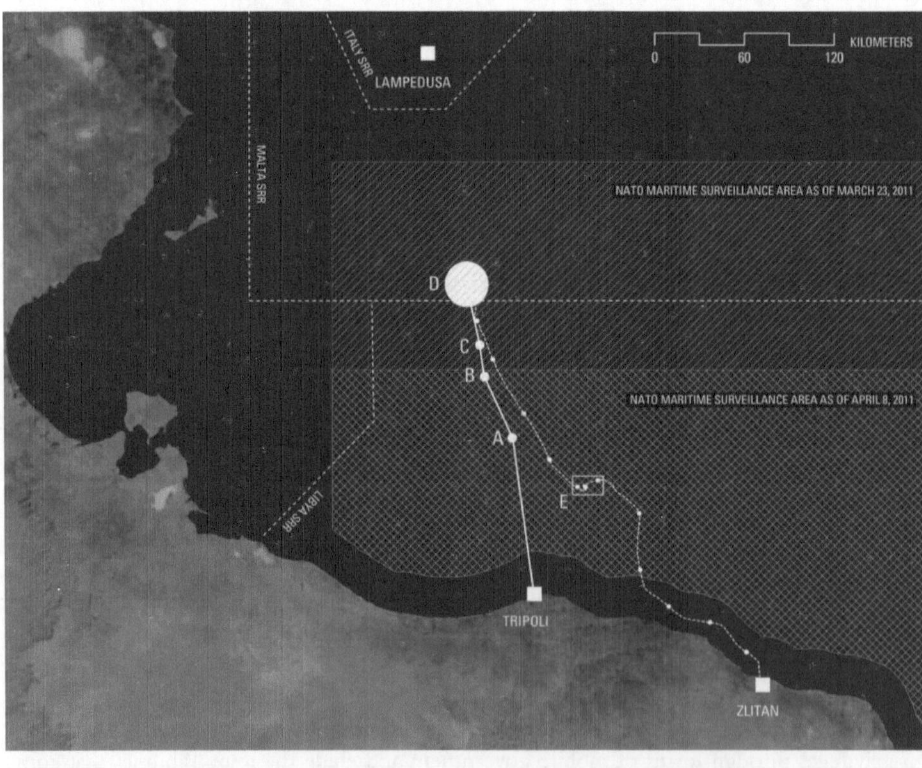


*Figure 20.1* Chain of events in the "left-to-die boat". For a detailed key to this map, see: www.forensic-architecture.org/case/left-die-boat/

*Source*: Forensic Oceanography and SITU Research, Report on the Left-to-Die Boat Case.

and how they complement and collide with each other, we now turn to a discussion of the 'humanitarianisation' of the border.

## The Mediterranean as a humanitarian border

In spaces like the Mediterranean "where it seems that the worlds designated by the terms Global North and Global South confront one another in a very concrete, abrasive way", William Walters has observed the emergence of a "humanitarian border" (2011: 146). For him (2011: 138), this "novel development within [the] history of borders and border-making" points to the complex entanglement of rationales of security and humanitarian care in practices of governing mobility. While rescue at sea has long been the humanitarian counterpart of the illegalisation of migrants, border control operations themselves have frequently been framed as *acts of saving*, blurring the notions of rescue and interception, as several scholars have observed (Garelli and Tazzioli, 2018; Heller and Pezzani, 2016; Moreno-Lax, 2018; Stierl, 2018). Border enforcement at sea then is often framed as a form of "humanitarian government", which Didier Fassin has described as "the administration of human collectivities in the name of a higher moral principle which sees the preservation of life and the alleviation of suffering as the highest value of action" (2007: 151). As this section will demonstrate, the humanitarian border is a highly unstable space as the logics of security and humanitarianism that operate within each actor and within the border regime as a whole, are always fraught with tensions, and the balance between them is in constant flux. While such fluctuations of security and humanitarian logics have left Europe's "restrictive migration and border regime" fundamentally unchanged over the last years (Cuttitta, 2018a: 649), the contradictions between these logics have opened a space of negotiation that migrants mobilise in the process of their unauthorised border crossings (Mezzadra, 2017).

In the central Mediterranean, the *humanitarianisation* of the border – a term we use in order to indicate a process, which, as we will see, is also reversible – became particularly visible after the shipwreck of 3 October 2013, when over 360 migrants died just a few hundred meters off the coast of the small Italian island of Lampedusa. This tragedy caused a public outcry that forced policy makers to position themselves. After his visit to Lampedusa on the 8 October 2013, Barroso, then President of the European Commission, declared: "[w]e in the European Commission [...] believe that the European Union cannot accept that thousands of people die at its borders" (European Commission, 2013). In the same speech, Barroso announced an increase in Frontex's budget and the launch of the European Border Surveillance System – that is, the continuation of a security approach to migration involving exactly the kind of measures that had prompted migrants to take deadly risks in the first place.

In the aftermath of this shipwreck, something did change, however. Italy launched operation *Mare Nostrum* (MN) days later, a military-humanitarian campaign involved in the rescue of about 150,000 people within a year. As Martina Tazzioli (2014, 2015, 2016) highlights, MN managed to focus public attention on the good 'scene of rescue', recasting the role of the state and the military as that of a merciful saviour. At the same time, however, this scene obscured other crucial aspects of the operation. First of all, by its official description, MN was a humanitarian *and* a security operation. In the frame of MN, saving lives and policing borders became one and the same thing. Not only did rescue operations lead to the arrest of 330 alleged smugglers, these operations also allowed for summary identification procedures to take place onboard the military ships, which for a time became floating detention centres, extending onto the high seas the biopolitical regime of identification normally applied on firm land. This, in turn, allowed for swift repatriation procedures for the nationals of countries with which

Italy held readmission agreements, in particular Tunisia and Egypt (Cuttitta, 2014). MN thus epitomised the blurring security and humanitarian practices characteristic of the humanitarian border. But the 'good scene of rescue' did not live up to its own image either: while more than 170,000 people were rescued that year, more than 3,000 deaths were also reported – MN did not make the crossing significantly less dangerous (Heller and Pezzani, 2016). Although MN assets were deployed close to the Libyan coast and thus came to operate as a "half-way bridge to Europe" (AEI, 2014: no page), it left untouched the political violence of the EU border regime which forced (and forces) migrants to resort to precarious means of crossing in the first place. It thus could not end the deaths of migrants at sea.

More than migrants' death at sea, what worried EU member states most was that the operation allowed migrants to arrive alive on EU shores, and that Italy was tacitly allowing them to continue their onward movement across the continent, in particular to Germany, Sweden or the UK, in violation of the Dublin regulation which provides that the first country in which migrants set foot are responsible for processing their asylum claims (Kasparek, 2015). In effect, the flip-side of Italy's extension of its operations to rescue migrants at sea was its retraction from its responsibility to fingerprint and process the asylum demands of the migrants once disembarked, thereby enabling their further movement across EU space. As a result of its perception as an Italian 'taxi service' to Europe, EU member states refused to 'Europeanise' MN as Italy requested, and Italy terminated the operation at the end of 2014. EU institutions and member states were keen to install, not a substitute, but a different presence at sea – a presence that would make migrant crossings more difficult and dangerous – and thus deter migrants from crossing towards the shores of Europe.

This operational shift was justified through a humanitarian discourse, exemplified by the UK Foreign Office Minister Lady Anelay's position when she stated "[w]e do not support planned search and rescue operations in the Mediterranean. We believe that they create an unintended 'pull factor', encouraging more migrants to attempt the dangerous sea crossing and thereby leading to more tragic and unnecessary deaths" (UK Parliament, 2014). With the launch of the Frontex operation *Triton*, which offered fewer European vessels patrolling areas much further away from the Libya coast, a deadly rescue gap was created (Heller and Pezzani, 2016; Stierl, 2018). Human rights advocates such as Amnesty International (2014) vocally denounced this policy of retreat, arguing it would not lead to fewer crossings but merely more deaths. Even Frontex (2014: 6), in an internal document, assessed that "the withdrawal of naval assets from the area, if not properly planned and announced well in advance, would likely result in a higher number of fatalities". This is effectively the reality that materialised in early 2015.

The week following 12 April 2015 saw what is believed to be the largest loss of life at sea in the recent history of the Mediterranean. About 400 people died on 12 April when an overcrowded boat capsized due to its passengers' excitement at the sight of platform supply vessels approaching to rescue them. On 18 April, a similar incident took an even greater toll in human lives, leading to the deadliest single shipwreck recorded by the UNHCR (2015) in the Mediterranean. Over 1,000 people are believed to have died when a migrant boat sank after a flawed manoeuvre led it to collide with a cargo ship that had approached to rescue its passengers (Pressly, 2020). As Médecins Sans Frontières (2015) commented at the time, these figures eerily resembled those of a war zone. On 29 April 2015, Jean-Claude Juncker, the President of the European Commission, admitted that "it was a serious mistake to bring the Mare Nostrum operation to an end. It cost human lives" (European Commission, 2015). However, the ending of MN and its (non-)replacement by Frontex's *Triton* operation, cannot adequately be described as a 'mistake' since it was a carefully planned policy, implemented in full knowledge of its deadly consequences. It was a policy of lethal non-assistance.

Regardless of admissions of misguided policy decisions, the humanitarian justification of the violent bordering of the Mediterranean continued in the shape of Frontex's *Triton* operation, as well as the launching of the EU's anti-smuggling military operation, EUNAVFOR MED – each of which was justified in the name of saving migrants' lives (Garelli and Tazzioli, 2018; Moreno-Lax, 2018). What these dominant narratives routinely occlude has been observed by critical scholarship for a long time: the causal relation between policies of closure and migrant deaths. Though by now a basic tenet in much of migration scholarship, European policymakers continue to reject the fact that restrictive migration policies have re-directed people on the move onto routes, such as those across the sea, that are often costlier, lengthier and far more dangerous than previous ones, often requiring the 'service' of human smugglers (Albahari, 2015, 2018; Mountz and Loyd, 2014). In order to maintain the Mediterranean as a space of humanitarian sentiment justifying security-oriented operations, radical de-contextualisation and de-politicisation are required. The spectacularisation of border control and migrants' deaths occlude the structural economic and political processes that have produced the Mediterranean's mobility conflict in the first place. As a result of this spatial and moral delimitation (Fassin, 2012: 253), the maritime borderzone is turned into a seemingly inherently exceptional and dangerous space requiring (EUropean) intervention.

## Contesting the humanitarian border

As the end of the *Mare Nostrum* operation demonstrates, the humanitarian border is a highly unstable and contested space. Here, a process of operational de-humanitarianisation led to large-scale deaths but was justified through the humanitarian discourse of saving lives. The inherent ambivalence of the humanitarian border means that it has remained a contested space also following the end of the MN operation. Confronted with this Mediterranean border, migrants instrumentalise its humanitarian dimension to transgress it, activists and researchers denounce its inherent violence, and NGOs attempt to mitigate its deadliness. Meanwhile, state actors and agencies have pursued the frontier's de-humanitarianisation ever-more forcefully in order to contain migrants on the southern shores of the Mediterranean. We address these different actors and tensions in turn.

One of the main factors that exposes European attempts to humanitarianise the border to "vulnerability, to reversal, to subversion, even to critical instrumentalization" (Butler, 2009: 10) are the main protagonists in the drama over movement and its control – migrant subjects themselves. In dominant public and policy debates, but also in many mainstream migration studies accounts in which the ongoing atrocities at sea are conceived as a 'migration crisis' (cf. New Keywords Collective, 2016; Stierl, 2020), the migrant is regularly reduced to a passive subject, a victim of smugglers or traffickers who lacks agency in the migratory process, simply being 'pushed' or 'pulled' around. Such portrayals deny the fact that migrants exercise agency and enact their freedom of movement, even in precarious conditions at sea, often generating new situations and dilemmas for Europe's border enforcers (New Keywords Collective, 2021; Stierl, 2019b). The Autonomy of Migration literature, associated with autonomous Marxism, has become one of the main scholarly interventions able to emphasise the subjective practices of migrants and the political dimension of their transgressive movements (Scheel, 2013). Migrant escape through an autonomist perspective is considered "a form of creative subversion capable of challenging and transforming the conditions of power" (Papadopoulos et al., 2008: 56).

Migrant boats, then, are "a site of political action" (Walters, 2015: 472), places of contestation that carry subjects who enact their right to leave, move, survive and arrive. These boats are mostly steered by 'migrant captains', who make use of the satellite phone merely after having moved a significant distance, regularly informing not (only) European coastguards, but (also) activists of their distress. Some boats do not want to be rescued but seek to reach European coasts independently, at times in order for their passengers to 'disappear' after arrival. There were other instances where dinghies have been intentionally deflated or people went overboard in order to force others to respond. Hundreds of precarious travellers have jumped into the sea in the presence of European forces to avoid being returned to Libya by Libyan militias. Through these risky acts of becoming shipwrecked, migrants force themselves into a regime of care that was never intended for them, thereby re-articulating European humanitarian narratives in a "reversed discourse" that is "parasitic on the 'dominant discourse'" (Baaz et al., 2017: 31–32). As Tazzioli and Walters (2016: 462) write, through such actions, migrants make themselves audible and visible "in terms of a politics of the governed in which migrants demand to be objects of humanitarian concern" and turn themselves into "a humanitarian problem".

Another force that has contested the Mediterranean border are the nongovernmental actors, which, since 2011, have transformed the Mediterranean into a laboratory for the innovation of new political practices. The Forensic Oceanography project, which was initiated in 2011 by two among us (Heller and Pezzani), developed new methods to document and contest the violence of borders. In our reconstruction of the 'left-to-die boat' case – discussed above – we assembled a composite image of the events by corroborating the survivors' testimonies with information provided by the vast apparatus of remote sensing technologies that have transformed the contemporary sea into a digital archive. While these technologies are often used for the purpose of policing illegalised migration as well as the detection of other 'threats', they were repurposed to find evidence for the failure to render assistance. We were able to model and reconstruct the drifting boat's trajectory and could account for the presence of a large number of vessels in the vicinity of the drifting migrant boat which did not heed its calls for help.

Through the 'left-to-die' case, we sought to put into practice a *disobedient gaze* that used some of the same sensing technologies of border controllers, but redirect their 'spotlight' from unauthorised acts of border crossing, to state and non-state practices violating migrants' rights. We conceived this gaze as

> [aiming] not to disclose what the regime of migration management attempts to unveil – clandestine migration – but unveil that which it attempts to hide, the political violence it is founded on and the human rights violations that are its structural outcome.
>
> *Pezzani and Heller, 2013: 294*

In addition to reconstructing events at sea by assembling their multiple traces, crucial to our project was the task of *spatialising* the practices of actors and inscribing them within the political geography of the sea. Over the last few years we have continued with these lines of investigation, and have contributed to uncovering the dynamics that led to the April 2015 shipwreck discussed above in our report 'Death by Rescue – The Lethal Effects of the EU's Policies of Non-assistance'.[2] Crucially, while the humanitarian border deplores migrants' deaths at sea but severs political responsibility for them, by combining case reconstruction with a *forensics of policies*, we could reconnect policy making and its lethal effects.

The WatchTheMed[3] monitoring platform, a network of activists, researchers, and NGOs that we helped found in 2012, intended to document human rights violations in the Mediterranean but also sought to intervene more directly at sea. In this endeavour, the Alarm Phone[4] project emerged, an activist hotline supporting boats in distress. Initiated by a coalition of freedom of movement, human rights and migrant activist groups – including WatchTheMed, Boats4People, Welcome to Europe, Afrique Europe Interact, Borderline-Europe, No Borders

Morocco, FFM and Voix des Migrants – the Alarm Phone was launched in October 2014, with the intention to respond to violent border 'protection' practices and the unabated mass dying in maritime spaces around Europe, and to offer travellers alternative ways to make their distress heard (Stierl, 2016, 2019a, 2019b). Thanks to a management software, the Alarm Phone can re-route distress calls to a vast number of volunteers operating shifts, situated in about 12 countries, thus ensuring that every call is attended to. In its first six years in existence, the activist phone project has gathered extraordinary momentum, and supported about 3,400 boats in distress in all Mediterranean regions (Schwarz and Stierl, 2019). Listening to and giving echo to the voices of those in the process of crossing maritime spaces allows the Alarm Phone to disobediently *observe* the Mediterranean Sea in the aim of supporting them in their exercise of their freedom of movement and mitigating the extraordinary risk they face for daring to do so. Besides supporting precarious human mobilities at sea, the wide solidarity network of the Alarm Phone, composed of about 200 activists and several connected groups, can exercise pressure in real time when there is a risk that a violation at sea may be perpetrated, such as cases of failing to render assistance or push-back (the illegal collective expulsion of 'aliens' from a country's territory). The story of the boats that left Libya in January 2019 but were intercepted and returned, garnered international attention only due to the collaboration between migrants at sea, and in detention, and activists on land.

Finally, after the ending of the *Mare Nostrum* operation, nongovernmental actors deployed their own rescue vessels to fill the lethal rescue gap the Italian operation had left in its wake and to denounce the violent (in)action of European states. Paolo Cuttitta (2017, 2018a) has referred to this moment as the 'non-governmental turn' in the Mediterranean, with first MOAS and later Sea-Watch, MSF, and several other NGOs sending SAR assets to the area off the coast of Libya. Also, Stierl (2018: 705) has considered the emergence of the nongovernmental "humanitarian fleet […] a new dimension in the humanitarian transformation of (maritime) borderzones" which would provoke conflict in the dominant humanitarian narratives constructed by EU institutions and member states. Rescue NGOs, then, simultaneously contest the partial de-humanitarianisation of the border at the operational level, and take the rescue practices that had come to be embedded in the humanitarian border into their own hands.

While, until now, we have discussed the practices of actors who have sought to contest the inherent violence of the humanitarian border and push its humanitarian dimension as far as possible, the humanitarian border has also continued to be contested by state actors, who, in the aim of sealing off the Mediterranean frontier, have sought to strip away any remnant of humanitarian practice and logic. As the EU had proven unable to deter migrants' from crossing the sea through its policies of non-assistance and militarised border control, since 2016 it has embarked on a policy involving the criminalisation of civilian rescue activities and the outsourcing of border control. This process has been particularly evident in Italy. While Cuttitta (2018b) has written of the 'end of the humanitarian turn' in Italy's policy, we prefer once again to think in more processual and variegated terms, and thus refer to this process as one of *de-humanitarianisation* of the border. We observe a process of de-humanitarianisation in several different respects: on the one hand, we see humanitarian actors and the humanitarian logics that are embedded in practices of border control progressively pushed to the side, and thus tending to leave only security actors and logics in operation at the EU's maritime frontier.[5] Furthermore, the collaboration of Italy and the EU with war-torn and politically fragmented Libya in the violent returning of migrants has been implemented despite full knowledge of their fate of detention, forced labour, torture and rape.[6] As Stierl and Sandro Mezzadra (2019) have argued: "[w]here we once spoke about a 'humanitarianisation' of the border, we now see viscerally the materiality and depth of inhumanity and a purity of violence that Europe is no longer able, or willing, to hide". Migrant deaths and the violence migrants are subjected to are increasingly accepted as a necessary evil to protect Europe against the 'threat' migrants are purported to constitute. In this phase, the lives of migrants have increasingly lost even their discursive and symbolic value as a justification for border control.

## Conclusion

One could say that while the current form of the territorial state on firm land is founded on an imaginary of sedentariness, the political form of maritime space is founded on movement and its management – the policing of the so-called 'freedom of the seas'. Movement is highly contested along the EU's Mediterranean frontier, one of the fault-lines of the world system in which profound inequality coincides with variegated cultures constructed as radically different. The EU's restrictive migration policies clash with the dynamics of migration from the Global South, leading to an enduring mobility conflict. The dialectic between illegalised migrants' precarious exercise of their freedom to move by boarding often unseaworthy boats in order to reach Europe and ever shifting militarised bordering practices has led to the harrowing loss of over 40,000 lives (UNITED, 2021).

In this chapter, we have pointed to several aspects of the mobility conflict that continues to play out in the Mediterranean. We have argued that illegalised and precarious sea migration is not a natural phenomenon but part and parcel of the interplay between the desire and need to move and attempts to govern and contain particular migratory subjects. The phenomenon of sea migration in the Mediterranean can be traced back several decades, when migration policies implemented by EU member states and institutions turned increasingly restrictive towards individuals and populations from the Global South. Instead of preventing migration to EU territory, such policies were productive of increasingly dangerous migratory projects, with increasing numbers of people seeing no other options than to cross the border via the sea. Despite Europe's concerted attempts to thwart and contain movements, migrants continue to struggle across the Mediterranean. Over recent years, hundreds of thousands of people have succeeded in claiming their right to survive and arrive.

Notwithstanding the structural violence that underpins the Mediterranean border regime, EU authorities have sought to turn the sea also into a space of humanitarianism, where the rescue of the distressed is depicted as expressions of fulfilling a moral duty despite it being combined with ever-more drastic interdiction and deterrence practices. What has emerged is a 'humanitarian border', which combines security and humanitarian logics. However, precisely because of its inherent contradictions, the humanitarian border is fraught with instability and conflict, pulled and pushed in different directions by migrants, nongovernmental actors and states. As we have shown, the sea's particular geography, its overlapping jurisdictional zones and conflicting norms, play a crucial role in shaping these conflicts, and are themselves a major stake. This Mediterranean is not simply the 'sea between the land' – a liminal space lying outside the realm of politics. Rather, through migrants' unabated movement and practices of nongovernmental solidarity, both of which contest the bordering of the sea, the Mediterranean has been turned into a fundamental space of international and transnational politics in its own right.

## Acknowledgements

We would like to thank Kimberley Peters for her excellent comments on a previous version of this chapter. Maurice Stierl would like to thank the Leverhulme Trust. Charles Heller would like to thank the Swiss National Research Fund.

## Notes

1 The term 'EUrope' seeks to problematise frequently employed usages that equate the EU with Europe and Europe with the EU and suggests, at the same time, that EUrope is not reducible to the institutions of the EU.
2 See https://deathbyrescue.org/
3 See http://watchthemed.net/
4 See https://alarmphone.org/en/
5 See https://blamingtherescuers.org/
6 See www.forensic-architecture.org/case/sea-watch/

## References


Africa Europe Interact (2014) *Mare Nostrum: Resistance From Below Forces Europe to Save People*. Available at: http://afrique-europe-interact.net/1205-1-Mare-Nostrum.html

Albahari M (2015) *Crimes of Peace: Mediterranean Migrations at the World's Deadliest Border*. Philadelphia, PA: University of Pennsylvania Press.

Albahari M (2018) From right to permission: Asylum, Mediterranean migrations, and Europe's war on smuggling. *Journal on Migration and Human Security*. https://doi.org/10.1177/2311502418767088

Amnesty International (2014) *Lives Adrift Refugees and Migrants in Peril in the Central Mediterranean*. Available at: www.amnesty.org/en/documents/eur05/006/2014/en

Andersson R (2012) A game of risk: Boat migration and the business of bordering Europe (Respond to this article at www.therai.org.uk/at/debate). *Anthropology Today* 28(6): 7–11.

Ataç I, Kron S, Schilliger S, Schwiertz H and Stierl M (2015) Struggles of migration as in-/visible politics. Introduction. *Movements: Journal für kritische Migrations- und Grenzregimeforschung* 1(2): 1–18.

Baaz M, Lilja M and Vinthagen S (2017) *Researching Resistance and Social Change*. London: Rowman and Littlefield.

Bakewell O and De Haas H (2007) African migrations: Continuities, discontinuities and recent transformations. In: de Haan L, Engel U and Chabal P (eds) *African Alternatives*. Leiden: Brill, 95–118.

Baldwin-Edwards M, Blitz B and Crawley H (2018) The politics of evidence-based policy in Europe's 'migration crisis'. *Journal of Ethnic and Migration Studies* 45(12): 2139–2155.

Bialasiewicz L (2012) Off-shoring and out-sourcing the borders of EUrope: Libya and EU border work in the Mediterranean. *Geopolitics* 17(4): 843–866.

Borutta M and Gekas S (2012) A colonial sea: The Mediterranean, 1798–1956. *European Review of History* 19(1): 1–13.

Boswell C and Geddes A (2010) *Migration and Mobility in the European Union*. Basingstoke: Palgrave.

Butler J (2009) *Frames of War, When Is Life Grievable?* New York, NY: Verso.

Clancy-Smith J (2010) *Mediterraneans, North Africa and Europe in an Age of Migration, c. 1800–1900*. Berkeley, CA: University of California Press.

Cuttitta P (2014) From the *Cap Anamur* to *Mare Nostrum*: Humanitarianism and migration controls at the EU's maritime borders. In: Matera C and Taylor A (eds) *The Common European Asylum System and Human Rights: Enhancing Protection in Times of Emergencies*. Den Haag: Asser Institute, 21–37

Cuttitta P (2017) Delocalization, humanitarianism and human rights: The Mediterranean border between exclusion and inclusion. *Antipode* 50(3): 783–803.

Cuttitta P (2018a) Repoliticisation through search and rescue? Humanitarian NGOs and migration management in the central Mediterranean. *Geopolitics* 23(3): 632–660.

Cuttitta P (2018b) *Pushing Migrants Back to Libya, Persecuting Rescue NGOs: The End of the Humanitarian Turn*. Available at: www.law.ox.ac.uk/research-subject-groups/centre-criminology/centreborder-criminologies/blog/2018/04/pushing-0

Crawley H, Duvell F, Jones K, McMahon S and Sigona N (2017) *Unravelling Europe's 'Migration Crisis'*. Bristol: Policy Press.

European Commission (2013) Statement by President Barroso following his Visit to Lampedusa. Available at: http://europa.eu/rapid/press-release_SPEECH-13-792_

European Commission (2015) Speech by President Jean-Claude Juncker at the debate in the European Parliament on the conclusions of the Special European Council on 23 April: 'Tackling the migration crisis'. Available at: http://europa.eu/rapid/press-release_SPEECH-15-4896_en.htm

Fassin D (2007) Humanitarianism: A nongovernmental government. In: Feher M (ed) *Nongovernmental Politics*. New York, NY: Zone Books, 149–160.

Fassin D (2012) *Humanitarian Reason*. Berkeley, CA: University of California Press.

Frontex (2014) *Concept of Reinforced Joint Operation Tackling the Migratory Flows Towards Italy: JO EPN-Triton*. Available at: https://deathbyrescue.org/assets/annexes/2.Frontex_Concept_JO_EPN-Triton_28.08.2014.pdf

Gammeltoft-Hansen T and Alberts T (2010) *Sovereignty at Sea: The Law and Politics of Saving Lives in the Mare Liberum*. DIIS Working Paper 18. Available at: https://pure.diis.dk/ws/files/43394/WP201018_sovereingty_at_sea_mare_liberum_web.pdf

Garelli G and Tazzioli M (2018) The humanitarian war against migrant smugglers at sea. *Antipode* 50(3): 685–703.

Grosz E (2012) Geopower. In: Yusoff K, Grosz E, Clark N, Saldanha A and Nash C (2012) Geopower: A panel on Elizabeth Grosz's Chaos, Territory, Art: Deleuze and the framing of the earth. *Environment and Planning D: Society and Space* 30: 971–988.

Heller C and Pécoud A (2018) Counting migrants' deaths at the border: From civil society counter-statistics to (inter)governmental recuperation. *International Migration Institute Network* 143: 1–20.

Heller C and Pezzani L (2012) Forensic oceanography: *Left-to-die* boat case. Available at http://migrantsatsea.files.wordpress.com/2012/04/forensic-oceanography-report-11april20121.pdf

Heller C and Pezzani L (2016) Ebbing and flowing: The EU's shifting practices of (non-)assistance and bordering in a time of crisis. *Near Futures Online 1 'Europe at a Crossroads'*. Available at: http://nearfuturesonline.org/wp-content/uploads/2016/03/Heller_Pezzani_Ebbing_2016.pdf

Heller C, Pezzani L and Stierl M (2017) Disobedient sensing and border struggles at the maritime frontier of EUrope. *Spheres, Journal for Digital Cultures*. Available at: http://spheres-journal.org/disobedient-sensing-and-border-struggles-at-the-maritime-frontier-of-europe/

Heller C, Pezzani L and Stierl M (2019) Toward a politics of freedom of movement. In: Jones R (ed) *Open Borders: In Defense of Free Movement*. Athens, GA: University of Georgia Press, 51–76.

IOM (2018) *Libya – Maritime Update Libyan Coast (1–15 December 2018)*. Available at: https://migration.iom.int/reports/libya-%E2%80%94-maritime-update-libyan-coast-1-15-december-2018

Kasparek B (2015) Complementing Schengen: The Dublin System and the European Border and Migration Regime. In: Bauder H and Matheis C (eds) *Migration and Borders Here and Now: From Theorizing Causes to Proposing Interventions*. London: Palgrave, 59–78.

Klepp S (2011) A double bind: Malta and the rescue of unwanted migrants at sea, a legal anthropological perspective on the humanitarian Law of the Sea. *International Journal of Refugee Law* 23(3): 538–557.

Klepp S (2010) A contested asylum system: The European Union between refugee protection and border control in the Mediterranean sea. *European Journal of Migration and Law* 12(1): 1–21.

Mann I (2016) *Humanity at Sea: Maritime Migration and the Foundations of International Law*. Cambridge: Cambridge University Press.

McMahon S and Sigona N (2016) Boat migration across the Central Mediterranean: Drivers, experiences and responses. MEDMIG Research Brief No. 3. Available at: www.medmig.info/wp-content/uploads/2017/02/research-brief-03-Boat-migration-across-the-Central-Mediterranean.pdf

Médecins Sans Frontières (2015) MSF calls for large scale search and rescue operation in the Mediterranean. Available at www.msf.org/article/msf-calls-large-scale-search-and-rescue-operation-mediterranean

Mezzadra S (2004) The right to escape. *Ephemera* 4: 267–275.

Mezzadra S (2017) Humanitarianism destroys politicality. An interview with Sandro Mezzadra by Davor Konjikusic. *Transversal*. Available at: https://transversal.at/transversal/1017/mezzadra-konjikusic/en

Moreno-Lax V (2018) The EU humanitarian border and the securitisation of human rights: The 'rescue-through-interdiction/rescue-without-protection' paradigm. *Journal of Common Market Studies* 56(1): 119–140.

Mountz A and Hiemstra N (2012) Spatial strategies for rebordering human migration at sea. In: Wilson T and Hastings D (eds) *A Companion to Border Studies*. Malden: Blackwell, 455–472.

Mountz A and Loyd J (2014) Constructing the Mediterranean region. *ACME* 13(2): 173–195.

Neal AW (2009) Securitization and risk at the EU border: The origins of FRONTEX. *Journal of Common Market Studies* 47(2): 333–356.

Neilson B (2010) Between governance and sovereignty: Remaking the borderscape to Australia's north. *Local-Global Journal* 8: 124–140.

New Keywords Collective (2016) *Europe/Crisis: New Keywords of 'the Crisis' in and of 'Europe'*. Zone Books, Near Futures [Online]. Available at: http://nearfuturesonline.org/wp-content/uploads/2016/01/New-Keywords-Collective_12.pdf

New Keywords Collective (2021) Minor keywords of political theory: Migration as a critical standpoint. A collaborative project of collective writing. *Environment and Planning C: Politics and Space*. https://doi.org/10.1177/2399654420988563

Papadopoulos D, Stephenson N and Tsianos V (2008) *Escape Routes*. London: Pluto Press.

Pezzani L and Heller C (2013) A disobedient gaze: Strategic interventions in the knowledge(s) of maritime borders. *Postcolonial Studies* 16(3): 289–298.

Pressly L (2020) 1,000 lost on one boat – this woman hopes to name them. *BBC News [Online]*. Available at: www.bbc.com/news/stories-55398000

Samers M (2010) *Migration*. Abingdon: Routledge.

Sassen S (2006) *Territory, Authority, Rights: From Medieval to Global Assemblages*. Princeton, NJ: Princeton University Press.

Scheel S (2013) Autonomy of migration despite its securitisation? Facing the terms and conditions of biometric rebordering. *Millennium* 41(3): 575–600.

Schwarz N and Stierl M (2019) Amplifying migrant voices and struggles at sea as a radical practice. *South Atlantic Quarterly* 118(3): 661–669.

Squire V, Dimitriadi A, Perkowski N, Pisani M, Stevens D and Vaughan-Williams N (2017) *Crossing the Mediterranean Sea by Boat: Mapping and Documenting Migratory Journeys and Experiences*. Available at: www.warwick.ac.uk/crossingthemed

Steinberg PE (2001) *The Social Construction of the Ocean*. Cambridge: Cambridge University Press.

Steinberg PE (2011) Free sea. In: Legg S (ed) *Spatiality, Sovereignty and Carl Schmitt: Geographies of the Nomos*. Abingdon: Routledge, 268–275.

Stierl M (2016) A sea of struggle: Activist border interventions in the Mediterranean Sea. *Citizenship Studies* 20(5): 561–578.

Stierl M (2018) A fleet of Mediterranean border humanitarians. *Antipode* 50(3): 704–724.

Stierl M (2019a) Migrants calling us in distress from the Mediterranean returned to Libya by deadly 'refoulement' industry. *The Conversation*. Available at: https://theconversation.com/migrants-calling-us-in-distress-from-the-mediterranean-returned-to-libya-by-deadly-refoulement-industry-111219

Stierl M (2019b) *Migrant Resistance in Contemporary Europe*. Abingdon: Routledge.

Stierl M and Mezzadra S (2019) The Mediterranean battlefield of migration. *Open Democracy*. Available at: www.opendemocracy.net/en/can-europe-make-it/mediterranean-battlefield-migration/?fbclid=IwAR3K5PTnnP2peI9lDXe9VhoIJQy-gZFiI57uOgJxpsoVLC970uMqGNS2IvI

Stierl M (2020) Do no harm? The impact of policy on migration scholarship. *Environment and Planning C: Politics and Space*. https://doi.org/10.1177/2399654420965567

Suárez de Vivero JL (2010) *Jurisdictional Waters in The Mediterranean and Black Seas*. Brussels: European Parliament.

Tazzioli M (2014) *Spaces of Governmentality, Autonomous Migration and the Arab Uprisings*. London: Rowman and Littlefield International.

Tazzioli M (2015) The desultory politics of mobility and the humanitarian-military border in the Mediterranean. Mare Nostrum beyond the sea. *REMHU: Revista Interdisciplinar Da Mobilidade Humana* 23(44): 61–82.

Tazzioli M (2016) Border displacements: Challenging the politics of rescue between Mare Nostrum and Triton. *Migration Studies* 4(1): 1–19.

Tazzioli M and Walters W (2016) The sight of migration: Governmentality, visibility and Europe's contested borders. *Global Society* 30(3), 445–464.

Tsing A (2005) *Friction: An Ethnography of Global Connection*. Princeton, NJ: Princeton University Press.

UK Parliament (2014) *Mediterranean Sea*. Available at: https://publications.parliament.uk/pa/ld201415/ldhansrd/text/141015w0001.htm

UNHCR (2015) *Mediterranean Boat Capsizing: Deadliest Incident on Record*. Available at: www.unhcr.org/553652699.html

UNHCR (2021) *Mediterranean Situation*. Available at: http://data2.unhcr.org/en/situations/mediterranean

UNITED (2021) *About the 'List of Deaths'*. Available at: http://unitedagainstrefugeedeaths.eu/about-the-campaign/about-the-united-list-of-deaths/

Vaughan-Williams N (2015) *Europe's Border Crisis: Biopolitical Security and Beyond*. Oxford: Oxford University Press.

Walters W (2011) Foucault and frontiers: Notes on the birth of the humanitarian border. In: Bröckling U, Krasmann S and Lemke T (eds) *Governmentality, Current Issues and Future Challenges*. New York, NY: Routledge, 138–164.

Walters W (2015) Migration, vehicles, and politics: Three theses on viapolitics. *European Journal of Social Theory* 18(4), 469–488.

Weber L and Pickering S (2011) *Globalization and Borders: Death at the Global Frontier*. Basingstoke: Palgrave Macmillan.

# SECTION V

# Ocean experiences, ocean engagements

# 21
# WRITING
## Literature and the sea

*Stephanie Jones*

### Introduction: Material and historical poetics

Visitors to the UK's National Maritime Museum (NMM) will walk past walls and over floors inscribed and posted with lines of poetry. Most prominently and repeatedly, they will read a phrase that is also the title of a poem by the Antillean poet, playwright and 1992 Nobel Prize winner, Derek Walcott: 'the sea is history'; 'The sea is History'; 'The Sea is History'. In one location, the words are materialised in smooth white laminate lettering on a dark blue wall; the font is elaborately serif; 'sea' is about a metre high, twice that of the other three words; and shafts of blue-gelled light lend a striking underwater effect. Alone on the wall, these words make a bold statement of material fact. In another place, words from the poem are constructed from a crisper font on brightly-lit planking that evokes the deck of a ship. Here, the first verse of the eighty-line poem is depicted:

> Where are your monuments, your battles, martyrs?
> Where is your tribal memory? Sirs,
> in that grey vault. The sea.
> The sea has locked them up.
> The sea is History.
>
> *Walcott, 1986: 364*

Extracted phrases and parts of poems by other writers from other centuries appear in varying fonts, heights, and depths of lettering across the rest of the wall. It is a collage that seems – rather refreshingly – to defy a linear literary historiography and – to a lesser extent – to challenge hierarchies of literary worth: an extract from a song by the rock band Sea Power is decked with Shakespeare, but few of the quotations are not by men. Behind all this, fainter but larger, individual words are depicted in plump outline lettering: 'Joy', 'Aggression', 'Pride', 'Anticipation', 'Love'. The 'S' of a great 'Sadness' palimpsests the 'ea' of Walcott's sea: 'The s[S]ea is History'; The s[Sadness]ea is History': The sea Sadness is History. This can be read as a critical direction, or as the creation of a new poem. And so, the wall encourages every visitor to become a sea-poet. In these many ways, it is wonderful to see poetry literally making meaning in the free public space of the NMM. But there is also an irony to this assembled dis-assemblage of poetry.

DOI: 10.4324/9781315111643-26

In a place of meticulously historicised objects, Walcott's poem – disaggregated and turned into a solid thing – is noticeably de-historicised. And this involves a further irony; because when reassembled, the poem does not really tell us that the sea is history.

In *The Sea is History*, Walcott rewrites the bible as epochs of Atlantic violence. In a middle passage, between old and new testaments, the poem states:

> … that was Lamentations -
> that was just Lamentations
> it was not History …
> *Walcott, 1986: 366*

That 'just' is both dismissive and exacting. Pitching between the two meanings of the word, it stalls the very idea of history between an only-merely 'just' and a judgement-making 'just'. The sea of this poem feels like the loss of historical justice; and at the same time, it feels like a craving for historical justice. But the possibility of finally apprehending the sea as such a history is not consolidated by the poem. This possibility is defrayed into the images – often wry and elliptical – that the poet draws from ocean space. The shark's shadows are 'like' benediction, the white cowries are 'like' manacles, the grouper is 'like' a bald queen (Walcott, 1986: 364–365). These are 'just' similes. They do not yield – not even metaphorically – history itself. The poem does not present a recuperative historiography of the Black Atlantic, but rather ends with "the salt chuckle of rocks/…/ like a rumour without an echo/of History, really beginning" (Walcott, 1986: 367). Walcott *is* writing about a specific history: but the *poetry* of the poem makes us feel ocean space as both the imperative and impossibility of a replete historiography of colonial and postcolonial Atlantic violence.

First published in 1979, the title of Walcott's poem has gained iconic status in the past 20 years. I have lost count of the books and chapters from across disciplines – from law to environmental science – that use 'The Sea is History' as an epigraph. For an academic visitor, the NMM's deployment of the poem is a *leitmotif* of the over-lapping spatial, material, and environmental 'turns' in literary studies over the past half century. More specifically, it symbolises the significance of literature – forms of narrative fiction as well as poetry – to the 'blue'/oceanic turn across the humanities and social sciences that has gathered such significant momentum in the twenty-first century. In this chapter, I trace the way ocean space has shifted as a literary object and subject in the decades since the publication of Walcott's poem. Most particularly, I identify twenty-first-century work on literatures in English that seem likely to become more influential and critically invigorating in the future. To begin, I identify how a growing concentration on ocean space has changed what is considered as literature: natural histories, ship's journals, and ocean ephemera have become established primary materials of the discipline. I then turn to a longer consideration of ocean space as a way of reading literature. I begin with the bold but not uncommon proposition that *the* ontology of English literature *is* the ocean, and that the history of literatures in English always takes us offshore: and so always takes us to narratives of unfreedom. I then outline some of the ways this approach is being redefined through closer attention to marine animals and matter. This leads me to other more material matters: to the curious relationship of ocean, desert and outer space as a literary concern, and then back to Earth and to geopolitics, and to the ocean as a geo for fantasising other forms of politics. I highlight new moves and future directions in literary approaches to Indian, Atlantic, Pacific and comparative regional sea-studies: and to a rising interest in the submarine, the deeper deep, and in icy seas. I end with some thoughts on the ocean as a future literary epistemology for the world.

## The ocean as genre: What we read

With other disciplines, the twentieth century saw literary studies moving towards a new and more expansive interest in labour at and by the sea. The 1930s saw the 'discovery' of Herman Melville's until-then obscure 1851 novel, *Moby Dick*. Full of sublimely gruelling and beatifically mundane scenes of labour at sea, the novel first gathered a readership as "a parable on the mystery of evil and the accidental malice of the universe" (Mumford, 1929: 184), then as an allegory of industrialisation that binds the great American novel to oceanic histories of destruction (Tanner, 1988), and more lately as a first text in prominent historiographies of 'the environmental imagination' as a slow move away from anthropocentrism (Buell, 1995). In these ways, the canonisation of Melville and *Moby Dick* has – oddly – not entailed an interest in labour at sea as a broader or eminent topic of literary investigation. But from the late 1980s, Peter Linebaugh and Marcus Rediker's rousing Marxist histories of seafaring can be credited with prompting an intensification of interest in labour at sea across disciplines (Linebaugh and Rediker, 2000; Rediker, 1987; see also Featherstone, this volume; Griffin, this volume). In English departments, this has involved both a re-assessment of the maritime adventure tale as a popular genre and the retrieval of book histories of other, less consolidated genres. Hester Blum (2008), for example, promotes a deep archive of seafarer authored work that includes life writings, logs, journalism. And Margaret Cohen reanimates once-popular, now-obscure narrative fictions that articulate the mariner's craft. In her work, attention to literary form is necessary to grasping the significance of ocean space to an understanding of the temporalities of labour (Cohen, 2010). Alert to discourses of imperialist nostalgia and racist disavowal that so easily and most particularly attach to later-nineteenth century narratives, such work reassesses and differently valorises canonised maritime authors: Joseph Conrad's famous literary mappings of South East Asian seas as spaces defined by an agon of masculine moral duty and disciplined skill – the 'craft' – of the European seafarer is an important and productively troubling reference point in this line of scholarship.

The scope of 'literature' is also expanding through attention to ocean work as well as labour. Rachel Carson's environmental work has long been feted for its lyrical as well as persuasive quality: but recent decades have seen a more specific and sustained surge in the realisation of her marine science/natural history as a form of creative literature (Carson, 1941, 1951, 1955). This reading is both cause and effect of the rise and consolidation of 'ecocritism' and 'ecopoetics' in the twenty-first century. (This second term is sometimes bound to experimental nature poetry, and sometimes more broadly deployed [Hume and Osborne, 2018].) These rubrics have now framed a critical mass of companions, handbooks, readers, special issues, and new journals: Routledge's own 2019 *Handbook of Ecocriticism and Environmental Communication* is one marker of the establishment of these ways of reading, which – necessarily – includes a chapter on Carson (Beudel, 2019). Carson's sea work is also being read as a pre-history of the burgeoning genre of 'new nature writing', which is linking academic work to a wider public understanding of the seas as a measure of the environmental losses of the Anthropocene: Philip Hoare's work is absorbing and exemplary in its combinations of personal and natural histories (2008, 2013). But Carson's work is also distinct from this self-conscious and ranging genre. Her style is defined by a combination of close description and rhetorical inhibition; she is notably unselfconscious in reaching for fairytale images or traditionally feminine domestic metaphor to define and embellish scientific observation of the non-human and more-than-human world. And her style is remarkable for its combination of determined subjectivity and scrupulous self-effacement. She is not interested in uninflected objectivity, nor in a language of discovery and mastery, nor in herself as a discoverer. The place of Carson in the humanities is therefore fraught: while essential to the history of environmentalism as a form of feminist activism, her writing can seem less a refusal of masculine perspective and more a diminishment of her own position (Gaard, 2018).

There are both continuities and contrasts between Carson's work and that of current feminist environmental activist-artists who are becoming so important for literary critical research and teaching of ocean space. In her writing on/with whales, Rebecca Giggs finds a tone that is subjective but also concentrated on giving-over subjectivity to the natural world (2015). Kathy Jetñil-Kijiner writes of the Marshall islands as they are impacted – but not inexorably defined – by a past of imperial racism, nuclear testing and a future of rising sea levels. Her work and her concentration on women's positions, viewpoints and bodies implicitly both critiques and segues with Carson's ocean (Jetñil-Kijiner, 2017).

## The ocean as method: How we read

*The Enchafèd Flood: Or, the Romantic Iconography of the Sea* collects three lectures delivered by the poet W. H. Auden in 1949: 'The Sea and the Desert', 'The Stone and the Shell', and 'Ishmael-Don Quixote'. The lectures take us from the Classical seas of a Homeric Southern Europe to the Biblical world-ocean to 'The Seafarer' – a key and inaugural work of Anglo-Saxon literature – to the 'purely negative' of Shakespeare's sea: 'The Enchafèd Flood' is a phrase that describes the sea in Act 2, Scene 1 of Shakespeare's *Othello*, and captures something of Auden's seductively certain thesis that the whole of pre-Romantic literature can be narrated through the ocean as a "state of barbaric vagueness and disorder out of which civilization has emerged and into which, unless saved by the efforts of Gods and men, it is always liable to relapse" (Auden, 1967: 11, 6). Auden is most interested in defining epochs of literature to prove a singular pre-history to the Romanic sea. In his interpretation, the Romantic writers of the eighteenth and nineteenth centuries – early Samuel Taylor Coleridge to later Lord Byron and including an American tradition to which Melville is key – present an aesthetically and existentially revolutionary ocean space. Under their influence, the ocean as an Anglo-American and European *affect* shifts from negative vagueness to *the* scene of self-knowledge: from a scene of barbarism to the siting of the individuated self, and of exquisite aloneness/loneliness. For Auden, the Romantics inaugurated the modern literary ocean because they understood that the sea is "where the decisive events, the moments of eternal choice, of temptation, fall and redemption occur. The shore life is always trivial" (Auden, 1967: 13). In Auden's analysis, the Romantics are most revolutionary in realising that this oceanic self-awareness doesn't resolve anything: the sea continues – constant and eternal, unstable and temperamental – and so the overwhelming understanding that *this* is the true condition of man both urgently matters and is inconsequential. (This is the ancient mariner's insistence, and Ahab's madness.) Read today, the lectures can seem arcane and overly-selective: an example of a tradition of literary history-making through the valorisation of writing by white men. But *The Enchafèd Flood* is more interesting as an expression of Auden's own post-war, late-modernist despair than as a construction of a canon of writers. He is arguably less entailed to imperialism (he is not really that interested in the ocean as 'frontier') than he is attuned to the de-territorialisations of the ends of empire-making. In this sense, Auden's book – particularly when read alongside his poetry of exile – is prescient because it asks us to read stories of the ocean as the stories that speak most urgently to our age. And in the decades since Auden gave his lectures, going offshore to find the beginning of literary histories has become a common – a classic, even – critical move.

Daniel Defoe's *Robinson Crusoe* (1719) is a classic beginning in accounts of narrative fiction in English in its modern form. This story of failed ocean crossing, ship-wreck, settlement and enslavement begins influential accounts of the novel genre as a binding force of modern capitalism and individualism (Watt, 1957). This book also starts classic accounts of the genre as the crucial imaginative force of imperialism: this novel and imperialism are, Edward Said persuades, unthinkable without each other (Said, 1993). But there is also a long tradition of bumping Crusoe from his starting position; not by bringing the novel back onshore, but by turning to the watery spaces of other writers. Aphra Behn's work – and particularly her classic of indigenous noble savagery, *Oroonoko* (1688) – can be understood as a more importantly nascent novel. And recent work on *The Interesting Narrative of Olaudah Equiano, or Gustavus Vassa, the African* (1789) presents another kind of challenge to the pre-eminence of Defoe's novel (Equiano, 2001). A story of capture, enslavement, manumission, seafaring, and conversion to Christianity, *The Interesting Narrative* combines elements of spiritual autobiography and Igbo life-writing, but also deploys novelistic strategies and oceanic imagery to abolitionist effect. As Laura Doyle argues, Crusoe's story of loss and recuperation of liberty does not properly belong to him, but to the enslaved persons of the Atlantic trade and their ancestors. She reads Equiano's narrative as retrieving this Atlantic liberty narrative and claiming ocean space as a contingent site of coming to 'freedom' through self-hood, labour, and trade (Doyle, 2008). These days, the novel genre is only becoming more embedded and influential: so to locate its generic beginnings in the writings of a freed slave is of more than narrow disciplinary interest, and connects to a longer history of arguing over the status and meaning of Crusoe beyond literary critical studies.

Crusoe is probably better known as a symbolic idea than as a literary invention. As Karl Marx noted in *Capital* (1867), the figure of Crusoe as an analogue of the self-made man has iconic resonance within classical economics (Marx, 1990: 169). He is also a key reference in debates about 'natural man', and even appears as an instructive model in the record of debates between drafters of the United Nations Declaration of Human Rights (Whyte, 2014). And Crusoe as an analogue of the sovereign refuses to leave the realm of political and legal philosophising (Derrida, 2011, 2017). Crusoe continues to inflect discourses that shape current hegemonies: so it isn't easy or always wise to dismiss him – boring, fastidious, and unsexy as he is – from an understanding of the literary ocean-scape. But Crusoe's energising and providential ocean world is itself a world away from the stories of crossings and ocean wreckage that are importantly shaping literary culture right now. Defoe's novel can help us to understand that these ocean spaces are part of an overwhelming and continuous political economy: but the fame of this novel can also over-burden literary analysis. Forgetting Crusoe's ocean space – his passage from exile to sovereignty – can be particularly important to reading ocean space within the literatures of twentieth- and twenty-first-century exile. Literature about and by refugees is presenting ocean space as a site of physical and existential loss that is radically discontinuous with Crusoe's story of suffering and redemption, loss and coming to self-hood.

In his story of a becalmed boat of Vietnamese refugees, *The Boat* (2008), Nam Le focuses the brute fact of bodily sensation at sea – thirst, exposure, sickness – to generate reader 'empathy': a word and idea that are of central and specific importance to his ambitions as a writer. At the same time, he insists on the strangeness of strangers, describing his stricken characters through distancing metaphors, and so refusing the global reader an easy relatability, an easy idea of universal humanity. In this way, he uses the taut discipline of the short-story form to recognise human suffering, and to invite this recognition as a political and not a merely a sentimental/self-indulgent act of reading. Other authors are finding it necessary to break and surpass expected of literary form in order to convey the experience of seeking and not-finding sanctuary across oceans. Over the past 30 years – and particularly since 2001 – the Australian government's offshore detention regime has generated a legal geography between Indonesia, 'Australian Indian

Ocean Territory', Nauru, and Papua New Guinea (PNG). This circumscribes a sub-region of ocean space that is juridically and politically stark, administratively complex, and enforcement heavy. Written from PNG's Manus Island, and opening with a harrowing account of his attempt to cross by boat from Indonesia to sanctuary in Australia, Behrouz Boochani's *No Friend but the Mountains* (2018a) brings to account the intricate cultures of violence that constitute and confine this heavily bound ocean space. The author has used "allegory" and "novel" to describe the book (Boochani, 2018b). His work has also been glossed as "prison literature", and as "poetic manifesto" (Flannagan, 2018; Surma, 2018). For Boochani's translator, the book is a "fusion of journalism, political commentary and philosophical reflection with myth, epic poetry and folklore": and so, he argues, *No Friend but the Mountains* must finally be understood as an "anti-genre" (Tofighian, 2018). Boochani's book connects to a long tradition of fusing and re-fusing literary genre in order to narrate oceans as spaces of unfreedom and freedom; in order to narrate ocean spaces as an enquiry into the "necropolitics" of the nation state, and what liberty might, can and can't mean (Mbembe, 2003).

Alexis Wright's *Carpentaria* (2006) also breaks the novel form to produce a new idea for the genre. This undoing and redoing is critically achieved by writing indigenous Australian geography as a complex relationship between fresh and salt water. Across five hundred pages, *Carpentaria* conveys shifting simultaneities of deep time, human history, and present tense through the lives of rivers, estuaries and oceans that are themselves capable of malicious whimsy or consoling intent or retributive justice. Terribly funny and epically sad, the violent floods, wreckages, lost-at-sea and washed-ashore storylines that traverse this vast calling-out of Australia's past and present of racialised and gendered violence confronts the coming-of-age beach narrative which has been such an alluring and reassuring aspect of the Australian literary scene in the post-war period. Although in the twenty-first century, ocean space in Australian literature has been more fully bound to stories of masculine violence than triumph, particularly through the popular and critically acclaimed novels of Tim Winton and Robert Drewe, both prolific authors and famed for writing about coastal and ocean spaces. Winton's surfing novel *Breath* (2008), for example, is significantly structured by the visceral experience of four different waves. This *bildungsroman* of young men set on conquering ocean space tells a story that defies a tradition of 'pioneering' settler suffering and triumph, and instead concentrates on the demands and mental and emotional debilitations of discourses of Australian masculinity. In Winton's work, the surf is both the euphoria and the damage of being a man. And across his oeuvre, Winton has increasingly used the novel form to join this kind of damage to environmental – particularly marine – destruction. His twenty-first-century novels are tuned to the impact of man on the natural environment: but his narratives are also often compelled by the idea of a connection to marine life as a means of personal and political, individual and community (if not national) redemption. For Winton, this narrative also involves a concomitant redemption of sharks from their status as the predator *par excellence* within narrative fiction: *Dirt Music* (2001), for example, includes an extraordinary scene of a man dance-wrestling with great whites off the North West coast of Australia.

## The inhabited ocean: Literary animals and other matter

Along with other disciplines and other arts, critical and creative literature is increasingly focused on the relationship between humans and animals. Literary studies is being redefined by a more concerted interest in the potential of fiction, poetry and drama to exceed this relationship and to reach beyond anthropocentrism. Recent readings and re-writings of one of the most canonised works of English literature, Samuel Taylor Coleridge's *The Rime of the Ancient Mariner* (1798/1834), has shifted from the "soul in agony" and the "LIFE-IN-DEATH" of the men and ghosts of the becalmed ship, and from the "a sadder and a wiser" interlocutor/everyman (Coleridge, 1999: 47, 41, 74–75). The poem is now being more attentively read through the albatross, the multitudinous "slimy things", and the shining "water-snakes" (Coleridge, 1999: 36–37, 46–47). More expansively, but also even more minutely, ocean studies as the study of race and biopower is being reimagined in conversation with an ecocritical concentration on the non-human and more-than-human; and on the human within the non-human. This is working alongside feminist commentaries on the sea – like Astrida Neimanis's *Bodies of Water: Posthuman Feminist Phenomenology* (2017) – which are compelled by the correspondence between the salt and water of the human body, and the salt and water of the sea. In his "black ecopoetics gone offshore", Joshua Bennett draws upon the increasing attention – in both literary criticism and creative writing – to ocean space as molecular: and thus as a continual integration and reconstitution of the matter of drowned slaves (Bennett, 2018: 103). In his readings, sharks both embody – literally – the potential of another way of human and non-human life, but also symbolise modernity's continuing threat to Black lives.

As the ocean's smallest non-human particles are becoming increasingly important to our understanding of ocean space, so too its largest are coming into more certain literary focus. A map of English literary representations of marine life would produce a dense and curious picture, including attempts to inhabit, to ventriloquise, but also to acknowledge the complete otherness of the animal perspective. This map would also noticeably demonstrate that there are many literary whales and other cetaceans that are not *Moby Dick*. Marine mammals are rising as a topos of literary creativity. They are also of increasing significance in the definition of regional ocean studies as a literary project. In new work on 'The Oceanic South' (2019), Charne Lavery and Meg Samuelson turn to novels about whales – including Witi Ihimaera's *The Whale Rider* (1987) and Zakes Mda's *The Whale Caller* (2006) – to take forward a new conception of an ocean region that exceeds and challenges the basins model, and offers a cultural formation that speaks to embedded political discourses of Northern wealth and Southern poverty. Amitav Ghosh's *The Hungry Tide* (2004), the elusive Irrawaddy dolphin is central to the apprehension of the tide country of the Sundarbans as a unique sea and fresh water space. In Tahmima Anam's *The Bones of Grace* (2016), ocean space is a desert in Pakistan that holds the fossils of an ancient whale, *Ambulocetus natans*. This narrative in some ways continues and in other ways revokes a tradition of writing about ocean space through reference to the desert. In Auden's thesis of the 1950s, the desert is abstracted into a scene of extreme deprivation, a sheer background to the man as poet, the man as deserted and desolate. In Anam's novel, the desert was – in fact – the sea; it is a living and inhabited site of both modern and archaic violence, and of intricate cross-cultural encounter. It embeds the woman as scientist, the woman as thinking and connected. This novel decolonises the desert by narrating it as an ocean space.

Like the desert, the ocean has been imagined through both metaphorical and material correspondence with space: the relation between the ocean *and* space is also a prominent literary trope. This connects to international legal constructions that impress the continuity between non-state waters and airspace. The laws of piracy are particularly intriguing on this point: as the United Nations Convention on the Law of the Sea (UNCLOS) tells us, acts of illegal violence, detention or depredation that occur on ships or on aircraft have the same legal status (UN, 1982: Article 101–07). This legal instruction meets an intriguing thread in the literary history of ocean space as a fervent writing and re-writing of pirate fictions. *Treasure Planet* (Clements and Musker, 2002) is a cult classic of this genre. This film reconceives Robert Louis Stevenson's popular classic of high seas and coastal island adventure, *Treasure Island*, to critical decolonising effect (Stevenson, 1883). The film draws out the moral and ethical complexity of Long John Silver and presents outer space as a scene of competing hegemonies and resistance movements. It invites the boy-girl viewer to be alert and responsive in ways that implicitly criticise the original story: or, arguably, draw out a submerged anti-imperial narrative. As scholars of Victorian literature have explored, the original *Treasure Island* seems most obviously to partake in a flurry of oceanic pirate adventures written for boys that provided an imaginative force to the newly violent consolidations of the British empire of the 1850s onwards. In this later nineteenth/early-twentieth century genre, ocean space appears as a playground: it is the scene through which boys can figure themselves as becoming good – and indefatigably brutally boyish – administrators of empire (Deane, 2011).

*Treasure Island* – alongside *Robinson Crusoe* – is perhaps one of the most re-written and filmed texts in literary history. Some of these re-writings are true to the plot but eschew the aesthetics of the original: for example, there is little room for ocean space in the scenes crowded with puppets in *Muppet Treasure Island* (1996). More generally, literary pirates are remarkable for the intensively decontextualised energy with which they sail around the global popular imagination, away from the stories that first brought them into being. The total and fast-paced mash-up of pirates re-drawn from historical record, epic, myth, fable, and fiction have shaped the successes of *The Pirates of the Caribbean* (Marshall, 2011; Ronning and Sandberg, 2017; Verbinski, 2003, 2006, 2007) films and *Black Sails* (Steinberg and Levine, 2014–2017). In these works, ocean space is often hyper-materialised: familiar figures from ocean lore – the Kraken, Davey Jones, Calypso, Blackbeard – are debunked and debauched as we are shown every sticky tentacle, barnacled hand, pore of watery skin, hair of bloody beard. The mix of is both funny and disturbing. This is not least because of the consistency with which ocean space itself is both hyper-materialised – monster waves and overly-bright rippling vistas – but also inadvertently unnatural. For all the wizardry of the film industry, it seems that CGI cannot quite fix a fictional sea to convincingly behave like a real sea. The ocean of fictional films is often narratively archaic, but also oddly technologically arcane. It is not as distant from the painted backdrops and awkwardly filmed waters of popular 'Swashbucklers' of the 1930s and 40s (Rennie, 2013). And this is not the only genre in which the ocean of pirate literature appears as an odd combination of the natural and unnatural. Pirates started to become significant to utopian fiction in the early eighteenth century with the publication of the influential *A General History of the Pyrates* (Johnson, 1724).

In the utopian genre, ocean space is the physical boundary and temporal absence that allows the construction of the island as a perfect form of governance. But then the crossing of the ocean by an interlocutor brings historical time back into play, and the utopia inevitably fails. Literary utopias are often less a dream of perfect governance, and more often a troubled exploration of ideology, human nature, and the meaning of equality. (Jameson, 2005). Through the many editions of a *General History* that have been published across the centuries, the story of the French pirate Captain Misson's 'Libertalia' has taken on a significant resonance beyond literary studies and in utopian thought. This is not unusual for this book: the version of famous Anglo and Anglo-American (and occasionally European) pirates that this compendium gathers has not only provided the key source for other writers of pirate fiction but has also directed historical and archaeological enquiry. As I have traced elsewhere, readings of 'Libertalia' have been particularly strong in utopian and anarchist thought, but often miss the critique, the irony, and – crucially – the denouement of this story. It is hard to make this narrative of a piratical community of radical racial and social equality and harmoniously shared property sustain a real utopian manifesto because – of course – visitors arrive, and the community fails. The utopia can't exist in history. However, this story has been strained to mean that if it *was* to exist it *might* exist in the Indian Ocean, where Misson's Libertalia is set (Jones, 2012).

## The Indian Ocean as/in literature

There is a strong European tradition of imagining Indian Ocean space as an arcadian as well as utopian past and potential. In the infamous novel *Paul et Virginie* (1788), Mauritius is a natural idyll that is both consummated and ruined in the figure of a girl offshore, drowning in the heavy weight of her clothes (Saint-Pierre, 1819). In a crazy little story first published by HG Wells in 1894, 'Aepyornis Island' (2000) is a small atoll in the Western Indian Ocean on which an environmental idyll between a man and a giant bird – an interspecies paradise that works on the Indian Ocean myth of the Roch, and the natural history of the *Aepyornis maximus* – turns sour and ends in a prescient story of species extinction. Such literary moments add up to an odd lineage of utopias, arcadias, dystopias that traces through a more general recognition of the Indian Ocean as a space defined by aspirations for alternative communities and connections, and new structures of governance.

As Isabel Hofmeyr helpfully summarises, "whether rooted in pan-Islamism, Hindu reformism, pan-Buddhism, ideas of Greater India, or ideals of imperial citizenship, the Indian Ocean offers a rich archive of transnational forms of imagination" (Hofmeyr, 2012: 585). But this ocean also bears its history as a scene of unfree labour and a 'penal zone', defined by 'a necklace' of island prisons stretching between South African's Robben's Island, India's Andamans, to Australia's Tasmania (Hofmeyr, 2012: 587–588). These histories of both expansive dreaming and incarcerating fear are been borne by a literary sensibility of accumulative and ambiguous 'belonging' (Hofmeyr, 2012: 589). This is most often accessed through the many globally circulating novels of Amitav Ghosh and Abdulrazak Gurnah. And the success of these writers advertises the Indian Ocean as itself a kind of novel genre of past and current work. This genre is deployed in important ways by women writers. Novels such as Lindsay Collen's tough and fragmenting *Mutiny* (2002) concentrate on the Indian Ocean as a continuing site of imprisonment and denial that impacts in specific ways on resistant women and women as political beings. Other writings – such as Yvonne Adhiambo Owuor's *The Dragonfly Sea* (2019) – find a lush and fluid prose in the oceanscapes off the East coast of Africa and acknowledges international dreaming as the purview and right of girls and women.

## The Atlantic Ocean as/in literature

Women's writing is also transforming the Atlantic as a creative text. Dionne Brand continuously turns and returns to the sea – as material space of her past and present, as emotional state, as symbol of lost family history. Experiences and images (sometimes sustained, sometimes fleeting) of the ocean segue literary criticism, poetry, and life-writing. In *A Map to the Door of No Return* (2001), the narrative non-fiction is both broken and culminated in her 'Ruttier for the Marooned in the Diaspora'. The idea of the ruttier – an oral chart of the ocean – turns into a poetics of orality, as sound and sibilance determine word choice, sailing becomes slipstreams, and the word 'marooned' gathers weight and meaning as both an historical and present tense (Brand, 2001: 215). Christina Sharpe also turns continually back to ocean space in her work on 'the wake' in all its meanings: as vigil, as sleeplessness, as the disruption of water that tails a ship. This etymological 'wake work' – this narrowing of ocean space into this idea/feeling/image – inspires a combination of life-writing, close critical reading, political commentary, and theoretical proposition that can feel like poetry, and sometimes like a manifesto. *In the Wake: On*

*Blackness and Being* (2016) surfaces gendered violence and the disenfranchisement of Black women as the endlessly repeating future of globalisation: one of her sharpest points of critique focuses on Sekula's dismissive representation of a "former mother" in his much-lauded film about maritime labour and the ocean of late capitalism (Sharpe, 2016: 26–34).

## The Pacific Ocean as/in literature

In the UK and the US, the Atlantic has often dominated the study of literary oceans. This concentration was given important new impetus through Paul Gilroy's *The Black Atlantic: Modernity and Double-Consciousness* (1993). His analysis of music, literature, painting and biography bound modernity to Atlantic studies: it made modernity unthinkable without the Black Atlantic. Similar moves in relation to other oceans have more recently gathered momentum. Epeli Hau'ofa's 'Our Sea of Islands' – also first published in 1993 – has been key to both reframing and expanding the reach of Pacific studies. Across the humanities and social sciences, this work has displaced Cook's navigations and French ethnographies.

Hau'ofa's essay and his overall sense of Oceania are crucially literary. In 'Our Sea of Islands', he often reaches for oral creation stories and fables to express what Bill Ashcroft has described as a 'utopianism'. Distinct from the utopia, 'utopianism' is a kind of descriptive/prescriptive temporality through which imagining what *was* becomes a statement of what *will be* (Ashcroft, 2012). In Jini Kim Watson's analysis, Hau'ofa's short stories – which concentrate on the hilarious and bitter ironies and oppressions of regional aid cultures – surface long entrenched alignments of territory and security that threaten utopianism as a Pacific modernity (Watson, 2015).

## Comparative sea studies as/in literature

The prominence of Atlantic studies is also being productively complicated by comparative sea studies. Paradigms developed in one sea space are crossing and inspiring literary configurations of other ocean spaces in ways that don't succumb to an easy idea of a global ocean, but rather bring alternative conceptualisations of regionality into clearer focus. As Elizabeth DeLoughrey's work across Atlantic and Pacific literatures evidences, Kamau Brathwaite's 'tidalectics' (a term worked up by the poet in the 1980s and 90s) helps us to read for both parallels and discontinuities between Caribbean and Pacific island poetics of indigeneity and diaspora. 'Tidalectics' is an encompassing poetics that nonetheless avoids universalisms (Brathwaite, 1983, 1999; DeLoughrey, 2010, 2017, 2019). It emerges in manifesto-like fragments and lyrical images across various. At times it is an oppositional idea: it reproaches a Eurocentric 'dialectics', although at the same time invokes and pays homage to a Caribbean Marxist intellectual history. At other times, it is an elliptical image – a woman sweeping sand from a step – that evokes repetition with difference, a circularity of habit and daily existence that is not about linearity and progress but is nonetheless dynamic. The 1980s and 90s also saw the development of *Antillanité* and then *créolité* by Jean Bernabé, Patrick Chamoiseau, Raphaël Confiant, and Édouard Glissant's These resolutely regional declarations initially arise from a rejection of *négritude's* discourses of universalising black culture and return to Africa. But in promoting a poetics of the local/obscure against monolithic languages, these ideas resonate through other geographies, and enjoin a poetics of *créolité* in other island chains and archipelagos, particularly in the Western Indian Ocean. Moving between abstract recondite philosophical statement, historical situation, and a materially marine language of rock, beach and sea, Glissant enacts his claim to a right to opacity across writings gathered into *Poétique de la Relation/ Poetics of Relation* (1990/1997). The book's epigraph is a quote from Brathwaite: "The unity is sub-marine".

Underwater, undersea, unfathomable: this is the material symbol of "nous réclamons le droit à l'opacitié" (Glissant, 1997: 189–194).

While for Glissant the sub of the marine is valued as an opacity, other literary work is bringing the 'sub' – from coral reef shallows to lightless trenches through the five-layered depths – into greater definition. And again, ground-breaking writing finds it necessary to combine creative and critical genres: James Hamilton-Paterson's *Seven-Tenths: The Sea and its Thresholds* (1992) is by turns environmental and geopolitical journalism, literary criticism, and fiction. It also reminds us of a longer history of popular writing that has been alert to the ocean as a primary scene of the devastations of the Anthropocene: this is, of course, not a twenty-first century revelation. It has been in writings aimed at a broad readership for longer than is often realised. Similarly, a fascination with the sea is often a fascination with ice, and this new focus is revealing a critical potential to re-read centuries of creative literature as an address to anthropocentric climate change (Dodds, 2018).

## Conclusion: Geography as literature

This is the final verse of Elizabeth Bishop's 1948 poem 'At the Fish Houses':

> If you tasted it, it would first taste bitter,
> then briny, then surely burn your tongue.
> It is like what we imagine knowledge to be:
> dark, salt, clear, moving, utterly free,
> drawn from the cold hard mouth
> of the world, derived from the rocky breasts
> forever, flowing and drawn, and since
> our knowledge is historical, flowing, and flown.
>
> *Bishop, 183: 66*

We might read this as an example of a dated anthropocentric approach to the sea, insufficiently attentive to the non-human and the more-than-human in its limited concentration on the poet's subjectivity. Alternatively, we might read it as a fine negotiation between ocean-space-as-material and ocean-space-as-abstraction that delimits our understanding of what constitutes knowledge, and of what knowledge feels like. After all, the reason the ocean is so common and powerful a metaphor is precisely because it is so intensively material. Read back into the poem's setting and alongside its central figure – a fishing port in Nova Scotia, 'an old man sits netting' – we might also consider how the poem connects the taste of knowledge to the ocean as an actual place in time. "Knowledge is historical"; but the poem understands this through an experience of knowledge as visceral and material (Bishop, 1983: 66). Once again we encounter literary oceans as literary time.

To approach literature as geography is to encounter long established genres. The meadow is 'the pastoral', the remote castle is 'the gothic', the nation-state is 'the novel'. 'The world' is also a long-established genre of literature. In early accounts, 'world literature' maps mostly-European works that have circulated widely, have entered the global imagination, and inflect everyday discourse. This idea of world literature winds back to the seafaring stories and figures of the Greek classical tradition, and their many re-writings. Over two millenia after it was first told, Homer's *The Odyssey* still has extraordinary currency. This story of a charismatic and resourceful, ethically obtuse, temperamental and distractible seafarer-king and his island crossings has activated some of the most famous writings in English.

James Joyce's *Ulysses* (1922) is also central to a growing interest in the modernist sea more generally (Feigel and Harris, 2011). (Focusing on Virginia Woolf's Atlantic or Katherine Mansfield's Pacific tells us that this literary history, can only be fully apprehended offshore.) And beyond literature, Odysseus has become an archetype of risky journeying, although this regular noun – 'odyssey' – brings an arcane sense to current situations. Refugee crossings of the Mediterranean are being framed as an 'odyssey' (Kingsley, 2016). The word is used to describe a gruelling and dangerous journey: but it also infers something that should not be happening in our time. Again, fiction eventually and inevitably takes us into the ocean as a site and scene of the unfree under global capitalism. And this is perhaps one way in which older and newer ideas of world literature join up.

Now, 'world literature' is being energetically reconceptualised as a decolonising idea. In recent work, literary critics such as Pheng Cheah (2016) are proposing that a world literature is more properly conceived as a practice of cultural production that both expresses and resists the temporalities of capitalism. It may not be a literature that circulates around the world or has entered a global consciousness: it is rather definitively a literature that is alert to the hegemonies of world circulation and capitalist time. As this chapter traces, ocean space as a literary project creates and expands, refutes and subverts worlds. It is a worlding of regions, depths, surfaces that brings interrogative strength to literature as a critical and creative practice.

## References


Anam T (2016) *The Bones of Grace*. Edinburgh: Canongate.

Ashcroft B (2012) Introduction: Spaces of utopia. *Spaces of Utopia: An Electronic Journal* 2(1): 1–17.

Auden WH (1967) *The Enchafèd Flood: or, the Romantic Iconography of the Sea*. New York, NY: Vintage Books.

Awuor YA (2019) *The Dragonfly Sea*. New York, NY: Knopf.

Behn A (1688) *Oroonoko, or, The Royal Slave. A True History*. London: Will Canning.

Bennett J (2018) Beyond the vomiting dark: Towards a black hydropoetics. In: Hume A and Osborne G (eds) *Ecopoetics: Essays in the Field*. Iowa City, IA: University of Iowa Press, 102–117.

Beudel S (2019) Rachel Carson. In: Slovic S, Rangaragan S and Sarveswaran V(eds) *Routledge Handbook of Ecocriticism and Environmental Communication*. London: Routledge, 265–276.

Bishop E (1983) At the Fish Houses. In: Bishop E (ed) *The Complete Poems 1927–1979*. New York, NY: Farrar, Straus and Girroux, 64–66.

Blum H (2008) *The View from the Mast-Head: Maritime Imagination and Antebellum American Sea Narratives*. Chapel Hill, NC: University of North Carolina Press.

Boochani B (2018a) *No Friend But The Mountains*. Sydney: Picador.

Boochani B (2018b) Manus prison poetics/our voice: Revisiting "A letter from Manus Island", a reply to Anne Surma. *Continuum: Journal of Media and Cultural Studies* 32(4): 527–531.

Brand D (2001) *A Map to the Door of No Return: Notes to Belonging*. Toronto: Vintage.

Brathwaite K (1983) Caribbean culture: Two paradigms. In: Maritini J (ed) *Missile and Capsule*. Bremen: Universität Bremen, 9–54.

Brathwaite K (1999) *Conversations with Nathanial Mackey*. New York, NY: We Press.

Buell, L (1995) *The Environmental Imagination: Thoreau, Nature Writing, and the Formation of American Culture*. Cambridge, MA: Belknap Press of Harvard University Press.

Carson R (1941) *Under the Seawind: A Naturalist's Picture of Ocean Life*. New York, NY: Simon and Schuster.

Carson R (1951) *The Sea Around Us*. Oxford: Oxford University Press.

Carson R (1955) *The Edge of the Sea*. Boston, MA: Houghton Mifflin.

Cheah P (2016) *What Is a World? On Postcolonial Literatures as World Literature*. Durham, NC: Duke University Press.

Clements R and Musker J (2002) *Treasure Planet*. United States: Buena Vista Pictures.

Cohen M (2010) *The Novel and the Sea*. Princeton, NJ: Princeton University Press.

Coleridge ST (1999) *The Rime of the Ancient Mariner*. London and New York, NY: Bedford St Martin's Press.

Collen L (2002) *Mutiny*. London: Bloomsbury.

Deane B (2011) Imperial boyhood: Piracy and the play ethic. *Victorian Studies* 53(4): 689–714.

Defoe D (1719) *The Life and Strange Surprising Adventures of Robinson Crusoe of York, Mariner*. London: William Taylor.

Derrida J (2011) *The Beast and the Sovereign, Volume I* (trans. G Bennington) Chicago, IL: University of Chicago Press.

Derrida J (2017) *The Beast and the Sovereign, Volume II* (trans. G Bennington) Chicago, IL: University of Chicago Press.

DeLoughrey E (2010) *Routes and Roots: Navigating Caribbean and Pacific Island Literatures*. Hawaii, HI: University of Hawaii Press.

DeLoughrey E (2017) Submarine futures of the Anthropocene. *Comparative Literature* 69(1): 32–44.

DeLoughrey E (2019) Toward a critical ocean studies for the Anthropocene. *English Language Notes* 57(1): 21–36.

Dodds K (2018) *Ice: Nature and Culture*. London: Reaktion Books.

Doyle L (2008) *Freedom's Empire: Race and the Rise of the Novel in Atlantic Modernity, 1640–1940*. Durham, NC and London: Duke University Press.

Equiano O (2001) *The Interesting Narrative of Olaudah Equiano, or Gustavus Vassa, the African*. New York, NY: Broadview Literary Texts.

Feigel L and Harris A (eds) (2011) *Modernism on Sea: Art and Culture at the British Seaside*. Oxford: Peter Lang.

Flannagan R (2018) Forward. In: Boochani B *No Friend but the Mountains*. Sydney: Picador.

Gaard G (2018) Feminism and Environmental Justice. In: Holifield R, Chakraborty J and Walker G (eds) *The Routledge Handbook of Environmental Justice*. London: Routledge, 74–88.

Ghosh A (2004) *The Hungry Tide*. London: Harper Collins.

Giggs R (2015) Whale Fall. *Granta* 133. Available at: https://granta.com/whale-fall/

Gilroy P (1993) *The Black Atlantic: Modernity and Double Consciousness*. London: Verso.

Glissant É (1997) *Poetics of Relation* (trans. B Wang). Ann Arbor, MI: The University of Michigan Press.

Hau'ofa E (1993) Our sea of islands. In: Waddell E, Naidu V and Hau'ofa E (eds) *A New Oceania: Rediscovering our Sea of Islands*. Suva: University of South Pacific, 1–16.

Hamilton-Paterson J (1992) *Great Depths: The Sea and its Thresholds*. London: Random House.

Henson B (1996) *Muppet Treasure Island*. United States: Buena Vista Pictures.

Hume A and Osborne G (eds) (2018) *Ecopoetics: Essays in the Field*. Iowa City, IA: University of Iowa Press.

Hoare P (2008) *Leviathan, or the Whale*. London: Fourth Estate.

Hoare P (2013) *The Sea Inside*. London: Fourth Estate.

Hofmeyr I (2012) The complicating sea: The Indian Ocean as method. *Comparative Studies of South Asia, Africa and the Middle East* 32(3): 584–90.

Ihimaera W (1987) *The Whale Rider*. North Shore: Ruapo.

Jameson F (2005) *Archaeologies of the Future: The Desire Called Utopia and other Science Fictions*. London: Verso.

Jetñil-Kijiner K (2017) *Iep Jāltok: Poems from a Marshallese Daughter*. Tucson, AZ: The University of Arizona Press.

Johnson C Cpt. (1724) *A General History of the Robberies and Murders of the Most Notorious Pyrates, and also, Their Policies, Discipline and Government*. London: T Warne.

Jones S (2012) Literature, geography, law: The life and adventures of Capt. John Avery (circa 1709). *Cultural Geographies* 19(1): 71–86.

Joyce J (1922) *Ulysses*. Paris: Sylvia Beach.

Kingsley P (2016) *The New Odyssey: The Story of Europe's Refugee Crisis*. London Guardian Faber Publishing.

Le N (2008) *The Boat*. Sydney: Penguin.

Linebaugh P and Rediker M (2000) *The Many-Headed Hydra: Sailors, Slaves, Commoners, and the Hidden History of the Revolutionary Atlantic*. London: Verso.

Marshall R (2011) *Pirates of the Caribbean: On Stranger Tides*. United States: Walt Disney Studios Motion Pictures.

Marx K (1990) *Capital: A Critique of Political Economy Volume I* (trans. B Fowkes). London: Penguin.

Mbembe A (2003) Necropolitics (trans. L Mientjes) *Public Culture* 15(1): 11–40.

Mda Z (2006) *The Whale Caller*. Johannesburg: Penguin.

Melville H (1851 [1994]) *Moby Dick*. London: Penguin Books.

Mumford L (1929) *Herman Melville*. New York, NY: The Literary Guild of America.

Neimanis A (2017) *Bodies of Water: Posthuman Feminist Phenomenology*. Sydney: Bloomsbury Academic.

Rediker M (1987) *Between the Devil and the Deep Blue Sea: Merchant Seamen, Pirates, and the Anglo-American Maritime World, 1700–1750*. Cambridge: Cambridge University Press.

Rennie N (2013) *Treasure Neverland: Real and Imaginary Pirates*. Oxford: Oxford University Press.
Ronning J and Sandberg E (2017) *Pirates of the Caribbean: Dead Men Tell No Tales*. United States: Walt Disney Studios Motion Pictures.
Saint Pierre B (1819 [1788]) *Paul et Virginie* (trans. HM Williams). London: John Sharpe.
Said E (1993) *Culture and Imperialism*. London: Vintage.
Samuelson M and Lavery C (2019) The Oceanic South. *English Language Notes* 57(1): 37–50.
Sharpe C (2016) *In the Wake: On Blackness and Being*. Durham, NC: Duke University Press.
Steinberg JE and Levine R (2014–2017) *Black Sails*. United States: Starz Original.
Stevenson RL (1883) *Treasure Island*. London: Cassell and Company.
Surma A (2018) In a different voice: 'A letter from Manus Island' as poetic manifesto. *Continuum: Journal of Media and Cultural Studies* 32(4): 518–526.
Tanner T (1988) Introduction. In: Melville H *Moby Dick*. Oxford: Oxford World's Classics.
Tofighian O (2018) Behrouz Boochani and the Manus Prison narratives: Merging translation with philosophical reading. *Continuum: Journal of Media and Cultural Studies* 32(4): 532–540.
United Nations (UN) (1982) *United Nations Convention on the Law of the Sea (UNCLOS)*. Available at: www.un.org/Depts/los/convention_agreements/texts/unclos/unclos_e.pdf
Verbinski G (2003) *Pirates of the Caribbean: The Curse of the Black Pearl*. United States: Walt Disney Studios Motion Pictures
Verbinski G (2006) *Pirates of the Caribbean: Dead Man's Chest*. United States: Walt Disney Studios Motion Pictures
Verbinski G (2007) *Pirates of the Caribbean: At World's End*. United States: Walt Disney Studios Motion Pictures.
Walcott D (1986) The sea is history. In: Walcott D *Collected Poems, 1984–1984*. New York, NY: Farrar, Straus and Giroux, 364–367.
Watson JK (2015) From Pacific way to Pacific solution: Sovereignty and dependence in oceanic literature. *Australian Humanities Review* 58: 29–49.
Watt I (1957) *The Rise of the Novel: Studies in Defoe, Richardson and Fielding*. London: Chatto and Windus.
Wells HG (2000) Aepyornis Island. In: Hammond J (ed) *The Complete Short Stories of H. G. Wells*. London: Phoenix Press, 54–63.
Whyte J (2014) The fortunes of natural man: Robinson crusoe, political economy, and the universal declaration of human rights. *Humanities: An International Journal of Human Rights, Humanitarianism, and Development* 5(3): 301–321.
Winton T (2001) *Dirt Music*. Sydney: Picador.
Winton T (2008) *Breath*. Sydney: Hamish Hamilton.
Wright A (2006) *Carpentaria*. Sydney: Giramondo Publishing.

# 22

# IMAGINARIES

## Art, film, and the scenography of oceanic worlds

*Greer Crawley, Emma Critchley and Mariele Neudecker*

### Introduction

Astrida Neimanis, Cecilia Åsberg and Suzi Hayes in their essay 'Posthumanist imaginaries' suggest that "the 'imaginary', which is gaining traction in debates on environmental politics, citizenship, and related socio-institutional practice […] can also refer to the imaginative space wherein we formulate – and enact – our values and attitudes towards 'nature'" (2015: 5). As they note,

> Within contemporary cultural research, imagination, or an imaginary, refers to that social domain of seeing, experiencing, thinking, fantasizing, discussing and enacting aspects of the material world. Imaginaries shape how we see ourselves in relation to certain phenomena, and our relations to others in the context of those phenomena.
>
> *Neimanis et al., 2015: 2*

Whilst the sea has long been captured in the imaginaries of writers (see Jones, this volume) and artists, in recent years the emergence of an 'imaginative space' for articulating relations with the sea has been reflected in numerous exhibitions devoted to artists' responses to the ocean.[1] For example, *Aquatopia – the Imaginary of the Ocean Deep* was an exhibition in 2013 at Nottingham Contemporary, UK. The curator Alex Farquharson said: "[i]n this imaginary […] the ocean is both a here and an elsewhere […] we become-other, crossing thresholds that in our terracentric lives present themselves as absolute frontiers" (Farquharson, 2013: 6–11). The ocean imaginary, then, becomes an access point for many in our landed lives, to a space most of us do not and cannot usually, or easily inhabit (Peters, 2016; Squire, 2016). We cannot so readily build 'on' the sea, or breath under the sea. Its material wetness, depth, character, rendering it a space 'apart' from the (apparent) stability of land and its ability to more readily host human life (Steinberg and Peters, 2015). The ocean imagined (Peters and Steinberg, 2019) enables blue space to reach beyond its liquidity to be engaged with, experienced, *felt*. This also enables engagements, experiments and visceral encounters with the meanings evoked in such imaginaries, including meanings related to our changing planet.

Other exhibitions have shown how the ocean is *re*imagined by artists from a perspective informed by science. The ocean scientist, Professor John Finnigan described how the artists in *Ocean Imaginaries* held in 2017 at RMIT University, Melbourne took

DOI: 10.4324/9781315111643-27

> patterns of ocean life that science has revealed, and they have made us see them anew. Through the creative tension they have built between science's cold equations and their own potent reimaginings, they have taken us one more essential step towards the time when sustainability will inform both our political choices and our personal behaviour.
>
> *Finnegan, 2017: 11*

Imaginaries, in short, are spaces – or locations – of representation. Yet representations do not present worlds as they are – they are spaces of presentation, negotiation and interpretation (Duncan and Ley, 1993). They allow us to reflect critically on a given phenomenon, event, place, time – in this case the ocean, its changing state, and the life within it.

This chapter addresses the construction and reception of ocean space through 'imaginaries'. It begins by setting the scene for the ways in which ocean space has historically been subject to imagining and the production of imaginaries, and the place of art in making sense of the geographies of our water worlds. The chapter then turns to explore the works of two artists, Emma Critchley and Mariele Neudecker, represented in the aforementioned *Ocean Imaginaries* exhibit. Here the chapter both visualises artistic engagements with, and interpretations of, ocean space whilst also bringing the artists into conversation about the work of ocean imaginaries. In their responses to the effects of technological, ecological and economic exploitation on the oceans, they demonstrate the claim by Neimanis et al. (2015: 11) that "[i]maginaries can be forged, in part through the power of art as a catalyst for new kinds of engagements".

## Histories of imaginaries

For centuries, the production and dissemination of oceanographic knowledges have depended on what was primarily a visual representation. It was on Captain James Cook's scientific voyages that professional artists were first included on board. Cook observed that "drawings and paintings… give a more perfect idea… than could be formed from written descriptions only" (in Percy, 1996: 17). Of course, representations are always far from 'perfect', being particular, positioned, windows to the world (Rose, 2016). Early artistic imaginations of 'distanced lands', it is important to note, did political work in the age of western exploration through colonial and imperial expansion and the construction of 'othering' (see, for example, Edmond [1997] on Cook and the Pacific).

Yet early artistic endeavour on Cook's voyages reminds us of the way art interprets or reframes the world through the imaginary. Bernard Smith in *Art as Information: Reflections on the Art from Captain Cook's Voyages* (1979), for example, shows how William Hodges, an artist on Cook's second voyage to New Zealand in 1772–1775 depicted the reflections of water in the interiors of the icebergs in *Ships Taking in Ice*. Smith comments on the artist's ability to capture the optical effects of weather and light (Smith, 1979: 83–125). He describes the paintings Hodges produced on his return for exhibition at the Royal Academy, as "imaginative recastings of visual information" in which the artist "presented his basic truths within conventional super-structures, constructed out of neo-classical, picturesque and romantic elements" (Smith, 1979: 98). It was proof of Smith's later claim that

> it would be profoundly misleading to assume that this increased use of art for the conveyance of relevant scientific information operated as a direct, unilinear process

> by which error and illusion were cast off and the *truth* progressively revealed – though that certainly was the way the scientific optimists of the day chose to regard it.
>
> *Smith, 1992: 39*

Rather, art worked as partial representation (Duncan and Ley, 1993) of ocean worlds and the spaces reached through western exploration/exploitation.

Michael Jacobs suggests that in the nineteenth century it was still "the ultimate goal of the intrepid travelling artist – to portray nature at her (sic) most strikingly unusual, and to do so with an overlay of metaphor, poetry and adventure" (Jacobs, 1995: 153). Yet Rosamunde Codling in 'HMS Challenger in the Antarctic: Pictures and Photographs from 1874' challenged this view, arguing that whilst the Artists on the Challenger[2] "recognized the adventure in their work, they had no need of overlays of metaphor or poetry" (Codling, 1997: 204). Rather, artists reflected an awareness of their scientific mission and the need for 'precise visualisation', which had become a well-established practice in the natural sciences by the time of the Challenger expedition. Nature was to be depicted without enhancement. Codling refers to a quote by John James Wild (1824–1900), the professional artist onboard the Challenger, who in his 1878 narrative of the expedition noted his intention of "representing… accurately … the natural scenery" (Codling, 1997). The expression of emotion and subjectivity was seen as incompatible with objective observation. Whilst there was a shift, then, to an idea of ocean imaginaries as reflecting environments accurately, it is nonetheless recognised in contemporary scholarship that any visualisation – however accurate – always remains a representation. Gregory and Walford note (in Barnes and Duncan, 2013: 2), texts and images can never be "mirrors which we hold up to the world, reflecting its shapes and structures immediately, without distortion. They are, instead, creatures of our own making". Nevertheless, where science and art combined, Wild and the other artists on the Challenger *wanted* to – and arguably did – illustrate their subjects including the deep sea with greater scientific rigour.

However, unlike terrestrial subjects, the deep sea was inaccessible to direct observation, so new diagrammatic approaches had to be adopted. The result was that the undersea world was represented primarily by charts, graphs, and tables of data (see Lehman and also Squire, this volume). Emma Zuroski in 'Imagining the deep sea: Modes of representation on the HMS Challenger expedition' describes one of the few images made of the deep sea. It is a cross-section showing a geometric arrangement of variant lengths of depth sounding ropes descending into the water from the HMS Challenger. The ship and the lines are the only elements depicted in an otherwise completely empty expanse of oceanic space. Zuroski points out the irony of this kind of representation saying:

> … one of the main goals of the Challenger naturalists was to disprove Edward Forbes' azoic theory of the ocean, which had asserted that life was not sustainable below three-hundred fathoms. While the azoic theory had already been challenged by others, the Challenger naturalists confidently disproved the theory by trawling specimens at depths up to ten times that. But in visualizing this achievement, they physically erased their findings. In essence, they depicted an azoic space in order to disprove the azoic theory.
>
> *Zuroski, 2018: 95*

It was apparent that there were neither the artistic nor the technological capabilities for observing and recording phenomena on the ocean floor. To acquire them would necessitate both the corporeal as well as the mechanical experience of being underwater.

Developing technologies would be crucial to the opening of ocean spaces to further imaginaries. One of the first artists to immerse himself fully into the ocean was the Austrian artist-explorer Eugen von Ransonnet-Villez (1838–1926). Using a diving bell made to his own design, Ransonnet observed and sketched the underwater world in colonial Ceylon (now Sri Lanka) in 1864/65. A lithograph from his 1867 *Sketches of the Inhabitants, Animal Life and Vegetation in the Lowlands and High Mountains of Ceylon: As Well as of the Submarine Scenery near the Coast Taken from a Diving Bell* shows the artist's legs protruding from the bottom of a small submersible attached to an air tube and weighed down by cannonballs. This arrangement enabled Ransonnet to move along the sea floor at a depth of five metres and to stay down for up to three hours "undisturbed and drawing with a soft pencil on greenish-coloured, varnished paper" (The Public Domain Review, 2020: no page). Ransonnet-Villez demonstrated his understanding of the experiential effects of his underwater immersion, when he wrote in 1868 that "one's normal sense of distance and size is completely lost. You soon realize that in the depths of the ocean you need not only learn how to move, but how to see and hear as well" (The Public Domain Review, 2020: no page).

Ransonnet used his underwater sketches as the basis for his oil paintings. The aesthetic of these works is described by Stefanie Jovanovic-Kruspel et al. (2017) as a combination of romantic-lyricism with scientific observation. Having established himself as an innovator in a new genre of painting in what he described as an "unprocessed field" (Jovanovic-Kruspel et al., 2017: 143), Ransonnet-Villez was to go on to experiment with colour photography and the use of an underwater telescope or reverse periscope for his later images.

Like Ransonnet-Villez, the French photographer Louis Boutan (1859–1934) adopted an innovative approach to capturing underwater images. In 1893, he dove to 50 metres wearing a deep-sea suit and carrying a heavy camera and glass plate negatives, igniting magnesium powder to illuminate the darkness. Ann Elias explains how Boutan's experience of the deep sea was 'visceral' and that crucial to his imaginaries was "the embodied, immersed perspective from below, looking through underwater space horizontally. He wanted to understand the authentic animal in an ecological context" (Elias, 2019: 41).

But photography remained difficult underwater because of the low light conditions and cumbersome technologies. Therefore, drawing and painting continued to be the primary forms of visualisation and in turn constructed the imaginaries of ocean space. When in the 1930s, biologist William Beebe (1877–1962) was able to descend to a depth of over 900 metres in a bathysphere, his sketches and transcribed descriptions were transformed into paintings by Else Bostelmann (1882–1961). The *National Geographic*, who was to publish her illustrations, described how although Bostelmann never went down in the bathysphere, "she often would put on a diving helmet, tie her brushes to a palette of oil paints, and drag her canvas underwater to paint and find inspiration" (Strochlic, 2020: no page) (see Figure 22.1). Bostelmann was particularly noted for her ability to suggest movement in her underwater scenes, and it was one of her paintings that featured on the front cover of Beebe's 1928 edition of *Beneath Tropic Seas*.

For Margaret Cohen, access to underwater space is "a fortuitous example of the congruence between the practical discovery of a new planetary environment and the history of the arts and epistemology […] clarity and distinction which were previously the hallmarks of reality came into question" (Cohen, 2014: 18). The aquatic perspective represented a new approach to visuality, modes of perception and in turn created new imaginaries of ocean space. The destabilisation created by underwater optics suggested an alternative reality and the artistic imagination was transformed by the possibilities offered by "immersing vision in water" (Cohen, 2014). The rendition of immersive effects was now the objective for underwater visualisations in art, photography, and film.

*Figure 22.1* A pen and ink by Bostelmann illustrating a Christian Science Monitor article dated 18 July 1935, and entitled "With an Artist at the Bottom of the Sea By Else Bostelmann, an Artist with Dr. William Beebe's Expeditions Into Undersea Wonderlands", The illustration shows and the article describes her method of painting underwater.

*Source*: Widder (2016: 176) by Creative Commons.

Between 1914 and 1932, the first live-action underwater films were made by the John Ernest Williamson (1881–1966) As well as documentaries, he made a number of feature length films including the underwater scenes for a 1916 adaption of Jules Verne's *Twenty-Thousand Leagues under the Sea*. Previously 'underwater' films were made on sets or using footage shot in front of aquarium tanks. Williamson, however, devised a 'photosphere' that was connected by cable to the bottom of a ship which allowed him to take cameras and cameramen underwater. Although this was an innovative technology, Jonathan C. Crylen in 'The cinematic aquarium: A history of undersea film' points out that the representation of the underwater scenes was still from the same land-based frontal perspective as those shot before an aquarium (Crylen, 2015: 25). This lack of full immersion by the cameramen, and the resulting perceptual and optical fictions would become exposed when divers could descend directly into the sea with their cameras.

It was in the 1940s and 1950s, that scuba-diving equipment and underwater camera technologies began enabling filmmakers to have a fully immersive experience in the depths of the ocean. This presented the possibility a new "environmental aesthetics of the marine" (Cohen,

2019: 158). Just as the artists on the European voyages of exploration found forms of visualisation for apparently previously 'unknown' seas, Jacques-Yves Cousteau and his fellow filmmakers and divers "had to invent the aesthetic contours of an environment that had never been seen before" (Cohen, 2019: 158).

From 1956 onwards, further innovations in submersible technologies opened up the deep sea to increasing visual documentation and visual exploitation through surveillance and spectacle. In 2008 Google Earth added an 'Explore the Ocean' function to its website, enabling viewers to go underwater to explore the seabed virtually. Remote imaging, satellite technology and digitalisation have transformed the perception and occupation of ocean space (see Lehman, this volume). But despite this apparent accessibility to the deep sea, the provider determines the limits to both the view and the experience. Representation of ocean spaces – and the imaginaries they produce – remain partial. Nonetheless, they allow us access points for critical reflection on water worlds.

And there are still hidden, occluded spaces – blind spots. The artists Mariele Neudecker and Emma Critchley bring perceptual and experiential lines of sight to these invisible places, *invisible ocean spaces*. Their oceanic perspectives and fluid practices create encounters and exchanges in an entanglement of virtual data and natural phenomena. The artists' works are instrumental in revealing the physical and conceptual realities that lie beneath the water and articulating affective relationships with the ocean.

## Introducing Emma Critchley

In *Slow Violence and the Environmentalism of the Poor* Rob Nixon suggests that artists have a role to play in exposing the " 'layered invisibility' of a world permeated by insidious, yet unseen or imperceptible violence" (Nixon, 2011: 16). Emma Critchley's imaginings – which traverse the surface and depths of the ocean – offer what Nixon calls a "different kind of witnessing" (Nixon, 2011: 15) of ocean space by making visible catastrophic events and phenomena through "devising iconic symbols that embody amorphous calamities as well as narrative forms that infuse those symbols with dramatic urgency" (Nixon, 2011: 10).

Critchley's *Frontiers* (Figure 22.2) featured in *Slow Violence*, an exhibition and symposium in 2017 on the long-term effects of environmental damage. Made in 2015, *Frontiers* was a response to China's expansion of its territories by moving the living seabed onto reefs in the South China Sea. The artist made film and photographic studies of a strange waterborne presence adrift in water. This spectral apparition of an anamorphic island or possibly wreckage of some mysterious disaster exists within a transitory unspecified oceanic space. Ghostly echoes come from sonars on the soundtrack for the film. Two sonar calls were played backwards to sound like a 'call and response' search in bad visibility. It was to emphasise the idea of roaming and searching the horizons of the unknown, trying to find meaning from the available information. The artist shows us that despite the increased field of vision in ocean space, there is less visibility and more phantasms.

Critchley's *Frontiers* demonstrates the power of the metaphor. Metaphor and understandings of ocean space have been an ongoing area of discussion in critical marine geographies (see Lehman et al., 2021; Peters and Steinberg, 2019; Steinberg, 2014, 2016). Whilst the ocean is material, metaphors and imaginings play important roles in meaning-making and understanding. In Critchley's work there is a suggestion that there is this place somewhere, "an edge of space and time: a zone of not yet" (Tsing, 2005: 28).

*Then Listens to Returning Echoes* (2016) (Figure 22.3) was a further film made by Critchley in Barbados and Portland, UK – two places that have a great number of shipwrecks because of the

*Figure 22.2* Frontiers I, Emma Critchley, photographic print, 40" x 30", 2015.
*Source*: Emma Critchley.

*Figure 22.3* Film still from *Then Listens for Returning Echoes*. Emma Critchley, 20 minute HD film, 2016.
*Source*: Emma Critchley.

treacherous waters surrounding both islands. It has been recorded that Carlisle Bay alone has lost approximately 200 ships since the seventeenth century primarily due to storms. Portland is a similarly perilous place due to a combination of 'The Race', a convergence of no less than seven tides coming together and the 'Dead line' to the west of the isle that has a 10-knot undercurrent. Drawing from the records and mythologies of each location and interviewing people from the surrounding area, Critchley's film focuses on the resonance with the local communities of these hidden burial grounds and the layers of histories and stories held within each wreck's carcass. The sonic approach to the film responds to the ritualistic searching and listening that has become embedded within each community. There are associations with warning, monitoring, responding, resonating, the throwing out of light and sounding the foghorn, the sending out of sonar waves and listening for returning echoes.

The artist said she was thinking about the wrecks as sonic chambers that contain histories and stories, lying dormant on the seabed. The sound was recorded in layers with cellist Lucy Railton who often uses the acoustic structure of places as a point for improvisation. They did multiple recordings, which were then layered to make the soundtrack. The first was inside a hyperbaric chamber where Railton did a pressure dive to 20 metres to be in similar conditions to Critchley's underwater filming (and still play a cello!). They also did recordings inside Portland Bill Lighthouse where the cello was tuned to the same pitch as the lighthouse's foghorn and another segment on a sonar boat reading the sonar mapping of the seafloor as a score. The first time it was screened the cellist played the final layer live, which was recorded and then added to the track. In this work as in other of Critchley's works "[t]o listen is to enter a spatiality in which time becomes space, located between past, present, and future and encompassing notions of the remainder – the trace" (Fischer, 2014: 16).

*Do You Know Nothing? Do You See Nothing? Do You Remember, Nothing?* (2018) was a site-specific installation made by the artist in the Nayland Rock Hotel on Margate Sands, UK (Figure 22.4a). The commissioned artists had been asked to respond to TS Eliot's *The Waste Land* (1922)[3] and Critchley said that initially she saw in the work a sense of critical exhaustion that corresponded to the feelings of resignation often felt by people when confronted with the effects of climate change on their lives. She chose to make her response as a conflation of present and future climatic conditions.

The artist has described how there was a room on the third floor of the hotel, where the sea had begun to come in through cracks in the walls and windows, and the wallpaper had started peeling away (an 'Ocean in excess' in Peters and Steinberg's words [2019], of an ocean that seeps and reaches on shore). She worked with this to create a room that appeared to flood at high tide and a fictional inhabitant who lived a makeshift existence in the room by adjusting the furniture so it was above the high-tide mark (Figure 22.4b). The sea was a strong presence in this interior scenography. The room was permeated with the smell of damp, salt, and seaweed. Conceived through weather and water, the installation was experiential and visceral.

In her constructed imaginative spaces, Critchley invokes situations, atmospheric conditions, and temporalities. Her works are liminal experiences; transitory states, where one space exists within another. The artist has said of her films that

> the surface of the screen is like […] a tenuous film that holds back the mass of water contained in the image behind. A boundary, created in a fraction of a second that remains so that we might see beyond its surface into the world that it contains.
>
> *Critchley, 2010: 94*

There is a realignment of perceptual expectations; everything shifts.

*Figure 22.4a* *Do You Know Nothing? Do You See Nothing? Do You Remember, Nothing?* Emma Critchley, Installation at The Nayland Rock Hotel, Margate, 2018.

*Source*: Emma Critchley.

*Figure 22.4b* *Do You Know Nothing? Do You See Nothing? Do You Remember, Nothing?* Emma Critchley, Installation at The Nayland Rock Hotel, Margate, 2018.
*Source*: Emma Critchley.

## In conversation with Emma Critchley: Perspectives on ocean space I

*To date, I have worked underwater for 20 years using photography, film, sound and installation. I've shot in waters across the globe with channel swimmers and aquaphobics, sea gypsies and deep-sea ecologists and trained with freedivers to explore our human relationship with this space; philosophically, ecologically and politically.*

*Being immersed in water is a powerful scenario that resonates not only with me as an artist but unites us all; it is something we have all experienced. The thick liquidity of the space brings to the fore our interconnected, symbiotic relationship with the environment we inhabit, we are part of, we are. Yet our relationship with the ocean is in crisis. Perhaps because to most the subaquatic realm is seemingly remote; any 'experience' is mediated through the mechanics of the camera's lens. Yet we are far from remote; the membrane of the ocean's surface masks a plethora of human activity such as mining, mapping and territorial dispute. Positioned between staged and documentary, my work interweaves constructed experimental scenes with archive material. The installations I create invite the viewer into constructed environments that enable questioning from an embodied, experiential perspective. Unfolding narratives reinforce historic complexities, whilst the surreal nature of the submarine becomes a psychological counterpart, to push boundaries of reality and provide a space for reflection. I am interested in the psychology of the human response to environmental change and the multiple forms of narrative that emerge as a process of sense-making; from scientific observation and explanation to lived accounts of experience, media reporting and cultural mythologies.*

*Unlike our occularcentric way of being, sound, is the primary sense to all inhabitants of the ocean and it is an essential element of the work I make. The installation and research project 'The Space Below'*

*(2020), made in collaboration with artist Lee Berwick, responds to the issue of acoustic sound pollution in our oceans. Every creature in the ocean can hear: The smallest larva listens to the reef to find where to settle, while the Blue Whale draws an acoustic map to navigate its way. Hearing is fundamental to communication, breeding, feeding and ultimately survival. But as humankind increasingly continues to explore and exploit the underwater world, so our sounds pollute and raise levels of noise to extreme levels. Transporting audiences into a space where ears rule over eyes 'The Space Below' is a multi-speaker sound installation that explores the global issue of underwater acoustic pollution as an underground walk-through experience. It was initially installed in the Greenwich foot tunnel, where audiences stand beneath the River Thames in London. Created from underwater sound recordings that have been made by scientists all around the world, the installation features both natural sounds made by a wide range of sea creatures as well as human-made sounds of boats, sonar, seismic surveys and acoustic deterrent devices.*

*The ocean is considered a location for mineral resources, territorial space, a carrier of goods and people, yet culturally it remains a place of fantasy. The film 'Common Heritage' (2019) (Figures 22.5a and 22.5b) is an urgent response to the rush of deep-sea mining for rare earth minerals, exposing how reverberant layers of industrialisation, colonialism and territorial claim have affected the way we relate to our environment. Highlighting the fantasies we construct and investigating the relationship between exploration and exploitation, the film draws into focus how these romanticised stages are in fact borders of conquest, annexed for geopolitical territory appropriation and mineral resources. In 1967, Arvid Pardo, the Maltese Ambassador to the UN gave a speech, which instigated the Common Heritage of Mankind principle and after 10 years of international conference and debate, bore the Law of the Sea Treaty. Pardo's speech, narrated by science fiction writer, Gwyneth Jones, is the provocation for the film. The story draws from a web of press conferences, interviews and speeches, which unfold to reveal the tensions and contradictions in the attempted governance of such a vast and powerful landscape and the continual disputes that probe the edges of law and territorial demarcation. Dystopian science fiction motifs are harmonised with a poetic montage of deep-sea exploration archive footage. This juxtaposition both questions our current state and our future engagement with this critical frontier.*

*The ocean is also ice. 'Witness' (2021) is a multi screen film installation premised on the examination of an ice core as the post mortem of a glacier. The piece was made during the Earth Water Sky residency with Science Gallery Venice, working with the Ice Memory Project – a global initiative building the first world archive of glacier ice, to preserve this invaluable scientific heritage for the generations to come. Through an enquiry into the incidents leading up to its death the work calls to account society's witnessing of such events. The film weaves together different forms of narrative taken from conversations with Professor Carlo Barbante and the Ice Memory Team, scientific papers and articles, interviews with people who live local to glaciers and my own thoughts and reflections, in order to analyse our complex, collective, non-linear past. Importantly, there is also the notion of the glacier itself as witness – an animate body, which has been subject to historic events, which are evident in its layers. The work considers different ways of knowing and different forms of knowledge, from scientific data to everyday experience. Similarly, the importance of acknowledging who is conducting the research, whose histories are being told and by whom. Unlike the layers in a core, history is by no means linear and a story very much depends on who is telling it and where it begins. The work has therefore become a collection of different forms of knowledge, bringing together scientific papers and historic events with interviews with people who are living with glacial retreat from Kilimanjaro and Cordillera Blanca on a daily basis in Kenya and Peru. This concept also informed the development and making process of the work itself, as we ran a series of workshops with dancers both in and out of water, where elements of the research were explored through movement, which eventually contributed to the choreography in the final film.*

*Figure 22.5a* Film still from *Common Heritage*. Emma Critchley, 25 minute HD film, 2019.
*Source*: Emma Critchley.

*Figure 22.5b* Film still from *Common Heritage*. Emma Critchley, 25 minute HD film, 2019.
*Source*: Emma Critchley.

## Introducing Mariele Neudecker

The images and concepts of Romanticism and the Sublime underpin the Neudecker's contemporary perspective on maritime facts and fictions. *Shipwreck* (1997) (Figure 22.6) – a tank work that is related to the *Shipwreck* (1760) painting by Philip James De Loutherbourg – and *Heliotropion* (Ship and Avalanche) (Figure 22.7) – a two-channel video piece (1997) referencing Loutherbourgh's *Avalanche in the Alps* (1803) and George Philip Reinagle's, *First Rate Man of War* (1826) – are early examples of the artist's engagement with her Romantic predecessors and the oceanic 'sublime'. Both render engagements with ocean space that open up imaginaries that explore the relations tensions between the 'romantic' and 'reality', a long-standing interest in critical geographic work on understanding space, landscape (or seascape) and its articulations (see Tuan, 2013).

Following from Critchley's underwater work, Neudecker's installations *The Great Day of His Wrath, Horizontal Vertical* (Figure 22.8), *Dark Years Away* and *It Takes The Planet 23 Hours and 56 Minutes and 4 Seconds to Rotate on its Axis* (all 2013) were filmed from a remotely operated vehicle (ROV) as part of a research project developed in collaboration with the University of Oxford marine biologist, Dr Alex Rogers, founder of the International Programme on the State of the Ocean (IPSO), a campaign to support the oceans.[4] In November 2011, Rogers had sent an automated submersible with cameras and robotic arms to take specimens in deep water trenches in the Indian ocean. Sixteen terabytes of video taken by cameras attached to a remotely operated vehicle (ROV) travelled to a depth of up to 3000 metres from the RRS James Cook (a British Royal Research Ship).

*Figure 22.6* *Shipwreck*. 1997, Mixed media incl. glass, water, food dye, light. 32.5 x 27.6 x 162 cm. Installation: SEDIMENT, LCGA, Limerick, Ireland. Photo: Benjamin Jones. Courtesy: Pedro Cera, Lisbon.

*Figure 22.7* *Heliotropion* (Ship And Avalanche), 1997, 2 looped 2'25" videos on monitors. Courtesy: the artist.

The deep-sea footage is tightly focused recordings of the ocean floor. The haze and dust raised by the ROV's scraping of the ocean floor together with the red guide lights of the recording equipment, the variable speed of the mechanical arm of the camera and the reversed footage of fish contribute to a perceptual disorientation which is increased by the accompanying soundscape. Neudecker edited the video footage of *Dark Years Away* adding a musical score by Pēteris Vasks and for *Horizontal Vertical*, sounds such as human heart beats, car traffic outside and ticking clocks in response to the eerie silence of the footage and the sounds in her domestic editing space. This soundscape together with the evidence shown in the videos of human-made debris and bits of machinery on the ocean floor emphasise the presence of humans in these remote marine environments. Whilst in the previous section we saw how art enables the sea to stretch onshore, this piece is a reminder of the stretching of human life to the very depths of deepest ocean space. Moreover, despite the implied sublimity of Vask's music, the *Dark Years Away* is not intended to illustrate a heroic expeditionary narrative, but considers the wreckage of a deep sea imaginary of an oceanic sublime (Figure 22.9).

Neudecker has said that in her collaborations with scientists and their methods and research her aim is "to make the immeasurable measurable, the invisible visible" (Neudecker, 2018: no page). Among the subjects of her investigations is a Giant Squid (*Architeuthis dux*). This elusive creature has long been associated with the Kraken the sea monsters of Scandinavian legend and was to figure in the new biological sublime that had emerged in the mid-nineteenth century when culturally, interest began to move beyond the sublime surface of the ocean to "the strange biologies lurking in the submarine" (Deam, 2019: 83). Phillip Hoare said of JMW Turner's *Sea Monsters and Vessels at Sunset* (watercolour and chalk on paper c.1845):

*Figure 22.8* *Horizontal Vertical, Net Fish* (1 of 5), 2013, video still, 5-channel video-installation, size variable. Photo: the artist. Courtesy: Galerie Thomas Rehbein and Pedro Cera.

> There's all this bubbling, strange stuff going on underneath. It's psychologically uncanny. It's about what's below, what's underneath the relentless movement towards progress in the 19th century: the wilderness. Below the ocean's skin, as Melville put it, everything is other – alien and unconquerable.
>
> *Sooke, 2016: no page*

*Figure 22.9* *Dark Years Away*, 2013, 6' & 180', looped, 1 single video projection with sound and 1 monitor. Sound: Peteris Vasks, Voices of Silence. Installation: *There is Always Something More Important*, Galerie Barbara Thumm, Berlin. Photo: Jens Ziehe. Courtesy: Galerie Thomas Rehbein and Pedro Cera.

In 2004, an *Architeuthis dux* was caught accidentally by a fishing trawler in the South Atlantic. This nearly complete specimen was subsequently frozen and transported to the Natural History Museum in London. Neudecker's *One More Time [The Architeuthis Dux Phenomenon]* (2017) comprises two video monitors running a twelve-meter tracking shot of the Giant Squid, out of synch to suggest its size (Figure 22.10a). The young female squid is just over eight and a half meters long and contained in a narrow 10-metre-long tank of formol-saline in the tank collection in the basement of the museum. This arrangement and location of the Giant Squid in the museum is significant when contrasted to the display of the 28.6-metre skeleton of the Wexford whale, renamed as 'Hope'. The whale skeleton was relocated in 2017 to become the centrepiece in the museum's main entrance hall and put into a new dynamic 'diving' posture (Williams, 2019: 170). Neudecker's slow scanning along the massive length of the squid's body emphasises the extent of the squid's containment and the anatomical examination of the decaying state of its pale body. The stately progress of the camera's digital tracing creates a melancholic elegy to this extraordinary creature revealing the Giant Squid to be "[T]he strange stranger […] at the limit of our imagining" (Deam, 2019: 83) (Figure 22.10b). Through her affective engagement with this 'stranger', Neudecker presents the 'body of evidence' for the vulnerability of the ocean and its inhabitants. This is the sublime as an artistic strategy which "calls for the creation of evidence, of perceptibility, of documents – the renderings of a fleeting world" (Ray, 2020: 15).

*Figure 22.10a* *One More Time* (The Architeuthis Dux Phenomenon), 2017, HD video loop on two monitors, duration: 2'35". Commissioned for Offshore by Invisible Dust in partnership with Hull Culture and Leisure with thanks to the Natural History Museum, London. Installation: Limerick City Gallery of Art. Photo: Benjamin Jones. Courtesy: Galerie Thomas Rehbein and Pedro Cera.

## In conversation with Mariele Neudecker: Perspectives on ocean space II

*My work, for many years, has been to seek new definitions for our understanding of the sublime today. Our perception and subjective experience of various layers of reality is fundamental to this ongoing quest.*

*What we see is always cropped and filtered through frames and lenses. From the tissues in the brain, via fibres and nerves, the lenses of our eyes, the frames of glasses, through cameras and all its lenses, out to 'the field of vision', out to the world … – or was it the other way round?*

*Subjectively we look out of our eyes, our windows to the world – which, in the case of the Deep Sea, is something that is only partly possible and facilitated by sophisticated methods and technologies; it becomes very specific in this limitation, and goes far beyond any measure of human territories, nationalities, experience and comprehension. This information and knowledge is captured as data, recorded and stored, and of course we are forced to make assumptions about the full view and bigger picture.*

*When I first saw the still and moving images from the Deep Ocean that I received from marine biologist Alex Rogers, the mediated nature of these representations from another world, in all their multiple layers of their capture, preservation and existence unfolded in front of me.*

*Silently another reality revealed itself.*

*A wilderness that was no longer pure or wild. If it ever was.*

*Starting on the seabed: visible to us only in light beams: captured by cameras mounted to the Remotely Operated Vehicle [ROV] steered by technicians on the ship, depending on strong lights to reveal themselves to us.*

*Figure 22.10b* *One More Time* (The Architeuthis Dux Phenomenon), 2017, Making – tracking camera in 'tank room', basement NHM, HD video loop on two monitors, duration: 2'35". Commissioned for Offshore by Invisible Dust in partnership with Hull Culture and Leisure with thanks to the Natural History Museum, London. Photo: Benjamin Jones. Courtesy: Galerie Thomas Rehbein and Pedro Cera.

*The ROV is attached with a thin cable to the RRS James Cook, the research vessel is at this point moving in the South West Indian Ocean near Antarctica. The camera operators control the movement of the ROV, the lights and cameras. Several laser-beams are being recorded throughout, to tell us that the camera is working and to measure the size of what is found and being filmed.*

*Live footage scrolls across the screens in the control room on the research vessel.*

*... bringing many previously invisible traces of other human impact, mostly abandoned fishing gear, up, for our eyes so see.*

*How long have those things been there?*

*Alex lets me use this recorded and preserved footage, for which I am very grateful.*

*I sit in my loft and edit a copy. All I can hear is my breathing, the clock ticking, the cars moving outside in the street, my heart beating, the computer and the world's humming .... I look into the screen of my computer, and see the silent videos I set out to edit for an installation, a 'piece of work'. The experience is one of heightened awareness of our limitations in not only seeing, but also in hearing – or: not hearing and imagining.*

*The sound of absent sound is a very particular kind of silence.*

*The witnessing of the continuously changing complexities of all this, brings new meanings to the word contamination. This silence is painfully impossible... and more and more people are listening.*

## Endings/Beginnings

The ocean imagined is not a separated realm from a felt material, liquid (or icy) ocean. Imaginaries are a way in which the ocean is "a space that is 'perpetually beyond itself', the ocean is not 'the ocean' (or 'the sea', a sealed unit) but is already and always in excess" (Peters and Steinberg, 2019: 297). As Kimberley Peters and Philip Stenberg write, the "materiality of the ocean in excess meshes together with human life in such a way that … [it] transcends liquid, 'wet' engagement" (Peters and Steinberg, 2019: 297–298). We can engage with, experience and feel the ocean as "the ocean's physicality extends beyond the material, reaching past its geophysical boundaries to facilitate imaginative transformation" (ibid., 2019: 304).

Through the lens of Emma Critchley's and Mariele Neudecker's work, this chapter has curated a discussion of the ways in which ocean space is imagined and how imaginaries – constructed and interpreted – enable us to grapple with historical, environmental, political, economic engagements with water worlds (and, of course, entanglements between these 'categories').

The history of imaginaries and the oceans has always been concerned with capturing and representing the ocean to know and understand it, in one way, or another (see Adler, this volume), to present it to varying and various audiences, but also to reimagine or redefine it in our imaginations. As charted in this chapter, imaginaries are *works* – they are not mirrors of a material ocean space 'out there' but are articulations of that 'out there' brought into being through artistic practice and form: paintings, photographs, installations, film and beyond. They merge equipment and materials, technique and technology. They do particular work through the very shape of those works. Past imaginaries can be deconstructed to tell us of colonial histories. Present imaginaries can be constructed to tell us of oceans in crisis.

Coming full circle, Astrida Neimanis, Cecilia Åsberg and Suzi Hayes note that:

> … imaginaries determine our orientations … and *in a very real and political sense, produce the world we seek to live in*. While imagination will not change the data issuing forth from scientific measurement, it will determine what we measure and how, what benchmarks we set, what policies and behaviours we adopt and what values come to orient all of the above.
>
> *Neimanis et al., 2015: 10*

In sum, then, imaginaries of ocean space matter.

## Notes

1 The 2003 *Liquid Sea* exhibition at the Museum of Contemporary Art Australia (MCA) explored the artistic imagination in relation to aquatic themes. *Art & Science: Envisioning Ocean Depths* was an online and traveling exhibition featuring multi-media works by seven artists and scientists from a 2012 field expedition to the Barbados volcanic seeps on the research vessel Atlantis. The artists were challenged to capture the essence of discovery as the scientists mapped the sea floor. *Oceanomania, Souvenirs from Mysterious Seas* held in 2011 at the Oceanographic Museum in Monaco and the National Museum (NMNM) was based on a concept by the artist Mark Dion and co-curated by Sarina Basta and Cristiano Raimondi featured over 24 artists' works. *Oceans Imagining a Tidalectic Worldview* curated by Stefanie Hessler in 2017 included new commissions by participants in the voyages of Thyssen Bornemisza Art Contemporary (TBA2 Academy) Dardanella research vessel with works by other artists whose practice concern the oceans, together with pieces from the TBA21 collection. The 2019 exhibition *The Art of Marine Sciences*, was the result of a collaboration between the Institute of Marine Sciences of the

National Research Council (CNR – ISMAR) and the Academy of Fine Arts of Venice and included work by artists and researchers curated by Francesco Marcello Falcieri Gabriella Traviglia.

2 The Challenger Expedition was a voyage that is said to mark the start of oceanography as a discipline. It ran from 1872 and 1876 and "was the first expedition organized specifically to gather data on a wide range of ocean features, including ocean temperatures seawater chemistry, currents, marine life, and the geology of the seafloor". John Murray and Charles Wyville Thompson led the expedition (Woods Hole Oceanography Institute, 2021).

3 Written about the aftermath of World War I, this poem is described as one concerned with breakdown and loss. Across five parts the piece speaks to a variety of spaces, including the ocean (Tearle, 2021).

4 These works were first shown in the basement of a Regency House at 'HOUSE 2013', Brighton Festival and then by Invisible Dust in St Thomas' Church at the British Science Festival, Newcastle September 2013 and Galerie Barbara Thumm, in Berlin in 2014. For the 'Science in the City festival, which took place as part of Euroscience Open Forum in June 2014, Neudecker's video installation *Horizontal Vertical* was located in unusual locations such as the living quarters on Dana, a Danish vessel used by scientists to investigate the effects of fishing on marine ecology and monitor ocean conservation.

## References


Barnes TJ and Duncan JS (eds) (2013) *Writing Worlds: Discourse, Text and Metaphor in the Representation of Landscape*. New York, NY and Abingdon: Routledge.

Codling R (1997) HMS Challenger in the Antarctic: Pictures and photographs from 1874. *Landscape Research* 22(2): 191–208.

Cohen M (2014) Denotation in alien environments: The underwater *je ne sais quoi*. *Representations* 125(1): 103–126.

Cohen M (2019) The shipwreck as undersea gothic. In: Cohen M and Quigley K (eds) *Aesthetics of The Undersea*. Abingdon and New York, NY: Routledge, 155–166.

Critchley E (2010) *Immersion*. Unpublished Masters dissertation, Royal College of Art.

Crylen JC (2015) The cinematic aquarium: A history of undersea film. Unpublished doctoral dissertation, University of Iowa.

Deam N (2019) The great melancholy mother Michelet's evolutionary ocean in *The Sea*. In: Cohen M and Quigley K (eds) *Aesthetics of The Undersea*. Abingdon and New York, NY: Routledge, 83–96.

Duncan JS and Ley D (eds) (1993) *Place/Culture/Representation*. Abingdon and New York, NY: Routledge.

Edmond R (1997) *Representing the South Pacific: Colonial Discourse from Cook to Gauguin*. Cambridge: Cambridge University Press.

Elias A (2019) *Coral Empire Underwater Oceans, Colonial Tropics, Visual Modernity*. Durham, NC: Duke University Press.

Farquarson A (2013) *Aquatopia: The Imaginary of the Ocean Deep*. London: Nottingham Contemporary and Tate Publishing.

Finnigan J (2017) Ocean imaginaries, introduction. In: Davies S (ed) *Ocean Imaginaries*. Melbourne: RMIT Gallery/RMIT University, 8–11.

Fischer B (2014) On the notion and politics of listening. In: Schuppli S (ed) *The Notion of Sound and Listening*. Luxembourg: Casino Luxembourg, 9–19.

Jacobs M (1995) *The Painted Voyage: Art, Travel and Exploration 1564–1875*. London: British Museum Press.

Jovanovic-Kruspel S Pisani V and Hantschk A (2017) Under water – between science and art – the rediscovery of the first authentic underwater sketches by Eugen Von Ransonnet-Villez (1838–1926). *Annalen Des Naturhistorischen Museums in Wien. Serie A Für Mineralogie Und Petrographie, Geologie Und Paläontologie, Anthropologie Und Prähistorie* 119: 131–153.

Lehman J, Steinberg PE and Johnson ER (2021) Turbulent waters in three parts. *Theory & Event* 24(1): 192–219.

Neimanis A, Åsberg C and Hayes S (2015) Post-humanist imaginaries. In: Bäckstrand K and Lövbrand E (eds) *Research Handbook on Climate Governance*. Cheltenham: Edward Elgar Publishing, 480–490.

Neudecker M (2018) There is Always Something More Important. POLAR 2018 Open Science Conference OSC. 19–23 June 2018.

Nixon R (2011) *Slow Violence and the Environmentalism of the Poor*. Cambridge, MA and London: Harvard University Press.

Percy CE (1996) To study nature rather than books: Captain James Cook as naturalist observer and literary author. *Pacific Studies* 19(3): 1–30.

Peters K (2016) Oceans and seas. In: Richardson D (ed) *International Encyclopedia of Geography*. Hoboken, NJ and Washington, DC: AAG-Wiley.

Peters K and Steinberg PE (2019) The ocean in excess: Towards a more-than-wet ontology. *Dialogues in Human Geography* 9(3): 293–307.

Public Domain Review (2020) *En Pleine Mer: The Underwater Landscapes of Eugen von Ransonnet-Villez*. Available at: https://publicdomainreview.org/collection/underwater-landscapes-of-eugen-von-ransonnet-villez

Ray G (2020) Terror and the sublime in the so-called Anthropocene. *Liminalities: A Journal of Performance Studies* 16(2): 1–20.

Rose G (2016) *Visual Methodologies: An Introduction to Researching with Visual Materials*. London: Sage.

Smith B (1979) *Art as Information: Reflections on the Art from Captain Cook's Voyages*. Sydney: Sydney University Press for the Australian Academy of the Humanities.

Smith B (1992) *Imagining the Pacific in the wake of the Cook voyages*. Carlton: Melbourne University Press at the Miegunyah Press.

Sooke A (2016) The monsters hidden beneath the sea. *BBC Culture [Online]*. Available at: www.bbc.com/culture/article/20160620-the-monsters-hidden-beneath-the-sea

Squire R (2016) Immersive terrain: The US Navy, Sealab and Cold War undersea geopolitics. *Area* 48(3): 332–338.

Steinberg PE (2013) Of other seas: Metaphors and materialities in maritime regions. *Atlantic Studies* 10(2): 156–169.

Steinberg PE (2014) Mediterranean metaphors: Travel, translation and oceanic imaginaries in the 'new mediterraneans' of the Arctic Ocean, the Gulf of Mexico and the Caribbean. In: Anderson J and Peters K (eds) *Water worlds: Human Geographies of the Ocean*. Farnham: Ashgate, 41–56.

Steinberg PE and Peters K (2015) Wet ontologies, fluid spaces: Giving depth to volume through oceanic thinking. *Environment and Planning D: Society and Space* 33(2): 247–264.

Strochlic N (2020) These women unlocked the mysteries of the deep sea. *National Geographic [Online]*. Available at: www.nationalgeographic.com/history/article/these-women-unlocked-the-mysteries-of-the-deep-sea

Tearle O (2021) The Waste Land by T. S. Eliot. Loughborough University website. Available at: www.lboro.ac.uk/subjects/english/undergraduate/study-guides/the-waste-land/

Tsing A (2005) *Friction: An Ethnography of Global Connection*. Princeton, NJ and Oxford: Princeton University Press.

Tuan YF (2013) *Romantic Geography: In Search of the Sublime Landscape*. Madison, WI: University of Wisconsin Press.

Widder E (2016) The fine art of exploration. *Oceanography* 29(4): 170–177.

Williams L (2019) Deep time and myriad ecosystems: Urban imaginaries and unstable planetary aesthetics. In: Cohen M and Quigley K (eds) *Aesthetics of The Undersea*. Abingdon and New York, NY: Routledge, 170–171.

Woods Hole Oceanographic Institute (2021) The Challenger Expedition. Dive and Discover website. Available at: https://divediscover.whoi.edu/history-of-oceanography/the-challenger-expedition/

Zuroski E (2018) Imagining the deep sea: Modes of representation on the HMS Challenger expedition. In: Abberley W (eds) *Underwater Worlds: Submerged Visions in Science and Culture*. Newcastle upon Tyne, UK: Cambridge Scholars Publishing. 90–107.

# 23

# SWIMMING

## Immersive encounters in the ocean

*Ronan Foley*

### Neither waving nor drowning

Swimming is having a zeitgeist moment. Whether on the pages of Sunday supplements or the growing breadth of academic writing, the experience of swimming, especially in the unpredictable but exciting wildness of the ocean, is an increasingly popular and commercially successful subject (Fitzmaurice, 2017; Heminsley, 2017; Hoare, 2017). That popularity springs from a range of emotional and embodied connections to the water linked to family, friendship, memory, loss, identity, socialisation and challenge/adventure that any discussion on swimming almost invariably mentions (Costello et al., 2019). Equally, the specific act of swimming in the sea has attracted researchers from the arts and humanities through to pure science (Parr, 2011; Swim England, 2017). This chapter, emerging from 'blue experiencescapes' (Doughty, 2019), provides a broad review of key recent writing on swimming in the ocean, from geography and beyond. My own positionality, as a culturally inspired health geographer and intermittent swimmer, specifically informs this account in relation to the health-enabling potential of oceans as therapeutic waterscapes and swimming as a practice (Bell et al., 2018; Denton et al., 2021; Duff, 2011; Foley, 2010, 2019). Finally, given Strang's suggestion that water has an almost limitless capacity to absorb metaphors, swimming draws from a range of different subjects and approaches that offer navigable empirical routes into recent theory within human geography (Andrews et al., 2014; Olive and Wheaton, 2021; Strang, 2004). Underpinning the chapter is a view of swimming in the ocean as an embodied, emotional and experiential immersive encounter, shaped by the fluid nature of the 'blue space' in which it occurs (Britton and Foley, 2021; Olive and Wheaton, 2021). For the purposes of this chapter, the focus is on the liminal setting of the near ocean/coast/shore, in contrast to inland, indoor or even tidal swimming (Ward, 2017); a self-imposed and necessary limitation in terms of scope and scale. Yet it is important to consider that any oceanic water space is always relational and becoming, shaped by the hydrological cycle, human and more-than-human ecologies and practices and multiple surface and bathymetric flows in, through and across those spaces to remind us that in any such research, we are always dealing with 'one blue' (Foley, 2015; Peters, 2010; see also chapters in Section VI, this volume).

In exploring a range of writing on the subject from multiple disciplines – geography, psychology, cultural studies, history, sport and leisure – swimming has become one of



DOI: 10.4324/9781315111643-28

several experiential strands within what might be termed a 'blue/oceanic turn' (Brown and Humberstone, 2015). Recent texts across human geography focus on active, contested and relational dimensions of living with the sea that draw from social, economic, cultural, political and environmental perspectives (Brown and Peters, 2018). Within that writing, specific activities like sailing, surfing, canoeing, diving and swimming act as useful nodes for connective and specifically immersive and co-produced engagements with the sea (Anderson, 2012; Merchant, 2016; Olive and Wheaton, 2020; Peters and Brown, 2017; Thompson and Wilkie, 2020). Within health geography, there has also been a similar turn to blue space, shifting traditional greenspace and nature-based research in that direction, with a growing attention to both hydrophilic and hydrophobic elements that have value for human health and wellbeing (Britton and Foley, 2021; Foley et al., 2019). In seeing relational engagements in and with place as a series of connections, Bowring et al. (2018) suggest those connections are spatial sutures; a series of small, temporary small stitches that bind bodies and spaces together (Anderson and Stoodley, 2019). In seeing swimming in the ocean as an act of suture, the connection can be temporary and lasting, always leaving behind a small scar or mark on the body-memory. The remainder of the chapter will introduce wider writing on swimming and explore themes linked to swimming bodies, spaces and experiences and the therapeutic potential of swimming, as well as suggesting some future directions for oceanic swimming research.

## New natarographers

Swimming in the ocean has always been an emotional and affective encounter that has lent itself, almost reflexively, to narration. There are many different narrative strands that reflect context and location, place reflections and specifically (auto)ethnographic accounts of its meaning and value for individual and communal lives (Gould et al., 2021; Moles, 2021; Sherr, 2012). As a starting point for such encounters, historical studies of the beach and the coast generally identified the ocean and its overlapping *limen* with land as an interaction-space for multiple societies, with direct impacts on leisure and medical histories and cultural and artistic associations (Anderson and Tabb, 2002; Lencek and Bosker, 1998; Mack, 2013). Of particular importance was its liminality and identity as a space 'away', where people relaxed or took time out/away from the stresses and worries of everyday life, but equally were able to discover themselves and enact roles and identities often suppressed in everyday life (Gould et al., 2021; Parr, 2011; Shields, 1991). It is no coincidence that many established LGBTI urban spaces – Brighton, San Francisco, Sitges, Key West, Provincetown, Sydney – are all on the coast (Hoare, 2017). As an activity that takes place in the sea/ocean, swimming is one important practice that cements that liminal identity, but also opens the body up to itself in physical acts of discovery and skill (Evers, 2015; Hoare, 2017; Straughan, 2012). This is evident in a rapidly expanding popular literature on nature engagement, within which the idea of 'wild' swimming (at least in British writing) features prominently (Atkinson, 2019). New 'natarographers' were initially inspired by Roger Deakin's *Waterlog* (2000), an account of his swimming journey across Britain that inspired other place-based, societal and individual 'wild' swimming narratives (Atkinson, 2019; Hoare, 2017; Parr, 2011; Sherr, 2012). In the public sphere, there has been a publishing boom in swimming biographies, often by women and regularly framed around affecting life events linked to active negotiations of illness/wellness (Fitzmaurice, 2017; Heminsley, 2017; Lee, 2017). Finally, there is a developing awareness by these natarographers of a respect for oceanic space and other more-than-human bodies and objects – cetaceans, jellyfish, plastics, vessels – that inhabit and share those same spaces (Hoare, 2017; see also Johnson, this volume; Squire, this volume). A common theme across these narratives is a deeper oceanic understanding specifically produced by repeated immersive encounters in the water.

While there is limited space to develop in depth, much recent writing on both the ocean and swimming have engaged with wider turns in cultural theory, especially broadly relational aspects of non-representational theory and assemblage as well as ongoing discussions around affect, embodiment and post-humanist thinking (Andrews et al., 2014; Broch, 2020; Duff, 2014; Foley, 2017b; Lorimer, 2008; Wetherell, 2015). The core ideas will be discussed below, but much of the literature takes a broadly relational approach, within which both swimmers' bodies (as affective intimate sensors), and oceanic swimming places/spaces (as mobile becoming environments), shape oceanic spatial outcomes (Anderson, 2012; Throsby and Evans, 2013). These outcomes are summarised below in relation to accretion, flow, mood and experientially framed (un)moorings that have social, cultural and wellbeing dimensions. In addition, critical wider human geography has addressed societal aspects of swimming and swimming spaces, specifically in relation to inequalities in access, gendered relations and bodily difference (Alaimo, 2012; Throsby, 2015; Wiltse, 2007).

Those same theoretical aspects also inform recent research in geographies of health and wellbeing, within which swimming has emerged, alongside surfing, sailing and other blue space activities, as providing important immersive accounts of how ocean spaces become both health-enabling and health-endangering (Foley et al., 2019). Foley's explorations of swimming and swimming places along the Irish coast is one starting point, concerned with the value of swimming to older and disabled bodies as well as a wider therapeutic accretion produced in place. Atkinson has also developed thinking around the sometimes contested and commodified idea of 'wild' swimming, which as a term captures a sense of adventure and challenge, but also acts as a synonym for more everyday and thankfully banal experiences of open-water swimming (Atkinson, 2019). Literatures of surfing, while documenting a different immersive practice and experience, have much in common with swimming; new research is combining the two in comparative explorations of 'place capture' (Britton and Foley, 2021). Other recent innovative work explores the possibilities of new 'in-situ' methodologies (Spinney, 2014) to expand exactly how to record that specific immersion in blue space; traditionally difficult to do, yet increasingly made possible by new technologies (action cameras, Fitbits, GPS) and accounts from directly within the water via 'swim-along' interviews (Britton, 2019; Denton and Aranda, 2020; Evers, 2015; lisahunter, 2018).

Finally, it is important to recognise that swimming in open water – the more so the farther out from shore one swims – can have considerable risks and negative health outcomes and experiences (Collins and Kearns, 2007; Pitt, 2018). Similarly, the near-ocean is also a site for pollution, collisions and in the complex third spaces of the beach, more unhealthy social encounters as well (Bell et al., 2018). As a very different strand, and one not considered in this chapter, there are also specifically scientific explorations of swimming bodies from sports science, physiology and psychology that focus on identifiable and measurable changes in bodily capacities and the benefits of physical exercise to improve human fitness and strength (Britton et al., 2018; Dugué and Leppänen, 2010; Leonard et al., 2018).

## Swimming bodies, spaces and experiences

In discussing embodiment, blue space and experiential outcomes as three core themes, the idea of an immersive encounter remains significant. All three themes are described separately yet clearly connect and overlap, building into a form of living assemblage to identify how swimming can be better understood and valued as an oceanic act. The specifics of swimming as a relational immersive encounter are explored around: accretions (flow and embodiment); moods (affective practice) and moorings/unmoorings (oceanic third spaces).

## Accretive flows and swimming bodies

For all swimmers, the action of ocean swimming requires them to get their bodies wet inside a natural and mobile bath of salt water. While historically, reluctant bodies were vigorously immersed by professional 'dippers' (Parr, 2011), the contemporary swimmer has considerable autonomy in how they negotiate their own immersive encounter with the ocean (Atkinson, 2019). The relational and material geographies of that embodied encounter are experienced by literal feeling/sensing bodies within which flow and accretion are important elements (Foley, 2017). The idea of flow features heavily in writing around swimming as a sporting and leisure activity, with competitive open-water swimmers (like surfers) talking about the ways in which different types of flow linked to their own bodies gliding through currents and swells bring them into a sort of encapsulated zen moment (Anderson, 2012; Csikszentmihalyi, 1997). For everyday swimmers, flow was also an important part of everyday experience and a way of challenging the body.

> Sometimes if you are against a current it's very interesting to challenge yourself … and I think that can be important too … where we swim … there are quite strong currents … and if you go with the flow you can cover a distance in three minutes … if you go against the flow that could take you 40 minutes to do against the flow. In summer when I am reasonably fit I quite like the challenge of working against the flow … course you know the thing is lots of people try swimming against the flow and forget that they can turn back if they can't make it.
>
> *Foley, 2017: 48*

The reference here to currents, as material flowing dimensions of ocean swimming in particular, was also important in terms of the immersive experience: for cold-water swimmers in Northern Europe and California, that current provided an embodied shock when they dived into it but also acted as a manager of flow in terms or risk or gain (Alaimo, 2012; Atkinson, 2019; Tipton et al., 2017).

Ongoing relationships between feeling/sensing bodies, body-memories, skill and time-specific engagements in the encounter were identified as important in building up resilient strength within a health-enabling accretion (Anderson and Stoodley, 2019; Foley, 2017). There are many debates in swimming in relation to differential experiences between bare skin and wetsuits, with the former often identified as a more authentic experience, depending on the warmth and condition of the water (Parr, 2011; Throsby, 2015). For swimmers who prefer the unmediated chill of the encounter in cold water, that chill is what makes it the visceral and accretive encounter it becomes, within which intimate individual bodies sense changing temperatures, seasons and shifts across life courses (Foley, 2017). For others, using wetsuits allows them to stay in the water longer and in competitive sporting terms, improved the speed of the flow (Knechte et al., 2020; Throsby, 2015). Even in a less direct embodied encounter, water still gets trapped between the skin and rubber and acts as both insulation and encrusted layering. Finally, body size and shape, especially for long-distance and endurance swimmers, also help to mediate a layered experience. Throsby's (2015) description of English Channel swimming identified the protective value of extra fat layers in a rare immersive encounter in deeper oceanic space. Yet the idea of a therapeutic accretion, linked to regular immersions in the ocean invoked oceanic processes like encrustation on an anchor or the development of a seed pearl, to suggest the swimming body bears traces of that regular embodied flow (Foley, 2017).

## Sea moods and immersive encounter

As well as the value to physical health of being-in-the-water, swimming as an immersive encounter has strong purchase within wider debates around affect, emotion and mood (Andrews et al., 2014; Foley, 2017). Much debate within non-representational research focuses on the separation of affect and emotion and the perceived 'unrepresentability' of the former as an almost spectral pre-cognitive force (Macpherson, 2010). Others argue for a more pragmatic take on what might be termed 'affective practice' (Bondi and Davidson, 2011; Foley, 2015; Wetherell, 2015). For Wetherell (a social psychologist slightly exasperated with geographers' hair-splitting), affect makes sense when expressed through a practice approach that; "positions affect as a dynamic process, emergent from a polyphony of intersections and feedbacks, working across body states ... entangled with cultural meaning-making and integrated with material and natural processes, social situations and social relations" (Wetherell, 2015: 139). In swimming terms, the whole process can be described as affective practice, from a sniff of the sea as an affective trigger, to being-in-the-water as an affective experience, to the post-swim 'glow' as an affective outcome in shared space (Denton et al., 2021). Using an affective practice approach brings together embodied, emotional and experiential dimensions that produce affective encounters through action/practice and a lamination of affect through the body in both pre-, during- and post-cognitive form (Foley, 2017).

In applying such an affective practice approach in relation to swimming, it is no harm to revisit mood as a connecting element. In considering the idea of a 'sea-mood' specifically, this can be considered in relation to both the swimmer and the sea, before, during and after a swim, emerging (literally) to translate mood into experience and memory-making. From earlier writing (Foley, 2017), it can be suggested that such affective swimming practices – always multiple and immanent becoming encounters – can be seen as a continuum of mood into experience. In its pre-phase, this can incorporate the smell or sight of sea as affective cue, a confluence of weather and time as internal affective push, decision and movement to the ocean as activated affect, leading to the decision to just do it (Fitzmaurice, 2017). Several swimmers talk about not feeling up to it – "I'm not in the mood" – but something ineffable inside them pushes them on (Foley, 2015; see also, Denton and Aranda, 2020) such that upon arrival at the ocean, the wider seascape and shared moods with other swimmers converts cue into action.

During the swim – the mood-in-experience if you like – there are a set of 'during affects' that become evident in decisive visceral ways. Swimmers describe the first unbuttoning of clothes as being a key step, with another related to an almost always recurring 'just-before' moment and then the entry into the ocean, variously quietly timid or splashily loud (Bell et al., 2018). Almost all senses are immediately engaged the minute one enters the water; sound, touch, sight, taste and smell, with many accounts identifying these as the moments the body and water become permeable objects that in the encounter, co-produce the experience (Anderson and Stoodley, 2019; Hartley, 2019). There are different responses depending on the swell and these are always uncertain and in-situ in their expression. Swimmers describe the experience of swimming some intense initial strokes out into the ocean and then looking back on the land and how that act alone shifts their own sense of affective emplacement and space orientation (Foley, 2015). Depending on whether one is swimming alone or with others, or for a leisurely bob or an intense training session, there are aspects of durability, physicality and age that trigger the length of the immersion, followed by a self-directed sense of when to come in. Sometimes in cold water that is a very short duration and experience, called 'teabag swimming' by some, or it can be a long session in warm and calm seas (Foley, 2017). Almost all swimmers talk about an affective sense of transformation when in the water, in terms of one's own physical and mental awareness and an always renewing sense of their own bodily capacities (Doughty, 2019; Hoare, 2017).

As a swim ends, and bodies leave the ocean, there are additional relational post-swim affects and mood-created-out-of experience as one winds down from the immersion; still part of the overall experience, yet a forerunner of the next one. While many swimmers talk of the glow of being back on land and being 'stung to warmth' (Regan, 2009), the lingering effects of what for some still feels like an out-of-body experience stays affectively present via a gladness drawn from a whole range of intermingled feelings and affects (Anderson and Stoodley, 2019). From participant observations there are a whole range of 'recovery sounds' – blowing, gasping, chattering, shaking, slapping – as the body warms up and clothes are put on (Andrews et al., 2014). The post-affective mood can often include relief, smiles and a sense of achievement, though there are sometimes other moods or disappointments if people didn't get in for long, or swim as much as they had planned, or indeed if they have had a fright, been stung by a jellyfish or banged a body part on a rock (Britton and Foley, 2021). Across all of the phases of a swim, what is being brought out are affects, emotions and moods that may have been previously hidden or buried for long or short time periods, that become enabled through a material practice that is specifically immersive and directly encountered. It is both the immersion and the encounter that affects us and makes what we do a form of meaningful leisure (Silk et al., 2017).

## Moorings/unmoorings and the role of place

In following on the mention of mood, the more-than-human moods of both the sea and the littoral places and spaces in which swimming happens are equally important components of immersive encounter and provide an important context for affective practice (Brown and Humberstone, 2015; Brown and Peters, 2018; Wetherell, 2015). Whatever about the swimmer, the sea too has its moods, its own emblematic energetic signal (Hartley, 2019) and as Philip Hoare noted, "the sea is never your friend" (Foley, 2015: 221). Another swimmer's passing description of the mobile colours of seawater, noted how mutable in form and mood it could be; "if the sea is agitated it could have sparkle" (Foley, 2017: 48). Equally that agitation could be violent and angry, reflecting both historical and recent accounts (Brown and Humberstone, 2015) and the ocean's ability to deter swimmers when a churning mood takes it. As Steinberg and Peters (2015: 255) note, "churnings of the ocean both enables and disrupts … earthly striations".

In linking bodies and places together, feeling/sensing bodies are always engaged in physical/emotional negotiations with different blue spaces (Broch, 2020; Hartley, 2019). These can include blurred, jagged edges of rocky coasts or smooth spaces of sheltered beaches and a range of man-made sutures – piers, platforms, diving boards – that bond the two together. One coastal swimmer from Brighton imagined their body as a small vessel, with the legs as an outboard engine that suggests a very specific assemblage sensibility and an unmooring from a static fixed position on the hard surface of the earth (Denton and Aranda, 2020). But equally there is a strong active role for the ocean in shaping swimming places and especially the mobile and subtle differences between cliffs, rocks, beaches and manmade built structures. Here those architectural notions of suture, re-orientation and permeable alignments of land and sea become important place components (Bowring et al., 2018; Ryan, 2012). They also align the immersive act of swimming as a reparative suture in the ongoing tension between culture and nature to moor spaces down. For those always uncertain built structures that enable swimmers to access the water, there are many accounts from cultural geography and wider natarographies around specific issues linked to access, loss, exclusion and rejection (Atkinson, 2019; Bell. et al., 2019; Wiltse, 2007). These can also include a loss of confidence, capacity or connection linked to other events in people's lives (Britton and Foley, 2021; Foley, 2019; Peters and Brown, 2017). In terms of swimming as a specific act of unmooring, a group of Cornish swimmers described their post-work evening swims as a form of erasure and wiping the everyday slate clean; being put back into shape by the swim (Silverstone et al., 2017). In suggesting swimming as a form of unmooring, it is also balanced by its moored settings, within which a range of important psychological benefits – stress-reduction, attention-restoration and social connection – also emerge (Doughty, 2019). This restoration of attention to both self and place (several swimmers talk about a renewed sense of perspective and purpose in the water) is a reflection of the act of re-balancing that takes place in the water, not unlike the realignment of chi/doshi (vital energies) in Ayurvedic and CAM practice (Bell et al., 2018).

Yet a full recognition of the ocean as a potentially dangerous space and its own liminal encounters with land, emphasise its immanent and open qualities linked to both land and sea (Macpherson, 2010). For swimmers these are most recognisable in the occasional dangers and risks involved in swimming in certain locations on land as well as the always open possibilities of encounters in the water with the more-than-human; sharks, jellyfish, dolphins, jet skis, boats, surfboards etc. (Brown and Peters, 2018). A recognition of swimming places as also being essentially unmoored applies to both. Swimming place like coves, beaches and rocks are, more often than not, free and open for all users and in essence become 'third spaces' where mixed encounters, of gender, generation, sexuality, class and indigeneity/race are acted out, with both positive and negative results in terms of how such spaces are managed and governed (Atkinson, 2019; Olive and Wheaton, 2021; Phoenix et al., 2020; Wiltse, 2007). Yet the unmooring is even more pronounced offshore in the swell of the waves, with no specific outcome guaranteed. Here that unmooring becomes a relational place encounter, where after casting off from the solid ladder or rock, individual place bathymetries' rhythms and flows play out. In this oceanic continuum, depth of experience seems to deepen as you go deeper off shore while to do so is to disconnect or unmoor oneself. This in turn is an act of becoming, developed through need, possibilities afforded by the mobile space (swell, weather, more-than-human co-residents), or one's own 'in the moment' physical and mental state (Anderson, 2013). As noted in many accounts, the importance of adventure and challenge emerges regularly in swimmer's accounts, as do the ideas associated with stretching bodily limits and the completion of a set of epic/wild swims (Lee, 2017; Sherr, 2012) despite everything from exposure and hypothermia to jellyfish stings and shark attacks (Throsby, 2015). Finally, several writers also correctly identify the capacity of oceanic waters to act as spaces of oblivion, often chosen deliberately for acts of self-harm for precisely that sense of the water closing over as a marker of rest and passage into an underwater/world (Foley, 2015; Hoare, 2017).

## Therapeutic waterscapes

Across all of the preceding themes there is an undercurrent – an affective sonic echo – that swimming is good for you. Across a range of literatures, especially environmental psychology and health geography, but also in biomedical and medical humanities work, there are implicit/explicit connections between swimming, health and wellbeing (Atkinson, 2019; Foley et al., 2019; Swim England, 2017). Given that swimming and the ocean are co-constituents of what we might term therapeutic waterscapes, that affective echo can also translate into an explicit dimension-shaping critical public health and environmental planning policy (Swim England, 2017). In addition, strong messages emerge from swimmers' narratives of an enhanced sense of place care and attachment that comes specifically from their immersions in it (Britton and Foley, 2020; Gould et al., 2021). This reflects an attentive oceanic vision in which 'blue space care' produces benefits for both the human and more-than human health of the ocean, from within a shared and sustainable vision (Brown and Peters, 2018; Olive and Wheaton, 2020). It also tallies with parallel work on green space and sustained immersion in nature, that has a mutually beneficial outcome for society (Hartig et al., 2014); while the suspended body in water reflects Alaimo's (2012) notion of trans-corporality that draws both from 'the sea within' and without, to produce a positive blue space connection.

In relation to the perceived health and wellbeing effects of swimming, elements of identity, ownership, autonomy and attachment have already been noted. Some of these findings have emerged from new methodologies that directly record in-situ encounters or 'place capture' in the water (Britton and Foley, 2021; Denton and Aranda, 2020; Foley, 2019). Again, using blue space research in the surf as an analogue (Anderson, 2012; Britton, 2019; Evers, 2015), attempts to use swim-along interviews, actions-cameras and emplaced (auto)ethnographic elicitations are part of a new wave of blue geo-narratives (Spinney, 2014; Bell et al., 2017). Denton and Britton's recent work in Brighton and Galway respectively are good examples, working with swimmers in the water and recording specific aquatic responses as well as those post-affective moments immediately after the swim. Here the voices are unmediated and attest to immediate feelings of achievement, worry, reflection, memory and, above all, a renewed appreciation of the potential of their own bodies. Several respondents in both locations speak of feeling like superwomen/men, when they go into work having already done a swim (Britton and Foley, 2019; Denton et al., 2021). Those direct accounts reflect moments of concern/fear in a natural space that is always an irregular and challenging environment. Equally the subjects of some recent interventions with swimmers and surfers have been young people with emotional health issues, for whom swimming (despite expressed concerns around self-harm) seems to act, once in the water, as both a literal and metaphorical salve and as an important source of resilience for a group not traditionally noted for it (Britton and Foley, 2019; Denton and Aranda, 2020). Perhaps, as in all good forms of therapeutic intervention, we must immerse ourselves in our fears to overcome them.

As a therapeutic space and practice, swimming draws from both biomedical and social-relational models of health. From specifically physiological and psychological perspectives there are considerable literatures from rehabilitation and recovery medicine, while open water swimming is beginning to gain acceptance as an acceptable therapeutic intervention for mental health (Gaunt and Mafulli, 2012; van Tulleken et al., 2018). Recent referrals for women with breast cancer to take up outdoor swimming recognise both the direct health (building pectoral muscle and overall strength) and indirect wellbeing (shared socialisation, group supports) benefits that enhance recovery (Costello et al., 2019; Foley, 2015). Equally importantly, in relation to people with chronic conditions or degenerative conditions, swimming can support aspects of illness management and recurrent treatment, to demonstrate its always ongoing value as a place-based coping mechanism. In a critical public health sense, it is that mixture of science and art that best promotes health and swimming is a good example of such an open and accessible intervention (Williams, 2010). An area of keen and developing interest are enabling relationships between swimming and various forms of disabled and othered bodies of difference, within which key findings note that swimming has the transformative capacity to make a disabled land body an enabled sea body (Bell et al., 2018; Foley, 2015; Throsby, 2015). Pitt (2018) also correctly notes that an enabled land body can equally become disabled in the water; both are useful embodied markers of essential oceanic dimensions of invertibility and churn that allow for multiple therapeutic outcomes (Steinberg and Peters, 2015).

## Oceanic 'blue bodies'

A wider dimension of the ocean as a cleanser and a space for renewal has always made it an attractive place for the swimmer to dive in to (Foley, 2019). There is a parallel machismo in the global trope of the leaping teenager on the first warm day of the summer; a test of courage and rite of passage that mark a crossing over from one group to another (Turner, 1969). In theoretical terms, one can see the crossing over from solid land into the flowing sea as a relational affective expression of a shift in the plane of immanence, to allow oneself to be immersed in a 'shiftspace' that brings one into an aquatic assemblage, from which one re-emerges always slightly altered (Duff, 2014; Spinney, 2014). Oceanic immersion is also an act of alterity, of a transformation and even a therapeutic accretion that comes from place as well as affective practice (Anderson and Stoodley, 2019). It taps into almost all our senses and emotions and is at its heart an embodied experience that, through a set of open connections across multiple subjects and settings creates space to develop an unbounded oceanic 'blue body' (Foley et al., 2019). While the frequency, regularity and shape of the encounter varies, it does provide an ongoing suturing of self, land and waterscape across the lifecourse. As a medical metaphor, the suture acts increasingly as a new form of stitching and repair into the skin, leaving a less visible but no less important trace on the body.

There are many shades and palettes also to what can sometimes be contested spaces and practices (Foley and Kistemann, 2015). Just as there are inherent divisions and snobberies in surfing, the swimming experience can also be contested, around clothing (bare skins or wetsuits), skill (experts and beginners), authenticity (year-round versus occasional swimmers) and gender (costume choice and bodily exposure choices). Despite this, there is a real sense that oceanic swimming spaces are generally open and within wider socio-spatial contexts and constraints, inclusive spaces. But there is also almost no research on either private pools by the ocean or private beaches that commodify the ocean in a visibly bounded way or indeed in a way that is potentially resource-damaging (Bell et al., 2019; Head and Muir, 2007; Phoenix et al., 2020). Several of the swimming spaces described here are former 'men-only' spots, while the relationships between sometimes naked or near-naked bodies in shared inter-generational and inter-gendered spaces can also be potentially disquieting (Caudwell, 2020; Hoare, 2017). Swimming is also an act carried out by bodies of all shapes and sizes, all of which have some capacity for buoyancy which is also relational in terms of time and place; linked to an often unknown personal capacity (until in the water), but also linked to memories of what the body did or could do, what it does or can do now, or what it might be able to do in the future (Throsby, 2015). Swimmers' accounts regularly attest to swimming as a particular practice, again because of its immersive and resolutely material physical characteristics, to really stretch their limits or begin to understand what limits mean for them (Britton and Foley, 2019; Sherr, 2012).

In suggesting directions for future research, there is an almost oceanic scale of potential; disability, gender, sexuality, regulation, inequality, class and ethnicity have all been flagged already. It is also true that much of the literature relates to cold or even icy water environments, murky and intermittently blue that by extension produce 'blue' swimming bodies (Knechte et al., 2020). Yet swimming tropes are global and there is a need, and one suspects, a readiness, to carry out meaningful research on warm water swimming. Such swimming might seem easy and uninteresting, but it would be valuable to identify different or parallel enabling experiences linked to effects of length of immersion and clarity of water (Foley et al., 2019). Similarly, in public health terms there are interesting new outdoor spaces using heated outdoor water and swimming spaces fed by the ocean to promote open-water swimming. These in turn can be linked to wider critical public health discussions around social prescribing and the positive health promotion or – in a more negative sense – governmentality of citizen bodies to provide a different strand to consider within marine spatial planning (Caudwell, 2020; Doughty, 2019). At the beach, one is often struck by the relative prevalence on public signage of regulatory health and safety and behavioural prohibitions, yet the ocean is always a space where such risk-averse geographies need to be balanced against adventure, autonomy, joy and the dignity of risk for bodies young and old (Phoenix and Orr, 2014). In finishing with a return to oceanic space as a limen, and swimming as a creative and healthy act within that space, one is always brought back to limits/limitations and the liminal potentialities of the sea, as a space for various gazes and deeper explorations of the swimmer as a gendered/different/becoming body across the lifecourse of both shared lives and shared spaces (Longhurst and Johnston, 2014).

## References

Alaimo S (2012) States of suspension: Trans-corporeality at sea. *Interdisciplinary Studies in Literature and Environment* 19(3): 476–493.

Anderson J (2012) Relational places: The surfed wave as assemblage and convergence. *Environment and Planning D: Society and Space* 30(4): 570–587.

Anderson J (2013) Cathedrals of the surf zone: Regulating access to a space of spirituality. *Social & Cultural Geography* 14(8): 954–972.

Anderson J and Stoodley L (2019) Creative compulsions: Performing surfing as art. In: Roberts L and Phillips K (eds) *Water, Creativity and Meaning: Multidisciplinary Understandings of Human-Water Relationships*. Abingdon: Routledge, 103–123.

Anderson S and Tabb BH (eds) (2002) *Water, Leisure and Culture*. Oxford: Berg.

Andrews G, Chen S and Myers S (2014) The 'taking place' of health and wellbeing: Towards non-representational theory. *Social Science & Medicine* 108: 201–222.

Atkinson S (2019) Wellbeing and the wild, blue 21st-century citizen. In: Foley R, Kearns R, Kistemann RT and Wheeler B (eds) *Blue Space, Health and Wellbeing: Hydrophilia Unbounded*. Abingdon: Routledge, 190–204.

Bell S, Foley R, Houghton F, Maddrell A and Williams A (2018) From therapeutic landscapes to healthy spaces, places and practices: A scoping review. *Social Science & Medicine* 196: 123–130.

Bell S, Leyshon C, Foley R and Kearns R (2019) The "healthy dose" of nature: A cautionary tale. *Geography Compass* 13(1) https://doi.org/10.1111/gec3.12415

Bell S, Wheeler BW and Phoenix C (2017) Using geonarratives to explore the diverse temporalities of therapeutic landscapes: Perspectives from 'green' and 'blue' settings. *Annals of the American Association of Geographers* 107(1): 93–108.

Bondi L and Davidson J (2011) Lost in translation. *Transactions of the Institute of British Geographers* 36(4): 595– 598.

Bowring J, Vance N and Abbott M (2018) Between seascape and landscape: Experiencing the liminal zone of the coast. In: Brown M and Peters K (eds) *Living with the Sea: Knowledge, Awareness and Action*. Abingdon: Routledge, 15–35.

Britton E (2019) *Dúchas*: Being and belonging on the borderlands of surfing, senses and self. In: Foley R, Kearns R, Kistemann T and Wheeler B (eds) *Blue Space, Health and Wellbeing: Hydrophilia Unbounded*. London: Routledge, Taylor and Francis, 95–116.

Britton E and Foley R (2021) Sensing water: Uncovering health and well-being in the sea and surf. *Journal of Sport and Social Issues* 45(1): 60–87.

Britton E, Kindermann G, Domegan D and Carlin C (2018) Blue care: A systematic review of blue space interventions for health and wellbeing. *Health Promotion International* 35(1): 50–69.

Broch TB (2020) Sensing seascapes: How affective atmospheres guide city youths' encounters with the ocean's multivocality. *Journal of Sport and Social Issues* 45(2): 161–178.

Brown M and Humberstone B (eds) (2015) *Seascapes: Shaped by the Sea*. Ashgate: Farnham.

Brown M and Peters K (eds) (2018) *Living with the Sea: Knowledge, Awareness and Action.* Abingdon: Routledge.
Caudwell J (2020) Transgender and non-binary swimming in the UK: Indoor public pool spaces and un/safety. *Frontiers in Sociology* https://doi.org/10.3389/fsoc.2020.00064
Collins D and Kearns R (2007) Ambiguous landscapes: Sun, risk and recreation on New Zealand beaches. In: Williams A (ed) *Therapeutic Landscapes.* Ashgate: Farnham, 15–32.
Costello L, McDermott, ML, Patel P and Dare J (2019) 'A lot better than medicine': Self-organised ocean swimming groups as facilitators for healthy ageing. *Health & Place* 60 https://doi.org/10.1016/j.healthplace.2019.102212
Csikszentmihalyi M (1997) *Finding Flow: The Psychology of Engagement with Everyday Life.* New York, NY: Basic Books.
Deakin R (2000) *Waterlog. A Swimmer's Journey Through Britain.* London: Vintage.
Denton H and Aranda K (2020) The wellbeing benefits of sea swimming. Is it time to revisit the sea cure? *Qualitative Research in Sport, Exercise and Health* 12(5): 647–663.
Denton H, Dannreuther C and Aranda K (2021) Researching at sea: Exploring the 'swim-along' interview method. *Health & Place* 67 https://doi.org/10.1016/j.healthplace.2020.102466
Doughty K (2019) From water as curative agent to enabling waterscapes: Diverse experiences of the 'therapeutic blue'. In: Foley R, Kearns R, Kistemann T and Wheeler B (eds) *Blue Space, Health and Wellbeing: Hydrophilia Unbounded.* London: Routledge, Taylor and Francis, 79–94.
Duff C (2011) Networks, resources and agencies: On the character and production of enabling places. *Health & Place* 17: 149–156.
Duff C (2014) *Assemblages of Health.* New York, NY: Springer.
Dugué B and Leppänen E (2010) Adaptation related to cytokines in man: Effects of regular swimming in ice-cold water. *Clinical Physiology* 20(2): 114–121.
Evers C (2015) Researching action sport with a GoPro TM camera: An embodied and emotional mobile video tale of the sea, masculinity and men-who-surf. In: Willard I (ed) *Researching Embodied Sport: Exploring Movement Cultures.* Abingdon: Routledge, 145–163.
Fitzmaurice R (2017) *I Found my Tribe.* London: Chatto and Windus.
Foley R (2010) *Healing Waters: Therapeutic Landscapes in Historic and Contemporary Ireland.* Farnham: Ashgate.
Foley R (2015) Swimming in Ireland: Immersions in therapeutic blue space. *Health & Place* 35: 218–225.
Foley R (2017) Swimming as an accretive practice in healthy blue space. *Emotion, Space and Society* 22: 43–51.
Foley R (2019) Mapping a blue trace: An intermittent swimming life. In: Roberts L and Phillips K (eds) *Water, Creativity and Meaning: Multidisciplinary Understandings of Human-Water Relationships.* Abingdon: Routledge, 87–102.
Foley R, Kearns R, Kistemann T and Wheeler B (eds) (2019) *Blue Space, Health and Wellbeing: Hydrophilia Unbounded.* London: Routledge, Taylor and Francis.
Foley R and Kistemann T (2015) Blue space geographies: Enabling health in place. Introduction to Special Issue on Healthy Blue Space. *Health & Place* 35: 157–165.
Gaunt T and Maffulli N (2012) Soothing suffering swimmers: A systematic review of the epidemiology, diagnosis, treatment and rehabilitation of musculoskeletal injuries in competitive swimmers. *British Medical Bulletin* 103(1): 45–88.
Gould S, McLachlan F and McDonald B (2021) Swimming with the Bicheno "Coffee Club": The textured world of wild swimming. *Journal of Sport and Social Issues* 45(1): 39–59.
Hartig T, Mitchell R, de Vries S and Frumkin H (2014) Nature and health. *Annual Review of Public Health* 35: 207–228.
Hartley J (2019) Waves as emblemata for knowledge. In: Roberts L and Phillips K (eds) *Water, Creativity and Meaning: Multidisciplinary Understandings of Human-Water Relationships.* Abingdon: Routledge, 124–135.
Head L and Muir P (2007) Changing cultures of water in eastern Australian backyard gardens. *Social & Cultural Geography* 8(6): 889–905.
Heminsley A (2017) *Leap In: A Woman, Some Waves and the Will to Swim.* London: Hutchinson.
Hoare P (2017) *RISINGTIDEFALLINGSTAR.* London: Fourth Estate.
Knechte B, Waśkiewicz Z, Sousa CV, Hill L and Nikolaidis P (2020) Cold water swimming – benefits and risks: A narrative review. *International Journal of Environmental Research and Public Health* 17(23): https://doi.org/10.3390/ijerph17238984

Lee JJ (2017) *Turning: A Year in the Water*. Toronto: Penguin.
Lenček L and Bosker G (1998) *The Beach. The History of Paradise on Earth*. London: Secker and Warburg.
Leonard A, Singer A, Ukoumunne OC, Gaze WH and Garside R (2018) Is it safe to go back into the water? A systematic review and meta-analysis of the risk of acquiring infections from recreational exposure to seawater. *International Journal of Epidemiology* 47(2): 572–586.
lisahunter (2018). Surfing, sex, genders and sexualities. London: Routledge.
Longhurst R and Johnston LT (2014) Bodies, gender, place and culture: 21 years on. *Gender, Place and Culture* 21(3): 267–278.
Lorimer H (2008) Cultural geography, non-representational conditions and concerns. *Progress in Human Geography* 32(4): 551–559.
Mack J (2013) *The Sea: A Cultural History*. London: Reaktion Books.
Macpherson H (2010) Non-representational approaches to body-landscape relations. *Geography Compass* 4(1): 1–13.
Merchant S (2016) (Re)constructing the tourist experience? Editing experience and mediating memories of learning to dive. *Leisure Studies* 35(6): 797–808.
Moles K (2021) The social world of outdoor swimming: Cultural practices, shared meanings, and bodily encounters. *Journal of Sport and Social Issues* 45(1): 20–38.
Olive R and Wheaton B (2021) Understanding blue spaces: Sport, bodies, wellbeing, and the sea. *Journal of Sport and Social Issues* 45(1): 3–19.
Parr S (2011) *The Story of Swimming: A Social History of Bathing in Britain*. Stockport: Dewi Lewis Media.
Peters K (2010) Future promises for contemporary social and cultural geographies of the sea. *Geography Compass* 4(9): 1260–1272.
Peters K and Brown M (2017) Writing with the sea: Reflections on in/experienced encounters with ocean space. *Cultural Geographies* 24(4): 617–624.
Phoenix C, Bell S and Hollenbeck J (2020) Segregation and the sea: Toward a critical understanding of race and coastal blue space in Greater Miami. *Journal of Sport and Social Issues* 45(2): 115–137.
Phoenix C and Orr N (2014) Pleasure: A forgotten dimension of ageing and physical activity. *Social Science and Medicine* 115: 94–102.
Pitt H (2018) Muddying the waters: What urban waterways reveal about bluespaces and wellbeing *Geoforum* 92: 161–170.
Regan N (2009) Forty foot Christmas. In: Howard A (ed) *Slow Dublin*. Melbourne: Affirm Press, 156–157.
Ryan A (2012) *Where Land Meets Sea: Coastal Explorations of Landscape, Representation and Spatial Experience*. Farnham: Ashgate.
Sherr L (2012) *Swim: Why we Love the Water*. New York, NY: Public Affairs.
Shields R (1991) *Places on the Margin: Alternative Geographies of Modernity*. London: Routledge.
Silk M, Caudwell J and Gibson H (2017) Views on leisure studies: Pasts, presents & future possibilities? *Leisure Studies* 36(2): 153–162.
Silverstone T, Robertson G and Tait M (2017) The new retirement. *The Guardian [Online]*. Available at: www.theguardian.com/society/video/2017/feb/13/wild-sea-swimming-in-my-60s-erases-problems-being-child-again-video
Spinney J (2014) Close encounters? Mobile methods, (post)phenomenology and affect. *Cultural Geographies* 22(2): 23–46.
Steinberg PE and Peters K (2015) Wet ontologies, fluid spaces: Giving depth to volume through oceanic thinking. *Environment and Planning D: Society and Space* 33(2): 247–264.
Strang V (2004) *The Meaning of Water*. Oxford, New York, Berg.
Straughan E (2012) Touched by water: The body in scuba diving. *Emotion, Space and Society* 5(1): 19–26.
Swim England (2017) *The Health & Wellbeing Benefits of Swimming*. London: Swim England.
Thompson N and Wilkie S (2020) 'I'm just lost in the world': The impact of blue exercise on participant well-being. *Qualitative Research in Sport, Exercise and Health* 13(4): 624–638.
Throsby K and Evans R (2013) 'Must I seize every opportunity?' Complicity, confrontation and the problem of researching (anti-)fatness. *Critical Public Health* 23(3): 331–344.
Throsby K (2015) 'You can't be too vain to gain if you want to swim the Channel': Marathon swimming and the construction of heroic fatness. *International Review for the Sociology of Sport* 50(7): 769–784.
Tipton MJ, Collier N, Massey H, Corbett J and Harper M (2017) Cold water immersion: Kill or cure? *Experimental Physiology* 102(11): 1335–1355.
Turner VW (1969) *The Ritual Process*. London: Penguin.

van Tulleken C, Tipton M, Massey H and Harper C (2018) Open water swimming as a treatment for major depressive disorder. *BMJ Case Reports* http://dx.doi.org/10.1136/bcr-2018-225007

Ward M (2017) Swimming in a contained space: Understanding the experience of indoor lap swimmers. *Health & Place* 46: 315–321.

Wetherell M (2015) Trends in the turn to affect: A social psychological critique. *Body & Society* 21(2): 139–166.

Williams A (2010) Therapeutic landscapes as health-promoting places. In: Brown T McLafferty S and Moon G (eds) *A Companion to Health and Medical Geography*. Chichester: Wiley-Blackwell, 207–233.

Wiltse J (2007) *Contested Waters: A Social History of Swimming Pools in America*. Chapel Hill, NC: University of North Carolina Press.

# 24
# SURFING
## The what, where, how and why of wild surfing

*Jon Anderson*

This chapter investigates the experience of surfing oceanic space. It looks at contemporary issues and future agenda for how this practice – that is both a leisure activity, a competitive sport, an artistic endeavour, a spiritual quest, and in many cases a colonising and environmentally destructive culture – are studied and understood. The chapter will outline the 'what' and 'where', 'how' and 'why' of surfing, before exploring key geographical research themes that assemble around the globalisation of surfing spaces.

### The what of surfing

As it is generally understood, "surfing is the deceptively simple act of riding a breaking ocean wave" (Kampion and Brown, 2003: 27). Surfing is a cultural activity which shares many of the defining features of all lifestyle sports (see Tomlinson et al., 2005; van Bottenburg and Salome, 2010). As Wheaton (2004) outlines, lifestyle, or 'whizz', sports refer to a range of cultural activities including rock climbing, sky-diving, skateboarding, and snowboarding. Defined as individualistic in nature (as opposed to team-oriented) these sports are understood to be participatory in nature (rather than spectator-focused). They centre on skill, risk, and hedonism, often display resistance to regulation and institutionalisation, and remain ambiguous in their relationship to competition (Wheaton, 2004: 12; although see Pill, 2019). Associated surfing cultures, including surf photography, surf film making, surf dress, fashionwear, advertising, and marketing may often be far more passive and consumerist than the act of surfing itself; however, these related cultures both mould and disseminate the cultural (b)orders of the practice (see Anderson, 2015a; Midol, 1993; Tomlinson, 2001; Rinehart and Sydnor, 2003; Stranger, 1999; 2011). The synergy between these cultures has led to surfing becoming an incredibly fast-growing and popular lifestyle sport. From surfing's modern incarnation, which was conceived in 1950s California, current participation is estimated to have grown from 10 million in 2002 (LipChain, 2018) to 35 million in 2014, with surfing now practised in over 160 countries across the globe (Ponting and O'Brien, 2014; see also LipChain, 2018).

DOI: 10.4324/9781315111643-29

## The where of surfing

As Kampion and Brown note above, surfing is broadly defined as an oceanic activity. More specifically, most accessible waves occur in the zone where the land meets the sea, in the "transitional, highly porous border between the primeval terrestrial and aquatic" (Barilotti, 2002: 34). Despite the overwhelming majority of surfed waves occurring in this littoral zone, surfing can also occur in other parts of hydrological cycle. For example, it is increasingly possible to go offshore and engage in deep-sea surfing, operationalised through boat charter. This practice often accesses larger waves and employs teams of riders on petrol-powered wave-skis to 'tow-in' board riders to wave crests. Beyond oceanic space, surf riding can also occur on naturally occurring estuarine bores, channelled river waves, and increasingly in artificial wave parks. The proliferation of wave parks challenges many aspects of surfing as it is currently known. These artificial surfing spaces change the geographical location of surfing (from the coast to inland, and potentially from outdoors to indoors), they alter the direct connection between waves and natural cycles (e.g. the generation of waves through solar heating, pressure change, wind generation, tidal and current influence, and associated erosion and deposition of rocks and sand on continental shelves), and they remove the skills necessary to identify where and when waves will break (as 'artificial' waves can now be designed, created, and timed to order). In short, wave parks usher in a new generation of surfing spaces and raise the need to become more sophisticated in the categorisation of the geography of this activity; it may be that littoral and oceanic surfing soon become known as 'wild' surfing (complementing the nomenclature of swimming in coastal, riverine, and ocean spaces; see Bottley, 2019, and for more, Anderson, 2022).

## The what and where of surfing: Research themes

Human geography could be characterised as exploring and defining the ongoing relations between people and place. With respect to the experience of surfing oceanic space, human geography's traditional research focus is somewhat realigned. In traditional western societies, the spatial identity of humans is predominantly oriented towards terrestrial spaces, however surfers are constitutively defined not simply by their co-ingredience with terrestrial places, but also to the spaces of the surf zone. It is in this zone that surfers feel they are alive and at home; and thus it is this mix of 'surf–shore' constitution that becomes the relation of interest for surf geographers (Anderson, 2014b, 2015b).

Surfers' identity is often rooted around a home surf zone and its associated local shores. Such 'bonds' are social, cultural, political, psychological and emotional in nature, and tie a place and its residents together in a constitutive co-ingredience that builds up over time and becomes part of a local culture and lifestyle. From this perspective, surfers' identity is, in the words of Cresswell, "actively territorialised" (2004: 110) not simply in the sea, but in the surf–shore assemblage. Clifton Evers highlights how this co-ingredience is felt by the surfing body, and how this surf–shore attachment becomes part of the surfer identity:

> Surfers form a sensory relationship with the local weather patterns, sea-floors, jetties and rock walls. Surfers' bodies intermingle with the coastal morphology, and it can be hard to tell where the local's body begins and the local environment ends. Knowing how to ride 'with' a wave at a particular spot is a clear marker that you're a local and works as a way to signal ownership of a space in an increasingly crowded surfing world. The environment and how it works becomes so ingrained that a local should

> be able to tell the different surf seasons by the way their body feels. We bond with th[is surf-shore] geographical turf.
>
> *Evers, 2007: 4*

As Evers outlines, the human body and the watery world interact together to form the assembled identity of the surfer. As Shields suggests, this interaction forms a provisional co-ingredience; for surfers "there is a tremendous complicity between the body and the environment and the two interpenetrate each other" (1991: 14). This emphasis of the body is vital to the practice and experience of study, and by extension it is also integral to the way in which surfing is studied.

## The how of surfing: Bodies and technologies

In the popular imagination, surfing is often reduced to a singular and homogenous activity: in this caricature, surfing is (simply) board riding. However, in practice, there are many approaches and technologies (Michael, 2000) that can be used to ride the breaking wave. Surfers can be categorised in terms of those who like to lie to surf, those who prefer to sit, and those who must stand in order to catch their waves. For each bodily approach there comes an array of styles that can be adopted (ranging from a 'glide' to a 'shred'), as well as an array of crafts that can be mobilised (including the body, a board, or a boat). With each position, style and technology, the place of the surfed wave changes for each of these surfing neo-tribes (to use Maffesoli's phrase, 1996).

For those surf riders who prefer to lie down to catch their waves, bodysurfing and body-boarding are the most popular surfing activities. Bodysurfing is the least mediated engagement with the surface swell, and involves floating then swimming before the cresting wave, and manoeuvring the human body (often using half-cut swim fins on the feet and/or hands) so the breaking wave can carry the human forward on its moving energy. Bodyboarding, in turn, involves lying prone on a small, torso-length, board, being well-positioned (initially with board and body facing to the shore as a wave approaches) and catching waves as they careen to shore. Bodyboarding is commonly undertaken in shallow water, using generally affordable boards which offer both buoyancy and refuge to the novice participant, and as such offers a relatively easy entry into surfing activity.

For those who prefer to sit down to surf there are a number of alternatives on offer. Surf-kayaking represents the modern equivalent of Polynesian canoes, kayaks and catamarans, the original pre-modern means to ride the surf. Surf-kayaks offer a sit-in hull, and require the use of paddles to generate the power and mobility to cross through the surf zone to beyond where the waves are breaking. Surf kayakers can then choose which waves to paddle for, coinciding their position and speed to guide their route down the breaking wave. Surf-skis are flat(-ter) alternatives to surf-kayaks; here the would-be rider sits on the ski (rather than in it) often strapped on with easy-release belts around the waist or thighs. Surf-skis are highly mobile when compared to surf-kayaks, but also highly unstable; sitting on rather than in the water they are liable to tip when stationary, but will glide effortlessly when propelled by the sea's momentum (or your paddling). Sit-on-top kayaks, by contrast, sit deeper in the water, have more stability, and less manoeuvrability. Often made from air-blown plastic, sit-on-tops offer greater control at stationary or slow speeds, but this stability trades off the craft's ability to reach high speeds and prime mobility in fast water. Both surf skis and sit-on-top kayaks are easy to 'bail out' from when they do capsize – meaning that riders do not need to 'Eskimo roll' a surf-ski or sit-on-top kayak, they can simply fall from their craft, and climb back on board when the wave has broken.

For those who wish to stand up to surf, the dominant technology is the popularly known surf-board. Boards come in a range of sizes with different aims of mobility and speed. Stand up paddle boards (or SUPs) are long, wide 'boards', often made from wood, polyurethane, or even inflatable plastic, which are ridden with an extended paddle which can help to generate momentum as well as steer the moving SUP. According to Fordham, SUPs are "ideal for easy wave-catching and smooth-flowing, style-conscious manoeuvres" (2008: 75). Short-(surf) boards are often known as 'thrusters' and perhaps the most well-known type of modern surf technology. Short-boards enable surfers to "turn more radically and to take aerial manoeuvres out over the lip of the breaking wave" (Fordham, 2008: 154). Long-boards are also widely ridden today, and have their origins in the birth of modern surfing in California. Long-boards, due to the length and weight, encourage a slower, more sedate surfing style, allowing the rider to move along to the front of the board to 'hang ten' (i.e. place all their toes over the nose of their board as the wave carries them). This board and this style is commonly perceived to be more elegant than the 'bmx-trix' style of the short-boarding, and is often associated with so-called 'soul surfers'. The largest and most durable stand up boards are known as 'guns', and are made for being towed-in to the biggest, offshore waves.

Different body positions, ridden-style, and technology shapes the experience of the surfed wave. The 'place' of the surfed wave (see Anderson, 2012) for a surf-*kayaker*, for example, is therefore subtly but importantly different from that of the surf*boarder*. Each technology enables a different 'dwelling-in-motion' on the sea (after Sheller and Urry, 2006) and each unique coming together of swell, fetch, geology, wind, surfer and riding-technology "shapes [the] experience of place, as well as shaping [the] place [of the surfed wave] itself" (after Price, 2013: 124). To paraphrase Edensor, the assemblage of the surfing body and riding technology "weaves a path that is contingent, and accordingly produces contingent notions of place as well as being always partially conditioned by the special and physical characteristics of place" (2010: 70).

## Carnal surfing and an actor-centred perspective

We have seen in this chapter that surfers are not defined by the traditional spaces of geographical enquiry. They mobilise and generate a surf-shore identity. We have also seen that 'the place of the surfed wave' is a not only materially liquid, but also metaphorically 'liquid' – in other words, it not only changes its visual appearance and specific geographical location on a moment to moment basis, but human engagement with it changes its constitution and how it is understood. The emergent and always changing reality of surfing space therefore offers a challenge to geographers who wish to study it. In order to meet this challenge, scholars have often turned to 'actor-centred perspectives' (after Jones, 2009), seeking to understanding the reality of the surfed wave by focusing attention on how humans "engage with the world tactually" and how this engagement then comes to "constitutes ... reality" (Lewis, 2000: 59). One key actor-centred perspective is the 'carnal' approach to geography.

Carnal geography emphasises the importance of the body to knowing. It is a fundamental challenge to the separation and privileging of the mind as the 'best' way of gaining insight into the world, and by implication, re-balances rationalisation, reflection, and post-moment mental cognition as simply one set of competing sources of onto-epistemological knowledge. Carnal geography raises questions regarding how bodily knowledge can be harnessed and communicated, how it relates to more traditional and restricted knowledge fragments, and, how we can effectively frame, approach and realise research questions that foreground embodied engagement in the world.

Carnal geography has its origins in Loic Wacquant's carnal sociology (2011, 2015). Wacquant's approach encourages the exploration of the body as a key means through which to engage with and understand the world (e.g. Cerulo, 2015; Dutkiewicz, 2015; Falaix, 2015; Pitts-Taylor, 2015). Promoting knowledge through embodied practice raises questions with regard to the nature of epistemology, the timing and action of method, and broad purpose and product of academic endeavour. In short, carnal geography invokes a radical challenge to how we know ourselves and our world. As such, this challenge is part of the destabilising and dissolution of what Latour (1993) described as the 'modern constitution' and has led to a re-embodied and newly relational framing of the world.

For carnal geographers, no longer are humans simply detached, isolated minds rationally reflecting on our distance and disconnection from the world, but rather we are spatial beings, with our very attachment to the world actively informing our mutual constitution. Such relational approaches (themselves informed by the geographies of Doel [1999] and Murdoch [2006] and the actor networks of Callon [1986] and Latour [1999]), mark the shift away from the independent conceptual categories of the modern constitution and towards an *inter* dependent epistemology where things are always acting and being acted upon by everything else. In sum, they mark a rejection of a static (phenomenological) ontology of 'being-in-the-world', and an embracing of a more emergent and emerging (post-phenomenological) ontology of 'becoming-in-the-world' (primarily associated with Deleuze, 1985, 1993).

With respect to the experience of ocean surfing, the actor-centred approach emphasises not simply the 'fluid' nature of surfer-identity (they are defined in part by their engagement with water worlds), but also that this identity itself is changeable, influenced in part by anticipating the moment of engagement, the moment of engagement itself, and its divergent legacies. This approach raises new questions with regard to what (surfing) humans do, why they do it, and what further actions are mobilised as a consequence. In the next sections we address some of these questions, starting with 'why surf'?

## The why of surfing

When approaching surfing from an actor-centred perspective, bodily affects are often cited as key motivations for surfing. Issues of bodily health, physical challenge, and skill enhancement are commonly referred to, but the dominant response often is the simple thrill experienced through the act of wave riding. This affect is often labelled by surfers as 'stoke'. In common parlance, stoke alludes to the rekindling and relighting of a fire that may be quietening, being dowsed, or dying. In the context of surfing, stoke refers to the re-energising of the human – the relighting of the fire and energy in their body and spirit – as a consequence of their engagement with the cresting energy of waves. Stoke is described by Duane as "the light joy of effortless, combustion-free speed" (1996: 11) and by Kampion as "the real-time neural stimulation and restorative prophylaxis … resulting in a stunning net profit on time and energy invested" (2004: 44). Stoke refers at once to the thrill of personal skill accomplishment (i.e. it refers in part to the successful catching of a wave and carving a route through its moving mass of water – see Csikszentmihalyi, 1990; and Anderson, 2012), as well as to the relational sensibility produced in the human through a sublime encounter (see Anderson, 2013) and convergent experience (Anderson, 2012) – or as one surfer simply puts it, stoke refers to the "immense feeling of being carried by the sea" (see Anderson, 2015b).

Identifying stoke as a key motivation for surfing is therefore a vital step in understanding the practice. However, it also raises broader questions with regard to how geographers, and indeed humans more generally, can communicate such carnal knowledge to others. If motivations for actions are rooted in sense, engagement, and bodily compulsion between people and places (in this case riders and the surfed wave), understanding the experience of oceanic surfing raises questions about the importance of sharing the "affective intensities [and] enduring urges" (Lorimer, 2005: 83) formed through engagement with the world, and, how can we successfully communicate them with a range of audiences.

Methodologically, this had led to scholars participating, observing, and filming phenomena, as well as requesting participants themselves to document events in real time (through photographs, film, narration etc.; see Garrett, 2011; Simpson, 2011). This had led to identifying how the bodily experience of riding the surfed wave prompts participants to smile, scream and talk of near-religious (see Anderson 2013a; Taylor 2007a, 2007b, 2007c, 2008) and mystical experiences. As Shaun Thomson, Surf World Champion in 1977, reflects:

> [w]hen you go into a deep barrel you certainly feel as if time's expanded. Life is slowed down. I felt as if I could curve that wall [of water] to my will. I really felt that. It's a magical, magical moment.
>
> *Thomson, quoted in Gosch, 2008*

Despite these insights, it remains difficult to communicate the 'beta' of a rock climb (see Dutkiewic, 2015) or the 'stoke' of a surfed wave to another climber, surfer, or human being/becoming. If embodied knowledge is central to understanding the experience of the surfed wave, can only those who have managed to engage with cresting water know how that changes one's sense of self and outlook on the world? Indeed, if each wave is a unique, emerging place, can only those who have experienced that particular coming together truly know what that event feels like? If the answer to this question is 'yes', then this is an isolating conclusion. This isolation is deepened when one considers how, in all situations, the experience and language of the embodied world often escapes the capacity of modern language to faithfully capture it. Even when we share encounters we are apparently trapped within the relatively crude vocabularies and representational instruments of language to communicate our experiences with others (see Bondi, 2005).

In this situation, exploring the experience of oceanic surfing raises questions concerning how we can find new spaces of understanding and collaboration to express our embodied encounters. How can we activate and encourage our own and our audience's experiences, however indirect or tangential, to create a currency of communicated lived experience, one that is freighted with empathic resonance, and with the ability to be moved, felt, and known? In relation to surfing practice, the power of surf writing has the capacity to move and trigger empathy amongst participants and spectators (see Anderson, 2014a). It is also possible that surf photography, from the shore, boat, ski, or board, has a similar capacity to recreate a moment and crystallise a brief encounter to anyone who experiences it. The advent of waterproof housing and miniaturisation of digital technology has also enabled researchers to capture surfer-eye-views of the green room, the wipe out, and the line-up, perspectives that were once the preserve of the surfer themselves. Similarly, it is now technically possible to replicate surfing virtually, 'the green room' via 'green screen' if you will, with reality augmented with mobile platforms, sound, temperature, and even water, to represent a particular break or individual wave. How can these technologies help share, and also help us understand, the nature of surfing, the threat of white water, and the appeal of the glide? Can emotional registers, captured through biomapping and electro dermal sensors also help us get closer to the effects of surfing on the body (see for example, Nold, 2009)? If reliable and waterproof devices become available (or indeed virtual surfing can be advanced to adequately stand in for 'standing' on a board), can neuro transmissions and the body electric be tapped into in order to chart the emotional journey of a ride? How can these graphs be contrasted to normal land life, and how can they be explained by the medical practitioner, social scientist, and surfer, in order to gain new understandings of the embodied experiences in the water world?

## The who of surfing

As we have seen, the urge to experience a range of bodily affects motivates the practice of surfing, and these affects are summed up in the apparently simple but incredibly complex word 'stoke'. As ocean space offers the unique forum in which surfers can access this bodily affect, surf zones become vital places in many individuals' lives. When a number of individuals seek to access stoke at one particular surf break, competition arises with respect to who can catch each wave. As access to waves effectively translates as access to stoke, "surfers have developed a complicated set of norms or rules that govern behaviour in the surf and priority over the waves" (Nazer, 2004: 656); in short, the why of surfing directly influences who can surf any particular surfing space.

As soon as more than one person seeks access to ocean waves, dominant cultural orders actively border that site (see Anderson, 2021). Surfing spaces are (b)ordered in numerous ways, initially with respect to how many people can surf a particular wave. Historically, waves were shared (indeed in contemporary times many surfers actively seek to share waves [see teamwavestorm, 2019]). However, in the modern version of surfing, waves have become individualised in nature – the dominant maxim is 'one surfer, one wave'. Who is permitted to 'own' a particular wave is often determined by an individual's *position* with respect to the breaking edge of that wave: the individual who is closest to the crest has priority and others have to defer. As Steve Bough (former editor of surfing magazine *Wavelength*) states, "surfing's primary rule is that the surfer who is already on the wave has right of way" (Wade, 2007: 86).

This rule, although dominant in most locations, is not absolute. Those who profess a strong surf-shore identity with a particular surf break sometimes feel that their local *provenance* outweighs any other's *position* with respect to the wave. This trumping of wave-position by surfer-provenance not only demonstrates the fluid and provisional nature of some surf (b)orders, but also how they are more likely to be flouted by particular 'tribes' of surfers than others. It is common for some locals to act on their strong surf-shore affinity by 'actively territorialising' a set of surfing (b)orders that benefit those who are regular participants in a surfing line-up over those who may be new or irregular visitors to that site. This risk of 'local rules for local surfers' is often enforced with a range of unfriendly tactics on both the sea and shore in order to intimidate and ultimately dissuade new surfers from competing for waves. In some cases, it is enough for the stories of 'localism' at particular sites to convince would-be surfers from competing for waves in the best conditions.

Identifying the (b)orders of position and provenance goes some way to demonstrating that surfing spaces are not free and open zones, but rather places that are explicitly and implicitly controlled by surf cultures. As we have also seen, there are a variety of ways to surf and further surf (b)orders ebb and flow with respect to not only the type of craft welcome at particular breaks, but also the type of surfing body that is 'in place' at certain sites (after Cresswell, 2004). (B)ordering the surfing body occurs with respect to the chosen riding position on a craft (e.g. lying, sitting, standing), as well as expertise of that body in the wave conditions. As surf is not graded in the same way as ski runs are (for example, green for 'learners', black for 'experts'), surfers themselves use 'localism' style tactics to dissuade those who they perceive to be 'out of their depth' on a particular series of waves from accessing and potentially wasting the scarce resource. Even more importantly, surfing bodies are not simply deemed in or out of place in terms of technology and expertise, but also in terms of gender, race, and (dis)ability (for more, see lisahunter, 2018; Nemani, 2015; Olive, 2019). Finally, it is becoming increasingly common for access to surfing spaces to be controlled by ability to pay as waves become privatised, enclosed, or artificially designed.

## The spread of surfing

The dominance of particular surf (b)orders in particular spaces opens up a range of new questions for human geographers. For example, how do particular (b)orders become dominant, in which places? How are they spread across the surfing globe, and how are they resisted? The global spread of surfing, and the associated cultures of surf fashion, reportage, and marketing increases the importance of these questions, and require us to explore how surf cultures may be becoming homogenised, in what ways, and by whom?

The (b)orders of modern surfing are increasingly influenced by the industries serving the practice. Serving surfers' needs has become a lucrative market; Surfer Today (2014) estimates that the industry is worth over US$20 billion, whilst companies such as Quiksilver report annual turnovers in the multiple billion dollars (e.g. US$1.81 billion in 2013, as cited in Surfer Today, 2014: no page). The surfing craft these manufacturers produce, alongside the wetsuits, boardshorts, neoprene boots, rashvests and thermal armour, become part of the surfer assemblage that engages with and in turn comes to (b)order the water world. Surfboards, for example, become amalgams of function and fashion, they are not simply purposeful by providing mobility to the wave, an island-like refuge whilst waiting for waves, or the ability to catch and ride waves when they occur, but are also "symbolic, even talismanic… of cultural, social and emotional meanings" (Warren and Gibson, 2014: 1). Surfboards are thus at once technologies which enable particular types of engagement with particular types of waves (e.g. the shortboard for aerial or acrobatic surfing on quick, fast and often messy waves; or the longboard for long, slow rides on larger, cleaner waves). They are also cultural objects that promise affective encounters with waves and present aesthetic meanings to the broader cultural world about the nature of the board-rider (see Featherstone, 1991). As Warren and Gibson confirm, surfboards "are now used to construct a personal identity as much as answer a utilitarian purpose" (2014: 10). Surfing craft, as well as other facets of surf dress, therefore become an integral dimension of being a surfer, for this neo-tribe "consumption for adornment, expression and group solidarity become not merely the means to a lifestyle, but the enactment of lifestyle" (Shields, 1991: 16). As a consequence of this expression and enactment, those who produce craft and clothes have a growing ability to influence the type of worlds surf spaces become, and which (b)orders dominate the markets in which they operate. It is through designing and advertising surf craft and clothes to the core surfing market, and surf-inspired shore-side clothes to the general public, that these surf companies attempt to "control the clothes and shorts that [surfers themselves] wear, and even the way we ride waves" (Surfer Today, 2013: no page).

The influence of the surf industry but also the surf media (including websites and magazines) are therefore crucial areas for geographic inquiry. As marketing campaigns by companies such as Ripcurl and Quiksilver suggest (see Quiksilver, 2013; Surfer, 2014), surf industries have identified and sought to profit from the growing trend of surfers to satiate their need for stoke through travel. Although surfers have a strong surf-shore identity, often tied to particular local spaces, an increasingly evident trait of the culture is the popularity of surf travel. Enabled by advances in transport, communication and surf forecasting, surfers can identify where ocean swells may be meeting particular shore lines across the globe and time their arrival to catch the waves. This 'trans-local' surfing activity creates tensions with particular local cultures, in many cases displacing the pre-existing surfing or other maritime (b)orders in that location, and substituting in their place those of the travelling surf tribe. Surfing therefore has significance beyond the actor directly involved in the activity, and has wider cultural, social, economic and environment impacts on the littoral spaces that host surf breaks. In a germinal piece of surf journalism, Barilotti acknowledges these wider impacts with respect to the village of Kuta on the island of Bali, Indonesia:

> Kuta Beach started out as a drowsy little fishing village in the 1930s, catering to a small number of vacationing European colonialists. Its surf potential was discovered by Australians in the mid-1960s. Since then, it has morphed into a fully-fledged surf ghetto on a par with Huntington Beach or the North Shore. ... In our blind zeal to set up insular surf enclaves, we parachute advanced technologies into third-world economies and set up brittle unsustainable infrastructures. The list of soiled third-world surf paradises [like Kuta] ... is long and growing.
>
> *Barilotti, 2002: 92*

Across the archipelagos of South East Asia, as well as other popular surf destinations across the globe, the creeping cultural colonialism that is present in Kuta is replicated, and is often more intense. Summarised by Barilotti, across South East Asia, "4000 years of ancient animistic squat culture [has] now smacked straight into Western techheavy materialism" (2002: 92). This has led to the cultural influence of western society colonising the traditions of regional and local cultures, as Barilotti explains with respect to Nias:

> the effect of surf tourism on the Niah, a proud, warlike tribe once notorious for their headhunting and elaborate costumed rituals, has sped the erosion and disappearance of traditional ways. Twenty-five years of cashed-up westerners tramping through Lagundi village has seduced the local youth with lurid Baywatch fantasies of the North American high life.
>
> *Barilotti, 2002: 93*

In Bali, as local journalist Eric reports, "for centuries, Balinese women have obediently carried the responsibilities assigned by tradition. Now, however, tradition is becoming increasingly compatible with modern life [spread by surf travel]. Women are beginning to ask questions about their own destinies" (2011: 30). Whilst western ideas of feminism may be seen to be progressive when compared to South East Asian patriarchies, the imposition of foreign ideas involves cultural imperialism and colonisation. Substituting long-held traditions for industrialised poverty or employment in hawking, prostitution, and casual labour, all in the service of surf tourism, appears to be a dubious advancement.

We can see, therefore, that the challenges posed by the spread of surfing and the satiation of stoke raises questions both for surfers and human geographers. How should surfing seek to sustain its own activity without destroying the societies of others? How can all surfing spaces, and associated shore cultures, be protected for the future?

## Protecting surf-shore spaces

Individuals who enjoy the experience of surfing the ocean are responding to these questions by translating their individual concern for stoke into collective campaigning to protect surf-shore spaces. This collective campaigning involves protests which respond to threats to specific surf breaks (see Save the Waves, 2019a), as well as ongoing activities which seek to limit environmental pollution in the oceans (see for example Wheaton, 2007). Surfers are also central in actively responding to the perceived absence of areas which protect surf-shore cultures by creating their own designations which identify, value, and sustain local social relations and visiting surfers to mutual benefit (see Save the Waves, 2019b). The nature of surfing (b)orders that these designations promote, combined with the advancement of artificial wave parks, offer human geographers two exciting new areas of critical inquiry. Is the future of surfing the construction of 'experience preserves', locations that – when the conditions are right – provide access to fleeting coming togethers that enhance life experience through embodied encounter? In such a scenario, how might access to these locations be regulated? Whose emotions are most valuable, most economically beneficial, or ethically sound, and who decides? The implications of carnal geography for how we consider the (surfing) world, our place within it, are astonishingly profound. For those of us who consider the water world to be integral to our identities, and vital to make our lives valuable, insights from the oceanic experience of surfing help us chart the potential that embodied and affective engagements can have in advancing our understanding of the relations between people and place.

## References


Anderson J (2012) Relational places: The surfed wave as assemblage and convergence. *Environment & Planning D: Society & Space* 30(4): 570–587.

Anderson J (2013) Cathedrals of the surf zone: Regulating access to a space of spirituality. *Social and Cultural Geographies* 14(8): 954–972.

Anderson J (2014a) Exploring the space between words and meaning: Understanding the relational sensibility of surf spaces. *Emotion, Space & Society* 10(1): 27–34.

Anderson J (2014b) Surfing between the local and trans-local: Identifying spatial divisions in surfing practice. *Transactions of the Institute of British Geographers* 39(2): 237–249.

Anderson J (2015a) On trend and on the wave: Carving cultural identity through water-side surf dress. Special Issue: Dress. *Journal of Leisure Research* 19(2): 212–234.

Anderson J (2015b) On being shaped by surfing: Experiencing the world of the littoral zone. In: Brown M and Humberstone B (eds) *Seascapes: Shaped by the Sea.* Farnham and Burlington, VT: Ashgate, 55–70.

Anderson J (2021) *Understanding Cultural Geography: Places & Traces*. London: Routledge.

Anderson J (2022) *Surfing Spaces*. Abingdon: Routledge.

Barilotti S (2002) Lost horizons. Surfer colonialism in the twenty-first century, *The Surfers' Path* 33: 30–39.

Bondi L (2005) Making connections and thinking through emotions: Between geography and psychotherapy. *Transactions of the Institute of British Geographers* 30: 433–448.

Bottley K (2019) Winter wild swimming as individual and corporate spiritual practice. *Practical Theology* 12:3 343–344.

Callon M (1986) Some elements in a sociology of translation. In: Law J (ed) *Power, Action, Belief.* Abingdon: Routledge, 19–34.

Cerulo K (2015) The embodied mind: Building on Wacquant's carnal sociology. *Qualitative Sociology* 38: 33–38.

Cresswell T (2004) *Place: A Short Introduction*. Oxford: Blackwell.

Csikszentmihalyi M (1990) *Flow: The Psychology of Optimal Experience.* New York, NY: Harper and Row.

Deleuze G (1985) Nomad thought. In: Alison D (ed) *The New Nietzsche.* Cambridge, MA: MIT Press, 142–149.

Deleuze G (1993) *The Fold.* Minneapolis, MN: University of Minnesota Press.

Doel M (1999) *Poststructuralist Geographies: The Diabolical Art of Spatial Science*. Lanham, MD: Rowman and Littlefield.

Duane D (1996) *Caught Inside: A Surfer's Year on the Californian Coast*. New York, NY: North Point Press.

Dutkiewicz J (2015) Pretzel logic: An embodied ethnography of a rock climb. *Space and Culture* 18(1): 25–38.

Edensor T (2010) Walking in rhythms: Place, regulation, style and the flow of experience. *Visual Studies* 25(1): 69–79.

Eric (2011) Women of Bali. *The Mag* 37 May: 29–32.

Evers C (2007) Locals Only! Proceedings of the Everyday Multiculturalism Conference of the Centre for Research on Social Inclusion, Macquarie University, 28–29 Sept. 2006, Sidney, Australia.

Falaix LK (2015) Aloha spirit. La vague habitee comme rampart a l'institutionnalisation de la culture surf. *Nature and Recreation* 2: 28–43.

Featherstone M (1991) *Consumer Culture and Postmodernism*. London: Sage.

Fordham M (2008) *The Book of Surfing: The Killer Guide*. London: Bantam Press.

Garrett B (2011) Videographic geographies: Using digital video for geographic research. *Progress in Human Geography* 35(4): 521–541.

Gosch J (2008) *Bustin' Down the Door*. New York, NY: Screen Media Films.

Jones M (2009) Phase space: Geography, relational thinking, and beyond. *Progress in Human Geography* 33(4): 1–20.

Kampion D (2004) *The Lost Coast*. Salt Lake City, UT: Gibbs-Smith.

Kampion D and Brown B (2003) *A History of Surf Culture*. Los Angeles, CA: General Publishing.

Latour B (1999) *Pandora's Hope: Essays on The Reality of Science Studies*. Cambridge, MA: Harvard University Press.

Latour B (1993) *We Have Never Been Modern*. Hemel Hempstead: Harvester Wheatsheaf

Lewis N (2000) The climbing body, nature and the experience of modernity. *Body & Society* 6(3–4): 58–80.

LipChain (2018) How many surfers are there in the world? *Facebook* 30 June Available at: www.facebook.com/1754237891332927/posts/how-many-surfers-are-there-in-the-world-according-to-ponting-and-obrien-the-late/1771274642962585/

Lisahunter (2018) *Surfing, Sex, Genders and Sexualities*. London: Taylor and Francis.

Lorimer H (2005) Cultural geography: The busyness of being 'more-than-representational'. *Progress in Human Geography* 29(1): 83–94.

Maffesoli M (1996) *The Time of the Tribes*. London: Sage.

Michael M (2000) *Reconnecting Culture, Technology and Nature: From Society to Heterogeneity*. Abingdon: Routledge.

Midol N (1993) Cultural dissents and technical innovations in the 'whiz' sports. *International Review for Sociology of Sport* 28(1): 23–32.

Murdoch J (2006) *Post-structuralist Geography: A Guide to Relational Space*. London: Sage.

Nazer D (2004) The tragicomedy of the surfers' commons. *Deakin Law Review* 9: 655–713.

Nemani M (2015) Being a brown bodyboarder. In: Brown M and Humberstone B (eds) *Seascapes: Shaped by the Sea*. Farnham and Burlington, VT: Ashgate, 55–70.

Nold C (2009) *Emotional Cartography. Technologies of the Self*. London: Space Studios.

Olive R (2019) The trouble with newcomers: Women, localism and the politics of surfing. *Journal of Australian Studies* 43(1): 39–54.

Pill E (2019) Waves of power. The spectacularisation of professional surfing. Unpublished doctoral dissertation, Cardiff University.

Pitts-Taylor V (2015) A feminist carnal sociology? Embodiment in sociology, feminism, and naturalized philosophy. *Qualitative Sociology* 38(1): 19–25.

Ponting J and O'Brien D (2014) Liberalizing Nirvana: an analysis of the consequences of common pool resource deregulation for the sustainability of Fiji's surf tourism industry. *Journal of Sustainable Tourism* 22(3): 384–402.

Price P (2013) Place. In: Johnson N, Schein, R and Winders, J (eds) *The Wiley-Blackwell Companion to Cultural Geography*. Oxford: Wiley Blackwell, 118–129.

Quiksilver (2013) Originals don't walk the path, they carve it. *Pinterest*. Available at: www.pinterest.com/pin/117304765266860651/

Rinehart R and Sydnor S (eds) (2003) *To the Extreme: Alternative Sports, Inside and Out*. Albany, NY: State University of New York Press.

Save the Waves (2019a) *Issues*. Available at: www.savethewaves.org/issues/

Save the Waves (2019b) *World Surfing Reserves*. Available at: www.savethewaves.org/programs/world-surfing-reserves/

Sheller M and Urry J (2006) The new mobilities paradigm. *Environment and Planning A* 38(2): 207–226.

Shields R (1991) *Places on the Margin: Alternative Geographies of Modernity*. Abingdon: Routledge.

Simpson P (2011) 'So, as you can see…': Some reflections on the utility video methodologies in the study of embodied practices. *Area* 43(3): 343–352.

Stranger M (1999) The aesthetics of risk: A study of surfing. *International Review for the Sociology of Sport* 34(3): 265–276.

Stranger M (2011) *Surfing Life: Surface, Substructure and the Commodification of the Sublime*. Farnham: Ashgate.

Surfer (2014) Maldives Issue. *Surfer* 55. July.

Surfer Today (2013) *The forces of power and influence in surfing*. Surfer Today website. Available at: www.surfertoday.com/surfing/9332-the-forces-of-power-and-influence-in-surfing

Surfer Today (2014) Red Bull is taking over surfing. Surfer Today website. Available at: www.surfertoday.com/surfing/10772-is-red-bull-taking-over-surfing

Taylor B (2007a) Focus introduction. Aquatic nature religion. *Journal of the American Academy of Religion* 75(4): 863–874.

Taylor B (2007b) Surfing into spirituality and a new, aquatic nature religion. *Journal of the American Academy of Religion* 75(4): 923–951.

Taylor B (2007c) The new aquatic nature religion. *Drift* 1(03): 14–23.

Taylor B (2008) Sea spirituality, surfing, and aquatic nature religion. In: Shaw S and Francis A (eds) *Deep Blue: Critical Reflections on Nature, Religion and Water*. London: Equinox, 213–233.

Teamwavestorm (2019) *Takeover Party Wave*. Available at: www.instagram.com/p/ByWd1KcgYfo/

Tomlinson A, Ravenscroft N, Wheaton B and Gilchrist P (2005) *Lifestyle Sports and National Sport Policy: An Agenda for Research*. Brighton: University of Brighton.

Tomlinson J (2001) *Extreme Sports: The Illustrated Guide to Maximum Adrenaline Thrills*. London: Carlton Books.

van Bottenburg M and Salome L (2010) The indoorisation of outdoor sports: An exploration of the rise of lifestyle sports in artificial settings. *Leisure Studies* 29(2): 143–160.

Wacquant L (2011) Habitus as topic and tool: Reflections on becoming a prizefighter. *Qualitative Research in Psychology* 8(1): 81–92.

Wacquant L (2015) For a sociology of flesh and blood. *Qualitative Sociology* 38(1): 1–11

Wade A (2007) *Surf Nation*. London: Simon and Schuster.

Warren A and Gibson C (2014) *Surfing Places, Surfboard Makers: Craft, Creativity, and Cultural Heritage in Hawai'i, California, and Australia*. Honolulu, HI: University of Hawai'i Press.

Wheaton B (2004) Introduction: Mapping the lifestyle sportscape. In: Wheaton B (ed) *Understanding Lifestyle Sports: Consumption, Identity, and Difference*. Abingdon: Routledge, 1–28.

Wheaton B (2007) Identity, politics, and the beach: Environmental activism in Surfers against Sewage. *Leisure Studies* 26(3): 279–302.

# 25
# SAILING
## The ocean around and within us

*Mike Brown*

### Preamble

I'm lying in my bunk and I can hear the boat 'talking'. Everything's alive – creaking and groaning.

I'm in constant motion. The boat moves and I move. The bunk boards flex under me.

Oh shut up!

Sleeping is difficult with this incessant racket.

I hear the swoosh of water as it rushes past my head. It's less than a foot away and I'm only separated from it by a few millimetres of fibreglass. I'm on the 'downhill' side of the boat. At times I'm actually under water as the boat rises and falls on the Atlantic swell. I'm conscious of every movement, every noise.

I'm desperately struggling not to vomit.

Five days later

Just after dinner we gybe onto starboard tack. I'm on watch from 23.30-02.30. I spend the first two hours focussing on adjusting to the wind coming over the other side of the boat. Initially I struggle to become accustomed to the new sensation of leaning the 'wrong' way. The wind is blowing in my right ear – this feels strange after days when my left ear was constantly exposed to the breeze.

Same boat, same wind strength – different motion.

No time for the iPod.

It's a new dynamic. I struggle to regain the intuitive feel – it takes me two hours of intense concentration to 'take in' my new bearings.

For the last hour I listen to music and relax.

The conscious becomes the unconscious once again.

These extracts, from a diary kept on a Trans-Atlantic crossing, recount the manner in which the movement of the ocean shaped the writer's experiences and ways of 'knowing' the ocean. We will return to these two accounts shortly.

DOI: 10.4324/9781315111643-30

## A brief overview of qualitative research on sailing

As a topic of inquiry long-term cruising sailors, people who give up shore-based life to go sailing, have attracted the most attention from scholars. Macbeth's foundational work (1992, 1998, 2000) focused on how cruising sailors came to share a subcultural ideology, the process of subcultural formation, along with why people participated in this lifestyle. He described the cruising life as a combination of freedom with challenge, where at best "results follow efforts and where one is confirmed existentially by surviving. At worst it is uncomfortable, insecure and sometimes frightening!" (Macbeth, 1992: 320). He found that cruising sailors were inspired by a utopian vision that sought an alternative definition of reality to that espoused in consumerist societies. His work also revealed that ocean sailing provided relief from the sense of alienation engendered in modern society by allowing participants to get closer to nature. Jennings's (1999) research also found that people went long-term ocean sailing as both an escape from their home society and in the pursuit of a more desirable lifestyle. These features included: freedom, control, adventure, challenge, and exposure to new cultures. Later research (Koth, 2013; Lusby and Anderson, 2008, 2010) reinforced how cruising sailors sought to both escape from the humdrum existence of consumerist land-based lifestyles and as a way to create new and enriching experiences. Koth referred to this as the quest for "positive freedom" where sailors could choose "*to* exert autonomous control and self-determination, in contrast to freedom *from* constraining structures" (2013: 148). The themes of freedom that emerge in these studies are arguably rooted in the European romantic movement, which draws on the discontent of industrialisation to herald a new attitude to the sea. Here the sea was conceptualised as a blank slate where one might be free (this idea is explored more later).

Through sea-based adventures a person might escape the banalities of shore-based life and its contaminating influences (Auden, 1951; Osborn, 1977). The sea, as a wilderness, became a place where you could re-create yourself; and as Ford and Brown (2005) argue, this motif of freedom continues to shape contemporary perceptions. A modern exemplar of this tradition is Frenchman Bernard Moitessier who famously withdrew from the first single-handed around the world yacht race (1968–69) when he was almost certain to claim the fastest time. After seven months at sea he sent a message to race officials simply stating, "I am continuing non-stop to the Pacific Islands because I am happy at sea, and perhaps also to save my soul" (Moitessier, 1974: 169). Three months later he arrived in Tahiti and his writings fired the imaginations of a generation of sailors who sought to escape the 'rat race' and to find peace at sea.

The notion of 'freedom of the seas' is, contends Steinberg (1999), based on the legal principle of *imperium* derived from the Roman control of the Mediterranean. As he pointed out in a subsequent work (Steinberg, 2001), what we take to 'be' the sea, and the extent of the freedom that might prevail are social constructions based on specific socio-cultural influences. Therefore, the previously mentioned studies that found cruising sailors viewed the sea as a site of freedom and escape is not surprising. What emerges is a combination of personal experience and the reiteration of existing metaphors. Freedom is not an inherent quality of salt water (whereas salinity is) – it is based in historical, economic and social traditions, which invariably advantage some people at the expense of others.

As briefly outlined above, research on sailors has been by scholars about *other* sailors. Until recently sailor/scholars have been hesitant to reveal aspects of their own engagement in this domain of social practice. This chapter provides an overview of writing from within human geography which has sought to provide first-person accounts of being with the sea. In doing so it opens up possibilities for new ways to express our relationship *with* the sea.

## Returning to the preamble

The two auto-ethnographic accounts that opened this chapter highlight the sentient aspects of being at sea and portray a different reality to images of freedom and escape. My 'freedom' to act was constrained, albeit temporarily, until I became attuned to my environment. The first account, written within 24 hours of departure on a 20-day trans-Atlantic sailing voyage, conveys the dis-harmony I experienced prior to becoming attuned to the environment onboard the boat. Seasickness inhibited my actions and impacted on the ability to perform tasks until such time as the continual motion was accepted as the new 'normal' and a sense of harmony with the environment was achieved. The second account shows the contingent nature of this attunement; altering course, with the boat now leaning the other way (what was the lower side of the boat is now the upper side), necessitated a process of reorientation and recalibration. This rebalancing was accomplished quickly and, is in fact, a frequent occurrence when sailing. Wave and wind patterns are constantly changing, the boat needs to alter course to maintain an optimum angle to the wind, and a process of constant adjustments are required to stay in harmony with the effects of different movement patterns.

## Auto-ethnography and 'knowing' the sea

Elsewhere Barbara Humberstone and I (Brown and Humberstone, 2015) have suggested that auto-ethnographic accounts by sailor/scholars, provide different perspectives on our relationship with the sea. Recent edited collections (see Anderson and Peters, 2014; Brown and Humberstone, 2015) have sought to provide new narratives of the human–sea relationship based on personal encounters that demonstrate the relational basis of human–sea interactions. Accounts of embodied experiences of the oceanic world – of being *with* the sea help to reveal how the sea shapes human experiences and contributes to a deeper understanding of how the sea permeates the very fabric of our being (Humberstone and Brown, 2015). This approach offers new ways of thinking about our relationship with the sea. Firstly, it centres the sailor/researcher's subjective experiences as data; the researcher is no longer investigating the experiences of the 'other'; secondly it provides a counter narrative to the portrayal of the sea in western thinking as unknowable or featureless. It has been argued that the positioning of the sea as featureless, unknowable or a void is a hallmark of western representations of the sea (Mack, 2011). Simon Winchester's popular, *Atlantic: A Vast Ocean of a Million Stories* reiterated this trope in its opening section when the author, standing on the deck of an Atlantic liner, stated "from here onward the sea yawned open wide and featureless, and soon took on the character that is generally true of all oceans – being unmarked, unclaimed, largely unknowable, and in a very large measure unknown" (2010: 8).

For the sailor the sea is neither empty nor featureless. On his Atlantic sailing trip, from the Caribbean heading towards Ireland, the Irish poet and sailor Theo Dorgan contemplated how descriptions of the sea as empty or featureless could be considered as "a crude and lazy shorthand, a way of saying 'I'm too busy to look, to see this as it is in itself'" (2004: 94). In the sections that follow I draw on a selection of recent literature, from within the broad field of human geography, written from the perspective of sailor/scholars, to show how our understanding of oceans can be enriched by close and attentive encounters *with* the sea. This approach is pertinent, for as leading figures in the field have pondered,

> How does our perspective change when we think not only from the sea, but with the sea? Over the past two decades, the sea has slowly crept into human geography.… Geography is 'earth-writing', and earthliness has been taken very literally in shaping the

> spaces in which geographical study has taken place. (Yet) as we have been arguing, new geographical knowledge can be unearthed when thinking from (and with) the sea…
>
> *Peters and Steinberg, 2015; in* Peters and Brown, 2017*: 2*

Autoethnographic accounts of sailors that are "attentive to the material conditions" of being at sea (Blum, 2010: 670) heralds an approach that extends our thinking about the sea in a manner which moves beyond "frameworks imported from existing discourses and takes the sea as a proprioceptive point of inquiry" (Blum, 2010: 671). The value of attending to the sea, through active engagement with it has the potential to not only alter how we think about our relationship with the sea but also to open up new approaches to expressing that relationship (Brown and Humberstone, 2015). The significance of auto-ethnographic accounts in understanding our engagement with the sea has considerable potential to enhance "empathetic forms of understanding" (Sparkes, 1999: 19) that can inform both the natural and social sciences. These sailor/scholar narratives of experiences with the sea enrich our understanding of thinking *with* the sea. This is significant for as Anderson and Peters (2014) note, we "understand and experience the sea as a 'place' with character, agency and personality" (2014: 9). These narratives have the "potential to challenge disembodied ways of knowing" that can take us "into the intimate, embodied world of the other in a way that stimulates us to reflect upon our own lives in relation to theirs" (Sparkes, 1999: 25).

Autoethnography enables each sailor/scholar to (re)present their lived experiences through the ways in which they describe their sensual and emotional relations with the sea. This approach to locating ourselves, through representations of lived experiences that portray our relationship with the sea, provides one version of a "different kind of 'map'" that Steinberg (2014: xv) calls for as we attempt to write off the sea as a non-objectified arena.

The use of autoethnographic accounts provides new opportunities to understand the sea. In *The Sea: A Cultural History*, John Mack argued that

> if we are to take seriously the observation that the understanding of the sea is predicated on an understanding of the people who inhabit the sea, accounts of mariners must clearly play a large part in what follows here … It [ethnography] has rarely reported on the experience of being on the seas; instead, to the extent that reference is made to the sea at all, it has almost always focused on the implications of being close to the sea, of having a relationship to it, not actually of being *on* it.
>
> *Mack, 2011: 23*

Thus, the autoethnographic accounts by sailor/scholars provide a way to understand how aspects of "understanding, knowing and knowledge" (Pink, 2009: 8) of the sea shapes both individual and collective identities.

## Examples of recent scholarship

In the section below I have outlined some recent contributions from sailor/scholars that may guide and hopefully encourage the reader to explore this mode of interaction more fully.

### *Sensations of a solid sea*

With extraordinary candour, and a great deal of humility, Kimberley Peters exemplifies the shift in perception that arises when new approaches are embraced. Peters reveals how her experiences of being on a four-day sailing trip enabled new ways of thinking *with* the sea in contrast to researching and writing about the sea from the land. It is worth quoting Peters at some length as her words capture a reappraisal of the 'liquid sea':

> I was ecstatic.
>
> But there was no denying. This boat *moved*. And it moved in a way that was alien to my legs, my arms, my sense of balance. Day 1, and in a bid to capture this world of full motion, I made two brief observations jotted by head-torchlight on my bunk on the *Steinlager 2*. But these weren't really observations. Sensations maybe, affects certainly.
>
> The first was the pitch. The sea – the wind against the sail – it threw that boat upwards, sideways and down. I'd written about a dynamic sea of angles – a more-than-horizontal world of shifts and verticals. But being at sea, being *with* the sea. It was different somehow. Now those words were just words on page. They were flat representations of a three-dimensional world. I was now *in* that three-dimensional world. And that three-dimensional world was in me, moving my limbs, moving me emotionally.
>
> The second was the slam. I still don't know what to make of this. But the sea lifted that boat up. And it dropped it right back down, with a thud. Before repeating – though never *exactly* repeating – that motion, Lift. Slam. Lift. Slam. These were elemental forces slapping against the carbon fibre hull. So it happened again. I thought: 'I've written about a motionful, liquid sea'. But something didn't fit. This sea was also solid.
>
> *Peters and Brown, 2017: 4*

Peters' engagement with the sea through the act of sailing, where the boat (and those on board) responded to the wind and waves, provides a different perspective to that of the scholar surveying the sea from land or from the perspective of the sea as 'other'. As I have suggested elsewhere (Brown, 2015, 2016) lived experiences in, and with, the sea allow us to understand how it "becomes part of us, just as we become part of it" (Ingold, 2000: 191).

Efforts to see ourselves as part of the world, by recognising the embodied visceral nature of our engagement with seascapes, has implications for how we relate to, and convey our relationship with the sea. For example, sailor/scholar, Peter Reason has drawn on environmental writers such as Gregory Bateson, Thomas Berry, and his own voyages under sail on the western edges of Ireland and Scotland to expound the need for a fundamental rethink of how we engage with the world (Reason, 2014, 2017). Reason's 'ecological pilgrimages' on his sailboat were experiential attempts to disrupt the binary between the human and non-human worlds. Whilst the scientific discourses of "evolution and ecology tells us we are also part of the community of life on Earth, we rarely feel that in our bones or our heart" (Reason, 2017: 9). Reason grapsed the value of being at sea, and taking time to work with its ebbs and flows, to bridge the human and more-than-human worlds.

Sailing, along with other embodied encounters with the sea (e.g., surfing, ocean swimming, SCUBA) provides opportunities to feel the sea through one's body (Peters and Brown, 2017; Throsby, 2015; Zink, 2015), to enhale it through one's breath (Moitessier, 1974), and to experience moments of grace (Reason, 2017). Reason's search for moments of grace, which he describes as times "when a crack opens in our taken-for-granted world, and for a tiny moment we experience a different world … . It is a world … no longer divided into separate things, but one dancing whole" (2017: ix), goes some way to capturing what has been referred to as 'ineffable encounters' (Peters and Brown, 2017) experienced by sailors. The interaction between wind, waves and a boat's hull shape results in unique 'signatures' of movement that give a boat a distinctive motion. This is exemplified in Peters' account of her experiences where the reader is invited to explore what thinking/writing with the sea might add to human geographers' understanding of the human–sea relationship when scholarship is based on/in embodied encounters.

While accounts of human encounters with the sea will always be distanced and partial (Steinberg, 2013), recent efforts by sailor/scholars (or scholars who have gone sailing as per Peters) have attempted to show how engagements with the sea can (re)shape new ways of thinking about our relationship with sea. Peters' narrative above highlights how the 'felt sea' differs from the imagined or metaphorical sea.

## *Skilful engagement: Constrains and affordances*

In *The Offshore Sailor: Enskilment and Identity* (Brown, 2016), I draw on the concept of enskilment – becoming skilful through active engagement – to investigate how a sense of identity, as an offshore sailor, is contingent upon being attuned to one's environment. Using auto-ethnographic accounts, such as those that opened this chapter, I highlighted the embodied practices that warranted claims of 'being' an offshore sailor. I detailed the process of enskilment, of gradually being at ease with the motion of the sea, to examine how I transitioned from a position of the sea as a 'constraint' (through seasickness) to becoming attuned to the environment that permitted me to '(re)inhabit' a particular identity. That paper attends to the temporal and contingent identity of 'being' an offshore sailor; an identity that is grounded in the practice of offshore sailing.

Both of these auto-ethnographic accounts (Brown, 2016; Peters and Brown, 2017) build on the embodied experiences of how the sea shapes a sense of identity, belonging, and connection that were explored in *Seascapes: Shaped by the Sea* (Brown and Humberstone, 2015). This edited collection included experiences of sailor/scholars, along with other accounts from 'sea-people' who articulated very personal engagements with the sea through the use of autoethnography. The chapters in this edited collection focus on various engagements with the sea, including open water swimming, sea kayaking, surfing and bodyboarding (see also Anderson, this volume; Foley, this volume; Waiti and Wheaton, this volume). Pertinent to this chapter are four contributions relating to experiences explored through the activity of sailing which are briefly discussed below.

## *Sensing the sea*

Barbara Humberstone promoted the importance of embodied narratives, the senses and subjectivities as a means to understand our relationship with the sea. She detailed how autoethnographic accounts can give rise to a greater appreciation of sensuous and embodied knowledge. For example, she draws on her experiences as a windsurfer to illustrate how sensuous encounters with the sea serve as the "seat of the senses" (2015: 30); as a way to connect with the environment:

> I feel the water rushing past my feet and legs. The wind in my hair.
>
> I sense the wind shifts in strength and direction and move my body in anticipation to the wind and the waves. I feel the power of the wind and the ability of my body to work with the wind and the waves. The delight and sensation when surfing down a small wave with the sail beautifully balanced by the wind.

> Seeing the sea birds and the fish jump delight further. The smell of salt and mud. The small seal that made its home on the tiny pebble spit.
>
> These are some of the beauties of windsurfing in this liminal space even with a monstrous power station chimney hovering in the distance and the occasional smell of sulphur from the large oil refinery when the wind blows from the north east.
>
> *Humberstone, 2010: 57*

She explained how such rich sentient experiences engender a strong sense of connection to a locale and the development of what Thrift (2008) refers to as 'kinetic empathy'. As she stated,

> how we learn to be in our bodies connects us to the wider world and, arguably for us 'sea-people', we are subtly connected to and intertwined with the energies of the waves, the sea and the universe. Who we are, and what we become is bound up with the dynamism of the sea.
>
> *Humberstone, 2015: 34*

She details how scholars from a variety of disciplinary areas (for example, cultural geography, sport and physical culture, tourism studies) have both recognised and engaged with embodied narratives to "elucidate the entwinement between the senses, the body and social thought" (Brown and Humberstone, 2015: 6). Humberstone's interdisciplinary approach provides a sound foundation upon which to further explore the ways in which we might know the water-world through our bodies.

## *What feels right*

Robyn Zink (2015) explored what makes sailing 'feel right'. Her reflections of sailing across Cook Strait (between the North and South Islands in New Zealand) provide a focal point for considering explanations of why she returns to the sea. Her explorations move beyond simplistic accounts of being 'free at sea' or Romantic notions of the sea as a wilderness where one might find one's true self (see earlier comments). Zink's work draws on, and extends thinking in relation to, the insights provided by phenomenology. She explores the work of Foucault and the writings of Deleuze and Guattari to better understand what it is that draws her back to sailing, and the sea, time and time again. Drawing on the notion of assemblage and affect, she considers how these concepts elucidate her lived experiences of being on watch on night passages. As she stated, "[t]he sea is a place where that feeling occurs in a way it does not anywhere else" (Zink, 2015: 81). Zink's insightful examination of what 'feels right' draws on her embodied *being* in an assemblage of relationships that are not divisible into individual elements. For Zink, it is the sea that affords opportunities to be in a relationship that has enduring appeal.

## *Moments of grace*

Sailing provided Peter Reason (2015) with the opportunity to reflect on Bateson's ideas concerning the errors of western epistemology. Reason draws our attention to issues of ecological sustainability through his voyages aboard his sailing yacht *Coral*. 'Slow' passages, where he worked with the elements (tides and wind), provided him with the opportunity to think about his, and humanity's, relationship with the sea. In the passage which informed his analysis, he draws from Bateson's idea "that human beings and human society are embedded in the general systemic structure of the natural world... [which are] self-organising and self-transcending" (Reason, 2015: 101). In reflecting on his own experiences, through the lens of Bateson, he revealed his preference to work with the elements (tides and wind) rather than pushing against or through them; an option he had via use of an engine. This preference, to avoid the mechanical noise and associated pollution, serves as a metaphor for how humanity might work with, rather than against, natural forces. He suggested that it is when one has experienced the difference between working with the tide, rather than against it, that the sense of working in harmony or rhythm becomes clear. Reason's reflections on what progress might actually mean and our choices about how we use technology (for example, a diesel engine) provided an example of the tensions that we all face in making decisions that are ecologically sustainable. Reason's challenge to all of us is to consider how we might think in new ways. Sailing may be one way to "rediscover the experience of grace" (2015: 108) that has the potential to reshape who we are.

## *The sea: Connection and separation*

Karen Barbour (2015) draws on feminist and phenomenological perspectives tracing multiple generations of family migrations by sea to Aotearoa New Zealand. In her writing she interweaves personal experiences and multi-generational family stories to show how personal and cultural identity is shaped by the sea and the voyages that sea travel entails. Her own narrative of a recreational sailing voyage to Fiji, along with her Grandmother and Mother's recollections of their voyages as passengers on commercial ships, formed part of a kaleidoscope of experiences that she has inherited and embodied in her own ways of being in relation to the sea. As a woman of the Pacific, Barbour's story encapsulates many elements that might resonate with those whose identity is shaped by seas; seas that both connect and separate us from the places of origin of our forebears. As she points out, familial stories, personal sensory experiences, and cultural myths serve to shape our subsequent experiences and choices throughout our lives (Barbour, 2011). We continually reframe and reinterpret our life stories, in relation to the sea, through our experiences and the insights gained as we better understand our individual and collective histories.

## *Encountering nature through sailing*

Pauline Couper's (2018) work provides an autoethnographic account of sailing, and boat ownership, in the Plymouth Sound area (UK). She explores the literature on the claimed health benefits of experiences of green and blue spaces and drawing on Merleau-Ponty's phenomenological perspective she argues that much of the literature on nature and heath "fails to recognise that 'nature', as a category in binary relation with 'culture' (or 'humans'), is a cultural construct" (2018: 285). She suggests that the act of sailing "entails different *embodied spatialities of being* from terrestrial urban life, and that this heightens a sense of nature as Other" (2018: 285). Through her accounts of exploring the Plym River and Plymouth Sound, Couper reflects on nature as Other – particularly what lies beneath the surface of the water which is 'hinted at' by navigation marks and subtle shifts in the boat's motion as tidal flows and wave patterns are influenced by the changing sea bed. As she eloquently states;

> The mysterious Other of the 'invisible beneath' is the most obvious and predictable; the presence of a lifeworld, a mode of being that, as human body-subjects, we cannot

> know. The invisible beneath always exceeds perception and yet must be attended to. But being on the water, coming to understand the space-in-motion of the boat-in-environment, means the presence of an Other can always be *felt*. The agency of water and weather is always there, in the unending motion, communicated through the boat and thus through the body.
>
> *Couper, 2018: 294*

Couper's experiences of sailing and being in/on blue spaces highlights the "potential for cultural geography to contribute to a much more nuanced interrogation of how people experience urban green/blue space" (285). However, she cautions that such endeavours need to be mindful of the complex and diffuse nature of such experiences; care needs to be taken to ensure that appropriate attention is given to the socio-cultural conditions that facilitate experiences of blue spaces.

## Concluding thoughts on a fluid field

This is unashamedly a clichéd heading but it does signify shifts in thinking about the sea and how we express our relationships with it. Efforts to write *with* the sea, to think of it as shaping who we are, and how we might understand our watery world, have opened up exciting opportunities for human geographers and other social scientists to think differently about the human–sea relationship. The authors mentioned in this chapter have sought to bring to life Lambert et al.'s request for greater consideration to be given to "the imaginative, aesthetic and sensuous geographies of the sea", opening up "new experiential dimensions and new forms of representation" (2006: 479).

Writing with the sea, based on personal experiences, provides new insights; from Peters' awareness of the solidness of the sea to Couper's invisible Other – that which lies below the surface but can be felt through the body – sailor/scholars are exploring new ways to express our relationship with the sea. This chapter has provided a brief overview of recent scholarship, where being with the sea, as "the seat of senses" (Humberstone, 2015: 30) is the phenomenological basis for enquiry. Technically, sailing places one 'on' the ocean in a buoyant vessel, yet much time can be spent below the water line (inside the boat), and being immersed or covered by waves or spray. The ocean deposits visible reminders of its presence (encrusted salt crystals), it shapes how we move (the sailor's swagger), and it permeates our bodies as we breathe salt-laden air. Hau'ofa's (1998: 408) assertion that "the ocean is in us" is not just metaphorical. If we are to take this maxim seriously we should explore further the role of direct experiences in developing connections with blue spaces, albeit mindful of the inherent complexity and diffusion (Couper, 2018). The importance of developing empathetic connections with the natural world via sailing, may encourage behavioural change, and is likely to gain greater importance as humanity faces changing climatic conditions (Nicol, 2015; Reason, 2014).

Sailor/scholars engagements with the sea and their articulation of these experiences adds to a growing body of literature which is emerging within geography. This is important for our understandings of the sea; for our relationships with it are vitally important as these impact on how we utilise and allocate resources, regulate its management, determine territorial authority, and work to preserve or deplete non-human life (Steinberg, 2013).

Sailor/scholars experiences allows us to understand the sea as "an alternative known world" (Raban, 1987: 220). This is something that the ocean sailor and mystic Bernard Moitessier knew when he wrote;

> A sailor's geography is not always that of the cartographer, for whom a cape is a cape, with a latitude and longitude. For the sailor, a great cape is both very simple and an extremely complicated whole of rocks, currents, breaking seas and huge waves, fair winds and gales, joys and fears, fatigue, dreams, painful hands, empty stomachs, wonderful moments, and suffering at times. A great cape, for us, can't be expressed in longitude and latitude alone. A great cape has a soul, with very soft, very violent shadows and colours. A soul as smooth as a child's, as hard as a criminal's. And that is why we go.
>
> *Moitessier, 1974: 141*

The sailor's sea may not be the sea that is experienced by other scholars, but this body of literature can enrich the broader discourses of sea scholarship through its focus on writing *with* the sea being a hallmark.

## References

Anderson J and Peters K (eds) (2014) *Water Worlds: Human Geographies of the Ocean.* Farnham and Burlington, VT: Ashgate.

Auden W (1951) *The Enchafed Flood, or the Romantic Iconography of the Sea.* London: Faber and Faber.

Barbour K (2011) *Dancing Across the Page: Narratives and Embodied Ways of Knowing.* Bristol: Intellect Books.

Barbour K (2015) In the middle of the deep blue sea. In: Brown M and Humberstone B (eds) *Seascapes: Shaped by the Sea.* Farnham and Burlington, VT: Ashgate, 109–124.

Blum H (2010) The prospect of oceanic studies. *Proceedings of the Modern Language Association* 125(3): 670–677.

Brown M (2015) Seascapes. In: Brown M and Humberstone B (eds) *Seascapes: Shaped by the Sea.* Farnham and Burlington, VT: Ashgate, 13–26.

Brown M (2016) The offshore sailor: Enskilment and identity. *Leisure Studies* 36(5): 684–695.

Brown M and Humberstone B (eds) (2015) *Seascapes: Shaped by the Sea.* Farnham and Burlington, VT: Ashgate.

Couper P (2018) The embodied spatialities of being in nature: Encountering the nature/ culture binary in green/blue space. *cultural geographies* 25(2): 285–299.

Dorgan T (2004) *Sailing for Home: A Voyage from Antigua to Kinsale.* Dublin: Penguin.

Ford N and Brown D (2005) *Surfing and Social Theory: Experience, Embodiment and Narrative of the Dream Glide.* Abingdon: Routledge.

Hau'ofa E (1998) The ocean in us. *Contemporary Pacific* 10(2): 392–410.

Humberstone B (2010). Youth and windsurfing in an 'urban context'. In: Becker P, Schirp J and Weber C (eds) *Water-Space for Experience: Youth and Outdoor Education in Europe.* Marburg: BSJ, 75–84.

Humberstone B (2015) Embodied narratives: Being with the sea. In: Brown M and Humberstone B (eds) *Seascapes: Shaped by the Sea.* Farnham and Burlington, VT: Ashgate, 27–39.

Humberstone B and Brown M (2015) Embodied narratives and fluid geographies. In: Brown M and Humberstone B (eds) *Seascapes: Shaped by the Sea.* Farnham and Burlington, VT: Ashgate, 187–189.

Ingold T (2000) *The Perception of the Environment.* Abingdon: Routledge.

Jennings G (1999) Voyages from The Centre to the Margins: An ethnography of long term cruisers. Unpublished doctoral dissertation, Murdoch University.

Koth B (2013) Trans-pacific bluewater sailors: Exemplar of a mobile lifestyle community. In: Duncan T, Cohen S and Thulemark M (eds) *Lifestyle Mobilities: Intersections of Travel, Leisure and Migration.* Farnham: Ashgate, 143–158.

Lambert D, Martins L and Ogborn M (2006) Currents, visions and voyages: Historical geographies of the sea. *Journal of Historical Geography* 32: 479–493.

Lusby CM and Anderson S (2008). Alternative lifestyles and well-being. *Loisir et Societe/Society and Leisure* 31(1): 121–139.

Lusby CM and Anderson S (2010) Ocean cruising – a lifestyle process. *Leisure/Loisir* 34(1): 85–105.

Mack J (2011) *The Sea: A Cultural History.* London: Reaktion Books.

Macbeth J (1992) Ocean cruising: A sailing subculture. *The Sociological Review* 40(2): 319–343.
Macbeth J (1998) Ocean cruising. In: Csikszentmihalyi M and Csikszentmihalyi I (eds) *Optimal Experience: Psychological Studies of Flow in Consciousness*. Cambridge: Cambridge University Press, 214–231.
Macbeth J (2000) Utopian tourists – Cruising is not just about sailing. *Current Issues in Tourism* 3(1): 20–34.
Moitessier B (1974) *The Long Way*. London: Granada.
Nicol R (2015) *Canoeing around the Cairngorms: A Circumnavigation of my Home*. Lumphanan: Lumphanan Press.
Osborn M (1977) The evolution of the archetypal sea in rhetoric and poetic. *The Quarterly Journal of Speech* 63(4): 347–363.
Peters K and Brown M (2017) Writing *with* the sea: Reflections on in/experienced encounters with ocean space. *cultural geographies* 24(4): 617–624.
Pink S (2009) *Doing Sensory Ethnography*. London: Sage.
Raban J (1987) *Coasting*. London: Picador.
Reason P (2014) *Spindrift: A Wilderness Pilgrimage at Sea*. Bristol: Vala Press.
Reason P (2015) Sailing with Gregory Bateson. In: Brown M and Humberstone B (eds) *Seascapes: Shaped by the Sea*. Farnham and Burlington, VT: Ashgate, 101–124.
Reason P (2017) *In Search of Grace: An Ecological Pilgrimage*. Winchester: Earth Books.
Sparkes A (1999) Exploring body narratives. *Sport, Education and Society* 4(1): 17–30.
Steinberg PE (1999) The maritime mystique: Sustainable development, capital mobility, and nostalgia in the world ocean. *Environment and Planning D: Society and Space* 17(4): 403–426.
Steinberg PE (2001) *The Social Construction of the Ocean*. Cambridge: Cambridge University Press.
Steinberg PE (2013) Of others seas: metaphors and materialites in maritime regions. *Atlantic Studies* 10(2): 156–169.
Steinberg PE (2014) Foreword of Thalassography. In: Anderson J and Peters K (eds) *Waterworlds: Human Geographies of the Ocean*. Farnham and Burlington, VT: Ashgate, xiii–vii.
Thrift N (2008) *Non-Representational Theory: Space, Politics, Affect*. Abingdon: Routledge.
Throsby K (2015) Unlikely becomings: Passion, swimming and learning to love the sea. In: Brown M and Humberstone B (eds) Seascapes: *Shaped by the Sea*. Farnham and Burlington, VT: Ashgate, 155–172.
Winchester S (2010) *Atlantic: A Vast ocean of a Million Stories*. London: Harper Press.
Zink R (2015) Sailing across Cook Strait. In: Brown M and Humberstone B (eds) *Seascapes: Shaped by the Sea*. Farnham and Burlington, VT: Ashgate, 71–82.

# 26

# DIVING

## Leisure, lively encounters and work underwater

*Elizabeth R. Straughan*

### Introduction

In early August 2017 I caught up with an old scuba diving buddy, Mel, who was working as an underwater photographer on the Great Barrier Reef (GBR) out of Port Douglas, Queensland. We met in Hemmingway's Brewery adjacent to the marina so that Mel didn't have to travel far from her work boat after a long tiring day on the water. As we chatted I indulged my curiosity as to what it might be like to work on one of the world's most famous reef systems. I asked questions such as "have you had any interesting experiences" and "what are the customers like"? One story sprang easily to Mel's mind, which I relay in narrative form below:

> A bikini clad woman ducks under the water in her snorkel gear pausing momentarily above a section of the Reef just long enough for Mel to take a crisp shot of her human subject and the spectacular coral behind her. Looking up from the digital camera's screen Mel saw the snorkeler signal to her and towards the surface. Obliging, she inflated her Buoyancy Control Device so that her head rose above the water where she was greeted with a request: "Can you make the fish come up? I want fish in my picture". Mel replied with a simple "no" and, grateful for being on scuba, sank back into the relative silence beneath the water's surface. Later back on the boat, Mel found herself next to the swimmer and asked if she's had a good time on the snorkelling trip. The lady explained that it had been OK, but really, she only came to get a photo of herself with "that fish in the brochure".

Mel's story outlines an experience of encountering a photographic desire set in motion by the hermeneutic circle where images of tourists posing alongside charismatic marine species are advertised and promoted, leading future tourists to seek their own re-production of the same image (see Alber and James, 1998).

In this chapter I take Mel's story as a provocation to thought and an invitation to reflect on the hermetic circle and associated issues of representation in the context of diving. Divided into three sections, the chapter draws out different foci implicit within this encounter on the GBR. The first section centres on diving as a recreational experience framed by the tourist industry, the second turns to consider non-human and human encounters in ocean space, while the third

 DOI: 10.4324/9781315111643-31

unpacks ocean space as a site in which diving bodies labour. In relation to each of these themes I trace through work that has already been done by researchers from various disciplines and suggest areas that are ripe for future research. Each section takes its lead from Mel's provocation and as such I now turn back to reflect on her story.

## Diving for leisure: From tourism management to experiential encounter

The comments made by Mel's snorkelling client sit uncomfortably against the wider context of recent coral bleaching events that signal ecological crisis in the face of global climate change. They also sat in contrast to the awareness of other tourists I encountered during a liveaboard holiday on the GBR completed a few days before meeting Mel. In this context scuba divers commented that their drive to experience the reef in 2017 was partly driven by wanting to see it 'before it disappeared'. But even comments such as these are problematic insofar as they are reflective of Shaw and Bonnet's contention that "there is a high level of awareness about environmental crises that are unfolding, but the capacity to meaningfully respond appears to be lacking, particularly in the context of consumer capitalism" (2016: 568).

Indeed, as a tourist sector and leisure pursuit scuba diving is recognised as a "multibillion-dollar industry and one of the world's fastest growing recreational sports" (Musa and Dimmock, 2012: 1). Further, the more recent uptake and growing popularity of freediving reflects a tendency for action sports to have potential commercialisation (Wheaton, 2010). Diving in the context of consumer capitalism, then, is big business. The growth of diving's popularity as a leisure pursuit and tourist activity has garnered attention from researchers interested in its wider environmental effects.

Recognising diving as "a strong force for marine tourism" (Cater, 2008: 233) and a tendency for tourists to travel from temperate regions to holiday in warmer climates (Cater, 2008) the impacts of the dive tourism industry are recognised to be significant. In this context, scrutiny has been paid to the environmental impacts on coral reefs. Such studies have considered impacts created by dive boat anchors dropped directly onto reefs (Hale and Olsen, 1993) and the management practices put into place to reduce damage to corals using mooring buoys (Dinsdale and Harriott, 2004; Hawkins and Roberts, 1992). Other studies have considered the overgrowth of algae resulting from pollutants such as oil, sewage, garbage and food discharged from dive boats (Harriott et al., 1997), which can lead to disease and subsequently effect species mortality (Kaczmarsky et al., 2005).

Researchers have also considered the physical proximity of diver bodies to coral reefs and the impacts this can have. In this instance researchers have examined how reefs can be damaged by holding on to coral or knocking it with dangling dive gear, poor finning, trampling, kneeling or removing it as a souvenir (Tratalos and Austin, 2001; Zakai and Chadwick-Furman, 2002). Studies have also found that proximity without physical impact can place coral under stress, increasing its susceptibility to disease and death (Hawkins et al., 1999). As such, researchers have observed that an increase of scuba divers to an area results in a reduction of living coral.

Looking beyond biophysical measurements of impact researchers have also sought to understand the role of diver education (Barker and Roberts, 2004) and experience (Dearden et al., 2007) in altering environmental perceptions. Educational efforts have been considered at all levels of the diving 'career' including training programmes for dive guides and instructors undertaken at popular tourism sites such as the GBR and Egypt's Ras Mohammed National Park to improve the efficacy of dive briefings in relation to environmental concerns (Marion and Rogers, 1994). Indeed, as an ecotourism activity, it has been argued that "with a strong educational component, strengthened links to reef enhancement and a clear focus on nature, diving would be one of the most effective ecotourism activities that clearly illustrates the potential benefits of tourism and conservation" (Dearden et al., 2007: 313–315).

Recognising a tendency for research to focus on management and policy initiatives, scholars within tourist studies have instead sought to understand diver motivations and rationales, a focus that has drawn attention to the subjective, embodied and sensuous experience of being underwater supported by technology (Cater, 2008). Such experiences, it is argued, are framed by individual motivations that direct the dive encounters sought, which Garrod (2008) has categorised in typologies such as adventure (for example, deep, wreck, drift cave, cavern or night diving), education (for example learning to dive through PADI's open water course) or the viewing of aquatic flora and fauna. These are typologies mediated by opportunities of access to different ocean spaces such as those located just off-shore and other deeper or more remote sites accessible only by boat. Extending this work, Dimmock and colleagues (Dimmock, 2009, 2010; Dimmock and Wilson, 2009) have unpacked scuba diving understood through a framework of 'Adventure' to argue that it takes place through a complex navigation of comfort, constraint and negotiation.

Taking up the motif of the embodied scuba diver researchers have built on to Cater's work in consideration of sensory, immersive experiences of scuba diving (Merchant, 2011a, 2011b; Straughan, 2012) and freediving (Adams, 2017; Strandvad, 2018). Geographic inquiries into the experiential qualities of ocean space have been influenced firstly by post-structural theories seeking to dislodge a focus on representational forms through consideration of practice and performance to emphasise the open-endedness of embodiment. Second, this body of work has been influenced by actor network theory and new-materialism's assessments of non-human agency enabling an associated appreciation of the human body's entanglement with environments (see also Anderson, this volume on surfing). Across both theoretical streams a need for methodological experimentation has emerged to tease out those aspects of embodied, underwater experience that are hard to talk about, especially in an environment where clear verbal communication is not possible.

Acknowledging the inability of verbal diver communication underwater Merchant used an experimental methodology on two counts. First, she learnt to dive so that she might develop an embodied understanding of this activity (2014). Second, she learnt to do underwater videography to record events and help elicit post-dive reflections from her research participants (2011a, 2011b). Focusing on the experience of novice divers on the island of Koh Tao (Thailand), Merchant used the already established practice of videography within dive tourism to carve out access to divers and space for group discussion within the context of PADI's entry-level Open Water Course. Following holiday makers learning to dive through their course and filming their encounters, Merchant was able to record movements and events that could be re-visited on land. Re-playing the footage, Merchant was able to elicit reflections from her audience on their experiences with aquatic life and the materiality of water which directs how sight, touch, hearing and the haptic senses are experienced underwater. Drawing out the sensory experiences of her research participants Merchant has demonstrated how the sensorium is re-arranged for divers, a re-arrangement that enables a "vastly different [experience] to the way we feel in/on land" (Merchant, 2011a: 231).

Further contributing to understanding on the sensory, embodied experience of scuba diving, I have focused on the sense of touch and its place within the haptic system to analyse the embodied experience of professional scuba divers (Straughan, 2012). These are individuals who guide and teach tourists, situating underwater encounters as an everyday practice enabling these divers to have a different embodied experience to Merchant's participants. Using my positionality as a dive guide learning to and then acting as an instructor I wove together empirics collected using auto-ethnography, participant observation and semi-structured interviews. Analysing embodied experience through the haptic system I unpacked how emotion was mobilised by internal sensations – kinaesthesia, proprioception and the vestibular system – effected by the materiality of the environment. For these experienced divers, ocean space facilitated a calming, meditative experience which led me to argue that the sub-aquatic environment is, for some, a therapeutic landscape.

Focusing on diving as an activity that enables tourists to visit specific sites of interest, Merchant's (2014) experiential, embodied account of diving the ship-wrecked SS Thistlegorm has drawn attention to underwater ocean space as something divers move *through*, positioning this activity as one with conceptual weight. That is, appreciation of embodied encounters with the ocean as a simultaneously horizonal and vertical space moved through by divers, highlights ocean space as a volume (Steinberg and Peters, 2015).

Unpacking the concept of volume, Squire (2017) has used her research on the geopolitics of experimental undersea living projects (Sealab I, II and III) of the Cold War. Squire describes her own pathway to learning how to scuba dive through acknowledgement of volumes as immersive states requiring "the researcher's body [to]... play a role in the process of constructing and interpreting ... spaces" (2017: 3) of volume. Learning to scuba dive provided her with a "sense of legitimacy" within the research process, but also to provide "insights into language and practices" of her research community (Squire, 2017: 11). As a result, Squire found "there was something significant about having been under the sea, about being part of a diving community, and in sharing a love and interest in inhabiting and moving through the water column that enabled an openness and conferred a legitimacy to my conversations" (2017: 11). On reflection, I suggest this work on the embodiment of scuba diving has highlighted a 'kinaesthetic culture' (Paterson, 2015), which draws attention to a shared understanding of movement somatically felt by divers entangled with ocean space.

In similar vein, research has also emerged on freediving. Adams (2017) has drawn attention to this as an immersive experience, which he describes as an activity characterised by simplicity and grace. He explains that freediving "mobilises the most powerful autonomic reflex known in the human body: the mammalian dive response" (Adams, 2017: no page) enabling humans to hold their breath and descend to depth. Unaided by technology, Adams relays his experience of learning and then doing freediving in both warm and cold-water environments. Highlighting the need to repress fear and encourage relaxation in that face of potential death, Adam's draws attention to freediving as a vehicle to reflect on personal and family traumas.

Elsewhere Strandvad (2018) has built onto Adam's work through an academic inquiry into freediving. Reflecting on Dewy and Tufts (1932) arguments, which used freediving as a case study to consider 'self-cultivation' as a process of continual interaction with the environment, Strandvad has unpacked the process of learning to freedive and considered this alongside the experiences of professional, competitive freedivers. Examining how freedivers move beyond fear to have a more meditative, spiritual encounter with ocean space, Strandvad has argued that freediving is a liminal experience as it enables divers to inhabit the edge between life and death.

A thread that emerges throughout these studies' consideration of vital, embodied 'doings' in ocean space are those qualities of dive experience that signal the diminishment of the human (Philo, 2017) as a diving body. A sensitivity to bodily limits emerges in Adams' (2017) reflections on his sons 'throat squeeze', as well as in Squire's decompression sickness (2017), suggesting avenues for future research. Those with an interest in corporeal finitude might, for example, build on Adey's (2016) interest in emergency mobilities to consider practices and process of dealing with pressure related dive injuries. For example, researchers might consider the geo-politics, mobilities, materiality and temporality of evacuating divers with decompression sickness to hyperbaric chambers. What, therefore, are the politics (macro and micro) of nitrogen saturation? Such lines of investigation might also include emotive and affective enquiry into subsequent trauma and anxieties around the same.

The space of the hyperbaric chamber, the politics of its use and the experience of the dry dive could open research out beyond an explicit focus on diving. For example, those with an interest in health geography and materiality might look to the practice of Hyperbaric Oxygen Therapy which turns oxygen under pressure into a healing treatment for ailments including server burns. In the UK such research would be timely insofar as cuts to the National Health Service (NHS) threaten two of ten chambers.

Further, diving as a popular leisure pursuit is ripe for enquiry by mobilities scholars' who might be tempted to consider the process of travelling for a dive holiday. This could include research into the techniques of travelling with heavy cumbersome dive equipment, or consideration of the liveaboard as a space which supports life for a couple of weeks while providing access to 'remote' dive sites. Researchers might look to the micro-politics of the dive boat (both day and liveaboard) where gear laden, cumbersome bodies negotiate the rhythms of both the ocean and dive operations.

## Encountering the non-humans of ocean space

In this section I return to the issue of representation highlighted by Mel's story at this chapter's opening. Reflecting on her experience on the GBR, I was interested in how else the hermetic circle might play out in ocean space. As such, I organised a semi-structured Skype with another old dive buddy, David, who was working as an underwater photographer for a whale shark tour operator at Coral Bay in Western Australia. During our conversation David outlined his role which required him to freedive whilst photographing whale sharks and tourists snorkelling alongside. He explained that tourists would often ask him to take their picture "holding a fin, like in a famous photo, like Ocean Ramsey's photos, but they want to do it with a whale shark instead".

Here David eludes to the promotion of images displaying shark conservationist and freediver Ocean Ramey hitching a ride through the water on the fin of a great white shark, an "exhibitionism [thought] to work in favour of pro-environmental behaviour" (Shaw and Bonnet, 2016: 527). Ramsey has explained the aim of her shark encounter expedition was to "collect video footage of their (great white sharks) natural behaviour, but also, if the opportunity arose and the conditions were right, to actually interact with them" (in Strege, 2013: no page), to counter the image portrayed by Hollywood "where you put a drop of blood in the water and the animals go crazy..." (Strege, 2013: no page). This was an act that sought to dislodge media representations depicting great whites as monstrous. Yet, as David's testimony suggests, the images circulated sit uncomfortably against best practice in ocean space which advocates a 'don't touch' ethos to limit damage and/or 'stress' to marine life (e.g. see Straughan, 2012). Scholars might consider, then, how non-humans that inhabit ocean space might be represented differently? Such a focus would build onto Bear and Eden's (2008) call to consider the agency of aquatic non-humans (see also Johnson, this volume; Squire, this volume). An appreciation of this agency might be gleaned through experiential accounts that relay feelings, emotion, affects and sensation which could help facilitate new understanding of human and non-human entanglements.

Scholars with an interest in diving have already started to do some work in this area. Merchant's (2014) chapter, depicting her experience of two dives on the SS Thistlegorm, is one such example as it relays the power of ocean processes to have material and atmospheric effects.

Building on a geographic understanding of ruins and haunting, Merchant's embodied experience of this wreck enabled her to reflect on how its temporality and materiality was affected by the "ocean's processual and ecological rhythms highlight[ing] that human made objects don't solely have social lives, but chemical and biological lives too" (Merchant, 2014: 126). Here Merchant acknowledges the agency of non-humans such as encrusting corals and sponges to create places that are at once "interesting, enchanting, chilling (literally and metaphorically) [both] structurally and atmospherically" (2014: 126). More recently Squire (2020) has considered the agency of non-humans in challenging the territorial ambitions of the US Navy's Sealab projects (1964–1969). Untangling records on encounters between fleshy human divers and non-humans from archival material, Squire reveals how aquatic marine life challenged the Navy's representations of the sea floor as uninhabited and actively shaped the Sealab projects.

It is, however, in consideration of sharks that most work has been done to examine interrelations between non-humans and humans in ocean space. This work has been galvanised by a spate of shark 'attacks' off Western Australia's coast in 2013–2014 and the ensuing controversial shark cull policy put in place by the Western Australian government, which mobilised a series of high-profile protests. Recognising the strength of geography's ability to unpack the cultures and politics of human and non-human interactions in ocean space, Gibbs and Warren (2014, 2015) investigated ocean users in Perth, Western Australia. Including the experiences of scuba and free divers in their study, Gibbs and Warren found that not only do ocean users frequently encounter sharks without harm "(including the three species considered most dangerous to humans)" (2014: 7), they also oppose shark culling. As such, these geographers recognised a rich area of future research and call for cultural geographers to illuminate the attitudes, knowledges and practices of ocean users who encounter sharks. Further, they have drawn attention to the role political and environmental geographers can play in unpacking the "policy, politics and governance of processes associated with regulating human-nonhuman interaction" in ocean space (Gibbs and Warren, 2014: 9).

Researchers with an interest in tourism management and marine environments have also considered shark–human encounters. Apps et al. (2016) have sought to understand the increase in demand for great white cage-diving tourism since the 1990s. Focusing on cage-dive experiences in South Australia Apps and colleagues also used surveys and found that within this space "education and the perceived naturalness of the experience" (2016: 231) were key drivers for scuba divers seeking to cage-dive. Further, emotions mobilised by this experience enhanced attitudes and knowledge (Apps et al., 2018) producing positive shark conservation outcomes. Perhaps unsurprisingly, however, these positive outcomes are threatened by debates around perceived threats bought about by cage-diving (Jøn and Aich, 2015). There is scope for researchers to move beyond the survey and consider other methods that might unpack the nuances and complexities of emotions and attitudes mobilised by orchestrated and serendipitous encounters with non-human marine life such as sharks.

Indeed, linguists Appleby and Pennycook (2017) have drawn on their experiences of swimming and scuba diving regularly with sharks to highlight the significance of embodied practice in challenging sociolinguist approaches to discourses of 'shark talk' such as the framings of nationalism, masculinity and risk which circulate in Australia. Drawing succour from vital materialism, ecological feminism and posthumanism these scholars argue for a "critical, embodied, positioned practice" (2017: 245) that is sensitive to the politics and ethics of de-centring the human through acknowledgement of an ethical inter-dependence with the more-than-human and non-humans of ocean space. Theirs is, then, a theoretical argument for future experimental and embodied methods to dislodge representations produced through language.

Such an approach might also enable scholars to re-materialise photographs of ocean space in recognition of the 'work' that can go into their production (Picken and Ferguson, 2014). Thinking beyond the shadow of the hermetic circle in Ocean Ramsey's great white expedition, we can find a carefully considered encounter as Amy Wilkes, Sydney Sea Life Aquarium's aquarist explained: "She [Ramsey] has gone to a great deal of effort to avoid threatening behaviour or scaring the sharks, so it's a calculated risk" (in Strega, 2013: no page).

These comments suggest an ethical attention to what Lorimer et al. (2019) call 'animal atmospheres', a concept that aims to recognise the affective lives of non-humans and highlights a future research direction. While this could mean a focus on affect in relation to shark encounters there is scope to take an embodied, practice-based approach to unpacking human encounters with other non-humans in ocean space. For example, researchers could explore processes of marine citizen scientists undertaking surveys and include the role of dive professionals (such as my protagonist David) and amateur underwater photographers' documenting marine life.

In addition, there is scope to consider atmospheres and entanglements between recreational divers and non-human marine life (and death) in ways that also attend to ocean space as one shared with other humans undertaking different recreational activities such fishing. In this context researchers might look to practices of hunting using scuba or freediving. Further, researchers could consider the liminal space of the jetty or sea wall where tensions and micro-politics between divers and fishers play out around perceived and real threats to fish and the environment. Such focus could interrogate the role of recreational divers in making visible issues of environmental concern that might otherwise remain hidden beneath the water's surface.

## Diving as working practice

In this final section I want to reflect on the positionality of my protagonists Mel and David. These are individuals for whom diving is not only a hobby or passion, it is also a working practice and as such they invite us to consider ocean space as a site of labour (see also Borovnik, this volume). Within our interview, David provided a window into his working practice by outlining his ideal angle for a photograph of tourist and whale shark, explaining "I will have the person between me and the whale shark and just get them to turn around, face me for a second". Highlighting that whale sharks, at the surface, offer a behaviour that produces a photographic backdrop, David went on to illuminate the role these non-humans' mobility plays in complicating his work. Given that whale sharks are animals on the move, he explained: "[s]ometimes you have to tow them [the tourists] to the front of the whale shark, if the whale shark is fast. You have to drag people up the front". David's labour, then, outlines some of the embodied challenges of working with materiality and non-human agency in ocean space.

The sea has a long been a space in which coastal and island communities dive as a fishing practice. For example, freediving has traditionally been used by the amas, or 'sea persons' of Japan who freedive to fish on the seafloor, a working practice performed today by women (Lim et al., 2012). Diving with technology is also undertaken in the fishing industry. For example, Purcel et al. (2016) have looked to sea cucumber harvesting conducted by small-scale artisanal fisheries where gendered dynamics of breath-hold and scuba diving are important considerations in managing coral reef exploitation. It is, however, recognised that dive fisheries can have serious health consequences (Barratt and Van Meter, 2004; Eriksson et al., 2012). Despite this prevalence, and notwithstanding Winker's (2016) work, links between dive behaviour and decompression sickness in the fisheries industry has been poorly understood. In this context, more work is needed to explore the role of place, economics, politics and culture in problematic dive fishery practices around the world.

As this literature suggests, diving can bring a body to its limits, indicating that a focus on how diving diminishes bodies should also be extended to consider working practices. Bodily limits have been a topic of interest for medical researchers interested in health effects for long-term professional, commercial and navy divers (Hoiberg and Blood, 1986). In this context researchers have found long-term diving can produce musculoskeletal symptoms (e.g. Hoiberg and Blood, 1986) and cognitive dysfunction (e.g. Todnem et al., 1990). Interested in the long-term impacts of such diving Ross et al. (2007) have considered the health effects of UK professional and off-shore (resource sector) divers using postal surveys, which found there were no major long-term health effects from a diving career. And yet, Grønning and Aarli (2011) have argued that long-term neurological effects from deep-diving (which takes place in the sea below 50 metres) are not yet conclusive.

Squire (2016) offers the most sustained consideration of deep-sea diving through a focus on the ocean as an immersive space for navy divers of the Cold War. Unpacking the geo-politics of the US Navy's Cold War project's Sealab I, II and III which sought to subdue ocean space into a battle ground, Squire's work draws attention to navy aquanauts living and working for around 15 days at 62 metres. Attending to the embodied and non-human challenges of working at depth, Squire positions ocean space as a terrain. Highlighting the time needed for aquanauts to decompress, she proffers food for thought around micro-politics at work in the time-space of decompression (2016). Further, her scholarship invites consideration of entanglements with materiality at depth such as those embroiled in activities carried out by commercial divers working on infrastructure, or Navy divers working with explosives (see also Squire, this volume).

Squire's work also draws attention to spaces used for diving several bodies to depth, a focus also considered by the anthropologist Helmreich (2007) who carried out ethnographic research into the working practices of marine micro-biologists, including their dives in a submersible. Relaying this immersive experience with a close focus on the soundscape, Helmrich's (2007) work draws attention to technologies that take people to ocean depths. Both Squire (2017) and Helmrich (2007) offer an invitation to look at spaces such as the submarine in which bodies labour and dive as a collection of bodies.

Another area where dive employment practices have been considered is the tourism industry where socio-economic inquiries have been made into the effects on local communities. For example, Daldeniz and Hampton (2013) have considered the capacity for dive shops to train and employ local staff, as well as for local individuals to own dive businesses. Looking across three Malaysian sites, these authors found the capacity for local training, employment and business opportunities was location dependent. Recognising a surprising dearth of research into socio-economic and development issues around dive tourism these authors have highlighted an area ripe for future research. Meanwhile social and cultural approaches could tease apart local pathways to becoming a dive professional in ways that may unpack best practice. As Daldeniz and Hampton's (2013) work suggests, dive labour is predominantly performed by migrants. As such, researchers could also consider the geo-politics of hiring and gaining visas, illegal work and how dive tourism fits within post-colonial contexts. Further, the enactment of guiding, instructing and videoing/photographing as both an immigrant and local dive professional could be unpacked to consider practices of care and responsibility as well as issues of burnout and precarity for seasonal workforces.

## Conclusion

In this chapter I have focused on diving in ocean space as a leisure pursuit with environmental implications, an activity carried out in proximity to non-humans and a working practice.

Highlighting work already done in these areas as well as possible future research directions, this chapter has drawn attention to the capacity for studies of diving to consider complex, vibrant, human and more-than-human and non-human relations and entanglements that occur within ocean space. However, as scholars have made clear, the ocean as a volume that is moved through by divers is only accessible for a duration of time because of technologies or considerable embodied and emotional skill. Therefore, diving is also an activity through which researchers might consider corporeal limits.

As Squire (2017), as well as Steinberg and Peters' (2015), work on volumes highlights, the diving imaginary offers liquid and gaseous metaphors and concepts to think with. I would like to end this chapter with the suggestion of other metaphors and concepts that pertain to accessing and managing ocean space as a diver. Following Peters' (2015) consideration of drift, questions could be asked around the *who*, *how* and *where* of compression and decompression, saturation, or acts of floating and hovering enabled by neutral buoyancy. The imaginaries offered by these concepts, which are central to diving in ocean space, might provide scholars with fruitful analytical frameworks.

## References

Adams M (2017) Salt Blood. *Australian Book Review [Online]*. Available at: www.australianbookreview.com.au/abr-online/archive/2017/208-june-july-2017-no-392/4100-2017-calibre-essay-prize-winner-salt-blood

Adey P (2016) Emergency mobilities. *Mobilities* 11(1): 32–48.

Albers PC and James WR (1988) Travel photography: A methodological approach. *Annals of Tourism Research* 15(1): 134–158.

Appleby R and Pennycook A (2017) Swimming with sharks, ecological feminism and posthuman language politics. *Critical Inquiry in Language Studies* 14(2–3): 239–261.

Apps K, Dimmock K and Huveneers C (2018) Turning wildlife experiences into conservation action: Can white shark cage-dive tourism influence conservation behaviour? *Marine Policy* 88: 108–115.

Apps K, Dimmock K, Lloyd D and Huveneers C (2016) In the water with white sharks (Carcharodon carcharias): Participants' beliefs toward cage-diving in Australia. *Anthrozoös* 29(2): 231–245.

Barker NH and Roberts CM (2004) Scuba diver behaviour and the management of diving impacts on coral reefs. *Biological Conservation* 120(4): 481–489.

Barratt DM and Van Meter K (2004) Decompression sickness in Miskito Indian lobster divers: Review of 229 cases. *Aviation, Space, and Environmental Medicine* 75(4): 350–353.

Bear C and Eden S (2008) Making space for fish: The regional, network and fluid spaces of fisheries certification. *Social and Cultural Geography* 9(5): 487–504.

Cater CI (2008) The life aquatic: Scuba diving and the experiential imperative. *Tourism in Marine Environments* 5(4): 233–244.

Daldeniz B and Hampton MP (2013) Dive tourism and local communities: Active participation or subject to impacts? Case studies from Malaysia. *International Journal of Tourism Research* 15(5): 507–520.

Dearden P, Bennett M and Rollins R (2007) Perceptions of diving impacts and implications for reef conservation. *Coastal Management* 35(2–3): 305–317.

Dewey J and Tufts JH (1932) *Ethics*. New York, NJ: H Holt and Company.

Dimmock K (2009) Finding comfort in adventure: Experiences of recreational SCUBA divers. *Leisure Studies* 28(3): 279–295.

Dimmock K (2010) CCN: Towards a model of comfort, constraints, and negotiation in recreational scuba diving. *Tourism in Marine Environments* 6(4): 145–160.

Dimmock K and Wilson E (2009) Risking comfort? The impact of in-water constraints on recreational scuba diving. *Annals of Leisure Research* 12(2): 173–194.

Dinsdale EA and Harriott VJ (2004) Assessing anchor damage on coral reefs: A case study in selection of environmental indicators. *Environmental Management* 33(1): 126–139.

Eriksson H, De La Torre-castro M and Olsson P (2012) Mobility, expansion and management of a multi-species scuba diving fishery in East Africa. *PLOS one* 7 https://doi.org/10.1371/journal.pone.0035504

Garrod B (2008) Market segments and tourist typologies for diving tourism. In: Garrod B and Gossling E (eds) *New Frontiers in Marine Tourism: Diving Experiences, Sustainability, Management*. New York, NY: Elsevier: 31–48.

Gibbs L and Warren A (2014) Killing sharks: Cultures and politics of encounter and the sea. *Australian Geographer* 45(2): 101–107.

Gibbs L and Warren A (2015) Transforming shark hazard policy: Learning from ocean-users and shark encounter in Western Australia. *Marine Policy* 58: 116–124.

Grønning M and Aarli JA (2011) Neurological effects of deep diving. *Journal of the Neurological Sciences* 304(1–2): 17–21.

Hale LZ and Olsen SB (1993) Coral reef management in Thailand: A step toward integrated coastal management. *Oceanus* 36(3): 27–35.

Harriott VJ, Davis D and Banks SA (1997) Recreational diving and its impact in marine protected areas in eastern Australia. *Ambio* 26(3): 173–179.

Hawkins JP and Roberts CM (1992) Effects of recreational SCUBA diving on fore-reef slope communities of coral reefs. *Biological Conservation* 62(3): 171–178.

Hawkins JP, Roberts CM, Van'T Hof T, De Meyer K, Tratalos J and Aldam C (1999) Effects of recreational scuba diving on Caribbean coral and fish communities. *Conservation Biology* 13(4): 888–897.

Helmreich S (2007) An anthropologist underwater: Immersive soundscapes, submarine cyborgs, and transductive ethnography. *American Ethnologist* 34(4): 621–641.

Hoiberg A and Blood C (1986) Health risks of diving among US Navy officers. *Undersea Biomedical Research* 13(2): 237–245.

Jøn AA and Aich RS (2015) Southern shark lore forty years after Jaws: The positioning of sharks within Murihiku, New Zealand. *Australian Folklore* 30: 169–192.

Kaczmarsky LT, Draud M and Williams EH (2005) Is there a relationship between proximity to sewage effluent and the prevalence of coral disease. *Caribbean Journal of Science* 41(1): 124–137.

Lim CP, Ito Y and Matsuda Y (2012) Braving the sea: The amasan (women divers) of the Yahataura fishing community, Iki Island, Nagasaki Prefecture, Japan. *Asian Fisheries Science* 25S: 29–45.

Lorimer J, Hodgetts T and Barua M (2019) Animals' atmospheres. *Progress in Human Geography* 43(1): 26–45.

Marion JL and Rogers CS (1994) The applicability of terrestrial visitor impact management strategies to the protection of coral reefs. *Ocean & Coastal Management* 22(2): 153–163.

Merchant S (2011a) Negotiating underwater space: The sensorium, the body and the practice of scuba-diving. *Tourist Studies* 11(3): 215–234.

Merchant S (2011b) The body and the senses: Visual methods, videography and the submarine sensorium. *Body & Society* 17(1): 53–72.

Merchant S (2014) Deep ethnography: Witnessing the ghosts of SS Thistlegorm. In: Anderson J and Peters K (eds) *Water Worlds: Human geographies of the Ocean*. Farnham: Ashgate, 119–134.

Musa G and Dimmock K (2012) Scuba diving tourism: Introduction to special issue. *Tourism in Marine Environments* 8(1–2): 1–5.

Paterson M (2015) On Aisthêsis, 'Inner Touch' and the aesthetics of the moving body. In: Hawkins H and Straughan ER (eds) *Geographical Aesthetics: Imagininging Space, Staging Encounters*. Ashgate: Farnham, 49–66.

Peters K (2015) Drifting: Towards mobilities at sea. *Transactions of the Institute of British Geographers* 40(2): 262–272.

Philo C (2017) Less-than-human geographies. *Political Geography* 60: 256–258.

Picken F and Ferguson T (2014) Diving with Donna Haraway and the promise of a blue planet. *Environment and Planning D: Society and Space* 32(2): 329–341.

Purcell SW, Ngaluafe P, Aram KT (2016) Trends in small-scale artisanal fishing of sea cucumbers in Oceania. *Fisheries Research* 183: 99–110.

Ross JA, Macdiarmid JI, Osman LM, Watt SJ, Godden DJ and Lawson A (2007) Health status of professional divers and offshore oil industry workers. *Occupational Medicine* 57(4): 254–261.

Shaw WS and Bonnett A (2016) Environmental crisis, narcissism and the work of grief. *cultural geographies* 23(4): 565–579.

Squire R (2020) Companions, zappers, and invaders: The animal geopolitics of Sealab I, II, and III (1964–1969). *Political Geography* 82 https://doi.org/10.1016/j.polgeo.2020.102224

Squire R (2016) Immersive terrain: The US Navy, Sealab and Cold War undersea geopolitics. *Area* 48(3): 332–338.

Squire R (2017) "Do you dive?": Methodological considerations for engaging with volume. *Geography Compass* 11(7) https://doi.org/10.1111/gec3.12319

Steinberg PE and Peters K (2015) Wet ontologies, fluid spaces: Giving depth to volume through oceanic thinking. *Environment and Planning D: Society and Space* 33(2): 247–264.

Strege D (2013) Free diver rides on back of a great white shark. *Grind TV*. Available at: www.grindtv.com/wildlife/free-diver-rides-a-great-white-shark-in-name-of- conservation/

Strandvad SM (2018) Under water and into yourself: Emotional experiences of freediving contact information. *Emotion, Space and Society* 27: 52–59.

Straughan ER (2012) Touched by water: The body in scuba diving. *Emotion, Space and Society* 5: 19–26.

Todnem K, Nyland H, Kambestad BK and Aarli JA (1990) Influence of occupational diving upon the nervous system: An epidemiological study. *Occupational and Environmental Medicine* 47(10): 708–714.

Tratalos JA and Austin TJ (2001) Impacts of recreational SCUBA diving on coral communities of the Caribbean island of Grand Cayman. *Biological Conservation* 102(1): 67–75.

Wheaton B (2010) Introducing the consumption and representation of lifestyle sports. *Sport in Society* 13(7–8): 1057–1081.

Winker N (2016) Diving dangerously: Exploring human health and resource trade-offs of extreme dive profiles in a Caribbean dive fishery. Masters dissertation, Dalhousie University. Available at: https://dalspace.library.dal.ca/handle/10222/72680

Zakai D and Chadwick-Furman NE (2002) Impacts of intensive recreational diving on reef corals at Eilat, northern Red Sea. *Biological Conservation* 105(2): 179–187.

# SECTION VI

# Ocean environments, ocean worlds

# 27

# DEPTH

## Discovering, 'mastering', exploring the deep

*Rachael Squire*

Once deemed an unstable and unfathomable abyss, home to sea creatures, and monsters yet to be discovered, the depths of the sea are increasingly being made 'known' to scientists, governments, and the public alike. Documentary series like *Blue Planet* have given mass audiences glimpses of the extraordinary, and often spectacular, everyday occurrences of the deep sea and sea bed. These images have been accompanied by a call to act, to prevent further destruction of the deep (see Brown and Peters, 2019). The series, narrated by David Attenborough, both explicitly and inexplicitly asked people to connect with the 'deep' and its inhabitants in ways centred upon awe and wonder, but also empathy and anger, at the effects of plastic pollution, climate change, and overfishing (Wilson, 2019). In this sense, then, the depths are increasingly visible, and so too are the myriad issues associated with practices of exploration and exploitation that are either taking place or are forecasted to take place within them. We have perhaps, as Hannigan suggests (2016 : 2), seen the transformation of the (deep) ocean as a half-known space of mystery and intrigue to "an emerging focus of global attention and concern".

In academia, the depths are also receiving increasing attention in what Deloughrey (2017: 32) would describe as an oceanic turn, or an important interdisciplinary "shift from a long-term concern with mobility across transoceanic surfaces to theorizing oceanic submersion… rendering oceanic space into ontological place" (Deloughrey, 2017: 32). Scholars are increasingly turning their attention to the sea, working to shed light on the myriad practices and processes that shape how the sea is lived, experienced, and understood. Increasingly we are being asked to reconfigure our understanding of the depths and its inhabitants too (Bear and Eden, 2008; Gibbs and Warren, 2014; Peters, 2010; Squire, 2020; Wang and Chien, 2020). The work of Childs (2020), Helmreich (2010), Merchant (2011), Peters (2020), Squire (2016), Starosielski (2015), and Straughan (2012), among others have demonstrated that the depths warrant our attention, generating important empirical insights alongside theoretical interventions (Peters and Steinberg, 2019; Steinberg and Peters, 2015).

Whilst offering alternative ways to understand and navigate the world, as Bremner (2015) highlights, the depths can still confound and obfuscate and there is still much to learn. The task, therefore, of writing a chapter on 'depth' is a difficult one. The possibilities for exploring such a theme are as vast as the ocean itself. To begin with, defining the term 'depth' is not straightforward. For some, the point at which 'depth' is reached may be a simple figure or striation, much

DOI: 10.4324/9781315111643-33

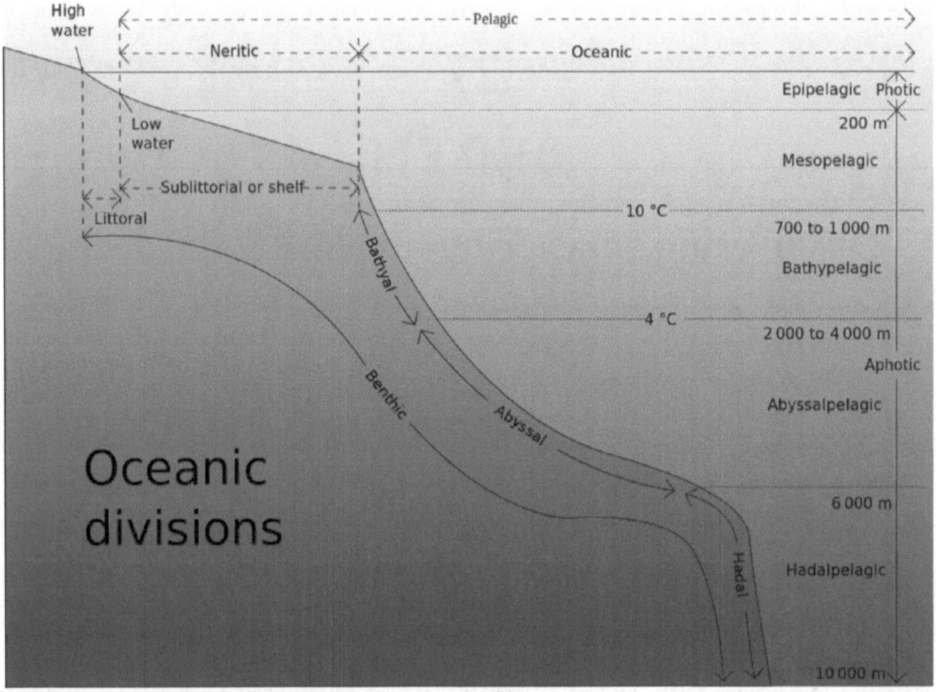


*Figure 27.1* Oceanic Divisions.
*Source*: K. Aainsqatsi at en.wikipedia, Public domain, via Wikimedia Commons.

like Figure 27.1 where the water column is divided by horizontal zones. Beyond 500 metres, for example, may be "referred to as the deep sea" (Houses of Parliament, 2015: 1; Sammler, 2020). The frame of reference for this chapter, however, refers, not to numbers and metrics, but to the complex, varied, spaces and contexts that exist beneath the surface of the sea. As Bear and Eden write, the

> straight lines and 90 degree angles… bear little relation to the coastline, the sea bed, the distribution and movement of fish or the fluidity of water itself… how far can strict cartographic boundaries deal with the essential fluidity of the seas and oceans?
>
> *Bear and Eden, 2008: 488*

All too often, argues Bremner (2015: 19), the oceanic volume is reduced to "a two-dimensional column of graded turquoise colour overlaid with… a selected collection of textual information".

Such an approach cannot account for the materiality of the oceans, its motion, undulation, and vast displacements of the watery volume across time and space, or the fact that water (both in its solid form as ice, and liquid form) can be "simultaneously encountered as a depth and as a surface" (Steinberg and Peters, 2015: 252; see also Bremner, 2015). As such, the chapter is less concerned with defining 'depth' as a metric or measurable line extending towards the seabed. Rather, it seeks to explore how the depths, the spaces beneath the surface of the sea, have been explored, made known, inhabited, and comprehended, and how they may be so in the future.

Whilst it cannot do so comprehensively, the chapter also seeks to point towards key thinkers and scholars in this field, and to touch on some of the key issues pertaining to 'depth'.

The chapter begins by providing some historical context to 'depth' in an Anglo-American context. It explores how understandings of the deep transformed from a sublime space of monsters and mayhem to a space that could be systematically measured and controlled for economic, cultural, and political gain. Moving forward in time, the chapter then turns to another period of exponential growth in oceanic innovation and 'deep' knowledge generation. Driven by geopolitical imperatives and super power rivalries, the Cold War saw unprecedented investment in anglophone oceanography. In addition, during this period, new governance strategies, such as the United Nations Convention on the Law of the Sea (UNCLOS), were brought into being. This section of the chapter will unpick these dynamics, exploring how the deep came to be configured during this period of geopolitical upheaval. The chapter finally turns to the contemporary issue of deep sea mining (DSM), exploring the importance of lived experiences and indigenous understandings of 'depth' in the debate surrounding DSM's practices and regulations, before finally concluding by pointing toward future 'uncharted depths'.

## Discovering the deep

Early understandings of the spaces beneath the surface of the sea were riddled with uncertainty and formulated through imaginative projections from land. At times, these projections took ghoulish forms, as sea monsters and mythical creatures were seen in cartographic representations emerging from this unknown watery entity (see Steinberg, 2009). Ortelius's oft cited 'Islandia' map (1590) is a prime example (Figure 27.2). The colourful, mountainous island of Iceland is surrounded by a flat, formless sea. Yet from this formless expanse emerge the beasts of the deep. These monsters are indicative of knowledges and understandings at the time. As Reidy and Rozwadowski highlight, much like their "mountain counterparts", the oceans "epitomised the sublime – a locus of anticivilisation, a void between developable, potentially civilised places" (2014: 338–339). Whilst appearing fantastical, these monsters reflected an effort by cartographers to accurately represent the sea's inhabitants (Waters, 2013). Reports from sailors and whalers provided the basis for artists and writers seeking to comprehend the depths of the sea and their reports of sea monsters, serpents and worms became a basis for "natural history texts and drawings on maps" (Waters, 2013). These drawings and depictions then, in turn, functioned to galvanise explorers, travellers and sailors who took to the water, aiming to confirm the existence of these strange and wondrous creatures.

Maps and imagery like that depicted in Figure 27.2 represent a fascination with the deep and a response to the alluring qualities of the space beneath the surface of the sea. It was one that could contain anything, a realm of unknowns, possibility and danger in equal measure, and a space that warranted further investigation. The late eighteenth century and nineteenth century in particular proved to be pivotal periods in generating understandings of depth, and its political, social, cultural, and economic significance (see Bravo, 2006). There was an energy and urgency underpinning this process. American oceanographer Matthew Fontaine Maury, for example, described the depths as a "sealed volume, abounding in knowledge and instruction that might be both useful and profitable to man" (Maury, 1855: 201). This seal, he asserted, was "of rolling waves many thousands of feet in thickness", but it was one that held promise – "could it not be broken?" (Maury, 1855: 201).

Breaking this surface bound 'seal' carried with it great societal significance at a time when speculation on the depths abounded. Indeed, there was a widespread belief that the deep sea was unfathomable and unreachable and that the water column was "weirdly populated by

*Figure 27.2* *Islandia* (Abraham Ortelius, 1590).
*Source*: Public domain, via Wikimedia Commons.

objects long forgotten" as they "found their depth" rather than sinking to the sea floor (Laloe, 2016: 125). A diverse range of actors were involved in dispelling these misconceptions (Bravo, 2006: 517). As Rozwadowski (2005: 39) highlights, whalers and sealers were some of the first to begin sailing away from previously established and 'long traced' sea routes. The hunt for new species to kill and the pursuit of waters where they could operate without competition drove exploration into new ocean spaces (Rozwadowski, 2005). In the process of whaling, new knowledges circulated throughout the industry (Bravo, 2006). As these enormous mammals pulled the harpoons down hundreds of fathoms through the water column, the experiences of whalers and the depths to which they were now being exposed travelled back to land with so-called 'fish stories' informing a land-based scientific community about the sea (Rozwadowski, 2005). Beyond the production of "virtuous knowledge about the oceans and the animal kingdom" (Bravo, 2006: 512), whaling, as Rozwadowski (2005) highlights, provided clear and concrete commercial and political imperatives to study the depths.

Of course, whaling was not the only practice driving further understanding of 'depth'. As Bravo highlights, in the nineteenth century the "oceans were spaces to be dominated by nations" (2006: 530). Within this framework, the ocean was deemed to be "unclaimed territory" (Rozwadowski, 2005: 40), with knowledge of its depths, volumes, and seafloor a means to bolster national authority (Reidy and Rozwadowksi, 2014). Across the Atlantic, Britain and the United States actively pursued knowledge about the oceans, "from the tides on its outer rim to the dark water at its greatest depths" (Reidy and Rozwadowksi, 2014: 340). They did so scientifically, rigorously, and systematically with unfolding imperial objectives, commercial

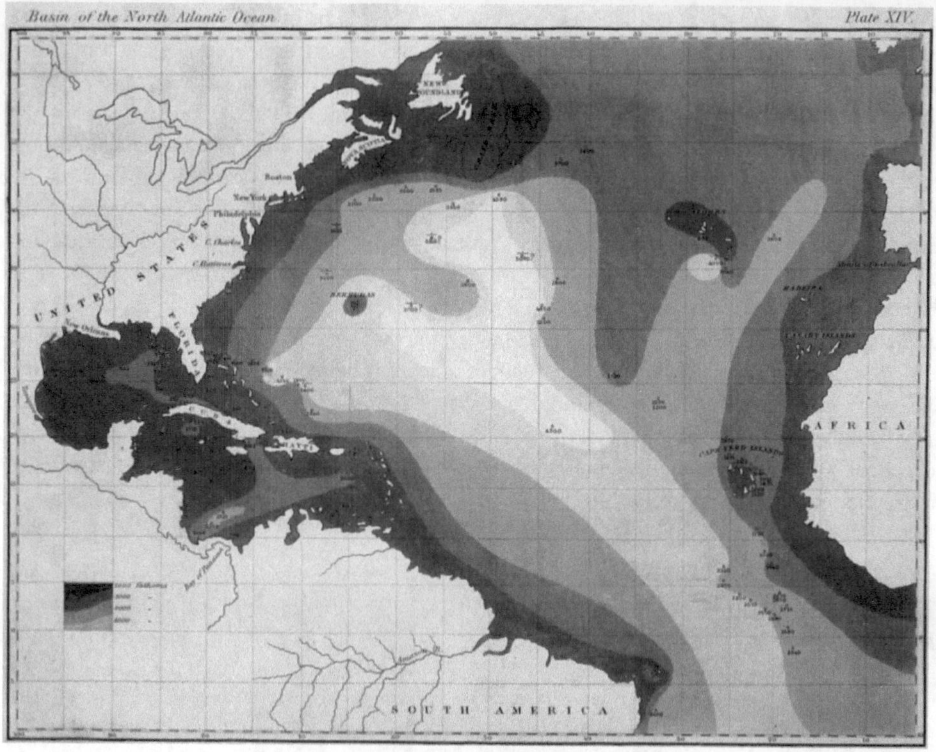


*Figure 27.3* Matthew Fontaine Maury's bathymetric map of the Atlantic Ocean basin (1853).
*Source*: Matthew Fontaine Maury, Public domain, via Wikimedia Commons.

activity, and new technologies providing the motivations and means to gain a synoptic view of the oceans (Reidy and Rozwadowksi, 2014: 340). This coincided with new representational practices and mapping techniques that enabled the communication of new observations and measurements to a wide audience (Bravo, 2006: 519). Figure 26.3 is a perfect exemplar of this, depicting Maury's bathymetric chart of the Atlantic, from 1853. It was the first ocean-basin map of its kind, using gradated shaded zones to mark ever deepening waters, bringing with it a new understanding of this crucial body of water (Reidy and Rozwadowski, 2014: 346).

The prospect of submarine telegraphy only energised the accumulation of knowledge of the deep and the sea floor. Believed to be a crucial tool in the maintenance of Empire, governments and commercial companies applied knowledge of the ocean to the cause of submarine communication infrastructure. With control over the extraction and distribution of the rubber-like gum from Malaysian trees used in the insulation of undersea cables (Starosielski, 2015: 33), Britain dominated the industry. The sea was transformed in the process, as a space capable of protecting vital infrastructure from interference, "potential colonial unrest, rival nations, and ships anchors" (Starosielski, 2015: 37). The "unexplored nautical spaces in between" land were suddenly taking on a whole new significance (Reidy and Rozwadowski, 2014: 339). Far from being a space to cross, the ocean was, for the "first time in Western history", a destination, a space to explore and dwell in and of itself (Reidy and Rozwadowski, 2014: 341–342).

Alongside state strategy and commercial imperatives the depths were being reimagined by publics too. Indeed, by the nineteenth century, the sea was an important part of Anglo-American culture. As rising literacy rates increased awareness of the depths of the sea, imaginations were projected from land and into the deep via maritime novels, a new appetite for maritime natural history, and the craze for home aquariums (Rozwadowski, 2005). Jules Verne's *20,000 Leagues Under the Sea* (1872) is often cited as being indicative of this trend (Cohen, 2018), not least because Verne himself was an avid sea traveller, writing detailed accounts of his journeys, including the crossing of the Atlantic on the *Great Eastern*. Whilst *20,000 Leagues Under the Sea* is clearly a fictional account, Verne drew on his experiences and the latest oceanographic knowledge of the deep sea at the time to inform his protagonist's journey into an undersea world. As Rozwadowski (2005) details, Verne penned this novel with Maury's *Physical Geographies of the Sea* beside him, with passages that mirror Maury's text. The Anglo-American discovery of the deep ocean was as cultural as it was political and scientific, an allure that perhaps stemmed in Verne's writing, at least, from the ocean's physicality and life-like qualities. Far from being inanimate, the ocean is given voice and character, with Captain Nemo, for example, proclaiming that: "The sea is everything! … Its breath is pure and wholesome… man is never alone, for he feels life pulsating all about him" (see Laloe, 2016). The sea and its depths are energised, living, and present limitless possibilities in the voluminous materiality (Laloe, 2016; Lambert et al., 2006).

*Figure 27.4* The Aquarium Craze: The carefully curated tank (1856), the glass was often absent to emphasise the connection with the sea.

*Source*: Internet Archive Book Images, No restrictions, via Wikimedia Commons.

The depths were also surfaced in other, but nonetheless extremely powerful ways during this period. The aquarium, as a prime example, provided not only an opportunity to extend the imagination into the deep sea but also to see its inhabitants first-hand and up-close. These encased spaces catered to, and boosted interest in the sea, and the saltwater 'aquarium craze' of the 1850s was indicative of the demand to experience something of the ocean's depths. Whilst natural historians extolled the scientific knowledge to be gained through the observation and 'befriending' of aquatic and marine creature (Hamlin, 1986: 132), for middle- and upper-class families, the aquarium brought the depths of the sea and its inhabitants into everyday contexts (Granata, 2018; see Figure 27.4). No longer the preserve of scientists and explorers, wealthy individuals were now able to "marvel at the intricate design of the invisible world" (Granata, 2018: 125). A sense of awe and wonder was implicit in this process, prompting Victorians to "dream of submarine journeys to a site of picturesque beauty, amazing novelty and variety" and inspiring them to want to achieve an "intimacy with the little people of the sea" through the space of the tank (Hamlin, 1986: 147).

Aquariums in private homes and public settings held romance and mystery, with the power to "fire the imagination with less threat than in previous centuries" (Rozwadowski, 2005: 21). At a time when interest in, and knowledge of, the sea was increasing at unprecedented speed, the aquarium offered a "reassuring sense of control" (Granata, 2018: 115). From the sublime spaces imagined prior to the nineteenth century, the tank represented peace and tranquillity, a miniature sea, or 'world in miniature' that could be contained, studied, and curated. As Hamblin (1986: 149) highlights, this had wider connotations, for while the "aquarium might represent 'a world in miniature' it was to be a civilized world", a world where order ruled chaos and where the recalcitrant depths could be mastered, arranged, and selectively utilised by 'man'.

## Mastering the deep

Whilst operating within a different and shifting geopolitical terrain, the Cold War also saw the extension of ocean knowledge into new depths. As with the nineteenth century, this was driven by the geopolitical imperatives of the time and the extension of the narrative that the depths were a space to be conquered and mastered in the pursuit of certain political, technological, and military objectives (Oreskes, 2003). The proliferation of submarine warfare, concerns about submarine rescue, and the salvage of atomic weapons in the sea animated the US military's desire to better "understand the environments through which men and machines might travel and communicate" (Oreskes, 2003: 699). Operating within these environments required knowledge and it is here that we see the significance of the relationship between the US Navy and oceanography as a discipline. Indeed, as Jacob Hamblin highlights, oceanography in the US, which remains a preeminent site of ocean science, owes its existence as a "mature discipline to financial and logistical assistance from the Navy" (2002: 15).

The Office of Naval Research, while established in 1946, came into its own during the Cold War with the aim of supporting scientific endeavours that would benefit both the naval services and scientific community. This was accompanied by an enormous influx of state funds to the geosciences in the 1950s and 60s to learn more about the global operating environment. Knowledge on the circulation of jet streams and ocean currents, for example, contributed to antiaircraft defence, whilst concerns over Soviet submarine activity were countered by vast investments in oceanography (Hamblin, 2002). It was felt that scientific research was an essential component of the strategy to "subdue the ocean environment, to make it a manageable and even advantageous battlespace" (Hamblin, 2002: 15). As Hamblin illustrates, whilst there were tensions, oceanographers wanted to conduct research and the Navy "wished to know everything there was to know about its own workplace – the sea environment" (2002: 3). Information garnered by Woods Hole Oceanographic Institution (WHOI) on underwater acoustics and thermoclines, for example, proved vital for the operation of US submarines (Doel, 2003: 636; see also Oreskes, 2003).

Alongside military applications, the 'intense thirst' for knowledge (Doel, 2003) underpinned a period of unprecedented scientific productivity in which scientists made numerous fundamental discoveries about the oceans, the sea floor, and the life associated with them (Oreskes, 2003: 699). As a prime example, Oreskes (2003) explores how the US Navy's interest in the deep sea led ultimately to the discovery of hydrothermal vents. With funds from the ONR to develop the Alvin, a deep-sea submersible (see Figure 27.5), scientists from WHOI discovered these hot springs venting on the seafloor, "supporting abundant biota under conditions previously thought impossible for life" (Oreskes, 2003: 699). Hydrothermal vents have since been found on the ridges of every major ocean, leading to the discovery of over 300 new species. As will be explored later, the knowledge generated from these sites had led to considerable interest in the seafloor from scientists, miners, and other commercial actors who believe that the vents may have many practical applications – from exploration for economic mineral deposits formed from submarine hydro-thermal fluids to the use of heat resistant organic molecules in synthetic organic chemistry and biotechnology (Oreskes, 2003: 699).

*Figure 27.5* The Alvin submersible (1978).
*Source*: Office of Naval Research, Public domain, via Wikimedia Commons.

Alongside scientific oceanographic studies, the Navy were also undertaking less conventional projects in the water column and seafloor. The Sealab projects (1964–1969), led by Capt. George Bond, are a prime example. Designed to enable 'man' to live and work on the seafloor in undersea habitats (see Hellwarth, 2012; Squire, 2016, 2018, 2021), the projects sought to increase the US's capability to "attack many significant oceanographic problems"; to better the Navy's ability "to live and to perform useful work under the sea" including salvage and rescue operations for sunken submarines, downed aircraft and atomic weapons; to expand the capabilities of the military on the continental shelf; and to test the feasibility of working productively with marine mammals (Pauli and Clapper, 1967: 17). The projects were undertaken with much publicity, seeking to ride the wave of 'ocean boosters' like Jacques Cousteau (1954) and Arthur C. Clarke (Rozwadowski, 2012), who dreamed of an era of development and exploration beneath the sea (Hannigan, 2016: 12). Similarly, artistic impressions of inhabiting the undersea environment played on an oceanic optimism that associated the sea with spatial transcendence and escape. In *Explorers of the Deep*, Cox, for example, declared with certainty that the day will come "when man will be able to live and work on the ocean floor with air of artificial gills" (1968: 90). Like Sealab, these outputs were pointing to a space in which geopolitical concerns unfold via human technologies and bodies beneath the waves, where the ocean could be consumed, subdued and put to work for the benefit of 'mankind'.

More broadly, these sub-marine projects, imaginaries, and discoveries had much wider significance and were taking place amidst concerns about ocean governance (Robinson 2020). These concerns took root in 1945 when US President Harry Truman made two proclamations – one of which, seen as a form of 'territorial enclosure', unilaterally asserted US jurisdiction and control over the US continental shelf and its resources (Steinberg, 2001: 140). Mexico followed suit and over the next five years, a series of similar proclamations ensued from countries including Chile, Costa Rica, Brazil, and Argentina. With no overarching governance framework to regulate these practices, fears about 'ocean grabbing' and hopes for the development of technologies "that might someday enable commercial extraction" of potentially lucrative manganese nodules from the seabed beyond the continental shelf abounded (Steinberg, 2001: 145). The seabed, like no other time in history, became the focus of international governments and legal practitioners alike. Amidst these "unilateral efforts to transform the rights of coastal states in ocean space, the UN convened the first international conference on the Law of the Sea (UNCLOS I)" (Steinberg, 2001: 143). Such a process, was believed to be, as articulated by Malta's ambassador to the UN, Arvid Pardo, "the only alternative by which we can hope to avoid escalating tension" (United Nations, 1998: no page). As such, few resisted his calls for an "effective international regime over the seabed and the ocean floor beyond a clearly defined national jurisdiction" (United Nations, 1998: no page). The culmination of a series of meetings and negotiations over this regime, as detailed by Steinberg (2001), was the signing of UNCLOS, containing some 320 articles, in 1982. A key component of UNCLOS saw the seas divided into zones. The first of these, territorial waters, gave states sovereign jurisdiction over the waters 12 nautical miles from the coastline. Pushing further out to sea, in what has been described as the "single greatest enclosure in human history" (Campling and Colás, 2018: 780), the Exclusive Economic Zone (EEZ) granted coastal states exclusive rights to economic and commercial activity with 200nm. Beyond this, the high seas were to be free to all and belonging to none.

The implications of the Cold War period on how the depths were being conceptualised, imagined, regulated, and inhabited cannot be underestimated. It was a time of technological innovation, where the potentiality of the sea as a space of consumption, of resources, and even possible future living space went unfettered. Concomitantly, however, anxieties underpinned moves to create a legal framework in which these activities could be governed, challenging the common view of the "global ocean as a lawless frontier" (Campling and Colás, 2018: 781). Moreover, as Hannigan (2016: 4) highlights, this era saw the emergence of a voluminous and volumetric understanding of the sea that confounded horizontal striation and measurement.

## Out of our depth?

The concerns and currents that swept through the Cold War period remain of significant interest today, both for thinking through contemporary issues relating to 'depth', but also when considering unfolding underwater futures. Resource exploitation in particular (see Thomas et al., this volume), remains a pressing concern, with the sea's inhabitants facing growing pressures and strains from practices including illegal fishing, whaling, acidification and temperature rise. The remainder of this section, however, will focus on DSM – a key challenge unfolding within the present-day context with potentially dramatic consequences in the future (for additional discussion of DSM, see Fawcett et al., this volume).

As highlighted by Hannigan (2016: 19), media reports of "an oceanic 'gold rush' have been appearing with increasing frequency" over the last decade or so. The sea is celebrated by some as having the "potential to save mankind" with new forms of energy, "low cost, high grade minerals, and miracle drugs" derived from new species and previously unknown seabed formations (Hannigan, 2016: 20). Whilst the deep seabed may have once seemed unreachable, for companies investing in seabed prospecting and the development of mining equipment, as well as a number of governments around the world, technology is making these spaces increasingly accessible and, as such, DSM "is being promoted as the next frontier of resource extraction" (Childs, 2020: 189). Advances in marine submersibles and mining technology are fuelling the redefining of this extractive frontier by national governments and commercial actors, who are intensifying their efforts "to explore and commodify" the metals, minerals, and phosphates imagined in the 'frontier' of the deep seabed (Childs, 2020: 190). Depth, as Childs notes, is a key variable here, with interests in DSM coalescing around three major categories of deposit; seafloor massive sulphides, found on hydrothermal vents (see Figure 27.6), polymetallic nodules, and cobalt-rich crusts (2018: 4). Around these three categories, commercial interests in exploration, mapping, sampling, logistics, and of course, the actual extraction of potential deposits abound. Not only does depth play a key role in the formation and subsequent potential value of these deposits, but it is also a determinant in the practicalities of DSM and in the possibilities imagined in the multi-dimensional sub-marine environment. The "fluidity, voluminosity, and dynamism" of the water column and seabed all matter in "shaping DSM's possibilities" (Childs, 2020: 190). For example, it is difficult to predict how the toxic materials contained within the plumes from hydrothermal vents may spread due to the nature of seafloor currents. Simultaneously, however, this dynamic topological, geological space can form, change and "move so slowly as to be inert, obdurate, even disconnected or bounded" (Childs, 2020: 190). It is perhaps these slow temporalities that "disengage nature from its previous ecologies", enabling the transformation of natural formations into corporate raw materials (Tsing, 2003: 5100).

The first effort to commercially explore the seafloor for massive sulphide systems, Nautilus Corporation's 'Solwara 1' copper-gold deposit project is under development in the Manus Basin, in the territorial waters of Papua New Guinea. Nautilus have been granted the lease needed for resource development at the site and it has plans to grow its holdings in the EEZ's and territorial waters of Papua New Guinea, Fiji, Tonga, and the Solomon Islands among others with the view to "build a pipeline of commercially viable projects" (Nautilus, 2019). The Solwara 1 site has yet to be mined commercially but extraction there is widely thought to be imminent (Childs, 2020; Houses of Parliament, 2015; Van Dover, 2011).

*Figure 27.6* 'White smoker' hydrothermal vents, Marianas Trench.
*Source*: NOAA, Public domain, via Wikimedia Commons.

The potentially catastrophic environmental consequences of DSM are well documented, including concerns about habitation destruction, loss of ill understood biodiversity and the 'exquisite organisms' that thrive around DSM sites, the lack of a framework to assess impacts (Van Dover, 2011), and the residual architectures of extractive landscapes (Bridge, 2009). Perhaps less well documented, however are the personal, spiritual, and cultural implications for those at the forefront of DSM practices (see Carver et al., 2020). For many, the 'deep' is not an abstract abyss or space of commercial exploitation. For communities in Papua New Guinea facing the mining of Solwara 1, the deep sea and its seabed are "intimately connected to humanity, despite the geographical distances involved" (Childs, 2019: no page). As Childs explores, for the people of the Duke of York Islands in Papua New Guinea, DSM "disturbs the spirits that inhabit their culture and beliefs" and even "a sense of who they are" (Childs, 2019: no page). As one clan chief asserted, "when they start mining the seabed, they'll start mining part of me". Feeling 'part' of nature, rather than external to it, is common across the Pacific Islands, alongside the belief that welfare is inextricably tied to the natural environment (Nunn et al., 2016). The dichotomy between nature and society that pervades many western understandings of the sea, and which facilitates and enables exploitation and extraction, does not exist here. Such perspectives offer a powerful rebuttal against eighteenth-century views of the ocean as an empty and external space that continue to characterise contemporary western ways of imagining and engaging with the sea (see also Waiti and Wheaton, this volume). As Theriault (2017: 114) illustrates, there is a need to "attend more carefully to the ontological multiplicity of forces that shape spatial practices and their regulation". Such perspectives and approaches to the sea and its depths "are no less significant in the (de)constitution of state power than many of the more directly observable agencies" that western scholars are more accustomed to tracing (Theriault, 2017: 114). As Childs (2019) highlights, this poses new questions about how we ethically, culturally, and humanely engage with the deep ocean and seabed, and how the human impacts of DSM can be foregrounded, further explored, and catalysed to destabilise dominant western capitalist abstractions of the sea. As it stands we are perhaps 'out of our depth' on this matter, with technology and commercial innovation outpacing research into the significant social, cultural, and spiritual impacts of DSM and the oceanic depths within it which it would take place.

## Conclusion: Unchartered depths

Whilst these challenges unfold through time and space, a recent study by NASA (see Weeman and Lynch, 2018), found that 'ocean rise' is accelerating. Conservative estimates forecast that if the current pace continues, the ocean will rise by 65 centimetres by 2100, "enough to cause significant problems for coastal cities" (Weeman and Lynch, 2018: no page). As Deloughrey states, "our planetary future is becoming more oceanic", and the sea and its depths will feature ever more prominently in everyday lives (2017: 33). For Deloughrey the potential of "subaquatic future worlds" looms large as the planet becomes "more submarine, more multispecies" and perhaps also more "unfathomable" (2017: 42). The questions raised by this are multifarious (see Jamero et al., 2017). How, for example, will UNCLOS adapt to potentially disappearing island spaces? If a low-lying island disappears, what happens to the state's territorial waters and Exclusive Economic Zone? On a human level, what happens to small island and coastal communities who simply do not wish to relocate amidst rising tides? Whilst relocation is often cited as the solution, as Jamero et al. (2017: 582) highlight, the reality is much more complicated. For one community on Ubay Island, in The Philippines, the water is central to the islanders' way of life and view of the world. Since an earthquake struck in 2013, causing the land to subside by up to one metre, the island has become either partially or completely flooded during high tides (Jamero et al., 2017: 582). According to Doherty, the situation on Ubay has created a "snapshot of what life might be like for many as sea levels continue to rise" and as the sea "quietly invades" the everyday life of islanders (2017: no page). Refusing a relocation programme, the islanders have used a combination of measures to adapt – from raising floors to coping with teaching lessons as fish and human waste circulate at the feet of the students (Doherty, 2019). For others, like, Nemo's Garden, a sub-marine project based in Noli, Italy, their response to climate change lies not in countering deepening waters but in utilising them. The project seeks to use the comparatively stable temperature of the undersea volume to grow crops in domed underwater greenhouses or 'biospheres' (see Squire et al., 2018), this, aiming to "make underwater farming an economically viable, long-term" (Nemo's Garden, 2018), sustainable and an environmentally friendly form of adaptation. The team have successfully grown basil, strawberries, and tomatoes under the sea, ushering into being the submarine futures that Deloughrey speaks of. More broadly, as Merrie et al. (2018) exemplify, the process of imagining ocean futures in the wake of deepening waters matter for their potential to help steer towards more equitable and socially and ecologically desirable futures. Using science-based storytelling, the Radical Ocean Futures Project (Merrie et al., 2018) offer four different visions of the future that include the oceans coming "back from the brink" due to efforts to protect submarine life, and creating "order from chaos" as a "pocket of humanity dwells in Earth's oceans". These radical narratives seek to prompt new ways of imagining the deep that disrupts the supposed 'objective', quantitative, abstracted, and 'value neutral' assumptions that are implicit in many scientific and policy-orientated projections seeking to grapple with submerging futures (Merrie et al., 2018).

Needless to say, the prospect of rising seas has led to a re-imagining of 'depth'. Much like the sea itself, understandings of this term and material, voluminous condition have undergone transformation. They fold and unfold, circulate, shift and are made mobile across vast distances. From the nineteenth century, through to the Cold War, the present day and into the future, 'depth' has been configured and reconfigured in ways that have profound societal, cultural, political, and economic implications. This process will only continue as oceans rise and we are confronted with the deep in new ways.

## References


Bear C and Eden S (2008) Making space for fish: The regional, network and fluid spaces of fisheries certification. *Social and Cultural Geography* 9(5): 487–504.

Bravo M (2006) Geographies of exploration and improvement: William Scoresby and Arctic whaling, 1782–1822. *Journal of Historical Geography* 32(3): 512–538.

Bremner L (2015) Fluid ontologies in the search for MH370. *Journal of the Indian Ocean Region* 11(1): 8–29.

Bridge G (2009) The hole world: Scales and spaces of extraction. *New Geographies* 2: 43–48.

Brown M and Peters K (2019) Introduction. In: Brown M and Peters K (eds) *Living with the Sea: Knowledge, Awareness and Action*. Abingdon: Routledge, 1–12.

Campling L and Colás, A (2018) Capitalism and the sea: Sovereignty, territory and appropriation in the global ocean. *Environment and Planning D: Society and Space* 36(4): 776–794.

Carver R, Childs J, Steinberg PE, Mabon L, Matsuda H, Squire R, McLellan B and Esteban M (2020) A critical social perspective on deep sea mining: Lessons from the emergent industry in Japan. *Ocean & Coastal Management* 19(3) https://doi.org/10.1016/j.ocecoaman.2020.105242

Childs J (2020) Extraction in four dimensions: Time, space and the emerging geo(-)politics of deep-sea mining. *Geopolitics* 25(1): 189–213.

Childs J (2019) Deep sea mining threatens indigenous culture in Papua New Guinea. *The Conversation*. Available at: https://theconversation.com/deep-sea-mining-threatens-indigenous-culture-in-papua-new-guinea-112012

Cohen M (2018) Adventures in toxic atmosphere. In: Abberley W (ed) *Underwater Worlds*. Newcastle: Cambridge Scholars Publishing, 72–89.

Cousteau J (1954) *The Silent World*. London: The Reprint Society.

Cox D (1968) *Explorer's of the Deep: Man's Future Beneath the Sea*. New York, NY: Hammond.

Deloughrey E (2017) Submarine futures of the Anthropocene. *Comparative Literature* 69(1): 32–44.

Doel R (2003) Constituting the postwar earth sciences: The military's influence on the environmental sciences in the USA after 1945. *Social Studies of Science* 33(5): 635–666.

Doherty B (2019) Enduring the tide. *The Guardian [Online]*. Available at: www.theguardian.com/world/2019/feb/01/enduring-the-tide-the-flooded-philippine-islands-that-locals-wont-leave

Gibbs L and Warren A (2014) Killing sharks: Cultures and politics of encounter and the sea. *Australian Geographer* 45(2): 101–107.

Granata S (2018) The victorian aquarium as miniature sea. In: Abberley W (ed) *Underwater Worlds: Submerged Visions in Science and Culture*. Cambridge: Cambridge Scholars Publishing, 108–129.

Hamblin J (2002) The Navy's 'sophisticated' pursuit of science: Undersea warfare, the limits of internationalism, and the utility of basic research, 1945–1956. *Isis* 93(1): 1–27.

Hamlin C (1986) Robert Warington and the moral economy of the aquarium. *Journal of the History of Biology* 19(1): 131–153.

Hannigan J (2016) *The Geopolitics of the Deep Oceans*. Cambridge: Polity.

Hellwarth B (2012) *Sealab: America's Forgotten Quest to Live and Work on the Ocean Floor*. New York, NY: Simon and Schuster.

Helmreich S (2010) Human nature at sea. *Anthropology Now* 2(3): 49–60.

Houses of Parliament (2015) POST Note: Deep sea mining. Available at: http://researchbriefings.parliament.uk/ResearchBriefing/Summary/POST-PN-0508

Jamero ML, Onuki M, Esteban M, Billones-Sensano XK, Tan N, Nellas A, Takagi H, Thao ND and Valenzuela VP (2017) Small-island communities in the Philippines prefer local measures to relocation in response to sea-level rise. *Nature Climate Change* 7: 581–588.

Laloe AF (2016) *The Geography of the Ocean: Knowing the Ocean as a Space*. Abingdon: Routledge.

Lambert D, Martins L and Ogborn M (2006) Currents, visions, and voyages: Historical geographies of the sea. *Journal of Historical Geography* 32(3): 479–493.

Maury MF (1855) *The Physical Geography of the Sea*. New York, NY: Harper & Brothers.

Merchant S (2011) Negotiating underwater space: The sensorium, the body and the practice of scuba diving. *Tourist Studies* 11(3): 215–234.

Merrie A, Keys P, Metian M and Österblom H (2018) Radical ocean futures-scenario development using science fiction prototyping. *Futures* 95: 22–32.

Nautilus (2019) *About Nautilus*. Available at: www.nautilusminerals.com/irm/content/overview.aspx?RID=252&RedirectCount=1

Nemo's Garden (2018) *The Project*. Available at: www.nemosgarden.com/the-project/

Nunn PD, Mulgrew K, Scott-Parker B, Hine DW, Marks AD, Mahar D and Maebuta J (2016) Spirituality and attitudes towards nature in the Pacific Islands: Insights for enabling climate-change adaptation. *Climatic Change* 136(3): 477–493.

Oreskes N (2003) A context of motivation: US Navy oceanographic research and the discovery of sea-floor hydrothermal vents. *Social Studies of Science* 33(5): 697–742.

Pauli DC and Clapper GP (1967) *Project Sealab Report: An Experimental 45-day Undersea Saturation Dive at 205 feet/Sealab II Project Group*. Washington, DC: US Government Print Office. Available at: https://archive.org/details/projectsealabrep00paul

Peters K (2010) Future promises for contemporary social and cultural geographies of the sea. *Geography Compass* 4(9): 1260–1272.

Peters K (2020) Deep routeing and the making of 'Maritime Motorways': Beyond surficial geographies of connection for governing global shipping. *Geopolitics* 25(1): 43–64.

Peters K and Steinberg PE (2019) The ocean in excess: Towards a more-than-wet ontology, *Dialogues in Human Geography* 9(3): 293–307.

Reidy M and Rozwadowski H (2014) The spaces in between: Science, ocean, empire. *ISIS* 105(2): 338–351.

Robinson S (2020) Scientific imaginaries and science diplomacy: The case of ocean exploitation. *Centaurus* 63(1): 150–170.

Rozwadowski H (2012) Arthur C. Clarke and the limitations of the ocean as a frontier. *Environmental History* 17(3): 578–602.

Sammler KG (2020) The rising politics of sea level: Demarcating territory in a vertically relative world. *Territory, Politics, Governance* 8(5): 604–620.

Squire R (2021) *Undersea Geopolitics: Sealab, Science, and the Cold War*. London: Rowman and Littlefield.

Squire R (2020) Companions, zappers, and invaders: The animal geopolitics of Sealab I, II, and III (1964–1969). *Political Geography* 82 https://doi.org/10.1016/j.polgeo.2020.102224

Squire R (2016) Immersive terrain: The US Navy, Sealab and Cold War undersea geopolitics. *Area* 48(3): 332–338

Squire R (2018) Sub-marine territory: Living and working on the seafloor during the Sealab II experiment. In: Peters K, Steinberg PE and Stratford E (eds) *Territory Beyond Terra*. London: Rowman and Littlefield.

Squire R, Adey P and Jensen R (2018) Dome, sweet home: Climate shelters past, present and future. *Nature*. Available at: www.nature.com/articles/d41586-018-07513-8

Steinberg PE (2001) *The Social Construction of the Ocean*. Cambridge: Cambridge University Press.

Steinberg PE (2009) Sovereignty, territory, and the mapping of mobility: A view from the outside. *Annals of the Association of American Geographers* 99(3): 467–495.

Steinberg PE and Peters K (2015) Wet ontologies, fluid spaces: Giving depth to volume through oceanic thinking. *Environment and Planning D: Society and Space* 33(2): 247–264.

Starosielski N (2015) *The Undersea Network*. London: Duke University Press.

Straughan E (2012) Touched by water: The body in scuba diving. *Emotion, Space and Society* 5(1): 19–26.

Theriault N (2017) A forest of dreams: Ontological multiplicity and the fantasies of environmental government in the Philippines. *Political Geography* 58: 114–127.

Tsing A (2003) Natural resources and capitalist frontiers. *Economic and Political Weekly* 38(48): 5100–5106.

United Nations (1998) *The United Nations Law of the Sea: A Historical Perspective*. Available at: www.un.org/Depts/los/convention_agreements/convention_historical_perspective.htm#:~:text=Pardo%20ended%20with%20a%20call,to%20continue%22%2C%20he%20said.

Van Dover CL (2011) Tighten regulations on deep-sea mining. *Nature* 470(7332): 31–33.

Wang CM and Chien KH (2020) Mapping the subaquatic animals in the Aquatocene: Offshore wind power, the materialities of the sea and animal soundscapes. *Political Geography* 83 https://doi.org/10.1016/j.polgeo.2020.102285

Waters H (2013) The enchanting sea monsters on medieval maps. *Smithsonian*. Available at: www.smithsonianmag.com/science-nature/the-enchanting-sea-monsters-on-medieval-maps-1805646/#ew2er1HZjLSEsPQU.99.

Weeman K and Lynch P (2018) *NASA Global Climate Science*. Available at: https://climate.nasa.gov/news/2680/new-study-finds-sea-level-rise-accelerating/.

Wilson HF (2019) Contact zones: Multispecies scholarship through imperial eyes. *Environment and Planning E: Nature and Space* 2(4): 712–731.

# 28

# LIFE

## Ethical, extractive and geopolitical intimacies with nonhuman marine life

*Elizabeth R. Johnson*

### Introduction

In 2013, at the age of 65, Diana Nyad swam 110 miles from Cuba to Key West. After 52 hours of being in the ocean, she stumbled onto the shore in Florida. Her body was swollen, stung, dehydrated. She had suffered jellyfish stings, sea sickness, vomiting, and delirium. It took a technologically enhanced, 35-person crew to keep Nyad supplied with the food, fresh water, emotional support, and shark surveillance necessary to keep her alive on her journey. Life may have emerged in the seas, but Nyad re-emerged from the water onto the shore of Key West evidencing the oceans' virulent inhospitality to our now terrestrial bodies.

Given the human body's inability to nakedly navigate the ocean waters for sustained periods, it is a small wonder that marine life often appears as 'alien'. The body plans of many marine organisms coordinate with the world on a register so different from our own, they often appear as 'otherworldly'. Take, for example, the *Vampyroteuthis infernalis*, 'the vampire squid from hell'. Vilem Flusser's 1987 philosophical treatise on the squid attempted to write through the being of a vampire squid, to "cross from our world into its", to "see with its eyes and grasp with its tentacles" (1987 [2012]: 38). In doing so, Flusser took on an impossible task. Human bodies cannot inhabit the same spacetime as the *Vampyroteuthis infernalis*. As Flusser describes, vampire squid live in an abyss so deep that we would be crushed by the pressure of the sea. There, eternal darkness is punctuated only by the squid's own bioluminescence. When humans have attempted to enclose the vampire squid in aquaria on land, Flusser reports, they commit suicide by devouring their own tentacles. We might conclude that the vampire squid is not of our Earth. Flusser's attempt to get inside the organism's head requires thinking outside the world we know with our human heads and human hands, thinking into an other-worldly body.

Encounters with marine organisms stretch the limits of what westerners have long believed constitutes life on earth, often inspiring equal parts fascination and horror. But acknowledging the radical alterity of many marine bodies does not mean that we hold marine organisms at a fixed distance. As Stefan Helmreich (2009, 2010) writes, our encounters with them – whether in the flesh or through the screens of televisions and laptops – are neither innocent nor transparent (see also Wilson, 2019). How we register the seemingly extra-terrestrial differences of marine organisms are part of "durable, multiple, and porous inheritances" (Helmreich, 2010: 2). In his


DOI: 10.4324/9781315111643-34

writing on coral reefs, Helmreich refers to marine organisms as "figures" – the products of both facts and fictions, material connections and metaphorical reverberations (see also Jones, 2020). These material and figurative practices mean that sea creatures are not merely encountered in the oceans; they are part of an elaborate and multi-modal dance of seduction and repulsion as well as distancing and intimate incorporations. Material and figurative recompositions bring us both nearer to and more distant from ocean organisms, often simultaneously.

The governance of marine life – whether for conservation, consumption, or a combination of both – work on and with what Philip Steinberg and Kimberley Peters (2015) refer to as the ocean's "turbulent materialities" (see also Lehman et al., 2021). The turbulent socio-material histories between land and sea are vividly captured, for example, by Donna Haraway's figure of our current eco-social milieu: the tentacular marine monster, *chthulu*. Haraway's name for our contemporary era – the Chthulucene – describes a world in which forms of life and their histories are incomprehensibly entangled. As she writes, marine organisms – corals, octopuses, squid, cuttlefish, etc. – are "good figures for the luring, beckoning, gorgeous, finite, dangerous precarities" of our present moment in which the world is at stake (Haraway, 2016: 55). For Haraway, the alterity of marine organisms, their "sheer not-us", requires that we consider the vast multi-species entanglements that make up life on our planet (Haraway, 2016: 55). As she writes, paying attention to coral and other organisms forces us to confront the dangers of our current economic and political moment that continues mining, drilling, and fracking the Earth, seemingly without end.

And we have increasingly been paying attention. David Attenborough's *Blue Planet* series and livestreams of the US National Oceanic and Atmospheric Agencies (NOAA) Okeanos Explorer dives have drawn deep sea life into closer intimacy with various publics (at least, those with internet and television access). Meanwhile, the growing academic fields of 'critical ocean studies' and the 'blue humanities' have brought marine life in from the margins of thought (see, for example, Braverman, 2018; Braverman and Johnson, 2020; Chen, et al. 2013; DeLoughrey, 2007, 2017; Elias, 2019; Gillis, 2013; Ingersoll, 2016; Mentz, 2009). In doing so, both popular and academic registers have played up the sublime nature of marine organisms to spark curiosity – and concern – about ocean life.

As this body of scholarship clearly shows, the *where* of marine life is a devilish question. In what follows, I describe some of the conflicting and contradictory ways that human bodies enter into material and symbolic relation with ocean life. I consider how, in the process, ocean lives become 'subjects of', 'subject to' and at times forceful actors within political practices (Hobson, 2007). I begin with the ethical question of moral exclusion and biopolitical governance. Marine life often plays a crucial role in governing the boundaries between what should be saved and what can be made 'killable' (Haraway 2008). In the second section, I explore how marine life is made a part of human (and nonhuman) metabolic systems through patterns of consumption and biomedical extraction. In the third, I contemplate how marine organisms have been drawn into geopolitical strategies and made a part of militarised sea space. I conclude by considering how engaging with ocean life might enliven alternative frameworks for rethinking human connections to marine ecologies. Throughout, I consider how stories of ocean lives regulate the proximity between marine life and human being, bridging but also mining gaps between the distant and familiar.

## Metaphorical and material intimacies: Biopolitics and ethics of marine life

Narratives of marine life bear on our understandings of ourselves. One of the most viral English-language news stories of 2018 was that of the orca, J35. Off the coast of British Columbia, she gave birth to a calf that died within hours. J35 proceeded to carry that dead calf with her for 17 days in the waters off British Columbia and the adjacent State of Washington. This so-called "tour of grief" (Simon, 2018: no page) fuelled a public outcry in solidarity with the orca and brought attention to ecological issues facing declining pod populations in the Pacific.

J35's moment in the social media spotlight follows a logic of proximity to and similarity with cultural value systems in ways that parallel wider trends in biopolitical governance. Contemporary biopolitical regimes are predicated on deciding what bodies can be made to flourish and, correspondingly, made grievable in their death. An increasing wealth of literature has examined how biopolitical apparatuses divide not only race and territory – as Foucault (1976) described – but also species from one another (Braverman, 2015, 2018; Chen, 2012; Wolfe, 2013), and life from non-life (Povinelli, 2016). The social intelligence of cetaceans renders them analogous to human life in its post-Romantic European forms. Narratives, like that of a mother's grief, project historically and socially contingent emotional states onto animal behaviours (Howard, 2018). In doing so, they bridge the distance between terrestrial and marine, redrawing the boundaries of ethical and biopolitical calculations.

Ethical frameworks built on similarity with certain human norms follow the twentieth-century revolution in animal rights. Whether cutting a line around an organism's capacity to suffer (Bentham, 1789; Singer, 1975), have a face (Lingis, 2003), or respond (Wolfe, 2013), such ethical frameworks and their associated stories require deciding on the differences and distances between humans and other organisms. We bring nonhuman animals like the orca close in language, culture, and metaphor to distance them from acts of cruelty and consumption. As Wolfe (2013) writes, humans are constantly required to define the conditions for incorporation and exclusion within spheres of biopolitical concern: those that can be made to matter can be distanced and, accordingly, saved. Those that are too distant (or too alien) to matter, however, are ironically brought nearer in the flesh as a material resource.

Such lines are not fixed, but transient. Contemporary narratives of cetaceans stand in stark contrast to those of the eighteenth and nineteenth centuries, when men pursued whales for fuel and financial gain (see Burnett, 2012). Then, whale bodies were objects of extraction, their flesh a currency that circulated into the intimate spaces of everyday life. Whale blubber fuelled bodies, lamps, and industry; pursuit of it also accelerated cartographic science and colonial expansion (see Korosy, 2020). Those colonial engagements with whales changed patterns of human life irrevocably. Whaling remains central to networks of cultural and material production in Japan and Norway. But, in the wake of nineteenth-century Romanticism and twentieth-century environmentalism, white westerners elsewhere now view marine mammals more as subjects of regard – and, therefore, part of discourses and legal frameworks for conservation – rather than objects of consumption.

Through analogies, marine mammals are drawn closer metaphorically into proximity with human life, motivating outrage against captivity and harvesting (see, for example, the documentaries *Blackfish* and *The Cove*). The social intelligence of marine mammals and a select few other organisms – the cunning of octopuses, for example – is part of what one might call the '*Us Weekly* ethical framework' based on the logic of 'Orcas: they're just like us'.

When it comes to terrestrial organisms – particularly cattle, pigs, and chickens – these frameworks have encouraged cultural change, not by engaging with agricultural or economic policy, but by resetting the boundaries of what we consider food versus "subjects-of-life" (Regan, 1985: 243). The primary site of concern according to this logic therefore rests on the dinner plate. When these ethical frameworks focus on consumption, moves to ethical exclusion ironically justify corporeal inclusion: if we recognise an organism as sufficiently alien, its incorporation within our own bodies can take place seemingly without question. The ethical framework that Peter Singer developed in *Animal Liberation* (1975), which was based on animals' capacities for suffering, notoriously excluded oysters and other filter feeders. Over the years, Singer and other animal advocates have continually waffled on whether bivalves should be granted entry into our circles of moral care. Meanwhile, other marine organisms exhibit body plans, life cycles, and behaviours so alien that we scarcely consider them animals. Or, unlike J35, their life courses do not fit within analogous stories of human life. These include crabs that, rather than mourning their dead offspring, eat their live ones; or even other orca mothers who have been observed committing infanticide (Gibbens, 2018). As Stacy Alaimo (2016, 2020) has written, the trope of the 'alien ocean' often fosters a kind of disconnect between ourselves and the sea that can lead to blatant disregard, rendering ocean life more easily 'killable' (Haraway, 2008; see also Gibbs and Warren, 2014; Schrader and Johnson, 2017).

Engaging more than superficially with marine life disturbs these conventional ethical framings that require deciding on the distance between humans and the sea. First, it is increasingly clear that ethical considerations of marine life extend well beyond patterns of consumption. Biodiversity loss, collapsed fisheries, the bleaching of coral reefs, ocean acidification, and the ubiquitous spread of micro-plastics are now near constant features of mainstream media reporting on the oceans (Adey, 2018; Bradley, 2018; Fox, 2019). As these anthropogenic phenomena affect not only marine life, but also national economies and human health, legislators at all scales of governance are working to author new policies that promote biodiversity, reduce pollution, and manage valuable fisheries (Briand, 2008a, 2008b, 2010, 2013, 2014; Cuttlelod, et al., 2009; Gross, 2015; Germanov et al., 2018; Margat and Vallée, 2000; Vié et al., 2009). And while knowledge of ocean life and the causes of its impoverishment remains slippery, there is growing interest in charting the circulatory feedback loops between human patterns of production and the recomposition of marine ecologies.

Moreover, while cetaceans and a selection of other organisms are easily drawn into analogous calculations, most marine lifeforms, as Elsbeth Probyn writes, "refuse to settle into a neat taxonomic order, to cuddle up to us" (2016: 20). The study of nonhuman marine species raises questions about the nature of interconnection as well as how we think the categories of the individual and the colony, the singular and the collective. The enmeshment of marine life and associated ethical and legal challenges has been a core concern of Irus Braverman's work on coral reefs and coral conservation science (2018). Braverman's research shows how corals themselves do not conform to the restrictions of international conservation law, which presupposes 'life' as individual and, most often, of the vertebrate variety. This has caused no end of issues for conservation scientists, who struggle to determine not whether corals 'count' as qualified life, but, given that they are colonial reef organisms, how to count them at all. Braverman's writing on the Crown of Thorne starfish also reveals a troubling biopolitical calculus. In attempts to preserve coral reefs, scientists actively kill the Crown of Thorne starfish that 'voraciously' dine on the reef (McDonald, 2018; McFarling, 2019). That they increasingly use autonomous robots to carry out the killing presents a curious displacement of decision-making over ecological governance away from humans and into algorithms (Braverman, 2020).

Other organisms fall outside of these frameworks entirely. Some forms of life that the ocean sustains are so radically different that we struggle to describe them, often reaching for awkward metaphors and ill-fitting comparisons. What would best make them live or let them die – sometimes even what counts as life and death – remains uncertain (see the discussion of jellyfish in Section II). Consider, for example, siphonophores like the Portuguese Man-O-War (named for its visual similarity to the sixteenth-century warships at full sail). Part of the phylum Cnidaria, siphonophores are related to coral and jellyfish. Scientists consider them colonial organisms, made up of differentiated 'zooids' that live collectively, performing separate functions in the collective animal's life cycle. The zooids emerge from a single embryo, but differentiate early on, similar to the organs collectively housed in the body of a mammal. But the siphonophore's component parts are more than parts of a whole; they are analogous to a solitary animal. In a 'Quick Guide' to siphonophores in *Cell*, evolutionary zoologist Casey Dunn compares them at various points to social insect colonies and conjoined twins. Their organismal coordination repeatedly appears as a "division of labor" (Dunn, 2009: 233). Yet, what makes these organisms 'work' as a collective remains something of a mystery.

The undecidability and strange ecologies of marine organisms make them well suited as emblems of 'queer' imaginaries and associated reconsiderations of ethical frameworks. Marine organisms collectively express the entire range of reproductive patterns evinced by life, including asexual cloning, parthenogenesis, hermaphroditism, and a multitude of sexual mating strategies. Confronting that diversity of reproductive pathways erodes many of the assumptions that underpin heteronormativity (see Alaimo, 2016; Griffney and Hird, 2016). Eva Hayward (2012a, 2012b) has written on how the visual presentation of captive jellyfish and coral creates new fields of sensation. These organisms, she writes, are encountered "viscerally rather than intellectually, sensuously rather than conceptually" (2012b: 184). Drawing on Haraway's notion of 'diffraction', Hayward explores how these organisms provoke a different kind of engagement with oneself as well as with jellyfish as others. Her encounters with jellyfish at the Monterey Bay Aquarium open up new ways of relating ethically, altering our sense of what does – and what might – come to matter in the world.

Astrid Schrader writes on marine microbes (2020) to show how conventional ethical frameworks elide the question of who – which humans under what conditions of knowledge about life – decide how we place difference and mark distance. In response to Cary Wolfe's writing on nonhuman biopolitics, Schrader questions his assumption that an ever more refined knowledge of nonhuman organisms will necessarily refine human capacities to determine the difference between qualified and unqualified life. Exploring the science of circadian rhythms and programmed cell death in cyanobacteria (known as apoptosis), Schrader shows how knowledge of marine microbes undermines long-held assumptions about the relationship between life and death as well as temporality. The internal clock of cells, she writes, is the product of intergenerational "haunting": "When the molecular mechanisms that stabilize the clock are internal to a cell and cannot be associated with interactions between the cells (as in eukaryotes)" scientists cannot distinguish "between an individual cell and a population" (Schrader, 2020: 261). As a consequence, Schrader argues that the conceptions of justice must be predicated on an expanded ethical framework that, following Derrida, she refers to as "abyssal intimacies" (Schrader, 2015) rather than the deciding lines of biopolitics.

Charting how to care for the circulation of marine organisms in the oceans and through our bodies on land – what Probyn refers to as "eating the ocean" – is clearly far from straightforward. In the next section, I explore how conservation and consumption are knots of economic, legal, ecological, and bodily relations.

## Resource intimacies: Political ecology and more-than-human assemblages

While human bodies cannot sustain extended exposure to the ocean, our bodies are kept alive by the stuff of the sea. Accounting for this intimate relationship between human bodies and marine life is a constant challenge.

Much of the geographic writing on marine life – like much of resource geography throughout recent decades – focuses on extraction. Unlike minerals and mines, however, living ocean resources are not bound to sites; they move through the layered legal jurisdictions of the sea. Accordingly, the material and economic entanglements between humans and marine life are tentacular. They extend well beyond networks of direct consumption and creep into nearly every sphere of production on land.

Contemporary western ways of knowing hold terrestrial and marine environments as distinct, often viewed through the lens of twentieth-century academic disciplines that cleave biology and ecology from political and economic formulations. But these conventional analytics are insufficient to examine marine organisms and their circulation as commodities. Alison Reiser's research on herring busses demonstrates that even early histories of maritime law are intimately connected with the life cycles of marine organisms (Reiser, 2020). Just as Korosy's work examines how whaling practices were central to the development of colonial expansion, Reiser's demonstrates how the migratory patterns of herring helped to co-produce Hugo Grotius's influential European legal doctrine, *Mare Liberum*, or the freedom of the seas. Grotius's legal writing, which serves as the foundation of the UN's Convention on the Law of the Sea (UNCLOS), was developed on an understanding of herring as an inexhaustible resource. Whether we view marine life as a common resource or privatisable set of commodities continues to create tension for ocean governance. Ever since Grotius's views were largely codified in European law, the West has viewed marine life as one of the few remaining resource commons. Without the prolific reproductive capacities of herring and their seasonal movement around the North Sea and Northern Atlantic, ocean space may have been subject to privatisation much earlier in history.

The precise dimensions of the socio-political networks that manage such a commons is dizzyingly complex, far from uniform, and increasingly subject to neoliberal policies. Neoclassical economic perspectives position fisheries as vulnerable to Garret Hardin's 'tragedy of the commons' and therefore in need of enclosure. The reproductive and migratory nature of fish populations, however, put the privatisation of wild fish stocks at odds with many in the fishing industry. Kevin St. Martin (2006), for example, has shown how fisheries off the coast of Maine in the United States are often managed through regional community networks. St. Martin describes fishing as a collective endeavour, rooted in regional communities and local knowledge networks. Accordingly, fisheries create the conditions for resource governance to emerge in accordance with networks of economic exchange that are not (exclusively) determined by capitalism's drive for accumulation and expansion (St. Martin, 2006; St. Martin et al., 2007).

Recent attempts to manage living marine resources differently – through, for example, emerging marine spatial planning policies (MSPs) – alternatively restrict and open up the capacities of communities to manage marine resources (Boucquay et al., 2016). Zoe Todd's work (2014) puts an even finer point on the challenges of managing fisheries by examining how Inuvialuit communities in Paulatuuq, Arctic Canada, account for the plural natures of fish in everyday life and Indigenous law. "Rather than treat fish as separate from humans or humans as separate from fish, fish are intimately woven into every aspect of community life" (Todd, 2014: 19). Todd and St. Martin examine the complexities of human–fish interactions and the institutions – formalised and informal – that govern those interactions.

Political ecologists have also focused on the circulation of fish commodities post-extraction, further complicating the deeply integrated network of law, history, affect, and the material properties of marine life. Becky Mansfield's writing on the fish paste surimi demonstrates the multiplicity of marine resources from the perspective of the commoditised product and its afterlives. Mansfield shows how surimi undergoes material transformations that are subject to changing legal and economic frameworks as well as the cultural and symbolic milieus through which they pass (Mansfield, 2003a, 2003b). Emphasising the way that extracted marine biomaterials are part of shifting spatial and economic relations, Mansfield describes patterns of production and consumption that simultaneously distance and entangle terrestrial and marine life. "Surimi is distanced from the actual fish and from Japan, to be entangled with more familiar seafood items", such as crab and lobster meat. Accordingly, "firms can sell surimi seafoods as precisely what they are not – or rather, sell them as something more than what they are" (Mansfield, 2003a: 181). The material fish that Mansfield tracks is part of a moving assemblage of economic, historical, and symbolic relations that coalesce to produce surimi as a commodity. Similarly, Amy Braun's work has mapped how promissory economies of algae production commoditise and compartmentalise life's components for patenting and privatisation. Like Mansfield, she has traced how algae's components multiply as actors in the Blue Economy (Braun, 2020).

Indeed, the more details of ocean life that emerge from 'Blue Economy' scholarship, the more entangled these networks seem. Marianne Lien's writing on the extraordinary intensification of aquaculture over the past half century introduces a further layer of complexity into the socio-material assemblages of the sea. She demonstrates how farmed Atlantic salmon is "systematically and simultaneously inscribed as a universal biogenetic artefact and a local brand commodity" (Lien, 2009: 65; see also Lien, 2015). Probyn's work similarly shows how blue economy strategies bring terrestrial and marine life into ever more intimate relation in ways that go well beyond the intentional act of eating seafood. Indeed, we often eat marine life without knowing it: 25 per cent of the global fish catch goes to non-direct food consumption. Much of that goes to feed vegetables, grains, and livestock on land. Some of it also ends up as fish oil, fortification for bread and milk, in cosmetics, and as fish food for aquarists and aquaculturalists. Ethical frameworks that presume we can decide what forms of life are fit for consumption fall apart on close consideration of marine creatures and their ties to terrestrial circuits of food production. As Probyn writes, "[t]he idea that you can solve such intricate and complicated human-fish relations by voting with your fork is deluded narcissism" (Probyn, 2016: 10).

The tight knit of more-than-marine ecologies makes solving ocean issues a thorny problem indeed. Recent efforts to engineer technological solutions to the vast rafts of ocean waste, for example, have been met with outcry from scientists who study little known neustonic organisms like blue buttons and blue sea dragons that float on the sea surface (Helm, 2019). As Rebecca Helm (2019) has written, plastics "mimic the neuston world—it's buoyant, surface bound, and rubbery" (2019: no page). Given that they both sail by ocean currents, garbage patches and neuston meadows often overlap. And while waste removal would likely benefit the neuston ecosystem in the long term, current techniques of plastic removal have yet to differentiate between life and trash.

Elsewhere, scientists and policy makers are attempting to control the populations of organisms that seemingly enhance ocean degradation. Over the past two decades, recent proliferations of jellyfish blooms in the Mediterranean and Black Sea have contributed to periodic reductions in commercial fish stock. Some blooms have been directly linked to the introduction of jellyfish species via the ballast water of commercial shipping. But the general causes of increased blooms are difficult, if not impossible, to pinpoint (Boero et al., 2013). Proposed solutions to expanding jellyfish populations have been limited. In 2013, South Korea famously innovated a fleet of unmanned robots known as the Jellyfish Elimination Robotic Swarm (or JEROS) in order to combat blooms responsible for crashing aquaculture fisheries and shutting down coastal nuclear power plants (see Johnson, 2015a). In the five years since, however, scientists have come to realise that shredding jellyfish is not an effective way of managing blooms; it only mixes their reproductive cells more effectively, promoting further blooms in the future (Horowitz, 2017). Even the installation of nets meant to protect beach goers from blooms increases the likelihood of stings, as nets often injure the jellyfish as they brush up against them, distributing (still stinging) tentacles throughout the water.

An alternative solution that has gained traction in the Mediterranean and, increasingly, the Atlantic, returns us to changing patterns of consumption. At an Expo Milano 2015 conference on nutrition for the Italian National Research Council, jellyfish were featured alongside algae and insects as the food of the future. That same year, the magazine *Fine Dining Lovers* featured an article on jellyfish that praised them as the "latest frontier" in "alternative seafood resources" (Saibante, 2015: no page). Jellyfish are therefore poised to transition from 'trash animal' to 'sustainable protein', praised for being low calorie, low fat, and high in collagen and B12. Last year, a team of Danish researchers reported that they had developed an ethanol preparation that would turn the "jelly" of jellyfish to "chips that have a crispy texture and could be of potential gastronomic interest" (Mathias Clausen, quoted in Del Bello, 2018: no page). All of these strategies are meant to facilitate the incorporation of jellyfish into human bodies in familiar ways.

In addition to eating away our environmental woes, the UN Food and Agriculture Organization also recommend harvesting jellyfish for biotechnology, marking them one of many marine organisms that hold promise for biomedical industries. Through the extraction of Green Fluorescent Protein and its use in biosensing, jellyfish are incorporated into a system of biological surveillance, in hospitals, laboratories, and field sites. Furthermore, jellyfish stem cells are uniquely pluripotent, offering incredible promise to regenerative medicines as well as cosmetic treatments. Through jellyfish, dreams of the fountain of youth and immortality are kept alive (one only need to think of Sergei Brin and other Silicon Valley obsessions with radical life extension).

The strategic growth of the Blue Economy is bringing marine life into ever greater proximity with human bodies. Even the most alien species have become central to human health. The blood of horseshoe crabs (Moore, 2018) and fluorescent proteins of crystal jellyfish, for example, are essential components of global biomedicine and biosecurity, protecting medical patients from biological harms like E. coli. And, as knowledge of marine biologies expands and intensifies, so too do the processes of enclosure that render biological life an economic resource. As Helmreich (2007) describes in his work on Blue Economy imaginaries in Hawai'i, biotech and pharmaceutical industries have supported the making of what he calls blue-green capital on the backs of an articulation of ocean ecologies as wide, free, and endlessly abundant. As Helmreich has written, "A vision of the ocean as endlessly generative mimes and anchors a conception of biology as always overflowing with (re)productivity" (2007: 289). Growing excitement over the Blue Economy brings marine life into the orbit of a wider logic of extractive governance and resource management alongside enclosures of deep sea minerals for mining (on proposals for deep sea mining and their potential effects on human and nonhuman life, see Childs, 2018; Havice and Zalik, 2018; Reid, 2020; Sammler, 2020; Zalik, 2018). Young Rae Choi has referred to this as the expansion of a governmentality, which fully justifies every "intervention toward utility, efficiency, and prosperity" (2017: 39; see also Bear 2017). Along with the seabed, these proposals render marine life a resource for development. These framings collapse biodiversity conservation and extraction into one another in a drive toward a singular ocean, brimming with planetary potential.

The expansion of this 'sustainable Blue Economy' reflects a classic, colonial mindset. It mistakes a regional, local way of viewing the world as totalising and universal. Such a viewpoint perhaps finds its apotheosis in the research and development aims of the US military. The Department of Defense (DoD) has long been at the forefront of harnessing marine biology for its potential utility. In the following section, I explore the US military's 'biological turn' and recent investments in bio-sensing to consider how attempts to harness life's capacities engender a vision of the earth as a site of planetary, geopolitical management.

## Geopolitical intimacies: Militarising marine life

On 29 April 2019, fishermen in Norway encountered a beluga whale wearing a harness with the words, 'Equipment' and 'St. Petersburg'. While unconfirmed, it is assumed that the whale had been trained into military service in Russia. The seemingly exotic notion that marine life might be harnessed for geopolitical interests sparked a media frenzy. Of course, as several journalists would later reveal, harnessing marine life for military aims is neither new nor exclusively Russian (Hu, 2019; Noack, 2019; Roache, 2019). The US Navy has been training dolphins and sea lions since 1959 to recover underwater mines, guard assets, and perform reconnaissance (Squire, 2021). The capacities of marine mammals to spend long periods underwater, navigate littoral environments, echolocate, and find and identify explosive devices are valued in an effort to expand the military's abilities in subsurface navigation (Moore, 1997; Squire, 2021).

Nonhuman life had long been utilised as test subjects in the development of military weapons, technologies, and pharmaceuticals. Beginning in the late-twentieth century, however, the US military newly turned to nonhuman life as technological inspiration. Like trained dolphins and sea lions, a range of nonhuman organisms became a resource to expand human capacities at sea as well as on land (see Johnson, 2010, 2015b). In the 1980s the US Navy began research into the development of biomimetic robots based on the forms and capacities of crustaceans and bivalves. The Defense Advanced Research Projects Agency (DARPA) followed suit and a suite of robotic marine organisms – lobsters, crabs, clams, and more recently jellyfish and manta rays – have since crawled and swam out of the DoD coffers (Johnson, 2015b). While very few have been taken up in service in the US military, these marine robotics programs seek to displace valued human life – and the lives of those aforementioned marine mammals – in dangerous ocean engagements. In this biopolitical calculus, the military attempts to short circuit 'killability' by rendering warfare a tournament of machines rather than war fighters (Johnson, 2020).

Marine life is now being used to enhance the capacities of military technologies more widely as well. In February of 2018, DARPA launched the Persistent Aquatic Living Systems program (PALS) in cooperation with Northrup Grumman, Raytheon, the Naval Research Laboratory and researchers at Florida Atlantic University and the University of Maryland (DARPA, 2019a, 2019b). The PALS program aims to harness the innate shrimp, plankton, and other marine organisms in order to sense and respond to environmental change. As one of the program leaders has noted, the goal of the project is "to leverage a wide range of native marine organisms, with no need to train, house, or modify them in any way" (DARPA, 2018: no page). By connecting organisms and their sensory capacities to a technological communication network, this investment in biosensing promises to enhance securitisation by turning organisms into a bio-technological platform.

Recent attention to verticality and volume in twentieth-century geopolitics (Elden, 2013; Graham, 2011) does not quite capture these projects that attempt to take hold of ecological processes and biological capacities to re-engineer the earth. PALS and other biosensing projects render marine space 'operational'. In doing so, they do not threaten to reduce or exhaust biological life, but amplify, fetishise, and reorganise its superabundance (see also Cooper, 2008). This recombination of abundant biologies is part of a geopolitical imaginary increasingly concerned not only with extensive territorial space, but intensive processes at the most intimate scales. Biology has thus become a repository of solutions to present military threats, anticipated future actions, and resurgent histories. This has been part of a transition in US military's strategy as it re-envisions its sites of engagement from battlefields to the Earth as a planetary 'solution-space' (Johnson, 2017, 2020).

The PALS program – along with much of the research on biosensing – reimagines the earth as a system capable of autonomous self-surveillance. That is, it engages with planetary management. The US military is not alone in this and their vision. Attempts to engender biological and technological solutions to environmental problems proliferate increasingly as part of regional and urban planning responses to ecological degradation and climate change (see, for example, Wakefield and Braun [2019] on oysters as climate mitigation infrastructure). Collectively, these efforts contribute to a logic of domination built on enrolling the superabundance of life in a single, universal trajectory of planetary management.

## Conclusion: Re-coordinating plural oceans and plural lives

Human life collides with marine organisms in ways both mundane and surprising. Entanglements with life in the seas condition political and ethical frameworks, mould structures of affect, and resonate with sensations beyond the everyday. Like the 'turbulent' matter of ocean waters and their currents, our relation to marine life ebbs and flows – sometimes in gentle waves, at other times in violent surges. Geographies of assemblage and relationally have better attuned western readers to these interconnections and the prospect of a 'one blue' world (Foley, 2017; Winder and Le Heron, 2017). Tracing those entanglements – between terrestrial and marine as well as human and nonhuman – are crucial as many of the aforementioned environmental issues facing today's oceans do not conform to Euclidean spatial coordinates. Highlighting the intimacies between human and marine organisms can also shake long-held assumptions in the western paradigm about the dichotomous nature of land and sea.

But there is reason to be wary: the proliferation of interconnections and their risks elide the powerful disconnections that are constitutive of our worlds and its multi-species "knots" (Haraway, 2008: 42). Indeed, at a current conjuncture that seeks to flatten forms of life into a singular vision of resource potential, dwelling on the wide gulf between the forms and patterns of life on the planet may be more necessary than ever. Thinking with 'alien' oceans and their inhabitants – perhaps through Schrader's abyssal intimacies – might call on us to rethink our ideas of connectivity altogether. Networked lives and assemblages are also part of fractured landscapes of power in which some forms of life are preserved while others abandoned. Such an approach might better align thinking on the seas with resistance to continued practices of colonial extraction that underpin many strategies in the blue economy – particularly around deep-sea mining – and the continued militarisation of ocean life. Studying marine life might thus help to show how to hold (rather than claim) space on the earth for others, no matter the nature of our connection to them.

## References

Adey J (2018) Building blocks of ocean food web in rapid decline as plankton productivity plunges. *CBC News [Online]*. Available at: www.cbc.ca/news/canada/newfoundland-labrador/ocean-phytoplankton-zooplankton-food-web-1.4927884

Alaimo S (2016) *Exposed: Environmental Politics and Pleasures in Posthuman Times*. Minneapolis, MN: University of Minnesota Press.

Alaimo S (2020) Afterward. In: Braverman I and Johnson ER (eds) *Blue Legalities: Life and Law of the Sea*. Durham, NC: Duke University Press, 311–325.

Bear C (2017) Assembling ocean life: More-than-human entanglements in the Blue Economy. *Dialogues in Human Geography* 7(1): 27–31.

Bentham J (1789) *An Introduction to the Principles of Morals and Legislation*. London: T. Payne and Son.

Boero F, Piraino S and Purcell J (2013) Jellyfish. *Vectors Fact Sheets*. Available at: www.marine-vectors.eu/factsheets/FS-26_jellyfish.pdf

Boucquey N, Fairbanks L, St Martin K, Campbell LM and McCay B (2016) The ontological politics of marine spatial planning: Assembling the ocean and shaping the capacities of 'Community' and 'Environment'. *Geoforum* 75: 1–11.

Bradley J (2018) The end of the ocean. *The Monthly [Online]*. Available at: www.themonthly.com.au/issue/2018/august/1533045600/james-bradley/end-oceans

Braun A (2020) Got algae? Putting life to work for sustainability. In: Braverman I and Johnson ER (eds) *Blue Legalities: Life and Law of the Sea*. Durham, NC: Duke University Press, 275–294.

Braverman I (ed) (2015) *Animals, Biopolitics, Law: Lively Legalities*. Abingdon: Routledge.

Braverman I (2018) *Coral Whisperers: Scientists on the Brink*. Berkeley, CA: University of California Press.

Braverman I (2020) Robotic life in the deep sea. In: Braverman I and Johnson ER (eds) *Blue Legalities: Life and Law of the Sea*. Durham, NC: Duke University Press, 147–164.

Braverman I and Johnson ER (eds) (2020) *Blue Legalities: Life and Law of the Sea*. Durham, NC: Duke University Press.

Briand F (ed) (2008a) *Climate Warming and Related Changes in Mediterranean Marine Biota*. Monaco: CIESM Publisher.

Briand F (ed) (2008b) *Impacts of Acidification on Biological, Chemical and Physical Systems in the Mediterranean and Black Seas*. Monaco: CIESM Publisher.

Briand F (ed) (2010) *Climate Forcing and its Impacts on the Black Sea Marine Biota*. Monaco: CIESM Publisher.

Briand F (ed) (2013) *Marine Extinctions – Patterns and Processes*. Monaco: CIESM Publisher.

Briand F (ed) (2014) *Marine Litter In The Mediterranean And Black Seas*. Monaco: CIESM Publisher.

Burnett DG (2012) *The Sounding of the Whale: Science & Cetaceans in the Twentieth Century*. Chicago, IL: University of Chicago Press.

Chen C, MacLeod J and Neimanis A (2013) *Thinking with Water*. Montreal: McGill-Queen's University Press.

Chen M (2012) *Animacies: Biopolitics, Racial Mattering, and Queer Affect*. Durham, NC: Duke University Press.

Childs J (2018) Extraction in four dimensions: Time, space and the emerging geo(-)politics of deep-sea mining. *Geopolitics* 25: 189–213.

Choi YR (2017) The Blue Economy as governmentality and the making of new spatial rationalities. *Dialogues in Human Geography* 7(1): 37–41.

Cooper M (2008) *Life as Surplus*. Seattle, WA: University of Washington Press.

Cuttelod A, Garcia N, Malek DA, Temple H and Katariya V (2009) The Mediterranean: A biodiversity hotspot under threat. In: Vié JC, Hilton-Taylor C and Stuart SN (eds) *Wildlife in a Changing World: An Analysis of the 2008 IUCN Red List of Threatened Species*. Gland: IUCN, 89–101.

Defense Advanced Research Projects Agency (DARPA) (2018) PALS Turns to Marine Organisms to Help Monitor Strategic Waters. Defense Advanced Research Projects Agency website. Available at: www.darpa.mil/news-events/2018-02-02

Defense Advanced Research Projects Agency (DARPA) (2019a) Five teams of researchers will help darpa detect undersea activity by analyzing behaviors of marine organisms. Defense Advanced Research Projects Agency website. Available at: www.darpa.mil/news-events/2019-02-15

Defense Advanced Research Projects Agency (DARPA) (2019b) Persistent Aquatic Living Sensors (PALS). Defense Advanced Research Projects Agency website. Available at: www.darpa.mil/program/persistent-aquatic-living-sensors

Del Bello L (2018) Jellyfish chips are the future of junk food. *Futurism*. Available at: https://futurism.com/jellyfish-chips-future-junk-food

DeLoughrey E (2007) *Routes and Roots: Navigating Caribbean and Pacific Island Literatures*. Honolulu, HI: University of Hawai'i Press.

DeLoughrey E (2017) Submarine Futures of the Anthropocene. *Comparative Literature* 69(1): 32–44.

Dunn C (2009) Siphonophores. *Current Biology* 19(6): R233–R234.

Elden S (2013) Secure the volume: Vertical geopolitics and the depth of power. *Political Geography* 34: 35–51.

Elias A (2019) *Coral Empire: Underwater Oceans, Colonial Tropics, Visual Modernity*. Durham, NC: Duke University Press.

Flusser V (1987 [2012]) *Vampyroteuthis Infernalis: A Treatise, with a Report by the Institut Scientifique de Recherche Paranaturaliste*. Minneapolis, MN: University of Minnesota Press.

Foley R (2017) One blue. *Dialogues in Human Geography* 7(1): 32–36.

Fox A (2019) "This is shocking". An undersea plague is obliterating a key ocean species. *Science Magazine*. Available at: www.sciencemag.org/news/2019/01/shocking-undersea-plague-obliterating-key-ocean-species

Germanov ES, Marshall AD, Bejder L, Fossi MC and Loneragan NR (2018) Microplastics: No small problem for filter-feeding megafauna. *Trends in Ecology & Evolution* 33(4): 227–232.

Gibbens S (2018) First case of orca infanticide observed. *National Geographic News [Online]*. Available at: https://news.nationalgeographic.com/2018/03/orca-killer-whale-infanticide-calf-video-canada-spd/

Gibbs L and Warren A (2014) Killing Sharks: Cultures and politics of encounter and the sea. *Australian Geographer* 45(2): 101–107.

Gillis J (2013) The Blue Humanities: In studying the sea, we are returning to our beginnings. *Humanities* 34(3) Available at: www.neh.gov/humanities/2013/mayjune/feature/the-blue-humanities

Graham S (2011) *Cities Under Siege: The New Military Urbanism*. London: Verso Books.

Griffney N and Hird MJ (2016) *Queering the Non/Human*. Abingdon: Routledge.

Gross M (2015) Deep sea in deep trouble? *Current Biology* 25(21): R1019–R1021.

Haraway D (2016) *Staying with the Trouble: Making Kin in the Chthulucene*. Durham, NC: Duke University Press.

Haraway D (2008) *When Species Meet*. Minneapolis, MN: University of Minnesota Press.

Havice E and Zalik A (2018) Ocean frontiers: Epistemologies, jurisdictions, commodifications. *International Social Science Journal* 68(229–230): 219–235.

Hayward E (2012a) Fingeryeyes: Impressions of cup corals. *Cultural Anthropology* 25(4): 577–599.

Hayward E (2012b) Sensational jellyfish: Aquarium affects and the matter of immersion. *Differences* 23(3): 161–196.

Helm R (2019). The Ocean Cleanup Project could destroy the Neuston. *The Atlantic*. Available at: www.theatlantic.com/science/archive/2019/01/ocean-cleanup-project-could-destroy-neuston/580693/

Helmreich S (2007) Blue-green capital, biotechnological circulation and an oceanic imaginary: A critique of biopolitical economy. *BioSocieties*: 2(3): 287–302.

Helmreich S (2009) *Alien Ocean: Anthropological Voyages in Microbial Seas*. Berkeley, CA: University of California Press.

Helmreich S (2010) How like a reef. *Party Writing for Donna Haraway!* Available at: https://drive.google.com/file/d/0BzmKs1Fz7m9uYmNhZjVmMjQtMmRkYi00OTljLWJmNTItZGE5MDQ2ZmQ3Yzcy/view?usp=embed_facebook

Hobson K (2007) Political animals? On animals as subjects in an enlarged political geography. *Political Geography* 26(3): 250–267.

Horowitz J (2017) Jellyfish seek Italy's warming seas. Can't beat'em? Eat'em. *The New York Times [Online]*. Available at: www.nytimes.com/2017/09/17/world/europe/jellyfish-climate-change-italy.html

Howard J (2018) The 'grieving' orca mother? Projecting emotions on animals is a sad mistake. *The Guardian [Online]*. Available at: www.theguardian.com/commentisfree/2018/aug/14/grieving-orca-mother-emotions-animals-mistake

Hu JC (2019) Why would the Russians use a beluga whale as a spy? *Slate Magazine [Online]*. Available at: https://slate.com/technology/2019/05/beluga-whale-spy-marine-life-russia-navy-dolphins.html

Ingersoll KA (2016) *Waves of Knowing: A Seascape Epistemology*. Durham, NC: Duke University Press.

Johnson ER (2010) Reinventing biological life, reinventing "the human". *Ephemera: Theory and Politics in Organization* 10(2): 177–193.

Johnson ER (2015a) Governing jellyfish. In: Braverman I (ed) *Animals, Biopolitics, Law: Lively Legalities*. Abingdon: Routledge, 59–78.

Johnson ER (2015b) Of lobsters, laboratories, and war: Animal studies and the temporality of more-than-human encounters. *Environment and Planning D: Society and Space* 33(2): 296–313.

Johnson ER (2017) At the limits of species being: Sensing the Anthropocene. *South Atlantic Quarterly* 116(2): 275–292.

Johnson ER (2020) Biomimetic geopolitics: The Earth, inside out. *Techniques and Cultures, Suppléments* 73. Available at: http://journals.openedition.org/tc/13832

Jones S (2020) The Seawolf and the sovereign. In: Braverman I and Johnson ER (eds) *Blue Legalities: Life and Law of the Sea*. Durham, NC: Duke University Press, 237–254.

Korosy Z (2020) Whales and the Colonization of the Pacific Ocean. In: Braverman I and Johnson ER (eds) *Blue Legalities: Life and Law of the Sea*. Durham, NC: Duke University Press, 219–236.

Lehman J, Steinberg PE and Johnson ER (2021) Turbulent waters in three parts. *Theory & Event* 24(1): 192–219.

Lien ME (2009) Standards, science and scale: The case of Tasmanian Atlantic Salmon. In: Inglis D and Gimlin D (eds) *The Globalization of Food*. Oxford: Berg Publishers, 65–81.

Lien ME (2015) *Becoming Salmon: Aquaculture and the Domestication of a Fish*. Berkeley, CA: University of California Press.

Lingis A (2003) Animal body, inhuman face. In: Wolfe, C (ed.) *Zoontologies*. Minneapolis, MN: University of Minnesota Press.

Mansfield B (2003a) Fish, factory trawlers, and imitation crab: The nature of quality in the seafood industry. *Journal of Rural Studies* 19(1): 9–21.

Mansfield B (2003b) "Imitation crab" and the material culture of commodity production. *Cultural Geographies* 10(2): 176–195.

Margat J and Vallée D (2000) *Mediterranean Vision on Water, Population and the Environment for the 21st Century*. Blue Plan for the Global Water Partnership/MEDTAC in the programme of the World Water Vision of the World Water Council White Paper. Available at: www.ircwash.org/sites/default/files/Margat-2000-Mediterranean.pdf

McDonald J (2018) When Crown-of-Thorns Starfish attack. *Jstore Daily [Online]*. Available at: https://daily.jstor.org/when-crown-of-thorns-starfish-attack/

McFarling UL (2019) Op-Ed: Scientists are borrowing from dystopian sci-fi in a last-ditch effort to save coral reefs. *LA Times [Online]*. Available at: www.latimes.com/opinion/op-ed/la-oe-mcfarling-coral-reef-solutions-20190619-story.html

Mentz S (2009) Toward a blue cultural studies: The sea, maritime culture, and early modern English literature. *Literature Compass* 6(5): 997–1013.

Moore LJ (2018) *Catch and Release: The Enduring Yet Vulnerable Horseshoe Crab*. New York, NY: New York University Press.

Moore PW (1997) Mine-hunting dolphins of the Navy. *Proc. SPIE 3079, Detection and Remediation Technologies for Mines and Minelike Targets II*. 3079: 2–6.

Noack R (2019) There's a new alleged Russian spy. It's a beluga whale. *Washington Post [Online]* Available at: www.washingtonpost.com/world/2019/04/29/norway-fears-alleged-russian-spy-whale-economists-wonder-if-kremlins-military-has-finally-peaked/

Probyn E (2016) *Eating the Ocean*. Durham, NC: Duke University Press.

Povinelli EA (2016) *Geontologies: A Requiem to Late Liberalism*. Durham, NC: Duke University Press.

Regan T (1985) *The Case for Animal Rights*. Berkeley, CA: University of California Press.

Reid S (2020) Solwara I and Sessile Ones. In: Braverman I and Johnson ER (eds) *Blue Legalities: Life and Law of the Sea*. Durham, NC: Duke University Press.

Reiser A (2020) Clupea liberum: Hugo Grotius, free seas, and the political biology of herring. In: Braverman I and Johnson ER (eds) *Blue Legalities: Life and Law of the Sea*. Durham, NC: Duke University Press.

Roache M (2019) A beluga whale is allegedly a Russian spy. There's a long history behind that. *Time [Online]*. Available at: https://time.com/5582694/russian-spy-whale-history/

Saibante CT (2015) In praise of the jellyfish, from menace to resource. *Fine Dining Lovers*. Available at: www.finedininglovers.com/stories/eating-jellyfish/

Sammler K (2020) Kauri and the whale: Ocean meaning and matter in New Zealand. In: Braverman I and Johnson ER (eds) *Blue Legalities: Life and Law of the Sea*. Durham, NC: Duke University Press, 63–84.

Schrader A (2015) Abyssal intimacies and temporalities of care. *Social Studies of Science* 45(5): 665–690.

Schrader A (2020) Marine microbiopolitics: Haunted microbes before the law. In: Braverman I and Johnson ER (eds) *Blue Legalities: Life and Law of the Sea*. Durham, NC: Duke University Press: 255–274.

Schrader A and Johnson ER (eds) (2017) LabMeeting – Considering killability: Experiments in unsettling life and death. *Catalyst: Feminism, Theory, Technoscience* 3(2): 1–15.

Simon D (2018) "Tour of grief is over" for killer whale no longer carrying dead calf. *CNN News [Online]*. Available at: www.cnn.com/2018/08/12/us/orca-whale-not-carrying-dead-baby-trnd/index.html

Singer P (1975) *Animal Liberation: A New Ethics for Our Treatment of Animals*. New York, NY: Harper Collins.

St Martin K (2006) The impact of 'community' on fisheries management in the US Northeast. *Geoforum* 37(2): 169–184.

St Martin K, McCay BJ, Murray GD, Johnson TR and Oles B (2007) Communities, knowledge and fisheries of the future. *International Journal of Global Environmental Issues* 7(2/3): 221–239.

Steinberg PE and Peters K (2015) Wet ontologies, fluid spaces: Giving depth to volume through oceanic thinking. *Environment and Planning D: Society and Space* 33(2): 247–264.

Squire R (2021) *Undersea Geopolitics: Sealab, Science, and the Cold War*. London: Rowman and Littlefield.

Todd Z (2014) Fish pluralities: Human-animal relations and sites of engagement in Paulatuuq, Arctic Canada. *Etudes/Inuit/Studies* 38(1–2): 217–238.

Vié JC, Hilton-Taylor C and Stuart SN (2009) *Wildlife in a Changing World: An Analysis of the 2008 IUCN Red List of Threatened Species*. Gland: IUCN.

Wakefield S and Braun B (2019) Oystertecture: Infrastructure, profanation and the sacred figure of the human. In: Hetherington K (ed) *Infrastructure, Environment, and Life in the Anthropocene*. Durham, NC: Duke University Press.

Wilson HF (2019) Contact zones: Multispecies scholarship through 'Imperial Eyes'. *Environment and Planning E: Nature and Space* 2(4): 712–731.

Winder G and Le Heron R (2017) Assembling a Blue Economy Moment? Geographic engagement with globalizing biological-economic relations in multi-use marine environments. *Dialogues in Human Geography* 7(1): 3–26.

Wolfe C (2013) *Before the Law: Humans and Other Animals in a Biopolitical Frame*. Chicago, IL: University of Chicago Press.

Zalik A (2018) Mining the seabed, enclosing the area: Ocean grabbing, proprietary knowledge and the geopolitics of the extractive frontier Beyond national jurisdiction. *International Social Science Journal* 68(229–230): 343–359.

# 29
# WAVES
## The measure of all waves

*Stefan Helmreich*

### Introduction

When Leonardo da Vinci, around the turn of the sixteenth century, wrote of "the numberless waves of the sea", (see Baskins, 2010) he articulated a vision of the ocean as an immeasurable expanse, a bounded surface that might contain the infinite. Western mariners and scientists later strove to bring this realm into the sphere of the accountable – though they were hardly the first, or alone. Ming dynasty mariner Zheng He from 1405 to 1433 created sailing charts outlining ocean winds between South India and East Africa (Pereira, 2012). Fifteenth-century Arab navigator and cartographer Ahmad ibn Mājid penned works on the currents, tides, and winds of the Indian Ocean, describing in 1490 a phenomenon later translated as the "wave of the Cross (Southern Cross)" which enabled trade across the monsoon ocean (Aleem, 1967). Sixteenth-century fishers on the West African coast of what is now Ghana designed dugout surf canoes (*ali lele* in the Fanti language) to ride waves safely into shore (Dawson, 2018). And wayfinders and surfers in the Pacific developed techniques of navigation and surfriding (*he'e nalu* in Hawaiian) pitched to a range of wave configurations (see Genz et al., 2009; Walker 2011). Such work brought wave worlds into a range of legibilities.

Later, over in the Atlantic world, in the eighteenth century, the British Parliament sought reliable methods for finding lines of longitude at sea, and, in recognising John Harrison's marine chronometer as the solution, came to envisage the gridded globe as a kind of clock face (Sobel, 1995). The United States, in the 1840s, saw Naval officer Matthew Fontaine Maury leading projects to map the oceans' currents and winds, seeking to create, in his words, "mile-posts … set up on the waves … and time-tables furnished for the trackless waste" (1860: 343), making ocean currents and winds into infrastructures that, to Maury, resembled the railroad tracks then beginning to cross-hatch the world (Hearn, 2002[1]) (Winds had previously been known by western mariners largely through the impressionistic frame of the c. 1805 Beaufort Wind Force Scale, which offered descriptors for local, not global conditions – e.g., "wind felt on exposed skin"). Bringing ocean waves into calculability came next, in the twentieth century, and just as longitude and oceanic wind tracks became thinkable through the techniques of their time (clocks, railroads), so did ocean waves become readable using conceptual schemes made available by twentieth-century technologies and media – from amphibious military landing craft to aerial photography to animated film to electronic communications devices to digital and

 DOI: 10.4324/9781315111643-35

virtual modelling. Apprehending waves through and as objects of transmission, inscription, and interface, scientists construed them as processes in time whose futures across space might be foretold, predicted.

I present here a compressed history of wave science, aware that an entry on waves for this handbook could as easily examine waves via cultures of surfing, Indigenous wave piloting in Micronesia, maritime storytelling about freak waves, or the rendering of waves in seascape painting, photography, and film (see Helmreich, forthcoming). I centre the scientific tale here because of the technocratic and institutional power wave science has had, through its measures and instantiations – in coastal engineering, in ship design, in weather and surf reporting – to contour a range of experiences (recreational, logistical) at shore and sea.

Natural philosophical investigations into waves have been undertaken for millennia, dating back to antiquity.[2] In recent centuries, before the rise of oceanography, researchers in fluid mechanics sought mathematically to describe waves as moving patterns of crests and troughs, characterised by wavelengths and periods. A roll call of European mathematicians including Newton, Euler, Lagrange, Cauchy, Weber, Russell, Airy, Stokes, and Kelvin created formal models of wave action – drawing from field observations, experiments they conducted in glass wave tanks, and mathematical models made possible by the calculus (Craik, 2004; Zirker, 2013). They determined that gravity is the force that works, over time, to restore waves to equilibrium; that the water beneath a wave surface traces out circles or 'orbits'; and that most ocean waves are not sinusoidal, but rather *trochoidal* – with crests pointy rather than rounded (which in turn corresponds to orbits 'drifting' slightly forward as waves travel). Many researchers centred attention on waves in canals rather than at sea (Green, 1839). In dialogue with scientists working on transmission in optics and acoustics, such thinkers came, too, to see waves as oscillations of energy relay prone to reflection, refraction, diffraction, and interference. By 1879, wave science had accumulated into a body of knowledge unified enough that Sir Horace Lamb could gather it up into a foundational textbook, *Hydrodynamics*.

Vaughn Cornish's 1934 *Ocean Waves and Kindred Geophysical Phenomena* moved decisively into the ocean context, seeking to anchor theory in observation through empirical, fieldwork-based corroborations of mathematical claims about the relation between, for example, wave height and period. Cornish's twentieth-century contemporaries and successors (e.g., Sir Harold Jeffreys and Peter Janssen) puzzled over how waves gathered strength from the forces of wind and pressure that rippled over the water's surface. Two major figures, Owen Phillips and John Miles, in 1957 brought intuitions to bear from work on turbulence over aircraft wings, thinking of ocean waves as interfaces between the aero- and hydrodynamic. Research on waves moved ever more towards thinking about *ocean space*, and about *swells*: wave forms traveling out from under the wind that initially generated them. Investigations fixed, too, on what happened when waves arrived at the shore to *break* as spillers (which sloppily collapse on themselves), plungers (which pitch over themselves to create the 'barrels' beloved of surfers), and surgers (which slosh up onto shore without generating whitewater) – processes that did not always map so neatly into analogies from light, sound, or electromagnetic wave theory. Wave research became the story of how waves unfurled across ocean space and time.

## Waves and war

The dominant narrative of twentieth-century ocean wave science usually gets going with World War Two, and the oceanographer Walter Munk. Together with his dissertation advisor at the Scripps Institution of Oceanography, Harald Sverdrup, Munk was tasked by the United States Army Air Corps with determining whether wave weather might be predicted in order to time Allied amphibious invasions on Axis-held beaches. Munk and Sverdrup's work became decisive for the landings of 'duck boats' at Normandy on 6 June 1944, D-Day. To generate their system for wave predictions – to improve upon the in-place Steere Surf Code, a 'rule of thumb' set of guidelines – Munk and Sverdrup first needed to gather spatial data. While there would eventually exist dozens of 'wave stations' around the world, from which observers would report incoming wave heights and periods (see Bates, 1949), much proof-of-concept work was done in La Jolla, California, at Scripps, which hosted a pier from which observations could be made. In one Scripps document, *Height of Breakers and Depth at Breaking*, from March 1944, authors report on a project in which "four or five [aerial] photographs were taken of a single well-defined wave as it advanced from the outer end of the pier to the point of breaking".[3] Such photos – laid out as a kind of flipbook – were later synchronised with underwater measurements of pressure, which could map the rise and fall of waves at surface. Munk argued that one could use photographically captured changes in wavelength and speed as waves travel to shore to infer the changing depth of the water beneath.

Recordkeeping – and standards for understanding what waves *were* – became essential to wave science. In *Proposed Uniform Procedure for Observing Waves and Interpreting Instrument Records* from 1944, Walter Munk was already offering a rubric for such features as *wave height*. He wrote that "the wave height shall be taken as the average of the highest one-third of the waves observed during a time interval of at least ten minutes", spelling out the measure he would later call *significant wave height*.[4] The measure emerged from Munk's work with Marine Corps steersmen, whose eyeballed estimates of wave height from small boats Munk sought to square with scientific measure; as Munk put it in an interview, "I made the policy decision that it would be easier to accept marine pilots' statistics than to try and teach them mathematics. So, I invented using *significant wave height* to accommodate the community that works with waves" (Author Interview with Walter Munk, 25 August 2015). Such measures – and the aspiration to fix waves with a kind of "mechanical objectivity" (an accounting for natural phenomena accomplished by self-registering instruments rather than through idealised human impressions) (Daston and Galison, 2007) – came increasingly to be embedded in wave monitoring and recording devices. (They also came, sometimes, to confuse modelers, as the "significant" in *significant wave height* does *not* refer to statistical significance [Carl Wunsch, personal communication, 8 February 2019].)

How were waves becoming knowable at that mid-twentieth-century moment? Through embodied knowledge (in boats) in the sea, to be sure, but also through aerial photography – that is, from *above*, as processes animated in time, by a kind of stepwise cinema (on the ways waves have sometimes been gendered as female through their treatment as fluid and fluxional entities to be objectified, see Rodgers 2016; Helmreich, 2017). The 'god's eye' vantage of the aerial (a view less immediate, but nonetheless also situated) has continued to be elaborated, as waves come, eventually, to be measured by satellite. *Significant wave height* continues today to be employed – by ships, oil rigs, coastal development projects – and although its military beginnings no longer necessarily shape the term's use, the word *significant* emphasises that waves are always measured with respect to projects *meaningful* to humans. Often, as with shipping and coastal engineering endeavours, that significance has to do with infrastructure – either with the stability of the built world or with waves themselves as an enabling infrastructure for maritime enterprise (including, to be sure, surfing[5]).

One of the more significant episodes in Cold War research came in 1946, when the United States detonated a 20-kiloton fission bomb in the Bikini Atoll. Scripps scientist Roger Revelle wrote that the Navy "wanted to learn about the waves that would be produced by the air and underwater explosions and about the dispersion of radioactive materials in the lagoon and ocean waters" (quoted in Shor, 1978: 380). In 1952, such investigations continued as the United States dropped a hydrogen bomb on Enewetak Atoll in the Marshall Islands. Walter Munk and his colleague Willard Bascom were tasked with measuring waves consequent upon the explosion (which they did, in bathing suits, standing on 3x3 floats in view of the explosion). If wave predictions for Normandy, during a hot war, sought to know a near-future wavescape, Cold War South Pacific 'tests' had the US military figuring nuclear-weapon shock waves as virtual proxies for not-yet-arrived waves of war, even as Munk was also "concerned that the H-bomb would trigger a tsunami with distant outreach" (quoted in von Storch and Hasselman, 2010: 26) (it did not). The time imagined by the Navy was the hypothetical time of future maritime nuclear combat. For Marshall Islanders, whose atolls were used as the test site, damage – and exile (euphemised as 'evacuation') – was far from hypothetical, as winds brought killing cancers over the next generations (see O'Rourke, 1985). Waves were not only formal patterns of energy, but also harbingers of a toxic future.

Waves and wave knowledge, in other words, exist in the wake of history. Consider the arguments of Christina Sharpe's *In the Wake: On Blackness and Being*, which presses readers to remember "the transverse waves of the wake" (2016: 57) of Atlantic slave ships. For Sharpe, following poet Derek Walcott, *the sea is history* – and waves, particularly ship-made waves – may be its deathly haunting inscriptions. We can say something resonant about waves in the wake of atomic testing in the Pacific. Marshallese poet Kathy Jetñil-Kijiner brings the atoll's story into the present, with a vignette of patients at an island hospital inundated by the waves of sea-level rise: "a nuclear history threaded into their bloodlines woke to a wild water world/a rushing rapid of salt" (2017: 78). As these perspectives on waves from beyond formal western wave science demonstrate, 'measures' of waves are always partial accountings; the *significance* of waves varies greatly across different communities' experiences of space and history.

## The wave spectrum

Wave observations during World War Two, and just after, were not yet animated by the now common sense that waves arriving into shore might be sorted out by frequency, into a *wave spectrum*, analogous to the sort of spectrum used to classify ranges of light, sound, or radio waves. Wave scientists had long known that waves were generated by ocean storms, which imparted energy to water and created swells that moved across the sea with definite wavelengths and periods. Rachel Carson summarised the dominant scientific wisdom (and observational practice) around 1951:

> As the waves roll in toward Lands End on the westernmost tip of England they bring the feel of the distant places of the Atlantic. ... As they approach the rocky tip of Lands End, they pass over a strange instrument lying on the sea bottom. By the fluctuating pressure of their rise and fall they tell this instrument many things of the distant Atlantic water from which they have come, and their messages are translated by its mechanisms into symbols understandable to the human mind. If you visited this place and talked to the meteorologist in charge, he could tell you the life histories of the waves that are rolling in... He could tell you where the waves were created by the action of wind on water, the strength of the winds that produced them, how fast the storm is moving... Most of the waves ... he would tell you, are born in the stormy North Atlantic eastward from Newfoundland and south of Greenland.
>
> *Carson, 1951: 113–114*

Such waves, it began to be surmised in the 1940s, might be ordered by their frequencies, into a spectrum. The intuition came from media technologies that were becoming part of scientists' everyday and professional lives.

British wave scientists, gathered by the British Admiralty during World War Two into a body called 'Group W' ('W' for waves), lit upon the idea that it might be possible to "measure the variation of frequency with time" (Ursell, quoted in Longuet-Higgins, 2010: 43). To this end, they focused on wave records from a Lands End observing station, seeking to infer something about the geographical origins of waves of varied frequencies. Twenty-minute records of wave pressure, taken by undersea gauges (the "strange instrument[s]" of Carson), would be relayed to a rotating drum. But where a more usual wave record might have been rendered by a pen wiggling lines on a scrolling cylinder, here researchers turned to a method inspired by the movies. The director of Group W, oceanographer George Deacon, recalled that colleague Jack Darbyshire,

> had a friend in the film industry and we learned that the Walt Disney film, 'Fantasia', which was then a recent success, had the sound part of the film as black and white wavy silhouettes along the side of the picture frames. It therefore occurred to us that we could do the same thing ... We could use [photographic] paper ... and instead of printing lines we would have a big block of light which would move about, generating a silhouette of the waves. We could then put this on a wheel and, as the wheel spun round, the variations in black and white could be detected by a photocell.
>
> *Darbyshire, quoted in* Longuet-Higgins, 2010*: 50*

Movies with sound were not the only media to inspire the spectral model. Oceanographer Willard Pierson in 1952 adapted the spectrum model to wave records from the work of Bell Laboratories statistician John Tukey, who had used it to examine the statistical properties of noise in electronic circuits (Pierson and Marks, 1952; Tukey and Hamming, 1949). Tukey rendered waves as populations, not individuals, similarly to how sociology at the time was coming increasingly to study people.

The formalism of the wave spectrum, then – derived from visualisations of sound waves, from electronics, from state-of-the-art statistics – reads ocean waves in the idiom of twentieth-century media; indeed, it makes ocean waves, materialised in the elemental *medium* of seawater, thinkable *as media* in a more technological sense (see Peters, 2015). As Melody Jue argues in *Wild Blue Media* (2020), oceanic phenomena, knowable through media (cameras, sonar, scuba, satellites, equations, more), can themselves be profitably theorised as media, forces that crystallise as well as query conceptions about how information manifests as *inscription*, across *interfaces*, and in regimes of *storage* and *transmission*.

The story of wave spectra came into focus in an ocean-spanning project during the summer of 1963 when Walter Munk led *Waves Across the Pacific*, a project aimed at tracking trains of waves from their origins in Antarctic storms across 10,000 miles of Pacific Ocean toward their eventual arrival on the shores of Alaska. Munk has reported that he was inspired by studies in radio astronomy, aimed at using diffraction patterns to locate the source of interstellar radiation (von Storch and Hasselman, 2010: 7). Munk, with eight other oceanographers, arrayed wave sensors at six sites across the Pacific. A 1967 documentary about the project described the full roster of locations:

> Cape Palliser Light in New Zealand, a rugged storm-battered point where the arrival of the great waves from an Antarctic storm could be expected

> Tutuila, one of the volcanic islands of Samoa, 2,100 miles to the Northeast
>
> the uninhabited equatorial atoll of Palmyra, 1,600 miles beyond Samoa, only two miles wide with no point of land more than six feet above sea level
>
> the easily accessible Kewalo basin in downtown Honolulu, selected for the central wave station and expedition headquarters
>
> the islandless North Pacific where the U.S. Navy's mobile island *FLIP* (FLoating Instrument Platform) was stationed at 45 degrees north and 150 degrees west
>
> the final recording site an Alaskan beach, the end of the line for the trains of waves.
>
> *From American Archive of Public Broadcasting, no date; and Dierbeck, 1967*

The task was to follow wave forms and energies as they travelled over a great arc spanning southern and northern hemispheres, investigating whether swells would be scrambled by crossing the equator (the answer was no). The project required observing stations that could permit scientists to position pressure sensors on a not-too-deep seafloor (or, for *FLIP*, on a portion of its stem just below the surface). Sensors would need to be sensitive enough to measure waves that could be a mile long and a tenth of a millimetre high. Sensors translated changing water pressure into electrical signals, which were relayed to spooling computer punch tape.

*Waves Across the Pacific* shows scientists unrolling punch-tape records onto graph paper and poking pencils through holes to make dots representing oscillating wave heights, producing graphs of the travel of waves. Such graphs permitted scientists to map patterns they hypothesised would become visible to stations later in the chain. Walter Munk, at a station on American Samoa, is shown radioing ahead.

The tape data was sent back to the Scripps Institution of Oceanography in La Jolla to be processed. As Munk put it in the film, spectral analysis could reveal how groups of waves travelled, since waves in the wild "are of all sizes and come from all directions. They are mixed and piled atop one another in lovely confusion". Translating punch tape into graphs, radio-ing, phone-ing, shipping, and more could generate spectra to sort it all out. The film shows the deployment of multiple sorts of media – tungsten wire, water, magnets, brass cylinders, globes, rulers, typewriters, paper, pencils, punch-tape, tape-computers, and telephones (many operated by women secretaries), all of which underscore the mediations through which waves come to be known. Taken as a whole, the *Waves Across the Pacific* project comes to *render the Pacific Ocean into a giant transmitter of wave signals*, the travel of its swells miming the relay of information by scientists from one station to another. This ocean is like a radio that stores and sends information about storms.

The Pacific, of course, is and was also a geopolitical medium. In *Waves Across the Pacific*, the script has Munk remarking that, for this project, "the sea itself was our laboratory" (in Dierbeck, 1967: no page). This laboratory was available to US scientists because, with the exception of New Zealand, most points along the great circle had come, since World War Two, to be under American rule, either as territories of the United States (Tutuila, on American Samoa, and Palmyra), or, more recently, as states (Hawai'i and Alaska). As historians of oceanography Michael Reidy and Helen Rozwadowski argue, "knowledge of the ocean was – and remains – inextricably connected to midcentury geopolitics and the growth of modern science" (2014: 351). They note, too, that "the expansion of empire enabled scientists… to study topics that required the accumulation and subsequent reduction, tabulation, and graphing of large amounts of observational data from all over the globe" (Reidy and Rozwadowski, 2014: 350). Ruth Oldenziel, writing of US island territories as an array of "naval nodes for the control of ocean space" (2011: 16), reads the Pacific Ocean as layering communications infrastructures, military outposts, and colonial enterprise (American, Japanese, more). This geopolitical context is subordinated to the nature-oriented narrative of *Waves Across the Pacific*. In fact, the "lonely coral atoll" (Dierbeck, 1967) of Palmyra was occupied by the US Navy from 1939–1959, during which time the Navy built a runway and dredged a ship channel, after which the island was used, in 1962, as a site from which the Department of Defense observed nuclear tests above nearby Johnston Atoll. The waves that "smash endlessly against the rock of high volcanic islands like Samoa" (Dierbeck, 1967) meet not just a shore but also a Marine Corps outpost. And the "distant coast" (Dierbeck, 1967) of the experiment is in Alaska, which had just in 1959 became a US state. When the script (written by freelance writer Harry Miles) has Munk casually say, about the experiment, "We preferred islands we could get to", it leaves to one side the historical contexts that made these places available.

*Waves Across the Pacific* represents island environments as zones of wild nature, emphasised by images of fieldworking oceanographers – all white men – often in shorts, wearing no shirts, sometimes sporting tropical bead necklaces. A scene of Munk in American Samoa – "along the exotic southwestern shore of Tutuila" (Dierbeck, 1967: no page) – shows him aided by shirtless local men placing a sensor underwater. He is shown working with tape-records in a *fale*, a "house … built entirely of coconut palm, an ideal place to work with waves", and his voice-over tells us, later in the film, as we see him reading a book on a veranda, "[d]ata taking became quite routine. My wife and some of the Samoans learned to operate the recording equipment". This positioning by the film's writers of local people (and American women) as ready help for scientists appears elsewhere; in Alaska, Gaylord Miller, working at a Coast Guard station, and "the only one of us who sometimes met a bear on his way to work", employs "a fishing boat operated by an Eskimo crew" (Dierbeck, 1967: no page) to aid in positioning a sensor. As if to underscore the tropical vacation that the South Pacific sites represented for white scientists, we see many with newly grown beards. The film gives us scientists and waves with lives (recall Carson's "life histories of the waves"), but not so much local or Indigenous people (though on-the-ground experience was likely more varied; Helen Raitt, wife of Scripps geophysicist Russell Raitt, wrote expansively of her time in Tonga and became an advocate for South Pacific causes [see Raitt, 1956]).

The spectral model got a boost in the later 1960s as wave scientists sought to find a spectrum applicable to oceans the world around, permitting more accurate wave forecasting. This time, however, the question would not be, as with *Waves Across the Pacific*, about the *decay* of swell across a seascape, but about the *generation* and *evolution* of a wavescape – about how, from a calm sea, different frequencies of waves might emerge and develop across a large "fetch" of water, perhaps into what wave scientists call a "fully developed sea" (in which waves and wind exactly balance each other out).

Physicist Klaus Hasselmann, at the Max Planck Institute for Meteorology at Hamburg, was inspired by his participation in *Waves Across the Pacific* to propose the Joint North Sea Wave Project, or JONSWAP. Laying out 13 observation stations (five were ships, connected by radio telephone) over a 160-kilometre stretch of the North Sea, heading northwest from the German island of Sylt, scientists from the United States, Germany, the Netherlands, and the United Kingdom in 1969 deployed pressure sensors, buoys, and wind-measurement instruments to gather 50 million data points over three months about North Sea winds and their corresponding wave heights and directions (Cartright, 2010; Hasselmann et al., 1973). The results, which determined that steady winds over a long fetch result in the highest and longest waves, were published in 1973. Hasselmann was also able to confirm his speculation that when waves fall into resonance with one another, they transfer energy from shorter to higher wave frequencies. That account depended upon Hasselmann using a mathematical formulation with so many integrals that the problem became *six-dimensional*. Leonardo da Vinci's infinite but bounded sea thus became legible through the infinitesimals of calculus. The JONSWAP spectrum is still in use by forecasters today who use fetch and wind speed to predict wave heights; whether it is truly universal across Earth's seascapes is an open question (indeed, the precise mechanism by which wind generates waves is still not known in its very finest details [Pizzo et al., 2021]).

Amid this search for universal laws of wave *systems* that reflect global scale forces and processes, wave scientists have long recognised that some waves, such as tsunamis, require analysis as individuals. Employing the JONSWAP spectrum, a number of wave scientists have devoted their attention to 'freak waves' (Draper, 1966), which, since 2000, have been known as 'rogue waves' (Rosenthal and Lehner, 2007). Their new name resonates not only with the 'rogue states' named by political scientists in 1994, but also echoes terms like *rogue elephant* or *rogue shark*, used by naturalists to refer to anomalous, wild, individuals. A rogue wave is statistically unexpected, twice the significant wave height of its surrounds. Such waves, once believed to be the fanciful imaginings of mariners, are now accepted as real. In 1995 came the first measured instance – a 26-metre wave, leaping out of a sea of 11-metre waves at a North Sea Norwegian gas pipeline-monitoring platform. What causes such waves? They may emerge from the superimposition of waves in 'crossing seas', or from wave–current interactions (Africa's Cape of Good Hope is notorious). As an increasingly recognised part of ocean space, rogue waves are finding their place in, for example, insurance calculations in the shipping industry.

## Putting wave science to use

This chapter has presented a narrative about wave science that examines how waves might be understood as media – and as media that bear the historical impress of those technologies and media that have been used to gather them into representation. Such media of technological capture have these days generated standardised measures of waves, measures that now also intercalate with wider meteorological, political, and legal systems. Mathematical models of waves animate devices that now remotely sense and report wave properties. Waves may be measured, for example, by buoys, created, owned, and operated by a collage of governments, companies, and other agencies. The world's most common buoy is made by a Dutch company, Datawell, which began its manufacturing in 1961 in response to the devastating North Sea flood of 1953. Their Directional Waverider uses an alchemical mix of liquids coupled with floating accelerometers to record significant wave heights, periods, and directions. The information such buoys collect is spooled out as strings of numbers formatted according to World Meteorological Organization codes. Scientifically defined measurements (e.g., 'spectral peak period') are codified into standardised forms that then circulate into other technical domains. Some are legal domains – take, for instance, Australia's Standing Council on Transport and Infrastructure – which, in its setting of national requirements for the safety of commercial vessels, offers measures of what count as 'smooth waters': "waters where the significant wave height does not exceed 0.5 metres from trough to crest for at least 90 per cent of the time" (2012: 8). Ocean space as wave space is, in addition to military, colonial, and recreational space, a space of law.

Wave codes, relayed to weather services, are made available in a range of formats, often through the Internet. One of the most important is the online portal of the United States National Weather Service's National Data Buoy Center (www.ndbc.noaa.gov/), which makes data readable through an interface that allows users to click on buoy icons to get updates on local wave conditions. The United States National Weather Service hosts the world's most extensive computer model of wave action, WAVEWATCH III, which takes input from buoys and satellites.[6] WAVEWATCH originated in the 1980s as the master's thesis project of Hendrik Tolman at the Technical University of Delft and has, since its inception, become a kind of Ship of Theseus, as new parameters and tweaks have been swapped into its FORTRAN infrastructure and as it has been coupled to larger weather models.

Surfers might consult predictions generated by WAVEWATCH, seeking to know when swells will come in, hoping for real waves to match the virtual waves of their imaginations. Such measures may attach to local spatial politics of surfing. Scholars of surfing have written about the politics of the surf line up – who is in and out by gender, race, and more (Comer, 2010; Nemani. 2015; Saldanha, 2007; Waitt, 2008,). Isaiah Helelkunihi Walker (2011), in *Waves of Resistance: Surfing and History in Twentieth-Century Hawai'i*, maps how politics of indigeneity and tourism tangle in Hawaiian surf. Here, the media of interface are surfboards.

Computational wave models are now being used in calibration with projects of indigenous maritime revival in the Marshall Islands. Marshallese navigators, American anthropologists, and oceanographers are revisiting the knowledge materialised in 'stick charts' – mnemonic diagrams, made of coconut fronds and shells, of wave and swell patterns around the islands of Micronesia (Genz et al., 2009; Hutchins, 1995). Anthropologist John Mack (2011: 118) writes that these charts, used as prompts for navigators to recall the feel of swell, might be "less representations of space" than "representations of experience of space" – though the implied contrast between western/objective and indigenous/subjective underplays how latitude-longitude charts, sextant-enabled navigation, and radar readings are also animated by technologically enabled perception and cognition. Both genres of wave knowledge are about meditation and media.

The becoming computational of wave forecasting now inflects the ocean with the logic of digital media (see Gabrys, 2016) and the *virtual.* Wave models like WAVEWATCH often demand *more* data points than there are buoys. The model can conjure proxy data points – virtual buoys created by interpolating between known data points. Waves, again, take the shape of the media representing them, becoming virtual forms layered on the real world. Ocean space as wave space is increasingly known through cyberspace. Waves in ocean space become, as they ever have been, hybrid forms that mix the phenomenological, mathematical, technological, legal, and more – conjunctures of watery media and media of representation.

All of this media-studies animated history of wave science, of course, is not to deny the persuasive use value of scientific formulations (though it may prompt us to ask for whom wave knowledge has been useful and for what), but rather to place these formulae in cultural history as well as in their conditioning epistemological, institutional, and geopolitical contexts, contexts that bear the marks of their Enlightenment, colonial, military, recreational, and Cold War history. The very latest wave science, tuned to matters of sea-level rise, the release of anthropogenic aerosols into the atmosphere from the action of breaking waves, and the wave-initiated break up of Arctic ice, is now opening a new chapter in how scientists understand the significance of waves in human and planetary histories and futures. The space and time that ocean waves are now preparing and proclaiming is still in the writing.

## Acknowledgements

I thank Melody Jue, Heather Paxson, Kimberley Peters, Nick Pizzo, Philip Steinberg, and Carl Wunsch for commentary on drafts of this paper. Portions of the material offered here also appear in Helmreich S. (2021) Flipping the ship: Ocean waves, media orientations, and objectivity at sea. *Media and Environment* 3(1): https://doi.org/10.1525/001c.21389

## Notes

1 And see Hardy and Rozwadowski (2020) on how Maury, who defected during the U.S. Civil War to the Confederacy, sought to use his maritime knowledge to expand the institution of slavery, which he advocated exporting to South America.

2 Plato (c 428 BCE–c 348 BCE), outlining his ideal city-state in *The Republic*, named "three waves" of desirable social change; some scholars have suggested that Plato here called upon ancient Greek knowledge about wave groups in the Mediterranean (Sedley, 2005). Aristotle (384 BCE–322 BCE) and Plutarch (46 CE–120 CE) each later pondered how wind generates waves. Still earlier, and more impressionistically, *The Iliad* (1260–1180 BCE) offered the word κύμα for wave. Jamie Morton (2001: 32) suggests that this drew upon an image of the sea as female: "[d]erived from κύω, to conceive or be pregnant, κύμα denotes something swollen".

3 *Height of Breakers and Depth at Breaking, Preliminary Report on Results Obtained at La Jolla and Comparison with South Beach State and B.E.B. Tank Results.* SIO Wave Project. Report No. 8 [11 March 1944], Walter Heinrich Munk Papers, 1944–2002, Accession Number 87-35, BOX 23, Scientific Papers, Manuscripts and Talks, Scripps Institution of Oceanography Archives.

4 *Proposed Uniform Procedure for Observing Waves and Interpreting Instrument Records.* Walter Heinrich Munk Papers, 1944–2002, Accession Number 87–35, BOX 23, Scientific Papers, Manuscripts and Talks, Scripps Institution of Oceanography Archives.

5 Surfer Kelly Slater's custom, machine-made breaks in his inland California wave pool represent a full realisation of the notion of waves as infrastructure (see Roberts and Ponting, 2018).

6 Satellites infer wave fields in part from the reflection of light, "sun glitter", a method first developed in connection with aerial photography (Cox and Munk, 1954).

## References


Aleem AA (1967) Concepts of currents, tides, and winds among medieval Arab geographers in the Indian Ocean. *Deep-Sea Research* 14(4): 459–463.

American Archive of Public Broadcasting (no date) Record – Spectrum #35; *Waves Across the Pacific*, Description: https://americanarchive.org/catalog/cpb-aacip_75-9995xh8c

Baskins W (trans.) (2010) *The Wisdom of Leonardo da Vinci.* New York, NY: Philosophical Library.

Bates C (1949) Utilization of wave forecasting in the invasions of Normandy, Burma, and Japan. *Annals of the New York Academy of Sciences* 51(3): 545–72.

Carson R (1951) *The Sea Around Us.* Oxford: Oxford University Press.

Cartright D (2010) Waves, surges, and tides. In: Laughton A, Gould J, Tucker T and Roe H (eds) *Of Seas and Ships and Scientists: The Remarkable Story of the UK's National Institute of Oceanography.* Cambridge: Lutterworth Press, 171–181.

Comer K (2010) *Surfer Girls in the New World Order.* Durham, NC: Duke University Press.

Cox C and Munk W (1954) Measurement of the roughness of the sea surface from photographs of the sun's glitter. *Journal of the Optical Society of America* 44: 838–850.

Craik ADD (2004) The origins of water wave theory. *Annual Review of Fluid Mechanics* 36: 1–28.

Daston L and Galison P (2007) *Objectivity.* New York, NY: Zone.

Dawson K (2018) *Undercurrents of Power: Aquatic Culture in the African Diaspora.* Philadelphia, PA: University of Pennsylvania Press.

Dierbeck D (directed) (1967) *Waves Across the Pacific.* McGraw-Hill Text-Films.

Draper L (1966) "Freak" ocean waves. *Weather* 21(1): 2–4.

Gabrys J (2016) *Program Earth: Environmental Sensing Technology and the Making of a Computational Planet.* Minneapolis, MN: University of Minnesota Press.

Genz J, Aucan J, Merrifield M, Finney B, Joel K and Kelen A (2009) Wave navigation in the Marshall Islands: Comparing Indigenous and Western scientific knowledge of the ocean. *Oceanography* 22(2):234–245.

Green G (1839) Note on the motion of waves in canals. *Transactions of the Cambridge Philosophical Society* 7: 87–96.

Hardy PK and Rozwadowski H (2020) Maury for modern times: Navigating a racist legacy in ocean science. *Oceanography* 33(3): 10–15.

Hasselmann KF, Barnett T, Bouws E, Carlson H, Cartwright D, Enke K, Ewing JA, Gienapp H, Hasselmann DE, Krusemnn P, Meerburg A, Müller P, Olbers DJ, Richter K, Sell W and Walden H

(1973) Measurements of wind-wave growth and swell decay during the Joint North Sea Wave Project (JONSWAP). *Ergänzungsheft zur Deutschen Hydrographischen Zeitschrift*, RA 12.

Hearn C (2002) *Tracks in the Sea: Matthew Fontaine Maury and the Mapping of the Oceans*. Camden, ME: International Marine/McGraw Hill.

Helmreich S (2017) The genders of waves. *Women's Studies Quarterly* 45(1–2): 29–51.

Helmreich S (forthcoming) *A Book of Waves: Science, Significance, Sea Change*. Durham, NC: Duke University Press.

Hutchins H (1995) *Cognition in the Wild*. Cambridge, MA: MIT Press.

Jetñil-Kijiner K (2017) Two degrees. In: Jetñil-Kijiner K *Iep Jaltok: Poems from a Marshallese Daughter*. Tucson, AZ: University of Arizona Press.

Jue M (2020) *Wild Blue Media: Thinking through Seawater*. Durham, NC: Duke University Press.

Longuet-Higgins M (compiler) (2010) Group W at the Admiralty Research Laboratory. In: Laughton A, Gould J, Tucker T and Roe H (eds) *Of Seas and Ships and Scientists: The Remarkable Story of the UK's National Institute of Oceanography*. Cambridge: Lutterworth Press, 41–66.

Mack J (2011) *The Sea: A Cultural History*. London: Reaktion Books.

Maury MF (1860) *The Physical Geography of the Sea*, and Its Meteorology. Being a Reconstruction and Enlargement of the Eighth Edition of "The Physical Geography of the Sea". London: Sampson Low, Son & Co.

Morton J (2001) *The Role of the Physical Environment in Ancient Greek Seafaring*. Leiden: Brill.

Nemani M (2015) Being a brown bodyboarder. In: Brown M and Humberstone B (eds) *Seascapes: Shaped by the Sea*. Farnham: Ashgate, 83–100.

Oldenziel R (2011) Islands: The United States as a networked empire. In: Hecht G (ed) *Entangled Geographies: Empire and Technopolitics in the Global Cold War*. Cambridge, MA: MIT Press, 13–41.

O'Rourke D (director) (1985) *Half Life: A Parable for the Nuclear Age*. Cairns: Camerawork Pty.

Pereira CJ (2012) Zheng He and the African horizon: An investigative study into the Chinese geography of early fifteenth-century eastern Africa. In: Chia LS and Church SK (eds) *Zheng He and the Afro-Asian World*. Melaka: Perbadanan Muzium, 248–279.

Peters JD (2015) *The Marvelous Clouds: Towards a Philosophy of Elemental Media*. Chicago, IL: University of Chicago Press.

Pierson WJ and Marks W (1952) The power-spectrum analysis of ocean-wave records. *Transactions of the American Geophysical Union* 33(6): 834–844.

Pizzo N, Deike L and Ayet A (2021) How does the wind generate waves? *Physics Today* 74(11): 38–43.

Raitt H (1956) *Exploring the Deep Pacific*. New York, NY: WW Norton.

Reidy M and Rozwadowski H (2014) The spaces in between: Science, ocean, empire. *Isis* 105(2): 338–351.

Roberts M and Ponting J (2018) Waves of simulation: Arguing authenticity in an era of surfing the hyperreal. *International Review for the Sociology of Sport* 55(3): 229–245.

Rodgers T (2016) Toward a feminist epistemology of sound: Refiguring waves in audio-technological discourse. In: Rawlinson M (ed) *Engaging the World: Thinking after Irigaray*. Albany, NY: SUNY Press, 195–214.

Rosenthal W and Lehner S (2007) Rogue waves: Results of the MaxWave project. *Journal of Offshore Mechanics and Arctic Engineering* 130(2) https://doi.org/10.1115/1.2918126

Saldanha A (2007) *Psychedelic White: Goa Trance and the Viscosity of Race*. Minneapolis, MN: University of Minnesota Press.

Sedley D (2005) Plato's Tsunami. *Hyperboreus* 11: 205–214.

Sharpe C (2016) *In the Wake: On Blackness and Being*. Durham, NC: Duke University Press.

Shor EN (1978) *Scripps Institution of Oceanography: Probing the Oceans, 1936–1976*. San Diego, CA: Tofua Press.

Sobel D (1995) *Longitude: The True Story of a Lone Genius Who Solved the Greatest Scientific Problem of His Time*. New York, NY: Walker Publishing.

Standing Council on Transport and Infrastructure (2012) *National Standard for Commercial Vessels*. Canberra: National Maritime Safety Committee.

Tukey JW and Hamming RW (1949) Measuring noise color 1. *Memorandum MM-49-110-119*. Murray Hill, NJ: Bell Telephone Lab.

von Storch H and Hasselmann K (2010) *Seventy Years of Exploration in Oceanography: A Prolonged Weekend Discussion with Walter Munk*. Dordrecht: Springer.

Waitt G (2008) Killing waves: Surfing, space, and gender. *Social & Cultural Geography* 9(1): 75–94.
Walker IH (2011) *Waves of Resistance: Surfing and History in Twentieth-Century Hawai'i*. Honolulu, HI: University of Hawai'i Press.
Zirker JB (2013) *The Science of Ocean Waves: Ripples, Tsunamis, and Stormy Seas*. Baltimore, MD: The Johns Hopkins University Press.

# 30
# HYDROSPHERE
## Water and the making of earth knowledge

*Jeremy J. Schmidt*

### Introduction

In his presidential address to the Geological Society of America, Victor Baker began by stating that, "Geology is both (1) a body of knowledge about Earth, and (2) a way of thinking about Earth" (1999: 633). This chapter argues that understanding the hydrosphere – the combined mass and movement of all water on Earth in all its forms – requires a similar disposition owing to how the body of knowledge about it has been shaped by geological ways of thinking. These, of course, are not the only ways of thinking that affect the hydrosphere. People push water around with all sorts of ideas in tow beyond its existence as solid, liquid, or gas, and in addition to its interactions with the atmosphere, biosphere, and lithosphere. Human forces alter how much water is in the atmosphere: they melt glaciers, treat rivers like nature's sewers, and pump out ancient aquifers at unsustainable rates. The cumulative effect of these anthropogenic forces is geological in scale (see Rockström et al., 2014; Vörösmarty et al., 2004). Humans have radically reorganised freshwater systems such that the mass of water impounded behind mega-dams has measurable effects on Earth's gravitational field and rotation (Chao, 1995; cf. Vörösmarty et al., 2015). Other anthropogenic forces require almost impossible mental images to fathom; a recent study found that the ocean absorbed 90 per cent of the heat gained by the planet between 1971 and 2010, which is the equivalent of 290 zettajoules, where 1 zettajoule equals $10^{21}$ joules (Zanna et al., 2019). When the UK newspaper *The Guardian* tried to relay these numbers in something approximating a human scale, they calculated it was the heat equivalent of 1.5 atomic bombs per second (Carrington, 2019). The challenge of thinking about the hydrosphere, however, is not only about large-scale impacts. Together with free energy, the hydrosphere animates the Earth system across kingdom and phylum. It is the medium that carries whale songs. Its annual excesses deposit silts on flood plains. It is a sign of life across the sacred, the secular, and the sought after – a precious endowment and a precarious resource (Johnson and Fiske, 2014). It is also the medium through which toxins in the milk of mothers pass to children. It is rising flood waters and coastal erosion. Torrential storms. Tidal waves. The list goes on. Owing to its multiple scales and sites, the hydrosphere is also the object of philosophical concern. UNESCO's World Commission on the Ethics of Scientific Knowledge and Technology recently argued that the interdependence of humans and all life with water "requires *a shift from an anthropocentric approach to a more ecocentric approach*" (2018: 4, original emphasis).

 DOI: 10.4324/9781315111643-36

There is no complete history of the hydrosphere, and this chapter will not provide it. It does not trace the disappearance of seas long forgotten as the Indian sub-continent drifted from the south end of Africa and squeezed out ocean waters. Or how it then inched upward into the Himalayas, which then became a water tower – the Third Pole – for the world's largest populations. Instead, this chapter outlines how the hydrosphere is entangled with both Earth and human history both as a body of knowledge and a way of thinking. To this end, it provides the contours for an account of the hydrosphere yet to come by identifying how links were forged among geology, hydrology, meteorology, and Earth system science. This strategy takes its leave from the substantive interest of Baker's presidential address, in which he argued that geologic reasoning runs the gamut from Earthly objects to thought itself. For Baker, it was the field of semiotics that best explained the "complex system of signs ... that is continuous from the natural world to the thought processes of geological investigators" (Baker, 1999: 633). The chapter takes semiotics – the study of signs – as a fillip for thinking about how different aspects of the hydrosphere may be brought together and thought together. It begins by considering why there is no fully satisfactory account of the hydrosphere yet. Part of the explanation is the variety of practical concerns that have shaped knowledge of the hydrosphere, from measuring cyclones to draining swamps. Other challenges are more recent, such as efforts to render the hydrosphere governable alongside other aspects of the Earth system it is integrated with, such as the lithosphere, the atmosphere, and the biosphere (Falkenmark et al., 2019; Rockström et al., 2014).

The second part of the chapter considers how different bodies of knowledge came together through international efforts to professionalise hydrology and to quantify the global hydrological cycle. Adjacent to these efforts, and eventually intersecting with them, geologists posed central questions about the hydrosphere, such as the problem of determining the origins of sea water in Earth's history. These were often accompanied by a healthy dose of humility. For instance, in his retiring address as president of the Geological Society of America, William Rubey (1951) humbly titled his influential efforts to understand the origins of sea water as an "attempt to state the problem". The third part of the chapter considers how Earth system science now syncretises different schools of thought and a variety of knowledge sources about the hydrosphere. These efforts in international collaboration, which started mid-twentieth century and accelerated alongside human impacts on the planet, now take an increasingly prominent role in reading the signs of the hydrosphere. These signs are complex and often foreboding. They suggest that to Baker's notion of geology as an object of knowledge and a way of thinking we might add that the signs linking objects and habits of thought do not only run from the "natural world to the thought process of geological investigators" (1999: 633). The signs run in several, crisscrossing directions as an outcome of human geology that now affects the functional processes of the Earth system.

## Geology, semiotics, and water

Although he is perhaps best known for his philosophical influence, the American polymath Charles S. Peirce spent many years employed by the United States Coast and Geodetic Survey in the late nineteenth century. He travelled to international conferences on geodesy and endlessly tinkered with instrumental devices to improve his work in calculating continental land masses through gravitational measurements. Peirce also thought carefully about space and applied his mathematical and logical skills to create a new map projection (a quincuncial projection). Amid these practical concerns, Peirce advanced philosophy and logic, and developed core tenets of American pragmatism (Misak, 2016). Among Peirce's most notable contributions is his theory of signs, or semiotics (see Hoopes, 1991; Liszka, 1996). In his writings, Peirce built a theory of semiotics that continues to inspire not only geologists like Baker, but also anthropologists interested in how the continuous and complex system of signs connecting Earth and human knowledge leads also to forms of non-human intelligence, such that anthropologists now research how forests think (Kohn, 2013).

It might not seem obvious straight away why a framework for interpreting geological signs was in high demand during Peirce's lifetime. But geology was in many respects consolidating its reign among late-nineteenth-century sciences, especially after the influential work of Charles Lyell was extended to biology by Darwin and Wallace. Indeed, Peirce (1901) vaulted geology in an annual report he wrote for the Smithsonian Institute in the United States at the turn of the twentieth century, in which he declared that America was geology's home (see also, however, Rudwick, 2007). For Peirce, semiotics was part of explaining knowledge in evolutionary terms. To this end, Peirce began by distinguishing three types of signs (see Buchler, 1955; Keane, 2003; Peirce, 1868; Short, 2007). The first type of sign denoted direct experiences of things that are alike one another. These signs are icons. For instance, to replicate a colour, such as that of wine dark seas, we must produce a likeness to experience directly. The second type of signs are indices. These signs have a physical connection to an object or to objects, such as how smoke indexes a fire, or how a weather vane points to the wind, or the ways lines index geometric shapes. Such signs are not restricted to humans. Many non-humans also use signs of this type. Indeed, it was Peirce's view that animals and plants "make their living … by uttering signs" (see Sebeok, 1990: 14). Peirce used a familiar word for the final type of signs, symbols, to describe words or phrases that convey meanings. Together, icons, indices, and symbols form a trio that anchors Peirce's account of semiotics. Signs, however, are just one aspect of semiotics. The other two aspects are the object(s) that signs refer to, and the interpretants who connect sign and object to habits of thought and action.

In the early-nineteenth century there were innumerable signs used to understand the hydrosphere. By then, monsters had (mostly) been removed from ocean maps, but there was no quantified understanding of global water flows. In Europe, nascent ideas of a global water cycle existed, but the logic underpinning these accounts was still largely explained through forms of thinking that relied on natural theology and which envisioned the water cycle in terms of an efficient and divine water economy (Tuan, 1968). The more pressing concerns at that time combined colonialism and state-building, which often found common cause in the need for knowledge conducive to ship travel as European hydrographers sounded coasts and mapped inland water ways from the United States to the Bay of Bengal to open up trade routes (Amrith, 2013; Bhattacharyya, 2018; Bray, 1970). Where these activities proceeded in relation to land there were obvious points of reference – a way to extend a system of signs outward through maps, measurements, and calculations. For many phenomena, however, building these connections took immense feats of dedication. Often it was a matter of sheer survival. For instance, calculations from the logs of two ships caught on either edge of a massive storm used the indexical relations of observational devices – barometers – to provide the first crude measurements of a cyclone in the Bay of Bengal (Amrith, 2018). Gaining knowledge of ocean dynamics, however, was laborious and slow. Where gaps in the emerging scientific view of the water cycle remained, there was still plenty of room for cosmological explanation. For instance, notions of 'moral meteorology' that took root in late imperial China had a long residency time before they were ultimately displaced (Elvin, 1998).

Different bodies of knowledge and ways of thinking about water posed difficult semiotic challenges. For one, throughout the eighteenth and nineteenth centuries there was a divergence between the sciences of hydrodynamics and flow versus the practical concerns of hydrology (Darrigol, 2005). The split wasn't unbridgeable, but it did reflect the differences between emerging scientific views and the predominance of state-building exercises that wielded hydrology for practical ends, such as draining marshes, damming rivers, building canals, and constructing irrigation works (e.g. Blackbourn, 2006; D'Souza, 2006; Pritchard, 2012). Very often, the methods and techniques of hydrology did not have scientific explanation as their end point but relied instead – at least initially – on forms of knowledge that had practical success in reinforcing colonial power relations (see Gilmartin, 2015; Mitchell, 2002). It is a common trapping among some social theorists to claim that the edifice connecting water, power, and state or colonial projects forced water into a box called 'nature' where it was stripped of its social aspects and allowed only one sign – $H_2O$ – that ultimately became the basis of the global hydrological cycle in the twentieth century (e.g. Linton, 2010). In fact, something quite different took place: throughout the nineteenth century water was increasingly interpreted geologically and with humans and their relationships to water viewed as part of those geological processes and their knowledge as an effective way of explaining Earth's evolution (Schmidt, 2017). For instance, when the climate scientist John Tyndall (1872) wrote his influential book *The Forms of Water*, he described the hydrological cycle by carefully tracing the geologic aspects of physical relationships. But he also did so in language full of references to social and industrial relationships to water, such as by explaining condensation and cloud formation using examples from steam-powered trains.

A full account of the hydrosphere will ultimately require numerous histories of how bodies of knowledge were linked to ways of thinking geologically. Consider, for instance, how water, Earth, and societies were thought of geologically (i.e. without the society–nature divide) in late nineteenth- and early twentieth-century America. At that time, William J. McGee worked among Washington DC's intellectual elite both for the United States Geological Survey and later as Director-in-Charge of the US Bureau of Ethnology. He co-founded the American Anthropological Association with Franz Boas and later held a key position in the administration of American president Theodore Roosevelt where he was one of the main architects of American natural resource conservation. McGee also sat on the board of *National Geographic* where, in a 1898 article, he outlined a theory of the three geospheres operating at and above Earth's surface: the atmosphere, the hydrosphere, and the lithosphere (it wasn't until decades later that Vladimir Vernadsky conceptualised the biosphere). At that time, McGee (1898: 439) described the atmosphere as an "aerial ocean" subject to the forces of different geospheres as they acted upon each other. The actions of the geospheres, in the view of McGee and indeed many of his fellow geologists, were what gave rise (degree by degree) to life, to societies, and ultimately to the scientific study of societies that shaped the geospheres themselves. The hydrosphere, in this account, was the primary agent of what McGee termed 'earth-making' because it bridged between life and non-life. For McGee and other geologists who held this view, such as the more famous John Wesley Powell, this meant that to manage and direct water was to direct planetary evolution itself (see Schmidt, 2017). The geospheres, in such accounts, were geological bodies of knowledge that came part-in-parcel with a way of thinking about Earth, water, and knowledge itself. McGee's way of thinking, like others that vaulted geology as the basis for totalising explanations of human and the planetary evolution, was highly Eurocentric in its dismissal of other ways of knowing water, such as those of Indigenous peoples. Nevertheless, it formed the cornerstone of key ideas of multi-purpose river basin planning and water resources conservation that the United States later exported on a global scale through international development programs and massive dam-building exercises in the twentieth century (Ekbladh, 2010; Schmidt, 2017; Sneddon, 2015).

There are many unsatisfactory aspects of how early notions of the hydrosphere sought to consolidate explanations of both human and non-human phenomena with respect to geology;

the tendency to use geological explanations to naturalise claims of cultural difference principal among them. The belief that one particular society held knowledge that justified controlling and rerouting the global water system in service to its own preferences is another. But uses of geology change over time, like any other domain of knowledge. To study or to use geology is not necessarily to endorse problematic uses in the past. We can accept, for instance, that Newton contributed significantly to understandings of ocean tides without accepting the cosmology he used to order relationships of measurement and meaning (see Schaffer, 2009). Similarly, Austrian climatologists in the nineteenth century worked amid projects of continental empire building when they developed explanations of eustatic shifts in global sea level and, later, approaches to scaling up from regional relationships of land-cover (i.e. forests) to meteorology and to early ideas regarding planetary water budgets (see Coen, 2018). In fact, the first meeting of the International Meteorological Congress in Vienna in 1873 set in motion a long and challenging effort to establish and standardise a global weather monitoring network; it was also the Austrian scientist Julius von Hann's 1896 book *The Earth as a Whole: Its Atmosphere and Hydrosphere* that helped to catalyse broader theoretical efforts to think about the planet as an interconnected system (see Edwards, 2010). In the case of combining multiple terrestrial, oceanic, and atmospheric sciences of water in an account of the hydrosphere, later efforts to distance the use of geology from cultural claims proceeded through appeals to objectivity shaped by the politics of international scientific collaborations in the twentieth century (see generally Wolfe, 2018).

## From global hydrology to geologic hydrosphere

William Rubey's "attempt to state the problem" regarding the geologic history of sea water was one that, he argued, "ramifies almost endlessly into many problems of earth history" (1951: 1143). The integrated nature of the hydrosphere with tectonic movement, the atmosphere, and biogeochemical cycles entangles it with the whole suite of practices through which Earth is understood. To unpack the history of the hydrosphere, then, is in some measure to unpack the deep history of Earth itself. Indeed, Rubey's attempt came only three years after the mathematician Henry Stommel (1948) made the key measurements and calculations connecting atmospheric and oceanic dynamics, such as wind-driven ocean currents (see also Dry, 2019). To tell such a history, however, requires both a body of knowledge – the measurements, standards, techniques – about the Earth and a way of thinking about the Earth such that the signs that index Earth history to the hydrosphere through multiple causal relationships are thought of as such. It involves much more than could be done here, including carefully tracing how local knowledge of the ocean could produce a planetary picture – a picture often framed by politics – alongside other efforts kickstarted by the International Geophysical Year in 1958–59 (Lehman, 2020; see also Collins and Dodds, 2008). Indeed, it was only shortly thereafter that another project of integration was launched by US hydrologists: the International Hydrological Decade, which operated from 1965–74 under the auspices of UNESCO and which brought together an international network of scientists, social scientists, and practitioners. It was during this period that hydrology both was consolidated as a modern science and became more closely connected to Earth sciences. A key output of this work was the first global water atlas in 1978, which subsequently became the basis for calculations that linked humans to the hydrosphere through ideas of water scarcity, international development, and the effects of 'industrial societies' on the global water system.

The International Hydrologic Decade (IHD) was launched during the Cold War and amidst the development of a particular social notion of scientific objectivity (cf. Reisch, 2005;

Wolfe, 2018). The IHD involved considerable cooperation across the geopolitical tensions that characterised American and Soviet sciences. Although the IHD was led by the American Raymond Nace, much of the global water atlas, for instance, depended on Soviet hydrology and the work of Kourzon et al. (1978) which had been published in Russian in 1974 under the auspices of the USSR Committee for the International Hydrologic Decade. The concerns of the IHD were introduced by Nace (1964) as those regarding how water cycled globally and how changes to land use impact the effective management of freshwater at the national level. Nace (1967) even described hydrology as a "global problem with local roots". Developing a global view of hydrology was key, at least for the American backers of the IHD, to establishing an objective view of hydrologic science that could inform a rational approach to managing freshwater (see Nace, 1961, 1969). The orientation of the IHD towards freshwater had numerous effects on global water policy: it formed the basis for rational planning and directly informed the first global assessment of world resources and water needs made by the geographer Gilbert White, which featured prominently at the first United Nations conference on water in Mar del Plata in 1977 (see Biswas, 1978). It was also important for establishing core principles of hydrology, notably independence and stationarity. These, respectively, treat interannual variability as statistically independent (e.g. a wet year this year is independent of what may happen next year) and also treat the parameters of hydrological variability as fluctuating within an overall envelope of stable statistical distributions (Milly et al., 2008; Serinaldi and Kilsby, 2015).

The orientation of the IHD towards concerns of national planning for freshwater scarcity also helps to explain why, at least at first, it was unclear how the objectives of its work fit with other international scientific efforts. For instance, when the World Meteorological Organization (WMO) reflected on its engagement with the IHD in 1974 – by which time it had jointly convened the end of decade conference for the IHD with UNESCO – it noted that for the decade's first five years the role of the WMO was "far from clear" (1974: 178). In fact, a bulletin seven years prior from the WMO (1967) contained only brief remarks about the IHD. Although the WMO had supported the IHD from the start, it was unclear how earlier efforts of the WMO would fit with the new international program. For instance, the statement of the Secretary-General of the WMO, David Davies, to the 'End of Decade Hydrological Conference' for the IHD notes how from 1943 the WMO (then known as the International Meteorological Organization) had already established the Commission for Hydrology as "one of its main technical commissions" (Davies, 1974: 2). Notwithstanding the slow start, however, by the end of the IHD the WMO was actively connecting atmospheric and oceanic data with national level concerns over freshwater availability and had produced dozens of reports and numerous technical manuals. A key turn took place at a mid-decade conference in 1969, when the IHD Co-ordinating Council invited the WMO to develop further work on: "meteorological and hydrological network design and operation; standardization of instruments, methods of observation and processing of data; hydrological forecasting of surface waters; and methodologies of computation of design data with inadequate basic observations..." (WMO, 1974: 179). Part of the foundation for this enhanced role was an earlier collaboration with the WMO's World Weather Watch program and a study of its implications for hydrology and water resources management led by the (late) Canadian James Bruce (Bruce and Nemec, 1967). A full history of the hydrosphere would require an exposition of how efforts to link meteorology to the practical concerns of hydrology proceeded. Even without it, however, it is evident that the standardisation of methodologies for computation of hydrology and meteorology became key to understanding the hydrosphere. Extending networks of observation and establishing shared practices provided a basis for calculations of large-scale water balances, such as those of nations and the planet itself (see 'Hydrology' in WMO, 1972).

As standardised measurements for sensing and calculating atmospheric vapour flux, precipitation, snow processes, and evaporation took hold, the last vestiges of a divine water economy and 'moral meteorology' were shuttered. The 1974 WMO report contains early calculations of ocean and hydrological flux on the climate system and plans to further integrate the WMO with the Intergovernmental Oceanographic Commission to more effectively govern remote sensing of the ocean. Reflecting on the IHD, and in which the WMO ultimately played a significant role, Nace (1980) argued it was during this decade that hydrology 'came of age' as a science. Significant challenges remained, and here I have only gestured at how a few were resolved. Oreskes (2021), in her opus on how the US military shaped ocean knowledge, shows the geopolitical dynamics operating in the background of the hydrosphere. One perennial challenge is that new forms of technology and new networks of observation began to alter the spatial imagination within which, and against which, the hydrosphere is understood. Today, the rise of enhanced remote sensing techniques (oceanic, atmospheric, and in orbit) increasingly shape the emerging field of Earth system science (see Lövbrand et al., 2010). These technologies, like their predecessors, are not free of non-scientific objectives, such as when LANDSAT was launched by the United States Department of Interior to map global resources (Black, 2018). But these technologies, and the international networks they are deployed within (both political and scientific), are also increasingly engaged with Earth and planetary sciences that began to converge on the idea of Earth as a single, integrated system.

## From hydrosphere to global water system

From the 1980s onwards, the hydrosphere took scientific form in relation to several initiatives that gave semblance to the Earth as an integrated system of which human activities were, through unequal social structures, an increasingly forceful part. For instance, the Scientific Committee on Problems of the Environment (SCOPE) had formed in 1969 and, through its first decade, published a series of reports on emerging understandings of global biogeochemistry. From 1976–1982, the president of SCOPE was the resource geographer Gilbert White, who worked across the natural and social sciences on issues of risk and global water needs from a pragmatist perspective inspired by John Dewey, who himself was influenced significantly by Peirce and other pragmatists (see Schmidt, 2017; Wescoat, 1992). White also picked up on the politics of the IHD, which Nace (1967) had previously identified in reference to hydrology's lack of integration with "industrial" society. As Figure 30.1 shows, by the end of the 1980s, White was advancing a quantitative approach to show how 'industrial societies' (both capitalist and communist) were having accelerated impacts on water resources. This acceleration, of course, was not unique to water, yet as calculations of the hydrosphere were refined from the late 1980s onwards, it became increasingly clear that treating the Earth as a single, integrated system – the Earth system – was no less political than efforts to provide objective assessments linking national resources to natural systems during the Cold War.

Among the most prominent efforts to position the hydrosphere in the field of Earth system science was work by the International Geosphere-Biosphere Programme (IGBP) launched in 1987. The IGBP ran until 2015 when a version of it (together with other aspects of the Earth System Science Project that convened the IGBP alongside other programs) morphed into what is now known as Future Earth. The IGBP did not arrive from nowhere, of course. It took much of its initial institutional model from the approach to Cold War scientific collaborations established through the International Institute for Applied Systems Analysis (IIASA) (Rindzevičiūtė, 2016). Through its links to IIASA and an international network of scientists, the IGBP was conceptualised from the start with an eye to the emerging field of

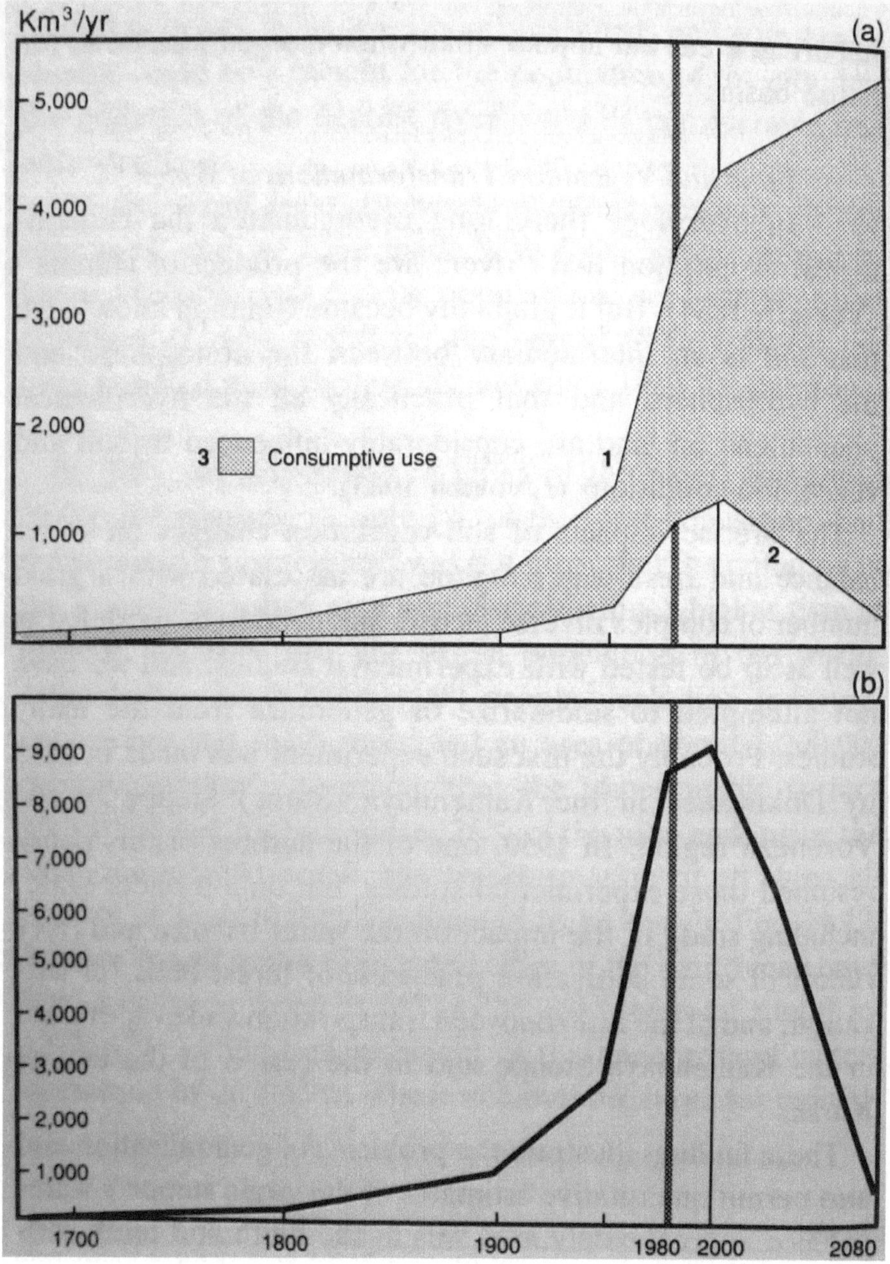


*Figure 30.1* The Chronology of water consumption for 400 years. Chart (a) shows (1) water intake from all sources; (2) waste water returned to streams, and (3) consumptive use. Chart (b) shows the volume of river runoff polluted by waste water. Both charts represent projections of all four parameters from 1680 to 2000 and from 2000 to 2080, assuming drastic measures are adopted to reduce waste-water discharge.

*Source*: Mark L'vovich and Gilbert White's *Use and Transformation of Terrestrial Water Systems* (1990: 248, fig 14.16). Reproduced with permission of the Licensor through PLSclear.

Earth system science. As such, the hydrosphere was increasingly treated as part of a complex and integrated Earth system and coupled closely with the biosphere, atmosphere, and lithosphere (National Research Council, 1988).

The IGBP had several phases and a comprehensive research agenda affecting hydrology, oceanography, and meteorology as it evolved. Under Phase I of the IGBP, several core programs developed scientific understandings of the hydrosphere, including on: (1) Biosphere Aspects of the Hydrological Cycle; (2) Global Ocean Ecosystem Dynamics; and (3) The Joint Global Ocean Flux Study. In Phase II, the IGBP pursued a set of complementary programs to further work on: (4) Future Earth Coasts and (5) Integrated Marine Biogeochemistry and Ecosystem Research. The IGBP's work also extended to joint projects, notably the Global Water System Project (GWSP), which launched in 2003 and which also provides a helpful example for understanding how quickly (and how much remains to be understood of how) the different contours of the hydrosphere converged towards an understanding consistent with Earth system science.

The GWSP's first report set out how thinking of human, physical, biological and biogeochemical components could move from old notions of integrating human–water relationships under "industrial societies" to one cognisant of "globalization" (GWSP, 2005: no page). As the GWSP laid out how to study the interactions of Earth-shaping dynamics of humans and non-humans, it incorporated the language of resilience and the non-equilibrium theories of complex systems from ecology and Earth system science (see Schellnhuber, 1999). The idea that the Earth system operated at a distance from equilibrium was developed, in part, through inspiration from Peirce and his insights into energy, entropy, chance, and order (see Prigogine and Stengers, 1984). Incorporating ideas of non-linearity and resilience into explanations of human impacts on the hydrosphere took shape in a series of studies at the turn of the millennium that conceptualised the dynamics of a global water system increasingly affected by human activities (e.g. Meybeck, 2003; Vörösmarty et al., 2004). By 2008, the extent of human impacts on the Earth system troubled stable notions of stationarity as anthropogenic climate change altered the outer bounds of 'natural' variability (Milly et al., 2008). In parallel, global water governance practitioners started linking Earth system crises to the risks and uncertainty of water security not only for 'industrial' societies but to a globalised economy (see Schmidt and Matthews, 2018).

In 2009, Rockström et al. (2009) introduced the powerful notion of planetary boundaries – a set of nine interacting aspects of the Earth system that defined a "safe operating space" for humanity with respect to the thresholds of planetary processes. Ocean acidification and global freshwater use figured prominently as two key boundaries, but by this point it was an unstated assumption that the hydrosphere – like all other geospheres – is integrated across an Earth system functionally altered by human impacts. Five years later, the Global Water System Project published the book *The Global Water System in the Anthropocene* (Bhaduri et al., 2014). Through Earth system science, the geologic force of humans signalled new forms of geologic thinking in which the stability of the Holocene had given way to an epoch in which human forces altered both stratigraphy and function of the Earth system (Röckstrom et al., 2014). The rapid conceptual and empirical developments that marked the shift towards the hydrosphere required in the Anthropocene were described by Vörösmarty et al. (2013: 539) as nothing short of a "transformation of a science".

Explanations of human impacts on the hydrosphere derived from Earth system science now shape the program of Future Earth and its sub-component: Future Water (see: www.water-future.org). These ongoing collaborations also shape the space of the hydrosphere in both systemic and semiotic terms – as a body of knowledge and a way of thinking that connects the hydrosphere to thinking geologically about human impacts on the planet, and also to broader agendas, such as those of the Sustainable Development Goals (e.g. Falkenmark et al., 2019; Rockström et al., 2014). The practices through which knowledge of human impacts on the Earth system are ordered, from times of deep Earth history prior to human existence through to the Anthropocene, now shape what Peirce (1868) might have termed the 'community' of interpretants seeking to get to grips with new realities in which the complex system of signs run in more directions than only from Earth to geological thought.

## Signs of the hydrosphere

The hydrosphere is closely linked to classic geology, in the sense of Earth history, and to more recent forms of human geology, in the sense of attempts to give integrated accounts of the Earth system and human forces on it. The hydrosphere in both of these senses entails a body of knowledge and a way of thinking. A satisfactory account of the hydrosphere, only some contours of which are outlined in this chapter, will ultimately need to traverse the semiotic bricolage through which the signs of the hydrosphere are observed, measured, and known by communities and networks. There is not likely to emerge a singular picture but rather a series of provincialised accounts owing to the complexity of water itself, and of social and scientific relations to it. As these accounts articulate with one another, the various ways in which knowledge of the hydrosphere has been produced for different ends, at different times, and with different effects, suggest that any Anthropocene 'transformations of water sciences' will also be a transformation of the signs of the hydrosphere as both a body of knowledge and a way of thinking.

## References

Amrith S (2013) *Crossing the Bay of Bengal: The Furies of Nature and the Fortunes of Migrants*. Cambridge, MA: Harvard University Press.

Amrith S (2018) *Unruly Waters: How Mountain Rivers and Monsoons Have Shaped South Asia's History*. London: Allen Lane.

Baker V (1999) Geosemiosis. *GSA Bulletin* 111(5): 633–645.

Bhaduri A, Bogardi J, Leentvaar J and Marx S (eds) (2014) *The Global Water System in the Anthropocene: Challenges for Science and Governance*. Springer: Dordrecht.

Bhattacharyya D (2018) *Empire and Ecology in the Bengal Delta: The Making of Calcutta*. Cambridge: Cambridge University Press.

Biswas A (ed) (1978) *United Nations Water Conference: Summary and Main Documents*. Oxford: Pergamon Press.

Black M (2018) *The Global Interior: Mineral Frontiers and American Power*. Cambridge, MA: Harvard University Press.

Blackbourn D (2006) *The Conquest of Nature: Water, Landscape, and the Making of Modern Germany*. New York, NY: WW Norton.

Bray M (1970) Joseph Nicolas Nicollet, geologist. *Proceedings of the American Philosophical Society* 114(1): 37–59.

Bruce J and Nemec J (1967) *World Weather Watch and its Implications in Hydrology and Water Resources Management* (WMO/IHD Project Report No. 4). Geneva: World Meteorological Organization.

Buchler J (ed) (1955) *Philosophical Writings of Peirce*. New York, NY: Dover Publications.

Carrington D (2019) Global warming of oceans equivalent to an atomic bomb per second. *The Guardian [Online]*. Available at: www.theguardian.com/environment/2019/jan/07/global-warming-of-oceans-equivalent-to-an-atomic-bomb-per-second

Chao B (1995) Anthropogenic impact on global geodynamics due to reservoir water impoundment. *Geophysical Research Letters* 22(24): 3529–3532.

Coen D (2018) *Climate in Motion: Science, Empire, and the Problem of Scale*. Chicago, IL: University of Chicago Press.

Collis C and Dodds K (2008) Assault on the unknown: This historical and political geographies of the International Geophysical Year (1957–8). *Journal of Historical Geography* 34: 555–573.

Darrigol O (2005) *Worlds of Flow: A History of Hydrodynamics from Bernoullis to Prandtl.* New York, NY: Oxford University Press.

Davies D (1974) Statement of Secretary-General, WMO, at the *UNESCO/WMO End of Decade Hydrological Conference (ENDEC/INF. 5).* Paris: UNESCO and WMO.

D'Souza R (2006) *Drowned and Dammed: Colonial Capitalism and Flood Control in Eastern India.* New Delhi: Oxford University Press.

Dry S (2019) *Waters of the World.* Chicago, IL: University of Chicago Press.

Edwards P (2010) *A Vast Machine: Computer Models, Climate Data, and the Politics of Global Warming.* Cambridge, MA: MIT Press.

Ekbladh D (2010) *The Great American Mission: Modernization and the Construction of an American World Order.* Princeton, NJ: Princeton University Press.

Elvin M (1998) Who was responsible for the weather? Moral meteorology in late imperial China. *Osiris* 13: 213–237.

Falkenmark M, Wang-Erlandsson L and Rockström J (2019) Understanding of water resilience in the Anthropocene. *Journal of Hydrology X* 2: 21–13.

Gilmartin D (2015) *Blood and Water: The Indus River Basin in Modern History.* Oakland, CA: University of California Press.

The Global Water System Project (GWSP) (2005) *The Global Water System Project: Science Framework and Implementation Activities.* Earth System Science Partnership. Available at: https://digital.library.unt.edu/ark:/67531/metadc11879/m1/

Hann JV (1896) *Die Erde als Ganzes ihre Atmosphäre und Hydrosphäre.* Prague: F. Tempsky.

Hoopes J (ed) (1991) *Peirce on Signs: Writings on Semiotic by Charles Sanders Peirce.* Chapel Hill, NC: University of North Carolina Press.

Johnston B and Fiske S (2014) The precarious state of the hydrosphere: Why biocultural health matters. *WIREs Water* 1(1): 1–9.

Keane W (2003) Semiotics and the social analysis of material things. *Language & Communication* 23: 409–425.

Kohn E (2013) *How Forests Think: Toward an Anthropology Beyond the Human.* Berkeley, CA: University of California Press.

Korzoun V, Sokolov A, Budyko M, Voskresensky K, Kalinin G, Konoplyantsev A, Korotkevich E and Lvovich M (eds) (1978) *Atlas of World Water Balance: Water Resources of the Earth.* Paris: UNESCO.

Lehman J (2020) Making an Anthropocene ocean: Synoptic geographies of the International Geophysical Year (1957–1958). *Annals of the American Association of Geographers* 110(3): 602–622.

Linton J (2010) *What is Water? The History of a Modern Abstraction.* Vancouver: UBC Press.

Liszka J (1996) *A General Introduction to the Semeiotic of Charles Sanders Peirce.* Bloomington, IN: Indiana University Press.

Lövbrand E, Stripple J and Wiman B (2010) Earth system governmentality: Reflections on science in the Anthropocene. *Global Environmental Change* 19: 7–13.

L'vovich M and White G (1990) Use and transformation of terrestrial water systems. In: Turner II BL (ed) *The Earth as Transformed by Human Action: Global and Regional Changes in the Biosphere over the past 300 years.* Cambridge: Cambridge University Press, 235–252.

McGee WJ (1898) The geospheres. *The National Geographic Magazine* 9: 435–448.

Meybeck M (2003) Global analysis of river systems: From Earth system controls to Anthropocene syndromes. *Philosophical Transactions of the Royal Society B* 358: 1935–1955.

Milly PCD, Betancourt J, Falkenmark M, Hirsch RM, Kundzewicz ZW, Lettenmaier DP and Stouffer RJ (2008) Stationarity is dead: Whither water management? *Science* 319(5863): 573–574.

Misak C (2016) *Cambridge Pragmatism: From Peirce and James to Ramsey and Wittgenstein.* Cambridge: Cambridge University Press.

Mitchell T (2002) *Rule of Experts: Egypt, Techno-Politics, Modernity.* Berkeley, CA: University of California Press.

Nace R and Panel on Hydrology (USA) (1961) A plan for international cooperation in hydrology. *International Association of Scientific Hydrology Bulletin* 6(4): 10–26.

Nace R (1964) The International Hydrological Decade. *Transactions, American Geophysical Union* 45(3): 413–421.

Nace R (1967) Water resources: A global problem with local roots. *Environmental Science and Technology* 1(7): 550–560.

Nace R (1969) *Water and Man: A World View*. Paris: UNESCO.
Nace R (1980) Hydrology comes of age: Impact of the International Hydrological Decade. *EOS* 61(53): 1241–1242.
National Research Council (1988) *Earth System Science: A Closer View*. Washington, DC: The National Academies Press.
Oreskes N (2021) *Science on a Mission: How Military Funding Shaped What We Do and Don't Know About the Ocean*. Chicago, IL: University of Chicago Press.
Peirce CS (1868) Some consequences of four incapacities. *Journal of Speculative Philosophy* 2(3): 140–157.
Peirce CS (1901) The century's great men in science. In: *Annual Report of The Smithsonian Institute For The Year Ending June 30, 1900*. Washington, DC: Smithsonian Institute, 693–699.
Prigogine I and Stengers I (1984) *Order Out of Chaos: Man's New Dialogue With Nature*. London: Verso.
Pritchard S (2012) From hydroimperialism to hydrocapitalism: 'French' hydraulics in France, North Africa, and beyond. *Social Studies of Science* 42(4): 591–615.
Reisch G (2005) *How the Cold War Transformed Philosophy of Science: To the Icy Slopes of Logic*. Cambridge: Cambridge University Press.
Rindzevičiūtė E (2016) *The Power of Systems: How Policy Sciences Opened Up the Cold War World*. Ithaca, NY: Cornell University Press.
Rockström J, Steffen W, Noone K, Persson Å, Chapin III FS, Lambin E, Lenton TM, Scheffer M, Folke C, Schellnhuber H, Nykvist B, De Wit CA, Hughes T, van der Leeuw S, Rodhe H, Sörlin S, Snyder PK, Costanza R, Svedin U, Falkenmark M, Karlberg L, Corell RW, Fabry VJ, Hansen J, Walker BH, Liverman D, Richardson K, Crutzen C and Foley J (2009) A safe operating space for humanity. *Nature* 461: 472–475.
Rockström J, Falkenmark M, Allan T, Folke C, Gordon L, Jägerskog A, Kummu M, Lannerstad M, Meybeck M, Molden D and Postel S (2014) The unfolding water drama in the Anthropocene: Towards a resilience-based perspective on water for global sustainability. *Ecohydrology* 7(5): 1249–1261.
Rubey W (1951) Geologic history of sea water: An attempt to state the problem. *Bulletin of the Geological Society of America* 62: 1111–1147.
Rudwick M (2007) *Bursting the Limits of Time: The Reconstruction of Geohistory in the Age of Revolution*. Chicago, IL: University of Chicago Press.
Schaffer S (2009) Newton on the beach: The information order of *Principia Mathematica*. *History of Science* 47(3): 243–276.
Schellnhuber H (1999) 'Earth system' analysis and the second Copernican revolution. *Nature* 402: c19-c23.
Schmidt JJ (2017) *Water: Abundance, Scarcity, and Security in the Age of Humanity*. New York, NY: New York University Press.
Schmidt JJ and Matthews N (2018) From state to system: Financialization and the water-energy-food-climate nexus. *Geoforum* 91: 151–159.
Sebeok T (1990) Indexicality. *The American Journal of Semiotics* 7(4): 7–28.
Serinaldi F and Kilsby C (2015) Stationarity is undead: Uncertainty dominates the distribution of extremes. *Advances in Water Resources* 77: 17–36.
Short T (2007) *Peirce's Theory of Signs*. Cambridge: Cambridge University Press.
Sneddon C (2015) *Concrete Revolution: Large Dams, Cold War Geopolitics, and the U.S. Bureau of Reclamation*. Chicago, IL: University of Chicago Press.
Stommel H (1948) The westward intensification of wind-driven ocean currents. *EOS, Transactions of American Geophysical Union* 29(2): 202–206.
Tuan Y (1968) *The Hydrologic Cycle and the Wisdom of God: A Theme in Geoteleology*. Toronto: University of Toronto Press.
Tyndall J (1872) *The Forms of Water in Clouds and Rivers, Ice and Glaciers [Fourth Edition, 1874]*. Akron, OH: The Werner Company.
Vörösmarty C, Lettenmaier D, Leveque C, Meybeck M, Pahl-Wostl C, Alcamo J, Cosgrove W, Grassl H, Hoff H, Kabat P and Lansigan F (2004) Humans transforming the global water system. *EOS, Transactions of American Geophysical Union* 85(48): 513–516.
Vörösmarty C, Pahl-Wostl C, Bunn S and Lawford R (2013) Global water, the Anthropocene and the transformation of a science. *Current Opinion in Environmental Sustainability* 5(6): 539–550.
Vörösmarty C, Meybeck M and Pastore C (2015) Impair-then-repair: A brief history & global-scale hypothesis regarding human-water interactions in the Anthropocene. *Daedalus* 144(3): 94–109.
Wescoat J (1992) Common themes in the work of Gilbert White and John Dewey: A pragmatic appraisal. *Annals of the Association of American Geographers* 82(4): 587–607.

Wolfe A (2018) *Freedom's Laboratory: The Cold War Struggle for the Soul of Science*. Baltimore, MD: John Hopkins University Press.

World Commission on the Ethics of Scientific Knowledge and Technology (COMEST) (2018) *Water Ethics: Ocean, Freshwater, Coastal Areas*. Paris: UNESCO.

World Meteorological Organization (1967) International Hydrological Decade. *WMO Bulletin* 16(4): 209–210.

World Meteorological Organization (1972) Hydrology. *WMO Bulletin* 21(1): 41–45.

World Meteorological Organization (1974) Hydrology. *WMO Bulletin* 23(3): 178–186.

Zanna L Khatiwala S, Gregory JM, Ison J and Heimbach P (2019) Global reconstruction of historical ocean heat storage and transport. *Proceedings of the National Academy of Sciences* 116(4): 1126–1131.

# 31
# ICE
## Elements, geopolitics, law and popular culture

*Klaus Dodds*

### Introduction

Sea ice is frozen seawater, and its genesis is quite different to glacial ice (which is formed via precipitation). It is commonplace in the Arctic and Southern Oceans, as well as the Baltic Sea and Bohai Bay in north east China. The Arctic Ocean is the main area of focus for this chapter, as it reveals well how sea ice expands, retreats and disappears depending upon ocean and wind currents and sea and air temperatures (Schmitt et al., nd; Turner et al., 2015; Yuan et al., 2017). According to the National Oceanic and Atmospheric Administration (NOAA), approximately 15 per cent of the world's oceans are covered with sea ice for some time of the year – around 25 million square kilometres (NOAA, 2021). Sea ice is most extensive in the winter months with, in the case of the Arctic Ocean, a maximum surface area in March and a minimum equivalent in September (Marshall, 2012: 106–107).

The initial section of this chapter considers sea ice properties, distribution and the web of ecological relationships it supports. The spotlight then shifts to the driving forces behind sea ice science from the 1950s onward. Initial efforts at understanding the geophysics of sea ice informed negotiations for the United Nations Convention of the Law of the Sea (UNCLOS) in the 1960s and 1970s. These efforts, in turn, have fed into contemporary enthusiasm regarding the emergence of the Arctic as a new shipping space and resource frontier (Dittmer et al., 2011). The penultimate section considers non-indigenous and indigenous Arctic experiences and meanings of sea ice (Fox Gearhead et al., 2017; Krupnik et al., 2010).[1] Arctic sea ice is the linchpin for this chapter and integral to debates over anthropogenic climate change, Arctic geopolitics, economic development, cultural meaning, environmental disaster, and indigenous resilience (for example, Dodds and Nuttall, 2016; Steinberg et al., 2015; Wadhams, 2017; Zellen, 2008).

Sea ice may be an exceptional form of ocean-space but it deserves greater attention because there is now a growing consensus amongst sea ice researchers that ice-free summers in the Arctic are likely to be a reality by 2035 (Voosen, 2020).

### Sea ice: Properties, distribution, and ecologies

Sea ice starts life as a very thin, almost delicate layer on the surface of the polar ocean and sea. It will thicken and extend only if currents and temperatures in the sea and in the air

DOI: 10.4324/9781315111643-37

facilitate growth. Its year on year dynamism is extraordinary, and reveals well the volatility of the frozen ocean.

Sea ice varies. Just in the western lexicon, there is an array of terms to describe the types of young and youthful sea ice. Frazil ice, for example, is composed of very fine sea ice crystals, which usually extends no more than 1 metre in depth. If conditions allow, frazil ice might thicken and accumulate in patches that are termed grease ice. The transition from frazil to grease ice is determined by the intersection of air temperature, wind speeds, and the strength of ocean currents. Grease ice is likely to be found in areas of the ocean experiencing cooler temperatures, stronger winds and oceanic currents. Grease ice can be transformed into pancake ice, if previously separate patches of grease ice collide with one another. The distinct qualities of pancake ice, such as rounded shapes with rim-like structures, are more likely to emerge in open water. And even this list does not account for other forms of sea ice such as nilas (similar to ice rind), which is defined as thin floating ice (around 5–10 centimetres thick) without hummocks (for further details see Marshall, 2012; Thomas, 2017).

As the upper layers of the ocean and sea warm in the spring months, sea ice melts. Every year, sea ice maximum and minimum will vary in the Arctic, Antarctic and other geographical regions such as the Baltic Sea, depending on water temperature, salinity, ocean currents, tides, and prevailing wind direction. As measures of sea ice extent, the maximum and minimum figures mark the point in the year when scientists gather an important snap-shot of the state of sea ice and draw conclusions about longer-term trends affecting ice-covered waters. The Arctic sea ice maximum tends to grab most scientific and political attention because sea ice in the summer months plays a crucial role in reflecting sunlight and contributes to regional weather patterns and global climate. It also informs the auditing of Arctic economic activity and geopolitical forecasting.

It is difficult to over-state the significance of sea ice. Sea ice, like glacial ice, has a high albedo affect. Any sunlight reaching sea ice is reflected away from earth. As sea ice shrinks and the ocean surface becomes darker, a higher percentage of solar radiation is absorbed, raising water temperatures and causing a feedback loop which leads to the melting of more sea ice, as well as further impacting the global climate. Worsening winter storms, an expanding 'underwater heat blob' and stronger ocean currents contribute to further disintegration, making the likelihood of multi-year sea ice enduring a remote prospect (Voosen, 2020).

The intersection of the submarine, surface and aerial produce a delicate form of 'vertical reciprocity' (Adey, 2010). Sea ice can and does act as intermediary between ocean and air, and underwater life has evolved to take advantage of its dynamic presence. Warmer waters bring about challenges for species adapted to life under and on top of sea ice. Algae is found underneath sea ice, and its presence enables others to feed on it such as zooplankton and fish. Copepods, nematodes and flatworms make the underneath of sea ice formations their homes. Reductions in these species' populations, brought about by the decline of sea ice, can have worrisome implications and legacies for biological systems, multi-species life and underwater ecologies (and speaks to wider agendas in blue humanities and volumetric territories scholarship; see, for example, Alaimo, 2019; Bille, 2020; Braverman and Johnson, 2020; and the work of the ICELAW Project, no date).

## Sea ice and the Arctic Ocean

Sea ice has been integral to the Arctic Ocean for millennia. For over 20,000 years, sea ice has covered vast areas of this marine basin. In comparison to more modern times, scientists estimate that there was twice as much sea ice maximum coverage (around 28–30,000 square kilometres, see Osborne et al., 2017). Recent paleo-climate research suggests that even with this extensive coverage of sea ice there were still pockets of open water (polynyas) that allowed marine life to endure and even flourish. Fossil remains of algae have been used to help that reconstructive process and indicate that polynyas appear, disappear and reappear over time. Over the last thousand years, sea ice extent has varied depending on eras of warming and cooling. The historic record is thus varied, and the remains of marine life provide vitals clues as to how it corresponds with shifts in planetary climate history.

Since satellite records began in the late 1970s, sea ice extent in the Arctic Ocean has been on an uneven but downward trend. The National Snow and Ice Data Center (2021) estimates that between 1979 and 2018 the sea ice extent minima (as measured each September) peaked at 1980 with over 7.5 million square kilometres to around 4.7 million square kilometres. The 2020 Arctic Ocean sea ice minima was the second lowest at 3.74 million square kilometres just behind the 2012, which remains the record low with a figure of 3.6 million square kilometres (Scott, 2020). Warming is detrimental to the perseverance of sea ice and contributes to climate forcing, but other localised factors can hasten its demise. One example is black carbon (soot) deposits from shipping and other activities. The darkening of Arctic sea ice contributes to heat absorption and further melting. The Arctic Council Expert Group on Black Carbon and Methane has called for radical reductions in emissions and proposed that they be reduced by around 25 per cent below 2013 levels by 2025 (see Arctic Council, 2021). They based their conclusion on information submitted by the eight Arctic Council member states and concluded that the main sources were diesel engines, oil and gas flaring and the burning of biomass.

Warmer waters will prove inviting to migratory species and it is highly likely that Atlantic and Pacific Ocean fish such as mackerel, cod, and haddock will move northwards, as their habitats change as well (Polyakov et al., 2017). Polar bears, walruses and seals depend upon firm sea ice to hunt, to reproduce and to simply rest and recover (Engelhard, 2016). As indigenous peoples of the Arctic have recognised for millennia, sea ice ecologies are capable of supporting a marine food-chain that makes living and travel possible. Due to recent declines in sea ice, predatory creatures such as killer whales arrive earlier in the Arctic and are now able to hunt fish and other marine mammals longer in the spring and summer seasons. Any significant removal of sea ice will unleash a series of feedback loops and chain-reactions, with mixed results for those who call ice-covered waters home. The Bering Sea Elders Group[2] have been vociferous in their warnings from what they describe as the 'front lines' of a fast-changing Bering Strait.

## Sea ice science and Cold War geopolitics

In his historical overview of sea ice and the Cold War, environmental historian Peder Roberts draws attention to a 1958 international conference funded by the US Office of Naval Research, where over 80 North American, Soviet and European scientists discussed the distribution and character of sea ice, the observation of sea ice, the physics and mechanics of sea ice, sea ice formation and disintegration, drifting sea ice and finally, prediction techniques (Herzberg et al., 2018; Roberts, 2014).

Conference proceedings reveal that the Soviet Union was considerably more advanced in its understanding of sea ice in comparison to the United States; a state of affairs that likely was due to the importance that the Arctic, and, in particular, the Northern Sea Route (NSR), had played in Soviet economic and security planning (McCannon, 1998). In 1932 Joseph Stalin decreed that, "[t]he Arctic and our northern regions contain colossal wealth. We must create a Soviet organization which can, in the shortest period possible, include this wealth in the general resources of our socialist economic structure" (cited in McCannon, 1998: 33). In the same year, the Glavsevmorput (also known as the 'Commissariat of Ice') was created for the explicit purpose of delivering on Stalin's vision. Making sense of the environmental qualities of the Soviet Arctic including sea ice and permafrost became a strategic priority. During the Cold War, some Russian writers even mused out loud about using nuclear bombs to destroy ice while others called for diverting warmer waters from the south to melt it (Fleming, 2010: 202).

The 1958 conference was deliberately aimed at encouraging information exchange, and the timing was significant (Roberts, 2014). Occurring in the midst of the International Geophysical Year (1957–58), US naval planners were mindful that they needed a more comprehensive understanding of sea ice behaviour. Infrastructural investment in the North American Arctic in the 1950s was the initial spur to this interest followed by concern that effective underwater surveillance in the Arctic Ocean and northern fringes of the North Atlantic depended upon a robust understanding of sea ice. As sea ice scientists were discovering, sea ice and pack ice were hazardous for submarine voyaging and disruptive of sonar recordings. Sonar operators found sea ice and whales being confused for possible submarines (Leary, 1999).

During the IGY, for example, research stations were established on drifting sea ice for the purpose of better understanding the circulatory patterns of the polar oceans. Although sea ice drift had been studied by Norwegian explorer Fridtjof Nansen in the 1890s, Soviet research institutionalised the effort in the 1930s through the establishment of North Pole-1, the first of a series of drifting ice stations designed to collect information of sea ice, meteorology, and oceanography (Althoff, 2007). In 1991, North Pole-31 marked the final iteration in this investment in drifting ice stations. After a break following the dissolution of the Soviet Union, Russia reinvested in floating ice stations again in 2003. The reintroduction of the floating ice station coincided with renewed interest in the Russian Arctic, and a desire to collect strategic knowledge on northern waters. Continuing a pattern from Cold War years, investment in sea ice has been strongly tied to wider resource and geopolitical priorities (Untersteiner, 1981). In December 2020, Russia launched a new $100 million floating research platform (Project 00903), which is designed to be autonomously controlled by operators elsewhere. The platform named the *North Pole* will be run by the Federal Service for Hydrometeorology and Environmental Monitoring and the Russian Arctic and Antarctic Research Institute and replaces existing North Pole stations based on drifting ice floes. The *North Pole* will deploy for up to two years and provide real-time reporting on sea ice and ocean conditions in the Central Arctic Ocean.

Recent innovations aside, sea ice prediction was, until the satellite era of the late 1970s, a scientific field where only a very few researchers were able to travel on military submarines patrolling under the Arctic sea ice (Wadhams, 2017). For long-term histories of past sea ice extent, researchers have had to deal with data collection gaps, although nineteenth-century sources such as whaling logs and newspaper reporting have proved noteworthy for building awareness.

Sea ice forecasting is considered to be a strategic imperative to many coastal states affected by its presence. Professional sea ice forecasting services exist in countries such as Canada and Russia as well as regional seas such as the Baltic Sea Ice Service.[3] Demand for such forecasting is varied, reflecting multiple stakeholders including commercial, military, scientific and indigenous/northern communities. Throughout the Cold War and beyond, despite national imperatives to collect and classify sea ice awareness and forecasting, there has also been plenty of evidence of international collaboration. Building on the innovative work of World Weather Watch established in 1963, Finland, Russia and the United States work together at the Tiksi International Hydro-meteorological Observatory (operational from 2010 onwards) working on sea ice monitoring as well as weather reporting.[4] Since 2003, the US and Canada have operated a joint North American Ice Service, sharing information on sea ice mapping on the Great Lakes. Arctic states have worked together to share information gleaned from their respective ice breakers on sea ice and the seabed below the Arctic Ocean.

## International law and sea ice

International law has made itself felt on sea ice, and sea ice has contributed to the manufacturing of rules, conventions, and standards pertaining to safety and security at sea (Rothwell, 1996). The sinking of the RMS *Titanic* in April 1912 by an iceberg off Canada provoked international parties to address safety through shared and enforceable standards, culminating in the Safety of Life at Sea (SOLAS) Convention, which was adopted in 1914 and re-adopted in 1929, 1948, 1960, 1974, and 1980. It addressed specific issues such as numbers and adequacy of lifeboats and initiated the creation of the International Ice Patrol. The latter was charged with warning shipping about the presence of ice-covered waters. Shipping accidents, not all of which involved sea ice and icebergs, triggered a series of amendments and led to the institutional development of specialist agencies such as the International Maritime Organization (IMO), which, decades later, developed the Polar Code for ships operating in icy waters (2017; see also IMO, 2021).

From the sinking of the *Titanic* to the unveiling of the Polar Code, international shipping in ice-covered waters has been the subject of legal intervention. The United Nations Convention on the Law of the Sea Convention (UNCLOS) has only one article devoted to ice-covered waters: Article 234. Strongly supported by two Arctic states, Canada and Russia, Article 234 grants authority to coastal states to sponsor and implement special regulations designed to manage activity in ice-covered areas of their exclusive economic zones (waters between 12 and 200 miles from territorial baselines). In effect, Article 234 endorsed Canada's Arctic Waters Pollution Prevention Act (AWPPA), which was adopted in 1970 in the face of concern about US transit shipping from Alaska to European and North American energy markets. Article 234 allows coastal states to act, proactively, to implement measures designed to negate 'major harm' in seas and waterways covered with ice 'for most of the year'.

What makes Article 234 notable is that ice-covered areas of the ocean are recognised as having properties that require additional intervention by affected coastal states. The impetus for Article 234 and the AWPPA lay not only with Canadian anxieties about the environmental consequences of international transit traffic through the Northwest Passage (NWP) but also discomfort about the legal status of those affected waters. International lawyers continue to debate whether the NWP is Canada's internal waters or akin to an international strait where third parties would enjoy rights of innocent passage and even transit passage for most military vessels (UNCLOS, Article 37). The US in particular has supported the international strait designation, while Canada holds that lack of historic usage as a transit corridor and long-term Inuit occupation and appropriation of land, sea and ice merit designation as Canada's internal waters. The legal status of the NWP and the Northern Sea Route continues to draw conjecture from advocates of both sides of the legal argument (see Byers, 2013).

Since the adoption of Article 234 in the early 1970s, global concern with the legal status of sea ice has continued to grow. In 2004 the Arctic Human Development Report (AHDR) had very little to say on the legal status of ice. However, in the second edition, published in 2015, Chapter 6 is entitled 'Legal systems' and addresses an array of topics relevant to polar law including ice-covered waters, the rights of indigenous peoples and the rights of coastal states and third parties. The new content in the report reflects an Arctic impacted by climate change, resource extraction and global geopolitical interests.

Three trends have made themselves felt. First, the five Arctic ocean coastal states (Canada, Denmark/Greenland, Norway, Russia and the United States – the A5) and the indigenous peoples of the Arctic have turned to UNCLOS and international indigenous law to adjudicate, consolidate and demonstrate their sovereignty and sovereign rights over Arctic land, sea, seabed, water and ice (Byers, 2013; Rothwell, 1996; Steinberg et al., 2015). Coastal state authority over extended continental shelves and ice-covered areas are addressed in Articles 76 and 234 respectively of UNCLOS, and have enabled the A5 to articulate a shared vision for a rule-based legal order for the Arctic Ocean, most notably in the 2008 Ilulissat Declaration.

Second, indigenous peoples of the Arctic have also turned to legal instruments such as the UN Declaration on the Rights of Indigenous Peoples (UNDRIP) as well as publicised declarative statements such as the 2009 Circumpolar Inuit Declaration on Sovereignty in the Arctic. The latter begins with an explicit mention of sea ice:

> Inuit live in the Arctic. Inuit live in the vast, circumpolar region of land, sea and ice known as the Arctic. We depend on the marine and terrestrial plants and animals supported by the coastal zones of the Arctic Ocean, the tundra and the sea ice. The Arctic is our home.
>
> *Inuit Circumpolar Council, 2009: no page*

The 2009 Inuit Declaration might transition to customary international law with implications for the legal status of sea ice as integral to indigenous life. As Article 14 of the International Labour Organisation's (ILO) Indigenous and Tribal Peoples Convention notes,

> The rights of ownership and possession of the peoples concerned over the lands which they traditionally occupy shall be recognized. In addition, measures shall be taken in appropriate cases to safeguard the right of the peoples concerned to use lands not exclusively occupied by them, but to which they have traditionally had access for their subsistence and traditional activities.
>
> *International Labour Organisation, 1989: no page*

Third, the rights of third parties moving through the ice-covered Arctic Ocean have also attracted more attention. While destination-based commercial shipping has occurred in the Arctic for decades, transit shipping is a more recent phenomenon. Sea ice shrinkage and thinning in the coming decades might enable, in the summer season, ever greater maritime traffic along the NSR. Improvements in professional polar training and ship design including icebreaker technology could make these the number of navigable days higher and even more attractive to operators looking for cost-savings between Europe and Asia. Greater usage brings opportunity but also concern to coastal states such as Canada and Russia about the ongoing legal status of those waters (NWP and NSR), and whether Article 234 will continue to enable extended control over ice-covered waters, in an Arctic that is losing sea ice (Lalonde and McDorman, 2015).

Physical state change raises question marks about the long-term viability of legal measures. For example, if ice-covered waters disappear then do the provisions adopted by the Canadian and Russian governments (using Article 234 as their legal authority) still hold? Does the NWP remain 'internal waters' if international maritime traffic continues to grow and exhibit 'transit'-like qualities? Coastal states and shipping states in the Arctic have interests and agendas that may not coincide, even if they are allies (e.g. Canada and the US). The relationship between Article 234 and the IMO Polar Code will continue to provoke further questions about the relationship between freedom of navigation and Arctic Ocean coastal state jurisdiction (see the essays in Kraska, 2014; Lalonde and McDorman, 2015).

Polar law is at a critical juncture. The uncertain state of sea ice captures well what is at stake: geophysical state-change and geopolitical transition. A melting, shifting and reconstituted Arctic is stress-testing a region where stakeholders, indigenous, northern, coastal state, shipping state, and other third parties are making their presence manifest. The identification of the Central Arctic Ocean (CAO), illustrates well the challenges ahead. While Arctic states are in the midst of negotiating their sovereign rights to extended continental shelves and coastal state jurisdiction, the CAO are high seas. The development of a regional fisheries management organisation is likely, and, coupled with intensifying maritime traffic, the Arctic Ocean will attract ever greater legal and geopolitical interest and investment (Dodds and Nuttall, 2019).

## Popular and Indigenous experiences and meanings of sea ice

For those who live in the Arctic region, sea ice is integral to a way of life that has endured for centuries and even millennia. And for those who did not live within it, there is plenty of evidence that sea ice has long fascinated and beguiled those who saw it (Bravo, 2018). Pytheas of Massalia, in the fourth century BC, is thought to have voyaged as far north as Iceland and witnessed a 'frozen ocean' in his pursuit of Thule (the furthest point north). To the far south, somewhat later in the eighth century AD, the Raratongan traveller Ui-te-Rangiara is credited with witnessing "white rock-like forms [growing] out of a frozen sea" (cited in Thomas, 2017: 5). Reports of sea ice (*Mare Concretrum*) from the seas around the Baltic, Iceland and Greenland began to become more numerous from the eighth and ninth centuries onwards (Marshall, 2012: 104; and more generally, Dodds, 2018).

Inuit living in the North American Arctic including Greenland tend to populate shoreline communities. In Canada, in particular, the vast majority of the residents of Inuit Nunangat (the Inuit homeland) are sustained by coastal and marine animals such as fish, whales, seals and walrus. These 'country foods' are integral to everyday life, and contribute to a form of sustainability that can be difficult to secure in the face of high food and fuel costs. Any visitor to a community supermarket or store will be horrified by the prices of staples such as bread and milk. Country food-sharing, as a well-established practice, is pivotal to the enhancement of community solidarity and hunters (often men) within Inuit communities are deeply respected for their skills and resourcefulness (Wright, 2014).

For six maybe even eight months, sea ice can be sufficiently robust to support overland travel by dog sledge or skidoo. Frozen water enables Inuit and other indigenous peoples to travel not only within local communities but also across sizeable distances, whether to hunt marine mammals that live close to the sea ice edge or to visit distant friends and relatives. Moving via sea ice enables migration and thus access to new hunting grounds depending on the change in seasons. The potential for fishing, hunting and picking varies, and communities across the Arctic have well-developed travel and resource networks. Knowledge and appreciation of those networks has then been communicated orally from generation to generation. Living memory is integral to many indigenous cultures in the northern latitudes. Fundamentally, sea ice invokes something quite different for those who inhabit these polar worlds than it does for those who are trying to pass through. When sea ice becomes less 'reliably frozen' however, it poses a far greater hazard to indigenous hunters. Drownings are not uncommon if hunters fall into the frigid waters (Sharma et al., 2020).

Geographers and anthropologists, working closely with northern communities, have been integral to co-producing knowledge and understanding about sea ice. Notably in Canada and

Greenland, studies have focussed on the importance of sea ice to the reproduction of northern lifestyles and communal identities. In the Pan Inuit Trails Atlas (n.d.), for example, one is afforded an extraordinary entrée into Inuit occupancy and mobility in the Canadian High North (Aporta, 2009, 2011). The Atlas shows not only the dense network of routes but also the rich tapestry of place-naming of land, water, ice, and coastal areas. Using historic accounts and oral memory, the Atlas reveals how Inuit use a network of trails and tracks to hunt, fish and socialise. Indigenous knowledge of sea ice reveals itself to be far more than simply knowing where and when it is safe to travel but also how an appreciation of sea ice is tied up to a wider cosmology of Arctic life. Formalising indigenous place-naming of land, water and sea ice not only creates a repository of indigenous knowledge. Recording and mapping the oral testimony of indigenous peoples contributes to the acknowledgement and recognition of Inuit presence in the Canadian Arctic, which, in turn, may be used to bolster Canadian sovereignty claims.

While these sorts of initiatives are not without critics, the identification of sea ice as integral to Inuit lifestyles and livelihoods might be viewed as a cultural corrective to how popular cultural forms such as film have represented sea ice. Cold War movies such as the *Bedford Incident* (1965), *Ice Station Zebra* (1968) and later *Firefox* (1982) depict sea ice as something that frustrates and ultimately enables the US armed forces (Shaw, 2007). In the case of submarine movies such as *Ice Station Zebra*, sea ice is an endemic hazard to the *USS Tigerfish*, threatening the safety of its crew and testing their skills to the limit. By contrast, in *Firefox*, the main protagonist is a retired US Air Force pilot who uses floating sea ice to land a stolen Soviet airplane in order to make a secret refuelling stop made possible by a US submarine. Despite the anxieties of pilot and submarine captain, the sea ice enables the United States to showcase its technical abilities and out-smart their Soviet counterparts.

As these examples suggest, the popular geopolitics of sea ice is, more often than not, complicit with militarisation and securitisation (Dodds and Nuttall, 2016). However, it can also be an opportunity to complicate those processes and outcomes. In two Russian-language films, *At the Edge of Russia* (2010) and *How I Ended the Summer* (2015), glacial and sea ice contributes to the making of the harsh northern frontier. For the men, serving either as border guards or meteorologists, the absurdity of their isolated situation is made manifest by the absence of human others and the toll that it takes on their mental health and physical welfare. The men in question don't travel across sea ice.

Movies made by indigenous Arctic directors, by contrast, tend to represent sea ice as integral to northern identities and lifestyles. In *The Fast Runner* (2001), directed by Zacharias Kunuk, the narrative of the film is grounded in Inuit oral history, shamanism and story-telling, and involves largely Inuit actors and film crew. While the film addresses social rituals and marriage customs, much of the film is literally grounded on the sea ice and tundra of the Canadian North. Dog-sledging is shown to be critical to mobility and connectivity between different communities. One of the most dramatic portions of the film features the main character Atanarjuat (also known as 'the fast runner') fleeing naked across the sea ice in order to escape the murderous intent of a love rival, Oki. Throughout the lengthy and slow-paced film, sea ice makes itself felt in multiple ways acting as a literal and affective platform for joy, fear, learning, resentment, and reflection.

Sea ice can and does enable other social dramas. In *On the Ice* (2011), written and directed by Andrew Okpeaha MacLean, the film's aesthetic is a more recognisably western action-drama. Filmed in Barrow, Alaska, the narrative arc is dominated by the trials and tribulations of three Alaskan Native teens who go on a seal hunt across the sea ice. Riding skidoos, they travel to the edge of the sea ice in pursuit of their prey. Unfortunately, the seal hunt is disastrous. One of their party is killed by accident during a fight, and the body hidden under the sea ice. Traces of blood on the snow are scooped up and hidden in a bag. The rest of the film is taken up by the consequences that follow when things are hidden, covered up and disavowed. The body of the dead teen is never found but the legacy of the death makes itself felt throughout the community. Although sea ice might have made for an ideal accomplice to the cover-up the film leaves open the possibility that summer melting might yet reveal the presence of the body.

There is a richer body of literary and visual work that can be cited here. From Mary Shelley's *Frankenstein* (1821) to the novel and then film *Miss Smilla's Feeling for Snow* (1993, and 1997 respectively), sea ice has proven a rich resource for polar fiction and melodrama (Spufford, 1997; Dodds, 2018). The tragic story of scientist Victor Frankenstein and his monstrous creation is one that climaxes on Arctic sea ice. At the start of the novel, Frankenstein's survival is made possible by sea ice (and he is rescued by a British ship heading towards the North Pole) but by the end of the novel sea ice has been transformed into something rather overwhelming and dreadful. As Frankenstein's rescuer Captain Robert Walton writes to his sister back in England, "[w]e are still surrounded by mountains of ice, still in imminent danger of being crushed in their conflict. The cold is excessive, and many of my unfortunate comrades have already found a grave amidst this scene of desolation" (quoted in Shelley, 1818). By the end, Frankenstein's death in the Arctic stands as a morality tale for those who are hell-bent on interfering in and with nature. As Walton and his crew return south from the Arctic, Frankenstein and his creation are left somewhere amongst the pack ice.

*Frankenstein* embodied an enduring European fascination with the Arctic, and the sublime qualities of ice; at once alluring but also terrifying and overwhelming. As the novelist Francis Spufford (1997) noted, the English were particularly obsessed with finding new ways through Arctic sea ice, and learned societies such as the Royal Geographical Society eagerly sponsored polar quests. Ideologically, the conquest of the Arctic and its material qualities of remoteness, coldness and iciness assumed considerable importance. This became most apparent in the aftermath of the disappearance of Sir John Franklin's expedition in 1848 (Craciun, 2010). The search for his lost party gripped the nation and led to countless expeditions dedicated to uncovering their whereabouts. What gave further urgency to the matter was the fear that, in their desperation, some of the survivors may have resorted to cannibalism. Had monstrous ice made men monstrous?

Separated by some 170 years, the Danish novel and later film *Miss Smilla's Feeling for Snow* depicts the quest of a Danish-Greenlandic woman, Smilla Qaaviqaaq Jaspersen, who seeks to discover what happened to a local Greenlandic boy killed in Copenhagen. The last movements of the boy were recorded in the snow on the roof of his apartment building. Did he fall or was he pushed? As part of her investigation into this mysterious death, Smilla takes the fateful decision to travel north to Greenland, and the story concludes aboard a Danish ship, the *Kronos*, as it weaves its way through sea ice. Smilia's 'sense of snow' is shown to be pivotal to her ability to uncover the conspiracy that led to the death of the boy and, at an earlier stage, his father in Greenland. Both father and son knew of the location of a meteorite buried in a glacier. The unsparing sea ice reveals the evil designs of others.

Aesthetically, images and feelings of ice, snow, frost and cold have proven extremely resourceful for Scandinavian film and novels. Nordic noir actively trades on the aesthetic economy of cold and ice, using it to embody expressions of trauma, tension and intrigue (Hansen and Waade, 2017). In German, the concept of *stimmung* (usually translated as mood or attunement) has been used by film scholars to explore how ice and cold are associated with particular moods and tones, revealing not only environments but also the inner lives of protagonists (Warner, 2017). In Henning Mankell's penultimate novel, *Italian Shoes* (2010), the noise of 'singing ice' awakens a retired surgeon from his slumber. The haunting sounds of sea ice straining against the rocks and islands of the Swedish archipelago trigger memories of a past life, which comes back to haunt him as a former lover trudges across the frozen sea desperate to be reunited with him.

Sea ice, in literature as in science, statesmanship, and practices of everyday life, enables governance, knowledge, physical and spiritual navigation and even survival in an uncertain future.

## Conclusion

Satellites such as Ice Sat 2, launched in 2018, offer the promise of greater resolution and coverage of the world's remaining sea ice, allowing the identification of the last 'hold-out' of multi-year sea ice to the north of Greenland. For human and non-human communities in the polar regions and elsewhere including the Baltic Sea and Great Lakes of North America, sea ice acts as a platform for mobility, reproduction, and habitat for marine life above and below the water surface. Whether sea ice is seasonal or year-round, its presence is far-reaching. For coastal indigenous communities sea ice is literally 'critical infrastructure', while others may view sea ice as a hazard that needs to be managed in order to facilitate commercial shipping, tourism, oil and gas extraction and fishing. Ice might be a hazard to shipping and resource extraction but it provides opportunities for Russian and Finnish ice-breakers to capitalise on its presence. Finally, the scope of criminal jurisdiction – when a crime is committed on sea ice or a floating iceberg in the international waters of the Arctic Ocean – will continue to attract attention (Wilkes, 1972).

Finally, let us end on a counter-institutive example brought to life by researchers at the University of Connecticut (Stephenson et al., 2018). They postulate that increased shipping in the Arctic Ocean could, on the one hand, contribute further black carbon particulates which, if and when they deposit themselves on sea ice, will alter albedo adversely. On the other hand, shipping brings with it increases in sulphur emissions which could encourage cooling and cloud formation. This in turn might mitigate the warming trend being recorded in the Arctic Ocean region. Shipping in the Arctic carries with it considerable environmental risks and sea ice will continue to be a hazard for the foreseeable future.

## Notes

1 See also northern and indigenous sea ice atlases available at: www.snap.uaf.edu/tools/sea-ice-atlas and https://sikuatlas.ca/index.html
2 For more information see: www.beringseaelders.org/
3 See for example www.bsis-ice.de and www.canada.ca/en/environment-climate-change/services/ice-forecasts-observations/latest-conditions.html
4 See www.esrl.noaa.gov/psd/arctic/observatories/tiksi/ for information on the Hydrometeorological Observatory of Tiksi, Russia.

## References

Adey P (2010) *Aerial Life*. Chichester: Wiley-Blackwell.

Alaimo S (2019) Introduction: Science studies and the blue humanities. *Configurations* 27(4): 429–432.

Althoff F (2007) *Drift Station: Arctic Outposts of Superpower Science*. Lincoln, NE: University of Nebraska Press.

Aporta C (2009) The trail as home: Inuit and their pan-Arctic network of routes. *Human Ecology* 37: 131–146.

Aporta C (2011) Shifting perspectives on shifting ice: Documenting and representing Inuit use of the sea ice. *Canadian Geographer* 55(1): 6–19.

Arctic Council (2021) Task forces and expert groups: Black carbon and methane expert group. Available at: https://arctic-council.org/en/about/task-expert/

Bille F (ed) (2020) *Voluminous States: Sovereignty, Materiality, and the Territorial Imagination*. Durham, NC: Duke University Press.

Braverman I and Johnson E (eds) (2020) *Blue Legalities: The Life and Laws of the Sea*. Durham, NC: Duke University Press.

Bravo M (2018) *North Pole: Nature and Culture*. London: Reaktion.

Byers M (2013) *International Law and Arctic*. Cambridge: Cambridge University Press.

Craciun A (2010) The frozen ocean. *PMLA/Publications of the Modern Language Association of America* 125(3): 693–702.

Dodds K (2018) *Ice: Nature and Culture*. London: Reaktion.

Dodds K and M Nuttall (2016) *The Scramble for the Poles*. Cambridge: Polity.

Dodds K and M Nuttall (2019) *The Arctic: What Everyone Needs to Know*. Oxford: Oxford University Press.

Dittmer J, Moiso S, Ingram A and Dodds K (2011) Have you heard the one about the disappearing ice? Recasting Arctic geopolitics. *Political Geography* 30: 202–214.

Englehard M (2016) *Ice Bear: The Cultural History of an Arctic Icon*. Seattle, WA: University of Washington Press.

Fleming J (2010) *Fixing the Sky: The Checkered History of Weather and Climate Control*. New York, NY: Columbia University Press.

Gearhead SF, Holm LK, Huntington H, Leavitt J and Mahoney A (eds) (2017) *The Meaning of Ice: People and Sea Ice in Three Arctic Communities*. Ottawa: International Polar Institute.

Hansen K and Waade A (2017) *Locating Nordic Noir: From Beck to The Bridge*. London: Palgrave.

Herzberg J, Kehrt C and Toma F (eds) (2018) *Ice and Snow in the Cold War: Histories of Extreme Climatic Environments* Oxford: Berghahn Books.

ICELAW Project (no date) ICELAW: Indeterminate and changing environments: Law, the Anthropocene and the world. Available at: https://icelawproject.org/

International Labour Organisation (ILO) (1989) Article 14. In: *C169 – Indigenous and Tribal Peoples Convention 1989 (No. 169)*. Available at: www.ilo.org/dyn/normlex/en/f?p=NORMLEXPUB:12100:0::NO::P12100_ILO_CODE:C169#A14

International Maritime Organization (IMO) (2021) International code for ships operating in polar waters (Polar Code). Available at: www.imo.org/en/OurWork/Safety/Pages/polar-code.aspx

Inuit Circumpolar Council (2009) Circumpolar Inuit launch Declaration on Arctic Sovereignty. Available at: www.inuitcircumpolar.com/press-releases/circumpolar-inuit-launch-declaration-on-arctic-sovereignty/

Kraska J (2014) Governance of ice-covered areas: Rule construction in the Arctic Ocean. *Ocean Development and International Law* 45(3): 260–271.

Krupnik I, Aporta C, Gearhead S, Laidler G J and Holm LK (eds) (2010) *SIKU: Knowing Our Ice*. Dordrecht: Springer.

Lalonde S and McDorman T (eds) (2015) *International Law and Politics of the Arctic Ocean* Leiden: Brill.

Leary W (1999) *Under Ice: Waldo Lyon and the Development of the Arctic Submarine*. College Station, TX: University of Texas Press.

Mankell H (2010) *Italian Shoes*. London: Vintage.

Marshall S (2012) *The Cryosphere*. Princeton, NJ: Princeton University Press.

McCannon J (1998) *Red Arctic: Polar Exploration and the Myth of the North in the Soviet Union*. Oxford: Oxford University Press.

National Oceanic and Atmospheric Administration (NOAA) (2021) Can the ocean freeze? Available at: https://oceanservice.noaa.gov/facts/oceanfreeze.html

The National Snow and Ice Data Center (2021) Scientific data for research. Available at: https://nsidc.org/

Osborne E, Cronin T and Farmer J (2017) Paleoceanographic perspectives on Arctic Ocean change. In: *NOAA Arctic Report Card*. Available at: https://arctic.noaa.gov/Report-Card/Report-Card-2017/ArtMID/7798/ArticleID/690/Paleoceanographic-Perspectives-on-Arctic-Ocean-Change

Pan Inuit Trails Atlas (no date) Available at: www.paninuittrails.org/index.html?module=module.about

Polyakov I, Pnyushkov A, Alkire M, Ashik I, Baumann T, Carmack E and Goszczko I (2017) Greater role for Atlantic inflows on sea-ice loss in the Eurasian Basin of the Arctic Ocean *Science* 356(6335), 285–291.

Roberts P (2014) Scientists and sea ice under surveillance in the early Cold War. In: Turchetti S and Roberts P (eds) *The Surveillance Imperative*. London: Palgrave, 125–144.

Rothwell D (1996) *International Law and the Polar Regions*. Cambridge: Cambridge University Press.

Schmitt, C, Kottmeier C, Wassermann S and Drinkwater M (no date) Atlas of Antarctic sea ice drift. Available at: https://data.meereisportal.de/eisatlas/
Scott M (2020) 2020 Arctic sea ice minimum second lowest on record. In: *NOAA Climate.gov Science and Information for a Climate Smart Nation*. Available at: www.climate.gov/news-features/featured-images/2020-arctic-sea-ice-minimum-second-lowest-record
Sharma S, Blagrave K, Watson SR, O'Reilly CM, Batt R, Magnuson JJ, Clemens T, Denfeld BA, Flaim G, Grinberga L and Hori Y (2020) Increased winter drownings in ice-covered regions with warmer winters. *PloS one* 15(11) https://doi.org/10.1371/journal.pone.0241222
Shelley M (1818 [2003]) *Frankenstein*. Harmondsworth: Penguin.
Shaw T (2007) *Hollywood's Cold War*. Edinburgh: Edinburgh University Press.
Spufford F (1997) *I May Be Some Time: Ice and the English Imagination*. London: Faber and Faber.
Steinberg PE, Tasch J and Gerhardt H (2015) *Contesting the Arctic: Rethinking Politics in the Circumpolar North*. London: IB Tauris
Stephenson S, Wang W, Zender C, Wang H and Davis S (2018) Climatic responses to future trans-Arctic shipping. *Geophysical Research Letters* 45: 9898–9908.
Thomas D (ed) (2017) *Sea Ice*. London: Wiley-Blackwell.
Turner J, Hosking J, Bracegirdle T, Marshall G and Philips T (2015) Recent changes in Antarctic sea ice. *Philosophical Transactions of the Royal Society A* 373(2045) https://doi.org/10.1098/rsta.2014.0163
Untersteiner N (ed) (1981) *The Geophysics of Sea Ice*. NATO ASI Series B: Physics Volume 146. Seattle, WA: University of Washington Press.
Voosen P (2020) Growing underwater heat blob speeds demise of Arctic sea ice. *Science Magazine [Online]*. Available at: www.sciencemag.org/news/2020/08/growing-underwater-heat-blob-speeds-demise-arctic-sea-ice
Wadhams P (2017) *A Farewell to Ice*. Oxford: Oxford University Press.
Warner R (2017) Orange is the warmest colour: Mood and chromatic temperature in Robert Altman's *McCabe & Mrs. Miller. New Review of Film and Television Studies* 15(1): 24–37.
Wilkes R (1972) Law for special environments: Ice islands and questions raised by the T-3 case. *Polar Record* 16(100): 23–27.
Wright S (2014) *Our Ice Is Vanishing / Sikuvut Nunguliqtuq: A History of Inuit, Newcomers and Climate Change*. Kingston and Montreal: McGill and Queens University Press.
Yuan N, Ding M, Ludescher J and Bunde A (2017) Increase of the Antarctic Sea Ice Extent is highly significant only in the Ross Sea. *Nature Scientific Reports* 7(1): 1–8.
Zellen B (2008) *Breaking the Ice: From Land Claims to Tribal Sovereignty in the Arctic*. Lanham, MD: Lexington Books.

## Filmography

*The Bedford Incident* (1965) Directed by James Harris
*At the Edge of Russia* (2010) Directed by Michał Marcza
*The Fast Runner* (2001) Directed by Zacharian Kunuk
*Firefox* (1982) Directed by Clint Eastwood
*How I Ended the Summer* (2015) Directed by Aleksey Popogrebskiy
*Ice Station Zebra* (1967) Directed by John Sturges
*On the Ice* (2011) Directed by Andrew Okpeaha MacLean
*The Hunt for Red October* (1990) Directed by John McTiernan
*Smilla's Sense of Snow* (1997) Directed by Bille August

# 32
# ISLANDS
## Reclaimed – Singapore, space and the sea

*Satya Savitzky*

### Introduction

According to a recent UN estimate, 40 per cent of the world's population now live within 100 km of a coastline. The world's most prominent cities are typically ports – many of them located on islands. As sea-levels rise, these settlement patterns present stark risks. Whilst the oceans have served as partners in supporting successive waves of globalisation, they now look set to play a much more adversarial role.

The island-city-state of Singapore is the product of a particularly *deep* relationship with ocean space. Singapore's location along the Malacca Strait, and close contact with the oceans, has enabled it to prosper even without land and resources. By first harnessing ocean-borne trade flows, and then becoming a connectivity and financial services hub, the 'tiny red dot' (as it is often referred to), has achieved an economic significance completely disproportionate to its size. Whilst Singapore ultimately owes its fortunes to its proximity to ocean waters, the latter are also a major source of insecurity, with most of the island no more than 15 metres above sea-level.

The chapter examines how the condition of being 'surrounded by sea', has shaped Singapore's development and will likely shape its future. The chapter documents the city's attempt to address its space constraints, through innovation with dense forms of urbanisation involving expansion into, under and above the surrounding oceans. Rather than examine land and sea as separate realms, the chapter explores relations between maritime and terrestrial processes showing how novel forms have emerged at their edges. Singapore's responses to space shortages – common to many islands –reveals the fluid character of physical territory in the Anthropocene.

### 'Place surrounded by sea'

Recent archaeological finds suggest that the island now known as Singapore was originally the site of a settlement known as Tamsek – 'place surrounded by sea' (Miksic, 2014: 181–182). In the nineteenth century, the British set out to create an entrepot in Singapore to challenge the Dutch maritime dominance in the region. As European maritime colonial powers expanded, they often established settlements on coastal islands, whose mix of territorial, defence and trade benefits historically made them ideal locations for establishing seats of government or trading

DOI: 10.4324/9781315111643-38

hubs. Mumbai, Hong Kong, Shanghai and New York provide other examples (Grydehøj, 2015: 429).

In 1818, statesman Stamford Raffles was sent by the British government to establish a trading presence in the strategically key Malacca region. Singapore – as it would come to be known – seemed a natural place for a new port. Positioned at the southern tip of the Malay peninsula, the island possessed a natural deep harbour, and crucially, provided access to the main trade route between India and China. Singapore also contained plentiful supplies of timber for repairing ships, and enough fresh water to supply the island's then low number of inhabitants. The entire island may have had a population of 1,000 including various 'tribes' and Orang Laut (sea gypsies). The island was nominally ruled by the Sultan of Johor, who in return for a yearly payment, granted the British the right to establish a trading post on the island (Rahim, 2010: 24). A formal treaty was signed on 6 February 1819 and modern Singapore was born. Singapore was made into a free port where fees (such as those that were often paid to the town, harbour, port and dock) were not collected. Ships from around the world were allowed to trade without custom duties, which were imposed only on select products, such as tobacco, opium, alcohol and petroleum. The combination of Singapore's strategic location and its 'free trade' policy, attracted large numbers of ships, and within five years of its establishment, the port of Singapore's had become a regional entrepôt (National Library Board, Singapore, 2017b: no page).

Along with the rise of the so-called 'Maritime Silk Road', from Dubai via Singapore to Shanghai, Singapore has since grown into the busiest container trans-shipment hub in the world, handling an estimated one-fifth of global container trans-shipment throughput (Ship Technology, 2019; see also Chua, this volume; Heins, this volume). Since achieving independence in 1965, Singapore has developed a hugely profitable maritime industry that includes shipbuilding, ship repair, oil rig construction, offshore engineering, and support services (Bennett, 2018). Singapore is the world's largest port for bunkering, or ship refuelling, a 'Texaco station of the high seas'. The total volume of bunkers lifted stands at over 42 million tonnes – the quantity to fill more than 17,000 Olympic-sized pools (Subramanian, 2017).

Built around its function as one of the world's great ports, Singapore has risen to become a hub of finance and services, a reminder "that the instantaneous globe of international finance… has always contained the ocean as its material substrate…" (Mentz, 2015: xxix; see also Savitzky and Urry, 2015). Starting from its initial export-oriented industrialisation in the 1960s, the economy has scaled up and 'deterritorialised' by way of significant overseas investments, joint ventures, and land purchases, including Cambodian farmland, deep-sea mines, and even Arctic shipping lanes (Bennett, 2018: 292).

Proximity to the oceans' is key to Singapore's success. Yet with the island just 719 square km in area, and most of it no more than 15 metres above sea level, the oceans have long presented it with an existential threat. A third of the island is only around 16 feet above sea level, and its port, airport and most of its business district lie less than two metres above sea level. It lacks both arable land and natural resources, including food and water. Its first ever Prime Minister Lee Kuan Yew famously stated that Singapore's 'raison d'etre' was its port. Often ignored in 'global cities' literature (see Boschken, 2013), five of the ten current 'command and control' nodes of the global economy (New York, Tokyo, Singapore, Hong Kong and Sydney) are ports, whilst two more (Amsterdam and London) were historically ports (Dawson, 2019: 126). Whilst the significance of ports in driving urbanisation has often been overlooked, so too has the fact that most of the world's ports are located on islands or archipelagos.

Eight of the world's ten busiest ports (Shanghai, Singapore, Tiajin, Rotterdam, Guangzhou, Ningbo-Zhoushan, Busan and Hong Kong) are located on islands, as are the largest and most densely populated urban centres in sub-Saharan Africa and the US (Grydehøj, 2015: 429).

Given that they provide ready access to the sea, islands invite the placement of ports, which then draw industry and populations around them. Grydehøj proposes that cities are "at their most city-like (densest) when circumscribed by water" (2015: 433). Furthermore, he argues, "agglomeration economies favour the spatially dense networks of industry, infrastructure, and knowledge that tend to arise in island cities in particular" (Grydehøj, 2015: 433).

The island-form has induced experiments in extending territory into the sea. As population numbers rise, and political and economic significance increases, so cities tend to expand in size. Cities located on the mainland usually expand by urbanising adjacent land. Where cities are restricted to islands or archipelagos, which Amir (2015) calls 'space scarcity', they tend to extend into – as well as extend above – surrounding waters, resulting in striking, extreme forms of urbanisation (Grydehøj, 2015).

## Making space: Outwards, upwards and downwards

So-called 'land reclamation', the process of creating new land from oceans, seas, river or lake beds, has been central to urban growth in several major global coastal cities, including Hong Kong, Tokyo and Mumbai. For instance, Mumbai, originally an archipelago of seven islands, has gradually been remade into a single peninsula (Perur, 2016: no page). Tokyo has added 25,000 hectares of land to its harbour since the seventeenth century. In Hong Kong, 6 per cent of the territory's land has been reclaimed as of 2011 (Graham, 2016: 297). China has embarked on a program of 'artificial island' construction in order to bolster its sovereignty claims over disputed areas of the South China Sea (Watkins, 2016). Dubai is home to the spectacular 'Palm Jumeirah' and 'World' archipelagos, self-consciously artificial islands built from an estimated 110 million cubic metres of dredged sand (Wainwright, 2018). Significantly, a port, the Port of Rotterdam in the Netherlands – which is now one of the world's largest container ports – was "a key laboratory of mass land-reclamation" in the 1970s (Graham, 2016: 296).

Around a quarter of modern-day Singapore was open sea when the nation came into existence in 1955 (Shepard, 2018). Making new land has therefore been central to Singapore's national strategy. Between 2004 and 2014, 120 square kilometres was added to the country – 20 per cent of its size at independence in 1965. The government is planning to add another 100 square kilometres of new land by 2030 (Graham, 2016: 298). It is used both to keep pace with population growth – which grew from 1.6 million in 1960 to 4.8 million in 2010 (Graham, 2016: 298) – as well as to anticipate future trends in the global economy. Faced with competition from other ports, its port facilities and cargo handling infrastructure must be continually upgraded for Singapore to maintain its pre-eminence as a shipping, refuelling and connectivity hub (Khanna, 2016: 238–39). This requires space.

The Marine Port Authority (MPA) have played a key role in Singapore's land reclamation activity. The MPA reclaimed land primarily to develop the Port of Singapore and Changi Airport, now one of the world's busiest airports. Its earliest project took place in 1967 when 23 hectares of land were reclaimed to build Singapore's first container terminal at Keppel Harbour. Between 1972 and 1979, some 61 acres of foreshore were reclaimed. MPA's reclamation works for Changi Airport began in 1975 when it supervised the reclamation of 745 hectares of land along Changi coast for the construction of the airport. In 1990, another massive reclamation was carried out for the expansion of Changi Airport as well as for mixed-use developments in the area (National Library Board Singapore, 2017a).

Land reclamation has also long underpinned Singapore's dramatic program of high-rise construction. Building upwards has been an important way through which Singapore has attempted to address its space constraints. The iconic five towers of the Marina Bay Financial Center are

*Figure 32.1* 'The Interlace' residential complex, Singapore.
*Source*: Mike Cartmell from Singapore, Singapore, CC BY 2.0 <https://creativecommons.org/licenses/by/2.0>, via Wikimedia Commons.

built on reclaimed land. This includes not only skyscrapers, but elevated roads, walkways gardens and villages. In 2015 Singapore won the world architecture prize for its 'vertical village' (Figure 32.1), a six-story luxury apartment block which contains 31 'stacked' residential blocks, complete with swimming pools, tennis courts, gardens and roof terraces (Blair, 2015).

The most ambitious, perhaps, of Singapore's vertical architectural innovations, reaches not overhead but deep underground (see Macfarlane, 2019, on underground geographies). To maintain its position as the 'Texaco station of the high seas', the island needs the capacity to store the millions of tons of fuel sold to ships each year. Off the southern coast of Singapore lies Jurong Island, fashioned from the 'merging' of seven previously separate islands (see Figure 32.2). The island landscape is described as the "a blur of brand names", completely given over to petrochemical industry names like Exxon Mobil, Vopak and BASF (Subramanian, 2017: no page). One of the island's most distinctive, although hidden features, are the Jurong Rock Caverns. Inaugurated in 2014, the 130-metre deep, 61-hectare (150-acre) caverns hold an estimated 126 million gallons of crude oil. The project invokes the prospect of Singapore expanding not only upwards, but downwards, below sea level. "Each time a dynamite explodes under the sea and the caverns become still deeper, Singapore expands its territory" (Duara, 2015: 44). Artist Charles Lim captured this astonishing penetration of the depths as the caves were being excavated, as part of his nine-part SEA STATE project, which explores Singapore's 'deep' relationship with the sea (Lim et al., 2015).

To facilitate this process of "radical vertical and horizontal expansion" (Graham, 2016: 298), stockpiles of sand are kept in the city, ready for use in the next reclamation projects. The areas of the city dedicated to storing sand and aggregates are heavily securitised and sometimes secret sites (Comaroff, 2014: no page). As existing reserves are used, sands are brought across growing distances – through legal and illegal means (Graham, 2016: 298). These 'sand mining' practices are known to devastate coastal and aquatic ecosystems at extraction sites, damaging local fishing and tourism industries. "This translates into a de facto transfer of territory from other countries" (Comaroff, 2014: no page). This has important implications for the geographical distribution

*Figure 32.2* Ariel view of Jurong Island.
*Source*: William Cho, CC BY-SA 2.0 <https://creativecommons.org/licenses/by-sa/2.0>, via Wikimedia Commons.

of climate risk. At sites of construction, these materials can be used to build defences against sea-level rises; at sites of removal their loss increases communities' vulnerability to flooding and erosion (Graham, 2016: 303).

## Swelling seas

Whilst prosperous coastal cities turn sea into land through land reclamation, global heating turns land into sea. Global mean sea levels rose by approximately 16–21 cm between 1900 and 2016 (USGCRP, 2018). Global warming is driving the thermal expansion of seawater, and the melting of land-based ice sheets and glaciers (see also Dodds, this volume). As parts of the Alps, Himalayas, Mount Kilimanjaro, and the glaciers of Alaska, Chile, Norway melt, large volumes of water are added to the oceans (Wadhams, 2017: 110). Antarctica has lost 3 trillion tonnes of ice in the last 25 years (Shepherd et al., 2018).

Of the world's 197 nation-states, 150 border an ocean and are therefore directly vulnerable to the gradual – or abrupt – swelling of the world's oceans. Trillions of dollars of economic assets are concentrated on the coasts and by large bodies of water, representing literally 'sunk' capital and infrastructure (Dawson, 2019: 59; Savitzky, 2018: 680). However, the effects of sea-level rises are expected to be most severe in low-lying, developing countries (Marshall, 2011: 235).

Small island states around the world are working to address these risks. Kiribati in the Central Pacific has bought 6,000 acres of land in Fiji, over a thousand miles away, with the aim of resettling some of its 100,000 population (Caramel, 2014). Climate refugees have begun to leave islands in the South Pacific, the Marshall Islands, and Micronesia (Subramanian, 2017). Five of the lowest Solomon Islands have already vanished (Klein, 2016). These islands are concentrated largely in the Pacific or in Asia, and depend on larger, wealthier states for financial aid and resources. Singapore, however, is placed fourth in countries ranked by per capita Gross Domestic Product.

Asia accounts for two-thirds of the world's urban population with almost three-quarters in low elevation coastal zones, less than 10 metres above sea-level (Hijioka et al., 2015: 1347). Asia is one of the most vulnerable regions in the world to climate change. Singapore's wealth and technological advancement provide it with a degree of insulation not afforded to its poorer neighbours. But its neighbours' vulnerability to sea-level rises has important implications for Singapore, which depends on parts of these places in important ways. Dependencies between food importing and exporting nations have increased in southeast Asia, especially since the establishment of the ASEAN Economic Community in 2015 (Khanna, 2016: 53). Low-lying rice fields in Thailand (as well as Indonesia) are especially important regions for Singapore's food supply (Chow, 2018). Singapore imports 90 per cent of its food, leaving it vulnerable to climate related disruptions and price hikes, resulting either from floods or droughts all over the world. Singapore imports half of its water from Malaysia, and the two (formerly conjoined) nations have a long-standing dispute over this supply, which could be exacerbated by sea level rises (Chow, 2018).

The average sea level around Singapore's coasts has risen steadily at a rate of between 1.2 mm and 1.7 mm per year and is projected to increase to about one metre by 2100 (Chow, 2018: no page). Today, over 70 per cent of Singapore's coastlines are fortified with hard structures and seawalls. In 2011, the minimum land reclamation level in Singapore was raised from three to four metres above the mean sea level. Coastal roads are being raised; a new airport terminal is being built 18 feet above sea level. In 2010, the Building and Construction Authority's (BCA) carried out shoreline restoration works to stabilise a section of the beach at East Coast Park. This consisted of large sand-filled bags, laid several metres into the ground to be level with the low tide, helping to reduce sand erosion (Tang and Lin, 2017). But whilst Singapore could cope with a rise of 50 cm to 1 metre, a rise of two metres would turn Singapore "into an island fortress", as it would involve constructing more and higher walls to protect against the sea (Fogarty, 2012). The costs of insulating Singapore from flooding and climate disruption are thought to be in excess of S$100 billion (Elangovan, 2019).

A government study identifies a stretch of land between East Coast, the City, and Jurong Island as the most vulnerable zones of Singapore's coastline (Elangovan, 2019). Sea levels rises will have major implications for property values, as well as for safety and liveability in these areas. The impact of flooding will, however, affect the whole island, as roadways and trainlines traverse low-lying areas, where many hospitals, schools and workplaces are located. The insularity and density of the island-city form means that effects spread quickly through its territory. "We cannot lose a big chunk of our city and expect the rest of Singapore to carry on as usual", one official said (cited in Elangovan, 2019: no page).

## Heating cities

The threat to Singapore is not only from water, but from heat. Island spatiality drives the densification processes which result in cities, due to replacing self-cooling vegetation with heat-retaining concrete buildings and bitumen roads, being on average 30 percent hotter than the countryside. This is known as the 'urban heat island' effect (Oke, 1982). Average daily temperature in tropical Singapore could increase by 2.7 to 4.2 degrees Celsius (4.9 to 7.6 degrees Fahrenheit) from the current average of 26.8 degrees Celsius (80.2 degrees Fahrenheit) by 2100 (Fogarty, 2012). As temperatures rise, demand for air-conditioning, and as a result also power, will increase, producing a vicious circle where more air-conditioning amplifies heating, and so on (Dawson, 2019: 130)

Indeed, Singapore is already one of the most energy intensive nations in Asia. Its carbon emissions increased to a record 37.1 billion tonnes in 2018, in order to power its airconditioned malls, industries and glass office towers (Global Carbon Atlas, 2018). Added to this, half of the emissions from global shipping are from vessels registered in one of only six states, Singapore being one. The heaviest concentrations of black carbon in the world are around Singapore and the Strait of Malacca (Olmer et al., 2015: v). Port cities – insofar as they have served as engines accelerating planet-warming forms of globalisation – are "drivers of the very processes that now threaten them with destruction" (Ghosh, 2017: 55).

Up until 2020, Singapore claimed exemption from the absolute, economy-wide emissions targets committed to by most developed countries under the 2015 Paris Climate Agreement. Singapore was pledged to a more modest goal of reducing emissions intensity by 36 per cent from 2005 levels by 2030, under a special UN Framework Convention on Climate Change article, which recognises country-specific limitations. Space constraints were cited as reasons preventing Singapore's widespread adoption of more ambitious emission targets and renewable energy programmes (Chin, 2019).

## Land into sea, sea into land

Singapore along with many other major cities around the world, exists precariously, at the water's edges. Whilst managing to leverage proximity to the oceans to harness global trade flows, growing population numbers and the continual infrastructural expansion required to maintain its status as connectivity hub, requires that Singapore constantly find more space. It has sought to do so by artificially extending its territory into the oceans and then building above and below its depths, in conjoined forms of horizontal and vertical expansion. The case of Singapore shows that: "the earth cannot be 'reclaimed' from the ocean by the magic of sovereign right; it needs to be brought from somewhere" (Comaroff, 2014: no page). Rather than examine land and ocean in isolation from one another, this chapter has detailed how forms emerge through their relations and at their edges. The 'island-city-state' of Singapore is one such form.

Singapore's founder Raffles famously said: "Our object is trade not territory". Yet with the large-scale purchase and import of sands for reclamation projects, a trade in territory itself has emerged. Physical territory – land itself – becomes 'fluid', as it is broken down and recomposed, extended and shrunk, in ways that challenge traditional, static conceptions of territory (Peters et al., 2018). "With the coastal earthworks that are under way throughout Southeast Asia and the Middle East… territory has acquired an unprecedented liquidity" states Comaroff (2014: no page). Such practices have significant implications given that "the geopolitical rubric of modernity is literally grounded in the idea… [that] territorial states can make claims to distinct portions of the Earth's terrestrial environment" (Dodds and della Dora, 2018: 1348). As a result of the so-called sand export-drive, it is feared that "national sovereignty claims are literally being stolen as material is used to bolster the claims of wealthy states and city-states elsewhere" (Graham, 2016: 299).

The practice of building artificial islands calls into question the distinction between nature and artifice (Dodds and della Dora, 2018: 1347). Intervening in the Earth's very geology in order to facilitate expansion and trade testifies to modern human actors' propensity to reshape the natural world to their designs. Yet this seeming ability to exert control over natural processes – in this instance 'reclaiming' land from the sea in ever-more ambitious projects – has been accompanied by a profound undermining of agency, as coastal populations and assets are increasingly imperilled by rising waters. The Anthropocene is marked by the simultaneous extension and diminution of human agency: cities, states and especially *islands* find themselves "humbled within the vast multidimensional forces of a rapidly changing planet" (Pugh, 2018: 96).

## Conclusion

Swiss-based research group Crowther Lab found that Singapore would be one of several cities experiencing major climate shifts by 2050 (Elangovan, 2019). The island will face harsher storm surges, flash floods, heatwaves, water and food disruptions. To address its food insecurity, a Singaporean company is experimenting with lab-grown meat, which uses stem cells from crustaceans such as prawns and crabs. Meat grown in Singaporean labs would reduce dependence on foreign food-supply, but more radically – dependence on (whole) animal bodies – freeing them of the enormous space requirements needed to hold and feed them (Lewis, 2019).

To decrease the impact of flooding, Singapore has implemented a system that reroutes rainfall into a river via a tank that can store 15 Olympic-sized swimming pools of stormwater. One of the main flows that Singapore will need to become adept at storing in the twenty-first century, is water. Faced with the prospect of submersion into the very waters that have served as its lifeblood, Singapore has sought assistance from the Dutch – commonly regarded as the world's greatest experts in holding back the sea. So successful in their struggle with the sea, the Netherlands have become global consultants on how to keep back rising tides, advising cities from Jakarta to New York. A Dutch firm is involved in helping Singapore convert its biggest river and marina into a huge downtown reservoir (Subramanian, 2017).

In the Netherlands, however, there is a realisation that they are caught in a 'control paradox': flood defences can encourage the development and habitation of risky areas (Dawson, 2019: 278). Defences are only built to withstand the kinds of flood events that are conceivable at the time of their building, a problem as projections continue to rise and forecasts are frequently found to underestimate the extent of sea-level rise. In the long term, no matter how high the seawalls and elaborate the flood defence systems, the oceans will, arguably, finally 'reclaim' island-cities from their human inhabitants. Large segments of humanity will then have to retreat not just from portions of the coastline, but from entire regions (Dawson, 2019: 280).

Ironically, the melting processes which threaten Singapore with submersion, also present it with commercial opportunities. Accompanying the collapse of glaciers, is the melting of Arctic sea-ice. Reductions in sea ice thickness and extent allows easier access to formerly frozen sea-routes and economically 'stranded' undersea hydrocarbons, whose burning will trigger further temperatures rises and ice loss, in a process referred to as the 'Arctic paradox' (Banerjee, 2012: 7). Singapore's state-linked Keppel Corporation has manufactured icebreakers for a Russian oil company and partnered with ConocoPhilipps in designing the first ice-capable jack-up rig. Singapore has also explored the possibility of establishing a foothold at the Port of Adak in Alaska's Aleutian Islands (Bennett, 2018: 303).

The island's significant overseas investments, including Arctic shipping lanes, farmland, and deep sea mines, "may suggest that even if low-lying Singapore were to disappear due to sea-level rise, it could still exert some form of regulatory power if not outright sovereignty over places beyond its borders" (Bennett, 2018: 298). This raises important questions regarding the relationship between states and their physical territories, namely how far and to what extent the former can exist without the latter (Dodds and della Dora, 2018: 1392). Singapore's 'space scarcity' may have been beneficial to its economic development, by compelling it to look overseas for investment opportunities (Bennett, 2018: 292).

The chapter has explored how islands – and particularly the city island of Singapore – has been shaped by the condition of being 'surrounded by sea'. How Singapore fends off the ocean will be of intense interest to many other islands, and cities, surrounded by sea. It is no accident, surely, that cities like Mumbai, New York, Boston and Kolkata were all brought into being through early globalisation, linked to each other as they were through patterns of trade, in circuits that expanded and accelerated the economies of Western Imperial centres. The history of habitation in these cities can be traced to the rise of capitalism over the past 500 years (Dawson, 2019: 72). Globalisation over the last few decades has only magnified the importance of such key nodes. Most of the world's megacities are in coastal areas, often on islands that concentrate truly astonishing numbers of people. These cities share significant traits, including "exceptional vulnerability to climate change" (Dawson, 2019: 73). As such, "their predicament is but an especially heightened instance of a plight that is now universal" (Ghosh, 2017: 54–55).

## References

Amir S (2015) Manufacturing space: Hypergrowth and the Underwater City in Singapore. *Cities* 49: 98–105.

Banerjee S (2012) *Arctic Voices*. New York, NY: Seven Stories Press.

Bennett MM (2018) Singapore: The "global city" in a globalizing Arctic. *Journal of Borderlands Studies* 33(2): 289–310.

Blair O (2015) This has just been declared the world's best building. *The Independent [Online]*. Available at: www.independent.co.uk/arts-entertainment/architecture/singapore-s-vertical-village-named-building-year-world-architecture-awards-a6725391.html

Boschken HL (2013) Global cities are coastal cities too: Paradox in sustainability? *Urban Studies* 50(9): 1760–1778.

Caramel L (2014) Besieged by the rising tides of climate change, Kiribati buys land in Fiji. *The Guardian*. Available at: www.theguardian.com/environment/2014/jul/01/kiribati-climate-change-fiji-vanua-levu

Chin N (2019) Singapore's greenhouse gas emissions top 50m tonnes: *TODAYonline*. Available at: www.todayonline.com/singapore/singapores-greenhouse-gas-emissions-top-50m-tonnes-report

Comaroff J (2014) Built on sand: Singapore and the new state of risk. *Harvard Design Magazine [Online]*. Available at: www.harvarddesignmagazine.org/issues/39/built-on-sand-singapore-and-the-new-state-of-risk.

Chow W (2018) How vulnerable is Singapore to climate change? *The Straits Times [Online]*. Available at: www.straitstimes.com/singapore/how-vulnerable-is-spore-to-climate-change

Dawson A (2019) *Extreme Cities: The Peril and Promise of Urban Life in the Age of Climate Change* (Reprint edition). London and New York, NY: Verso.

Dodds K and della Dora V (2018) Artificial islands and islophilia. In Baldacchino G (ed) *The Routledge International Handbook of Island Studies*. Abingdon: Routledge, 392–415.

Duara P (2015) Island territoriality. In Lim C (ed) *Sea State*. Milan: Skira, 44–46.

Elangovan N (2019) Explainer: Why climate change should matter to Singaporeans and what the Govt is doing about it. *TODAYonline*. Available at: www.todayonline.com/singapore/explainer-why-climate-change-should-matter-singaporeans-and-what-government-doing-about-it

Fogarty D (2012) Singapore raises sea defences against tide of climate change. *Reuters*. Available at: www.reuters.com/article/uk-climate-singapore-idUSLNE80Q00J20120127

Ghosh A (2017) *The Great Derangement: Climate Change and the Unthinkable*. Chicago, IL and London: University of Chicago Press.

Global Carbon Atlas (2018) CO2 Emissions. *Global Carbon Atlas*. Available at: www.globalcarbonatlas.org/en/CO2-emissions

Graham S (2016) *Vertical: The City from Satellites to Bunkers*. London and New York, NY: Verso.

Grydehøj A (2015) Island city formation and urban island studies. *Area* 47(4): 429–435.

Hijioka Y, Lin E, Pereira JJ, Corlett RT, Cui X, Insarov G and Surjan A (2015) Asia. In: *Climate Change 2014: Impacts, Adaptation and Vulnerability: Part B: Regional Aspects: Working Group II Contribution to the Fifth Assessment Report of the Intergovernmental Panel on Climate Change*. Cambridge: Cambridge University Press, 1327–1370.

Khanna P (2016) *Connectography: Mapping the Future of Global Civilization*. London: Random House.

Klein A (2016) Five Pacific islands vanish from sight as sea levels rise. *New Scientist*. Available at: www.newscientist.com/article/2087356-five-pacific-islands-vanish-from-sight-as-sea-levels-rise/

Lewis P (2019) Lab-grown and plant-based meat can help us move towards a more sustainable food system. *TODAYonline*. Available at: www.todayonline.com/commentary/lab-grown-and-plant-based-meat-can-help-us-moving-towards-more-sustainable-food-system

Macfarlane R (2019) *Underland: A Deep Time Journey*. London: Hamish Hamilton.

Marshall SJ (2011) *The Cryosphere* (Illustrated edition). Princeton, NJ: Princeton University Press.

Mentz S (2015) *Shipwreck Modernity*. Minneapolis, MN: University of Minnesota Press.

Miksic JN (2014) *Singapore and the Silk Road of the Sea, 1300–1800*. Singapore: NUS Press.

National Library Board Singapore (2017a). *Land From Sand: Singapore's Reclamation Story*. Available at: www.nlb.gov.sg/biblioasia/2017/04/04/land-from-sand-singapores-reclamation-story/

National Library Board Singapore (2017b) *Port of Singapore*. Available at: http://eresources.nlb.gov.sg/infopedia/articles/SIP_2018-04-20_085007.html

Oke TR (1982) The energetic basis of the urban heat island. *Quarterly Journal of the Royal Meteorological Society* 108(455): 1–24.

Olmer N, Comer B, Roy B, Mao X and Rutherford D (2015) *Greenhouse Gas Emissions from Global Shipping, 2013–2015*. Available at: https://theicct.org/publications/GHG-emissions-global-shipping-2013-2015

Perur S (2016) Story of cities #11: The reclamation of Mumbai – from the sea, and its people? *The Guardian [Online]*. Available at: www.theguardian.com/cities/2016/mar/30/story-cities-11-reclamation-mumbai-bombay-megacity-population-density-flood-rise.

Peters K, Stratford E and Steinberg PE (eds) (2018) *Territory Beyond Terra*. London: Rowman and Littlefield.

Pugh J (2018) Relationality and island studies in the Anthropocene. *Island Studies Journal* 13(2): 93–110.

Rahim LZ (2010) *Singapore in the Malay World*. Abingdon: Routledge.

Savitzky S and Urry J (2015) Oil on the move. In: Birtchnell T, Savitzky S and Urry J (eds) *Cargomobilities: Moving Materials in a Global Age*. Abingdon: Routledge, 180–198.

Savitzky S (2018) Scrambled systems: The (im)mobilities of 'Storm Desmond'. *Mobilities* 13(5): 662–684.

Shepard W (2018) Cities from the sea: The true cost of reclaimed land. *The Guardian [Online]*. Available at: www.theguardian.com/cities/2018/may/02/cities-from-the-sea-the-true-cost-of-reclaimed-land-asia-malaysia-penang-dubai

Shepherd A, Ivins E, Rignot E, Smith B, van den Broeke M, Velicogna I, Shepherd A, Ivins E, Rignot E, Smith B, Van Den Broeke M, Velicogna I, Whitehouse P, Briggs K, Joughin I, Krinner G and Nowicki S (2018) Mass balance of the Antarctic Ice Sheet from 1992 to 2017. *Nature* 558(7709): 219–222.

Ship Technology (2019) *Port of Singapore*. Available at: www.ship-technology.com/projects/portofsingapore/

Subramanian S (2017) How Singapore is creating more land for itself. *The New York Times [Online]*. Available at: www.nytimes.com/2017/04/20/magazine/how-singapore-is-creating-more-land-for-itself.html

Tang F and Lin X (2017) As sea levels rise, Singapore prepares to stem the tide. *The Straits Times [Online]*. Available at: www.straitstimes.com/singapore/as-sea-levels-rise-singapore-prepares-to-stem-the-tide

US Global Change Research Programme (USGCRP) (2018) *Climate Science Special Report*. Available at: https://science2017.globalchange.gov/chapter/12/

Wadhams P (2017) *A Farewell to Ice: A Report from the Arctic*. London: Penguin.

Wainwright O (2018) Not the end of the world: The return of Dubai's ultimate folly. *The Guardian [Online]*. Available at: www.theguardian.com/cities/2018/feb/13/not-end-the-world-return-dubai-ultimate-folly

Watkins D (2016) What China has been building in the South China Sea. *The New York Times [Online]*. Available at: www.nytimes.com/interactive/2015/07/30/world/asia/what-china-has-been-building-in-the-south-china-sea-2016.html

# INDEX

*Note*: Page numbers in *italics* indicate a figure and page numbers in **bold** indicate a table on the corresponding page. Page numbers followed by 'n' indicate a note.